INDUSTRIAL MAINTENANCE

THIRD EDITION

AMERICAN TECHNICAL PUBLISHERS
ORLAND PARK, ILLINOIS 60467-5756

Denis Green
Jonathan F. Gosse

Industrial Maintenance contains procedures commonly practiced in industry and the trade. Specific procedures vary with each task and must be performed by a qualified person. For maximum safety, always refer to specific manufacturer recommendations, insurance regulations, specific job site and plant procedures, applicable federal, state, and local regulations, and any authority having jurisdiction. The material contained is intended to be an educational resource for the user. American Technical Publishers, Inc. assumes no responsibility or liability in connection with this material or its use by any individual or organization.

American Technical Publishers, Inc., Editorial Staff

Editor in Chief:
    Jonathan F. Gosse
Vice President—Production:
    Peter A. Zurlis
Art Manager:
    James M. Clarke
Technical Editor:
    Julie M. Welch
Copy Editor:
    Jeana M. Platz
Cover Design:
    Jennifer M. Hines

Illustration/Layout:
    Thomas E. Zabinski
    Samuel T. Tucker
    Melanie G. Doornbos
Multimedia Manager:
    Carl R. Hansen
CD-ROM Development:
    Gretje Dahl
    Nicole S. Polak
    Daniel Kundrat
    Hannah A. Swidergal

3 4 5 6 7 8 9 – 10 – 9 8 7 6 5 4 3

Printed in the United States of America

ISBN 978-0-8269-3641-7

This book is printed on recycled paper.

atp made in USA

Acknowledgments

The author and publisher are grateful to the following companies and organizations for providing information, photographs, and technical assistance.

- Alemite Corp.
- Alnor Instrument Company
- Ansell Edmont Industrial, Inc.
- Automated Logic Corporation
- Browning; Emerson Power Transmission
- Carrier Corporation
- Chemgrate
- Cleaver-Brooks
- Computational Systems Incorporated
- Cone Drive Operations Inc./Subsidiary of Textron Inc.
- Cooper Industries Bussmann Division
- Copeland Corporation
- Crane Valves
- Crosby Valve Inc.
- Cutler-Hammer
- Datastream Systems, Inc.
- DPSI (DP Solutions, Inc.)
- Dunham-Bush
- Exxon Company
- Factory Mutual Engineering Corporation
- Fenner Drives
- Fluke Corporation
- The Foxboro Company
- The Gates Rubber Company
- General Electric Company
- Gilian, Inc.
- W.W. Grainger, Inc.
- Hand Held Products
- Henry Valve Co.
- Heidelberg Harris, Inc.
- Honeywell's MICRO SWITCH Division
- I.T.W. Devcon
- John Snell & Associates
- Justrite Mfg. Co.
- Kewanee Boiler Manufacturing Co., Inc.
- Kop-Flex, Inc.; Emerson Power Transmission
- Lab Safety Supply, Inc.
- Lau, a Division of Tomkins Industries
- Lennox Industries Inc.
- The Lincoln Electric Company
- Loctite Corporation
- Lovejoy, Inc.
- Ludeca, Inc., representative of PRUEFTECHNIK AG
- Mastercool Inc.
- McDonnell & Miller
- Miller Equipment
- Mine Safety Appliances Co.
- National Fire Protection Association
- New Pig Corporation
- Omron Electronics, Inc.
- Pacific Circuits
- Panduit Co.
- Parker Hannifin Corp.
- Parker Motion & Control
- Power Team, Division of SPX Corporation
- Predict\DLI
- Ranco Inc.
- Robinair Division, SPX Corporation
- Rockwell Automation, Allen-Bradley Company, Inc.
- SEW-Eurodrive, Inc.
- Siemens Corporation
- SKF Condition Monitoring
- Spirax Sarco, Inc.
- Spokane County Facilities Maintenance Department
- Sporlan Valve Company
- SPX Robinair
- SSAC, Inc.
- Super Radiator Coils
- Tecumseh Products Company
- The Timken Company
- UE Systems, Inc.
- U.S. Filter
- Yellow Jacket Div. Ritchie Engineering Co., Inc.

Contents

Interactive CD-ROM Contents

Using This CD-ROM
Quick Quizzes®
Illustrated Glossary
Maintenance Forms
Flash Cards
Media Clips
Link to ATPeResources.com

Introduction

Maintenance personnel are responsible for an ever-expanding range of industrial processes and building systems. These technicians and engineers are expected to be prepared for many different types of maintenance tasks, and able to recognize complex problems caused by different parts of a system. They are also expected to understand the operating principles of various systems and develop the necessary skills for effective troubleshooting. Also, maintenance departments are key participants in facility-wide energy-efficiency and waste-reduction programs.

Industrial Maintenance, 3rd Edition, is a comprehensive textbook covering fundamental maintenance and troubleshooting principles, procedures, and practices. Maintenance management is featured through the application of preventive and predictive maintenance, including the role of computerized systems to promote efficiency. The textbook is arranged in a system-based format. Common industrial systems covered include electrical, electronic, PLC, refrigeration, boiler, HVAC, mechanical, and fluid power systems. Troubleshooting principles are applied within each chapter and detailed further in a capstone chapter with case study examples. Safety, service, and repair principles are also explained through common industrial examples. The glossary defines over 840 technical terms used throughout the textbook.

This edition includes new material that reflects some of the current trends in maintenance management, energy efficiency, and occupational health and safety. New and expanded topics include the following:

- Lean principles
- Energy efficiency and green buildings
- Safety regulations and standards
- Advanced multimeter functions and features
- Voltage drop measurement
- Motor winding insulation testing
- Refrigerant charge verification
- Boiler furnace control sequence
- Building automation systems
- Energy audits
- Indoor air quality issues

The CD-ROM included at the back of the textbook includes Quick Quizzes® for each chapter, an Illustrated Glossary, Maintenance Forms, Flash Cards, Media Clips, and link to ATPeResources.com. The Maintenance Forms are a collection of common forms used by maintenance personnel, which can be printed for use in the field. Information about using the *Industrial Maintenance* CD-ROM is included at the back of the book.

The *Industrial Maintenance Workbook* complements the textbook, testing comprehension with review questions and challenging the student to apply critical-thinking skills with activities based on real-life maintenance problems.

To obtain information on additional related training products, visit the American Tech web site at www.go2atp.com.

The Publisher

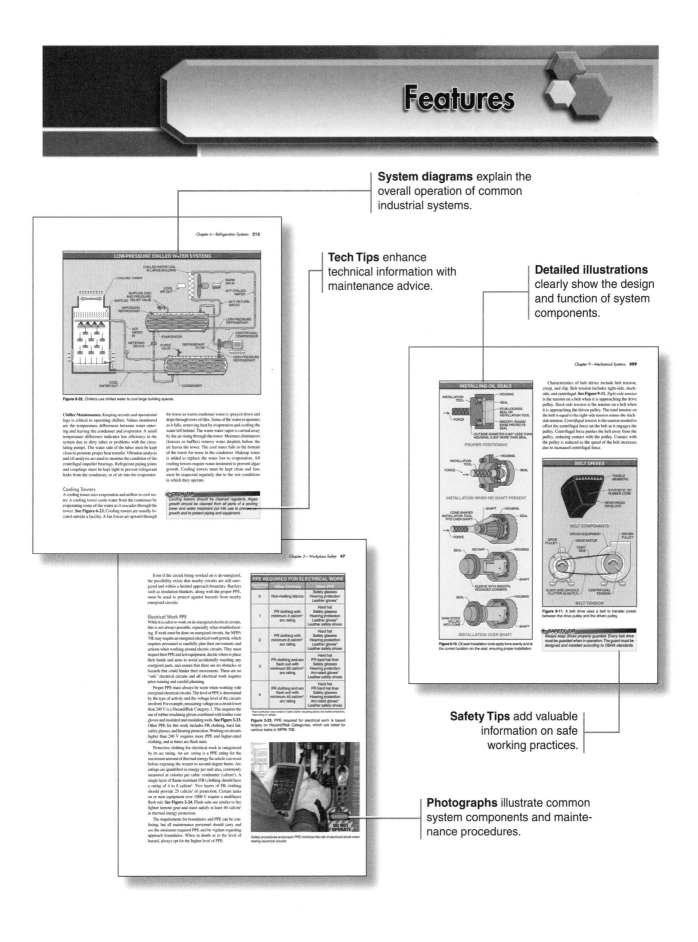

Features

System diagrams explain the overall operation of common industrial systems.

Tech Tips enhance technical information with maintenance advice.

Detailed illustrations clearly show the design and function of system components.

Safety Tips add valuable information on safe working practices.

Photographs illustrate common system components and maintenance procedures.

Maintenance Principles

Maintenance personnel specialize in the maintenance, troubleshooting, and repair of building system and process equipment in a wide variety of industrial facilities and commercial buildings. Maintenance personnel use preventive and predictive maintenance skills to minimize equipment malfunctions and failures and maximize efficiency in industrial facilities. Preventive maintenance is regular maintenance work required to keep equipment in peak operating condition. Predictive maintenance is the monitoring of wear conditions and equipment characteristics to detect potential for future malfunctions or failures. Troubleshooting skills are used to identify the cause of an equipment malfunction or failure.

MAINTENANCE PRINCIPLES

Maintenance personnel operate building systems in order to optimize facility resources and reduce costs through energy efficiency, minimized downtime, and waste reduction. General maintenance principles outline philosophies and methods for achieving these goals. Efficiency of resource use is a critical aspect of facility operations that directly involves maintenance personnel.

Lean Principles in Maintenance

Maintenance departments in modern production and commercial facilities have been significantly influenced by quality-improvement philosophies. Lean manufacturing principles were developed by the Japanese auto industry and are based on the goal of gradual, continuous improvement in all aspects of plant operation. While these principles were originally developed for manufacturing industries, many of the same goals can be applied to other areas, including facility maintenance. Maintenance personnel can apply these concepts to their maintenance, troubleshooting, and repair tasks in order to continually improve plant operations and efficiency.

Lean principles focus on the elimination of waste in seven key areas: transportation, overproduction, waiting, inventory, defects, extra processing, and movement. **See Figure 1-1.**

- Transportation waste is the unnecessary moving of parts or materials during a process. Effective task planning allows all parts and tools to be gathered and moved only once, which avoids personnel having to go back and forth between the shop and the worksite.

- Overproduction waste is work that is done beyond what is required. This type of waste is usually applied to product manufacturing, but it can also apply to maintenance activities. For example, replacing filters in air conditioning systems based on a time schedule, regardless of whether they are actually exhausted, can be considered overproduction.

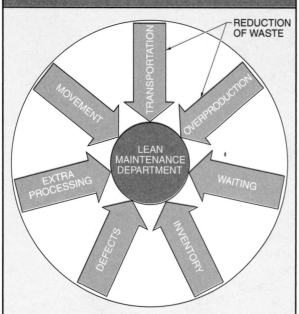

Figure 1-1. Lean principles identify seven areas of waste that reduce the efficiency of a business or organization. Targeting these areas for reduction produces a lean organization.

- Waiting is unproductive time due to needed parts or materials being unavailable. This is a common problem when making repairs. Having the correct parts and materials available at the locations they are needed reduces this type of waste.

- Inventory waste is the stocking of unnecessary parts or materials that consume space, contribute to disorganized inventory stocks, and contribute to unnecessary spending. Inventory control of maintenance parts and supplies is integral to most modern maintenance operations.

- Defects are a form of waste in which work must be repeated to correct or replace the defective product or task. In maintenance work, tasks completed incorrectly or with faulty materials are examples of defects that can be eliminated with proper planning.

- Extra processing is waste due to unnecessary steps performed on each product or task. This waste can be reduced by eliminating unneeded maintenance work and accomplishing the remaining work as efficiently as possible.

- Movement waste is waste due to the inefficient motion of persons or material handling equipment when performing a particular task. Motion studies are used to identify movement sequences that are faster, easier,

and safer. For example, maintenance personnel can reduce movement waste by stocking commonly used materials where they are easier to reach. This type of waste is different from transportation waste, which refers to inefficiencies of moving material between tasks throughout a process, rather than within an individual task.

To eliminate waste, a lean maintenance department sorts, straightens, scrubs, systematizes, and standardizes all maintenance processes. Sorting requires the elimination of all unnecessary parts, tools, and supplies. The remaining items are then organized in an easily accessible system. Scrubbing refers to keeping all work areas, tools, and equipment clean. Systematizing requires that processes be devised so that the organization and cleanliness continues to be an ongoing activity. Standardizing improves the efficiency of all workplace processes.

Effective maintenance personnel work to improve all aspects of the services offered through the maintenance department. This work lends itself to continuous learning and improvement. Maintenance personnel can find and prevent problems and can suggest improvements based on their observations and analysis of daily facility operations.

Energy Efficiency

Reducing energy consumption through efficient facility operations is an increasingly important concern. Maintenance personnel are essential to improving energy efficiency through their regular maintenance, troubleshooting, and repair activities. **See Figure 1-2.**

Regular maintenance activities keep equipment operating at optimal conditions, which minimizes energy use. For example, electric motors require less energy to overcome the resistance of rotating bearings when the bearings are well-lubricated. Regular maintenance also reduces the amount of maintenance needed to address avoidable breakdowns or failures. Continuous monitoring of systems, such as boiler systems, for unusual energy-use profiles is also now a common aspect of regular maintenance.

Effective troubleshooting reduces the impact of failures and breakdowns and may uncover inefficiencies that waste resources. Diagnosing and remedying a problem as quickly as possible minimizes costs and waste associated with downtime. For example, the length of time HVAC equipment is down for repair directly affects the extra energy needed later to return the building to its correct temperature.

ENERGY EFFICIENCY THROUGH MAINTENANCE		
Regular Maintenance	Troubleshooting	Repair
• Optimizes equipment performance • Avoids breakdowns • Monitors energy use • Distributes maintenance work evenly	• Diagnoses problems quickly • Indentifies areas of improvement • Uncovers hidden wastes	• Extends life of equipment • Selects efficient equipment and materials • Improves equipment performance

Figure 1-2. Improving energy efficiency in a facility through regular maintenance, repair, and troubleshooting activities has many benefits.

Troubleshooting may identify further maintenance that could be used to prevent future problems. For example, if an electronic component inside an enclosure fails due to overheating, maintenance personnel may install a cooling fan in the enclosure when replacing the board. To avoid failures in similar enclosures throughout the facility, fan installation or other cooling measures are taken at these locations as well.

Maintenance personnel are often uniquely qualified to address energy-efficiency issues due to their familiarity with a facility's spaces and systems. This is particularly important when troubleshooting problems in rarely seen areas or when building modifications are made. For example, personnel performing maintenance in the space above a false ceiling may discover leaks in the building envelope that were previously unknown. Dirt marks discovered at ductwork joints indicate that conditioned air is leaking from the ducts and not reaching the intended spaces. These leaks allow conditioned air to escape outside, resulting in energy waste. Additional maintenance to remedy these problems could significantly improve energy efficiency.

The repair portion of maintenance work also presents many opportunities to improve plant operations and to reduce waste. A repair strategy must account for factors such as energy use, waste generation, and emissions, both at the time of repair and in the future. For example, maintenance personnel must properly size a replacement for a failed motor. A higher horsepower motor may be needed, but if the new motor is too large it will use more power and may contribute to power quality problems in the facility. However if the new motor is too small, it might fail again, increasing downtime, costs, and waste. Choosing equipment, materials, and procedures may involve balancing many factors.

Energy Auditing

A plan for improving energy efficiency in a facility typically includes an energy audit. An *energy audit* is a comprehensive review of a facility's energy use and a report on ways to reduce the energy use through changes to buildings, equipment, and procedures. The audit gathers detailed information about every use of a particular resource, such as electricity, and suggests strategies to reduce the consumption, which increases efficiency. These changes may include improving building insulation, replacing equipment with more efficient models, implementing more sophisticated system controls, changing maintenance procedures, repairing leaks, and changing occupant activities. The primary goal is the reduction of costs through lower energy use, but other benefits may include improved safety, increased occupant comfort, reduced maintenance, and reduced downtime.

Energy audits are targeted at areas that are expected to yield the largest improvements, such as a specific department, but may include an entire facility. Similarly, the audit may survey the use of one or more types of energy or resources, such as electricity, natural gas, water, steam, compressed air, and even solid-waste disposal. Maintenance personnel are integral to all phases of an energy audit. Energy audits involve gathering system information, measuring energy use, developing reduction strategies, choosing the most cost-effective plan, implementing changes, and verifying results.

Green Buildings

To help promote energy efficient facility operations, some organizations have developed rating systems and achievement levels to award building owners and operators for their environmentally friendly practices. One of the most prominent programs is from the U.S. Green Building

Council®. Their Leadership in Energy and Environmental Design® (LEED®) Green Building Rating System™ is the nationally accepted standard for the design, construction, and operation of green buildings. **See Figure 1-3.** LEED serves as a guide for building owners and operators in realizing the construction and performance of a green building. The LEED certification process addresses sustainable site development, water savings, energy efficiency, materials selection, and indoor environmental quality throughout the construction and operation of the building.

The LEED certification program is focused primarily on commercial buildings but is applied to industrial facilities as well. Most categories are intended for new construction, as many energy efficiency measures require planning at the design phase, but there is a special category for existing buildings. Both categories consider the ongoing operations and maintenance of a building, which are directly affected by the maintenance programs and personnel. Therefore, maintenance personnel may be involved in the implementation, maintenance, and improvement of energy efficiency measures.

MAINTENANCE PERSONNEL

Companies or institutions commonly employ their own personnel to perform maintenance tasks in a facility. Duties and responsibilities of these workers are determined by the structure of the maintenance organization. Maintenance organizations range in size from one individual in a small organization to several individuals in a large organization.

Over the years, the maintenance trades have changed with the needs of industry. Maintenance personnel in the past were trained and hired to perform maintenance in a specific craft. For example, welding repairs were performed only by welders and machines were installed only by millwrights. In some organizations, especially larger ones, this is still the case. **See Figure 1-4.** However, the responsibilities of some trades have expanded into additional areas. For example, as the use of steam for heating and powering industrial processes has changed, the stationary engineer has become responsible for all heating, cooling, and industrial process equipment, in addition to operating the boilers. Therefore, maintenance personnel are broadly classified as specialists or multiskilled.

Specialist maintenance personnel have a great amount of expertise in one craft. Multiskilled maintenance personnel have expertise in several crafts. The structure of the maintenance organization is determined by the classification of maintenance personnel. Maintenance organizations with specialist maintenance personnel have a lead person for each craft. Maintenance organizations with multiskilled maintenance personnel have a lead person for each shift or crew.

✚ SAFETY TIP

Maintenance personnel should hold monthly safety meetings to address problems and accidents that have occurred. Discuss safety regulations or changes in equipment and methods to be adopted for safety reasons.

LEED® GREEN BUILDING RATING SYSTEM™	
Key Measurement Areas	**Rating Categories**
• Sustainable Sites	• New Construction
• Water Efficiency	• Existing Buildings: Operations & Maintenance
• Energy & Atmosphere	• Commercial Interiors
• Materials & Resources	• Core & Shell
• Indoor Environmental Quality	• Schools
• Location & Linkages	• Retail
• Awareness & Education	• Healthcare
• Innovation in Design	• Homes
• Regional Priority	• Neighborhood Development

Figure 1-3. The LEED® Green Building Rating System™ addresses the efficient use of resources through measurements in several key areas. This system can be used to rate the operations and maintenance of existing buildings.

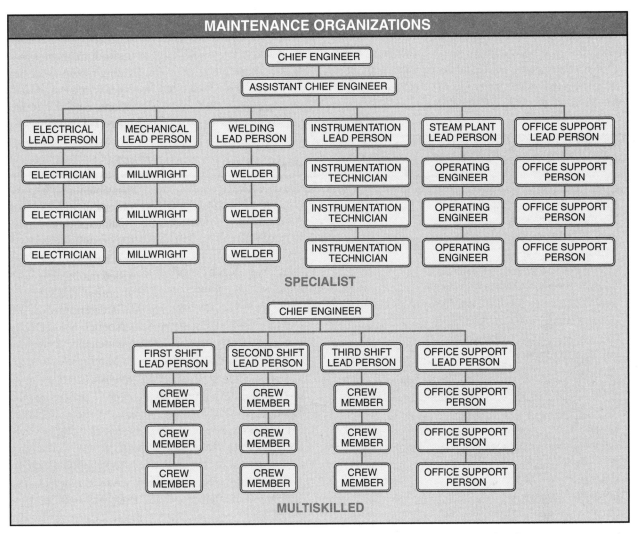

Figure 1-4. The structure of the maintenance organization is determined by the classification of maintenance personnel.

Multiskilled maintenance engineers are becoming more popular in industry because they often provide flexibility and cost effectiveness in maintenance organization. With more individuals qualified for each task, it is often easier to assign maintenance work and have necessary maintenance staff available when needed. Additionally, multiskilled maintenance personnel may be more effective at troubleshooting. For example, an electrical specialist investigating a boiler malfunction may be familiar with only the electrical system and overlook a mechanical problem with the water-level controller. However, a multiskilled maintenance engineer that is familiar with many aspects of boiler operation may identify the problem more quickly.

Even among those with a variety of skills, it is common for each engineer to have strengths in certain crafts. For example, one member of a maintenance crew might specialize in electrical work while another might be a better welder. However, each member is qualified to perform a variety of maintenance tasks. If a critical system malfunctions, each must be able to address the problem quickly to minimize downtime.

Various maintenance-personnel jobs are defined by the federal Standard Occupational Classification (SOC) system and identified by job titles and numbers. **See Figure 1-5.** This classification system is also used by the U.S. Department of Labor's O*NET® database, which details the knowledge, skills, abilities, and tasks required of each job.

Maintenance personnel are employed in a variety of building types, including commercial and industrial. Commercial buildings include schools, office buildings, retail stores, and hotels. Industrial buildings include facilities for food processing, manufacturing, refining, mining, assembly, and other heavy processes. Much of the maintenance work found in these two settings is similar, though industrial buildings often require larger or more sophisticated equipment.

SELECTED MAINTENANCE JOB TITLES	
Job Title	SOC Number*
First-Line Supervisors/Managers of Mechanics, Installers, and Repairers	49-1011
Electric Motor, Power Tool, and Related Repairers	49-2092
Electrical and Electronics Repairers, Commercial and Industrial Equipment	49-2094
Control and Valve Installers and Repairers	49-9012
Heating, Air Conditioning, and Refrigeration Mechanics and Installers	49-9021
Industrial Machinery Mechanics	49-9041
Maintenance and Repair Workers, General	49-9042
Maintenance Workers, Machinery	49-9043
Millwrights	49-9044
Riggers	49-9096
Helpers—Installation, Maintenance, and Repair Workers	49-9098
Installation, Maintenance, and Repair Workers, All Other	49-9099
Stationary Engineers and Boiler Operators	51-8021

* 2000 Standard Occupational Classification (SOC) system

Figure 1-5. Various maintenance-related jobs are defined in the Standard Occupational Classification (SOC) system.

While all maintenance work should be performed promptly, this can be especially critical in industrial facilities. For example, if an HVAC system in a commercial office building fails workers may become uncomfortable, but can usually continue to work until it is repaired. In an industrial facility, however, downtime due to equipment maintenance or breakdown can be extremely costly because it stops production. This results in a loss of worker productivity and possible spoilage of products. Downtime costs are often estimated at thousands of dollars per minute, so a failure in an industrial setting must be corrected immediately.

Safety is a critical aspect of maintenance tasks, as much of the work involves potential electrical shock, burning, chemical, noise, flying debris, or other hazards.

Maintenance personnel must be trained in all personal protective equipment (PPE) and safety procedures appropriate to the tasks. In the course of their work, maintenance personnel should also be aware of any potential safety problems with other facility equipment or procedures. These problems should be reported immediately so that they can be remedied in the appropriate manner.

UNSCHEDULED MAINTENANCE

Unscheduled maintenance is impromptu service that is required due to a failure. Unscheduled maintenance should be minimized, though in many cases it cannot be entirely avoided. This type of work may also disrupt regularly scheduled maintenance, as some unscheduled tasks must be handled immediately. Unscheduled maintenance includes emergency work and breakdown maintenance, which differ based on whether the failed equipment receives regular maintenance.

Emergency Work

Emergency work is unscheduled service to correct an unexpected failure on equipment that receives regular maintenance. The regular maintenance, such as cleaning or lubrication, is intended to prevent most failures. However, problems can still occur due to accidents, changing conditions, or previously undetected abnormalities.

Emergency work orders are issued to repair damaged equipment immediately. **See Figure 1-6.** Keeping a log of emergency work provides information that can improve maintenance procedures or equipment design by identifying common equipment problems.

Breakdown Maintenance

Breakdown maintenance is unscheduled service on failed equipment that has not received any regular maintenance. Breakdown maintenance is the least sophisticated maintenance work, but is often appropriate for equipment that is inexpensive and noncritical to facility operations. For example, light bulbs are serviced using breakdown maintenance because it is less costly to replace a bad bulb than to predict a bulb failure by testing.

However, if applied to the wrong equipment, breakdown maintenance can be expensive. For example, centrifugal pump bearings should be maintained and replaced according to manufacturer's recommendations because excessively worn bearings that are allowed to operate until failure can result in costly shaft assembly replacement and/or pump damage.

EMERGENCY WORK ORDER		No. 821463

MACHINE ___Box Former___   Date ___3/12___

Time Called ___2:10___   (AM) PM   Mechanical ☒

Time Arrived ___2:15___   (AM) PM   Employee ☐

Time Finished ___2:22___   (AM) PM   Other ☐

Machine Downtime ___12 min___

Production Time Lost ___12 min___

Number of Employees affected by shutdown _____

Reason for Stoppage ___Vacuum grip failed due to bad solenoid.___
___Loose connection. Grip also out of place.___

Maintenance Technician ___Jean Smith___
Production Supervisor ___Gerald Brown___

Figure 1-6. Emergency work orders are issued to repair damaged or malfunctioning equipment immediately.

PREVENTIVE MAINTENANCE

Preventive maintenance (PM) is scheduled work required to keep equipment in peak operating condition. When in peak operating condition, the equipment operates as designed, producing high-quality output at maximum efficiency. PM minimizes equipment malfunctions and failures and maintains optimum production efficiency and safety conditions in the facility. This results in increased service life, reduced downtime, and greater overall plant efficiency. PM tasks and frequency for each piece of equipment are determined by the manufacturer's specifications, equipment manuals, trade publications, and worker experience. PM work is requested and scheduled by plant personnel using work orders.

Scheduled Maintenance

Scheduled maintenance is work that is planned and scheduled for completion. Scheduled maintenance is performed to minimize emergency work and ensure reliable and efficient operations. Work orders are used to organize, plan, and monitor scheduled maintenance tasks. A *work order* is a document that details specific maintenance tasks to be completed. Work orders commonly include time, date, name and location of the equipment, work description, approximate time to complete the work, and safety requirements. Some work orders also list the steps for completing the task. Work orders are often printed on paper forms so they can be easily brought to the worksite for reference. **See Figure 1-7.**

Work order forms may be handwritten, but many facilities use computer systems to automatically generate and print work orders. **See Figure 1-8.** The printed work orders are based on maintenance intervals and other information stored in the system's database. All work orders have a scheduled completion date. Work that cannot be completed by this date is listed in a delinquent work order report until the work is completed. **See Figure 1-9.**

A work order is generally required for all scheduled maintenance. Scheduled maintenance work includes periodic maintenance, corrective work, and project work.

Periodic Maintenance. *Periodic maintenance* is work completed at specific intervals. Periodic maintenance is scheduled based on time, such as daily, weekly, monthly, quarterly, or according to equipment operating hours. Periodic maintenance tasks commonly include the following:

• inspection of equipment for conditions that indicate potential problems such as unusual noise, leaks, or excessive heat

- lubrication of equipment at scheduled intervals
- adjustments and parts replacement to maintain equipment's proper operating condition
- checking the electrical, hydraulic, and mechanical systems of operating equipment

⬡ TECH TIP

As the size and complexity of a maintenance organization increases, the extent to which maintenance work planning must be formalized and scheduled for efficiency also increases.

PRIORITY	☐ IMMEDIATE ☐ 24 HOURS ☒ TIMELY	MAINTENANCE REQUEST FORM	WO #: _1556_ REQUESTED BY: _Chris K._

| DATE: *7/28* | TIME: *9:23* ⓐⓜ pm | AREA: *Packing Area* | EQUIPMENT: *Overhead Lights* |

ASSIGNED TO: *Susan W.*

WORK DESCRIPTION:

SAFETY
☐ TAGOUT REQUIRED
☐ LOCKOUT REQUIRED
☐ HOTWORK PERMIT REQUIRED
☐ CONFINED SPACE ENTRY
   PERMIT REQUIRED

Replace fluorescent lamps.

Check lighting ballasts.

START DATE _____

START TIME _____

COMPLETION DATE _____

COMPLETION TIME _____

TECHNICIAN INITIALS _____

☐ DOWNTIME: ___ DAYS ___ HOURS ___ MONTHS
☐ PARTS ORDERED - REQUISITION #:

Figure 1-7. A work order includes information required to complete specific maintenance tasks.

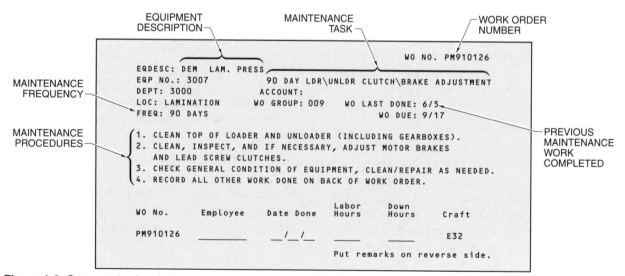

Figure 1-8. Computerized maintenance systems can automatically generate work orders when scheduled maintenance is due.

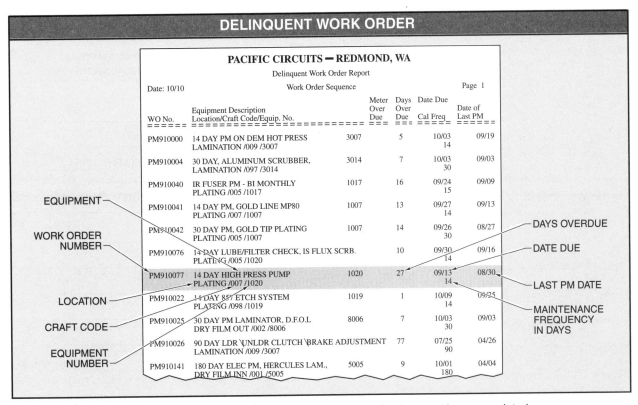

Figure 1-9. A delinquent work order report lists scheduled work orders that have not been completed.

Periodic work orders specific to one piece of equipment or several pieces of equipment are scheduled at specific intervals throughout the year. If a computerized system is used, a master schedule can be created with work orders projected automatically by day, week, month, or year. **See Figure 1-10.**

Corrective Work. *Corrective work* is the repair of a known problem before a breakdown occurs. Corrective work is requested after a problem is discovered during periodic inspections or while performing other maintenance tasks. Information on the work completed, supplies used, cause of problem, costs, and time for completion is recorded in the PM system when corrective work is completed. **See Figure 1-11.**

Project Work. *Project work* is work on long-term projects that require advanced planning and more time than typical maintenance tasks. Project work commonly includes rebuilding or modifying equipment, renovating structures, or installing new equipment.

Work Priority

Work priority is the order in which work should be done based on importance. The most important work is done first, followed by less important work. Work priority is indicated on the work order. Though work priority methods vary from plant to plant, a three-level priority method is commonly used. The highest priority is for work relating to safety, downtime, and production efficiency. The medium priority is for periodic maintenance. The lowest priority is for long-term projects. Some plants list work priority by completion time, such as "emergency", "within two weeks", or "timely".

The size of the plant and the number of available personnel dictate work-priority procedures. For example, some plants list PM tasks as the highest priority with only designated workers responding to emergency calls. Additionally, many maintenance engineers are "on-call" during their lunch and breaks. This allows quick response to high-priority maintenance calls.

Regardless of work priority procedures, maintenance personnel are always alert for signs of potential maintenance problems. For example, a change in the sound of a motor may indicate potential failure. The smell of hot electrical insulation requires immediate investigation. A glance at critical gauges can identify a problem.

WORK ORDER PROJECTION - TASK BY WEEK

RF Industries

Week 2

Date	Task No.	WO No.	Equipment No.	Cost Center	Expense Class	Hours
7/12	BEARING-REPLACE	-	MOTOR-EXTRUD1-LN1	7001	MECH	2.0
7/12	BEARING-REPLACE	-	MOTOR-EXTRUD1-LN1	7001	MECH	2.0
Total						4.0

Week 3

Date	Task No.	WO No.	Equipment No.	Cost Center	Expense Class	Hours
7/19	BEARING-REPLACE	-	MOTOR-EXTRUD1-LN1	7001	MECH	2.0
7/19	BEARING-REPLACE	-	MOTOR-EXTRUD1-LN1	7001	MECH	2.0
7/19	EXTRD-MTR-BELT-3M	-	MOTOR-EXTRD2-LN2	7001	MECH	2.0
7/19	EXTRDSCREW-BRNG-6M	-	SCREW-EXTRD1-LN2	7001	MECH	4.0
7/19	EXTRDSCREW-BRNG-6M	-	SCREW-EXTRD1-LN2	7001	MECH	4.0
Total						14.0

Week 4

Date	Task No.	WO No.	Equipment No.	Cost Center	Expense Class	Hours
7/26	BEARING-REPLACE	-	MOTOR-EXTRUD1-LN1	7001	MECH	2.0
7/26	FKLFT-PM-1M	-	-MULTITASK-		COMB	6.0
7/26	BEARING-REPLACE	-	MOTOR-EXTRUD1-LN1	7001	MECH	2.0
7/26	DIE-CLEAN	-	DIE-LN2	7001	MECH	2.0
Total						12.0

Datastream Systems, Inc.

Figure 1-10. A computerized maintenance system can compile lists of future maintenance tasks, which helps schedule personnel needs.

WORK ORDER-CORRECTIVE

```
Work Order:   COR 345              Description:  Repair damaged chain guard
Asset/Equip:  Conveyor #2
Procedure:  Repair damaged chain guard
Requested by:  Packing Lead
Telephone:  Ext. 556
Asset Shutdown:   X   Plant Shutdown
Model:   NA
Serial No:   NA
Location:   Packing Room
Skill:   Level 2
Sch Date:   Timely
Priority:   2
Shift:   NA
Status:   Open
Labor      Craft     Labor Description    Est Hrs. REF OVE DOUBL
   Maintenance      (Name printed in)
Tasks:   ID COR 345 PR 2
   Notify Packing Lead Person about repair.
   Lock out power to packing line #2.
   Repair damaged chain guard or fabricate new guard on
      drive motor for conveyor #2 in packing area.
   Remove Lockout.
   Notify Packing Lead Person when completed.
```

Comments: 2 hours.
Damage caused by forklift.
Rebuilt guard from mounts out.
No unusual costs.
Need to order more 6010 rod.

Figure 1-11. A corrective work order specifies repair work needed before a breakdown occurs.

Logbooks

A *logbook* is a book or electronic file that documents all work performed during a shift and lists information needed to complete further work by maintenance personnel during other shifts. **See Figure 1-12.** Maintenance personnel begin each day by reviewing the logbooks from previous shifts. Based on information in the log and the quantity and type of work orders issued, the maintenance engineer makes a list of all work to be completed. The work is then prioritized. Emergency or high-priority work requests are handled as they occur during a shift. In large facilities, a supervisor schedules daily activities. In smaller shops, maintenance engineers do their own scheduling.

Some maintenance organizations have specialized logs. For example, a maintenance department complaint log records all complaints regarding conditions in the facility and how each condition was addressed. In some cases, a personal log is used as a secondary means for documenting work completed.

When removing replacement or repair parts from inventory, the part numbers and quantity should be recorded in the inventory control system.

MAINTENANCE LOG

DATE: 3/5   NAME: Pat Williams   SHIFT: 4 PM to 12 PM

TIME	TASK	COMMENTS
1. 4-6	PM of oven #3, conveyors #5, 6, 7, & 8, pan washer.	No unusual conditions.
2. 6-7	Assisted Jean with mixer #6 motor replacement.	Motor overheated when mixer locked up.
3. 7-7:20	Dinner	
4. 7:20-9:30	Repair chain guard on conveyor #2 as per work order 35-556.	Used portable welder. Reorder 6010 welding rod.
5. 9:30-12	Repaired pie filler with Jean & Chris.	Retimed conveyor and pie trays with pie indexer. Worn parts. Ordered new conveyor chain, On-Off switch.

Graveyard Shift - Check timing of indexer.
Parts will arrive about 3 AM.
Filler keeps going out of time.
Check side mounted limit switch.

Figure 1-12. Logbooks are used by maintenance personnel to record tasks completed during a shift.

Preventive Maintenance Systems

Modern maintenance operations require more than just basic preventive maintenance activities designed to prevent major problems. A *preventive maintenance (PM) system* is a system used to record and organize maintenance information, which is then used to make the decisions required to maintain the facility and equipment. **See Figure 1-13.** This information includes items such as equipment data, maintenance costs, consumables, time on task, and breakdown resolutions. With a consistent, accurate flow of operation information, PM systems can help increase efficiency, reduce costs, and minimize health and safety problems. PM systems can also be used to document compliance with environmental and health and safety regulations. The PM system may include the entire facility, or only the departments that are expected to see the greatest benefit from improved maintenance practices.

PM systems can be implemented as paper-based systems, but many maintenance organizations use computerized maintenance management systems. A *computerized maintenance management system (CMMS)* is a software package that organizes preventive maintenance information and automatically generates reports, work orders, and other data for implementing and improving future maintenance activities. **See Figure 1-14.** CMMS software provides quick access to maintenance information, issues and tracks work orders, determines the costs of maintenance activities, schedules maintenance work, manages maintenance inventories, and assists in troubleshooting.

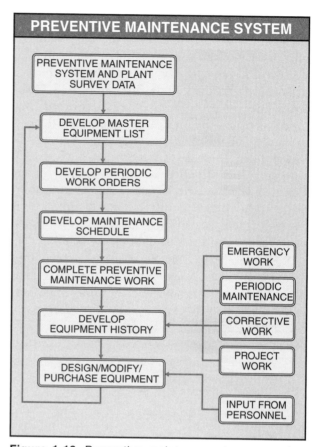

Figure 1-13. Preventive maintenance is an on-going system that generates data that is useful for refining plant operations.

CMMSs also help analyze the frequency and type of PM work and make adjustments in order to keep equipment in peak operating condition with minimal cost. Excessive PM work increases maintenance costs, but inadequate PM work also results in high maintenance costs due to an increase in breakdowns. Data compiled from the PM system is also used for assessing plant performance, equipment service life, energy costs, equipment purchase needs, insurance costs, and personnel and plant budget decisions.

Plant Survey. A plant survey is the first step in implementing a PM system. A *plant survey* is a complete inventory and condition assessment of a facility's equipment and structure. Data from the plant survey is entered into the PM system to create a master file for each piece of equipment. This file lists manufacturer, vendor, serial and model numbers, other identifying codes, parts suppliers, equipment location, and complete service history.

Fluke Corporation
Preventive maintenance involves scheduled testing and recording of certain equipment parameters.

COMPUTERIZED MAINTENANCE MANAGEMENT SYSTEM (CMMS)

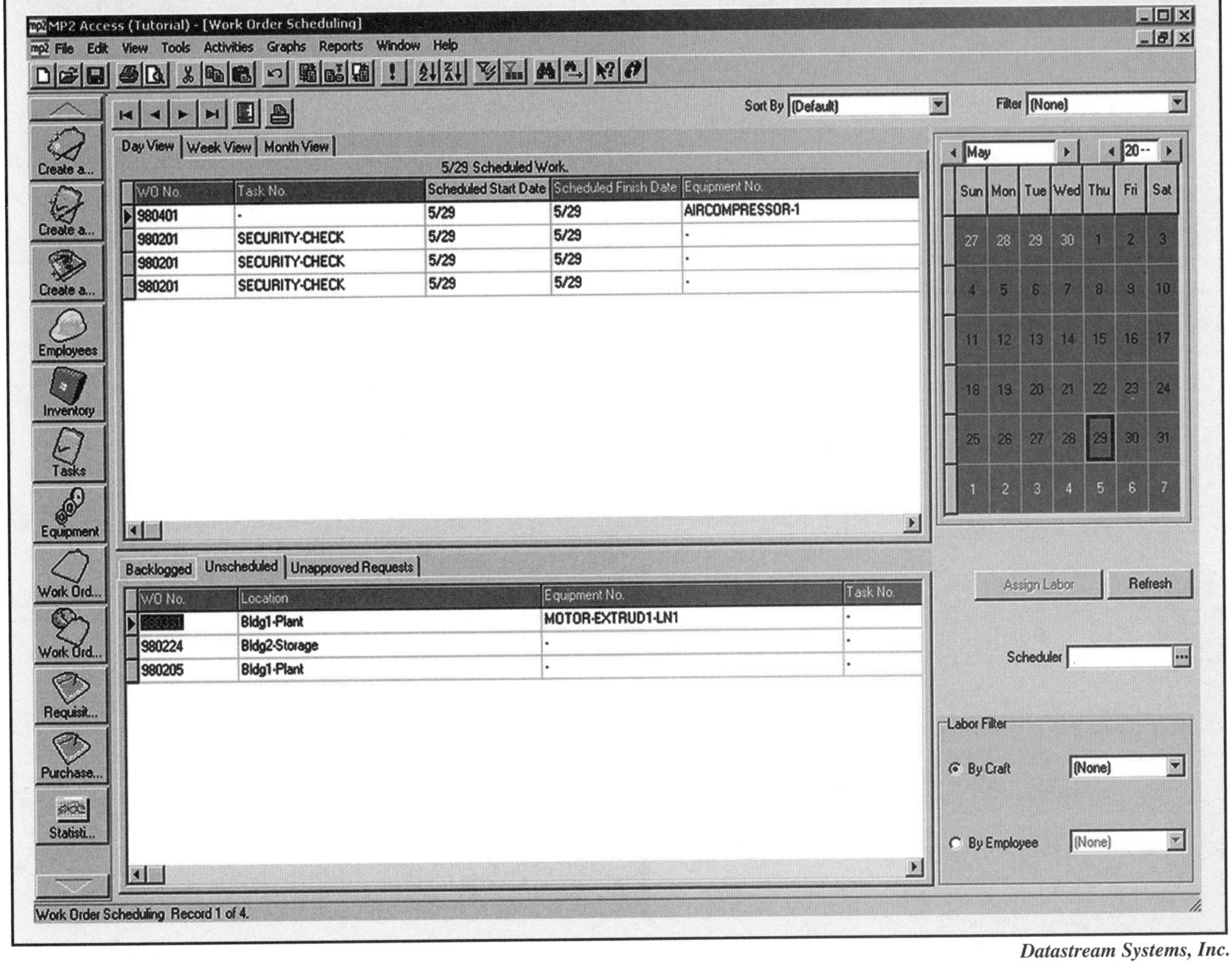

Datastream Systems, Inc.

Figure 1-14. A computerized maintenance management system (CMMS) takes the entered information for a preventive maintenance program and helps organize, schedule, and record future maintenance activities.

The plant survey is completed by in-house personnel or outside contractors. As part of the plant survey, the equipment is carefully inspected and analyzed. **See Figure 1-15.** Its condition is noted, as are any repairs needed to return the equipment to peak performance. These repairs become the first corrective work orders in the new PM system.

The plant survey typically uses paper forms to collect information that is later entered into a computer. Information can also be entered directly into a hand-held computer and transferred to a main computer later.

A master equipment file expands with newer information as the equipment is maintained and repaired. The data is used to identify possible problems when troubleshooting and quantify the equipment's reliability. The equipment reliability and total cost of operation information then influences future purchasing decisions.

TECH TIP

Some CMMSs can be integrated with building automation systems. This allows the maintenance system to access the building information being generated and shared by the building automation system controllers. This system information can be used for maintenance applications, such as ensuring optimal system settings, monitoring energy use, and tracking equipment operating times.

PLANT SURVEY FORM

EQUIPMENT IDENTIFICATION: Furnace #1 - FUR0001.00

LOCATION: 4247 Piedmont Building 2, Floor-1 Room 23 CL15F

DATE OF PURCHASE: 08/17

VENDOR: Lewis Systems, Inc.

1862 Erie St. Cleveland, OH

VENDOR PHONE: 216-555-1340

MANUFACTURER: Brown Boveri, Inc.

MODEL #: IT6P

SERIAL #: OP2810C2B

EQUIPMENT DATA: (List all parts and part numbers important to the operation of the equipment.)

PLANT SURVEY FORM

Description:	Furnace #1	Voltage:	440	Warranty ID:	IT6P882
Asset ID:	FUR0001.00	Amperage:	600	Warranty Date:	02/17
Asset Type:	Furnace Systems	Wattage:	264000		
Parent ID:		Phase:	3	YTD Labor Hr:	45.00
Priority:	8 Active:☒	Elec Line:	10	YTD Downtime:	22.00
Manufacturer:	Brown Boveri, Inc.	Air Area:	COMP 6		
Model:	IT6P			TD Labor Hr:	740.00
Serial Number:	OP2810C2B			TD Downtime:	493.00
Vendor:	Lewis Systems, Inc.	Counter UOM:			
Vendor Address:	1862 Erie St.	Current Counter:	1	YTD Labor Cost:	724.00
	Cleveland, OH 55117	Counter Rollover:	0	YTD Misc Cost:	1231.22
Vendor Phone:	216-555-1340	Meter UOM:		YTD Part Cost:	7701.19
Asset Tag:	00509	Current Meter:	1234550	Total:	9656.41
Location:	4247 Piedmont Building 2				
	Floor-1 Room 23 CL15F			TD Labor Cost:	9620.60
Department ID:		Meter Rollover:	0	TD Misc Cost:	5631.80
Cost Center:	Fixed Asset Repair	Purchase Date:	02/17	TD Part Cost:	14,942.00
Supervisor:	Jones, Fred	Install Date:	08/05	Total:	30,194.40
		Retire Date:			
		Install Cost:	97000		
		Replacement Cost:	97000		

Comment: Manufact. warranty extremely strict. Document all hours worked and parts used.

Report Totals:					
	YTD Labor Hr: 45.00	YTD Labor Cost: 724.00	TD Labor Cost: 9620.60
Assets: 1	YTD Downtime: 22.00	YTD Misc Cost: 1231.22	TD Misc Cost: 5631.80
	TD Labor Hr: 740.00	YTD Part Cost: 7701.19	TD Part Cost: 14,942.00
	TD Downtime: 493.00	Total: 9656.41	Total: 30,194.40

DPSI (DP Solutions, Inc.)

MASTER EQUIPMENT FILE

Figure 1-15. A plant survey and data from the PM system are combined to create a master equipment file for each piece of equipment.

Plant Documentation. All equipment documentation is gathered while assembling the master equipment files for the plant survey. This includes construction prints, wiring diagrams and schematics, replacement parts data, and installation, operating, maintenance, and troubleshooting manuals. **See Figure 1-16.** System documentation may be a mixture of paper-based documents or electronic files. Maintenance personnel must be able to locate information quickly in either paper or electronic formats.

Figure 1-16. Plant documentation consists of prints, operating manuals, and other documents required to operate the facility and equipment.

Paper-based documents require two copies. A reference copy, often the original document, is stored in a secure and permanent location. The working copy is used for daily tasks. The reference copy is used to make a new working copy if the working copy is lost or damaged. Any changes must be noted on both sets of documents.

Electronic files are becoming more popular for storing maintenance information, as they are easy to store, organize, and access. Electronic files of large, complicated drawings are sometimes easier to use than traditional paper documents because specific sections can be enlarged on screen to display details. The files are stored on shared computers or networked hard drives so that it is available to all maintenance personnel. Although electronic files are considered to be the reference copy, some organizations print paper copies to record changes and to use as backups in the event of computer problems.

Program Implementation. Once all the data is gathered, the amount and type of PM work is determined. Working from the manufacturer's recommendations, maintenance and inspection routines are formulated, along with the expected costs.

Maintenance costs and manufacturer's warranties determine the maximum amount of work a piece of equipment should receive. **See Figure 1-17.** It is wasteful to spend more money on maintenance for a piece of equipment than its replacement costs, unless the unit is critical for operations or involves health or safety issues. For example, most electric motors receive little or no preventive maintenance, but a motor that is critical to production might receive regular insulation testing to anticipate its possible failure so that it can be replaced before it fails.

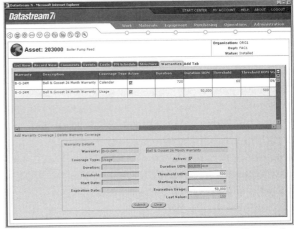

Datastream Systems, Inc.

Figure 1-17. Entering information about equipment warranties into a preventive maintenance system helps determine the most cost-effective solution to equipment breakdowns.

Changes to the formulated routines are made as needed to assign the appropriate amount of preventive maintenance to each piece of equipment, then the work is scheduled and assigned. Routine lubrication and inspections might be assigned to equipment operators, while work requiring more skills is assigned to qualified maintenance personnel. Efficient scheduling generates an even workload throughout the year and ensures that equipment is ready when needed.

Maintenance personnel periodically review maintenance routines and their actual costs, adjusting tasks as needed to balance equipment availability and costs. For example, if a piece of equipment is performing well, the maintenance interval might be lengthened. The breakdown rate is monitored and the frequency of work is changed if necessary. Other factors affecting the type and frequency of preventive maintenance are energy usage and the quality of work by the equipment.

Inventory Control. *Inventory control* is the organization and management of commonly used parts, vendors and suppliers, and purchasing records in the PM system. Each part type is assigned a unique number that is used to track its quantity on hand and associate the part with data such as supplier, equipment use, purchase date, and cost. The most effective use of an inventory control system is one that integrates this information with a CMMS.

To expedite data entry and tracking, the part number label is often printed in a computer code that can be read automatically. Barcodes have been used for this purpose for decades, as they are inexpensive and reliable. A *barcode* is a code consisting of a group of variously patterned bars and spaces that represent a certain number. **See Figure 1-18.** Barcodes can be scanned quickly and easily with an optical barcode reader, which inputs the identification number into software.

Scanning barcodes as parts are entered and removed from inventory allows the computer to keep up-to-date records of available parts. When quantities of parts fall to certain levels, the system is triggered and generates a purchase order for the number of parts required. This helps maintain an adequate inventory to support operations without excessive parts and material.

> ⊕ **SAFETY TIP**
>
> *Inventory control systems can also be used to keep track of consumable safety supplies, such as earplugs, respirator filters and cartridges, and safety glasses.*

BARCODES

Hand Held Products

Figure 1-18. Barcodes are often used for inventory control, which requires the organization and management of parts commonly used for maintenance tasks.

> **TECH TIP**
>
> *Paper-based organization systems for preventive maintenance programs are less expensive, but do not provide the capabilities of a CMMS. For example, in paper-based systems, cross-referencing files requires more time because files must be searched by hand.*

PREDICTIVE MAINTENANCE

Predictive maintenance (PDM) is the monitoring of wear conditions and equipment operation characteristics for comparison against a predetermined tolerance to predict potential malfunctions or failures. The data is periodically analyzed for trends in equipment performance and to check whether values are within acceptable tolerances. **See Figure 1-19.** Corrective maintenance is required if the data falls outside these tolerances. The equipment is closely monitored after maintenance is performed. If the problem reoccurs, the equipment application and design are analyzed and changes are made as required.

Since PDM requires a substantial investment in training and equipment, it is most commonly used on expensive or critical equipment. The equipment monitoring may be random, scheduled, or continuous. *Random monitoring* is unscheduled equipment monitoring as required. *Scheduled monitoring* is equipment monitoring at specific time intervals. *Continuous monitoring* is equipment monitoring at all times. Common PDM procedures include visual and auditory inspection, vibration analysis, lubricating-oil analysis, thermography, ultrasonic analysis, and electrical analysis.

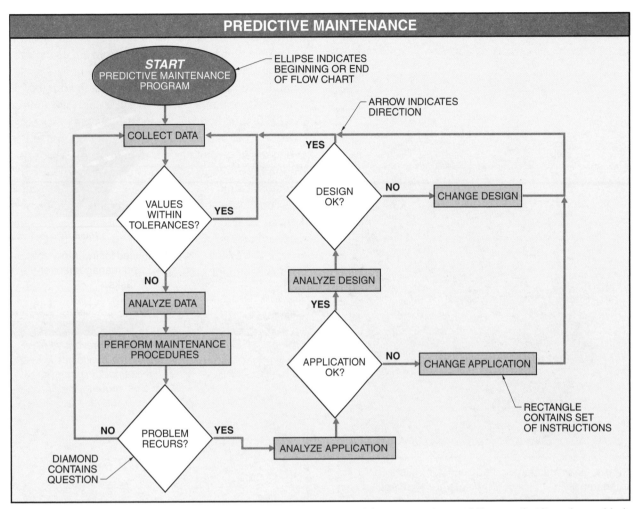

Figure 1-19. Predictive maintenance uses tests and procedures to anticipate an equipment failure, so that it can be avoided. Since it is a significant investment, predictive maintenance is most commonly used on expensive or critical equipment.

Visual and Auditory Inspection

Visual and auditory inspection is the analysis of the appearance and sounds of operating equipment. **See Figure 1-20.** Visual and auditory inspection is the simplest PDM procedure performed in a facility and requires no tools or equipment. Visual and auditory inspection is most effective where a potential problem is obvious to trained maintenance personnel. Unusual operating characteristics are noted and the equipment is scheduled for the required maintenance.

Visual inspection can be supplemented with processes such as dye penetrant testing to locate fine surface metal fractures. With dye penetrant testing the metal is completely cleaned and sprayed with dye that collects in small fractures

or pits on the surface of the metal. Excess dye is removed to reveal small cracks or pits below the surface.

Vibration Analysis

Equipment failure is frequently preceded by excessive vibration or noise, which can be isolated and measured using vibration analysis. **See Figure 1-21.** *Vibration analysis* is the monitoring of individual component vibration characteristics to analyze the component condition. Vibration in equipment may indicate gear wear, bearing wear or failure, imbalance, misalignment, or loose parts. Vibration analysis is the most common form of monitoring technique that uses tools or equipment.

VISUAL AND AUDITORY INSPECTION

Figure 1-20. Maintenance personnel routinely check the appearance and sound of operating equipment through visual and auditory inspection.

VIBRATION MEASUREMENT

UE Systems, Inc.

Figure 1-21. Vibration measurement records the displacement and frequencies of noise and vibration produced by equipment.

Vibration is oscillating motion in response to a force. **See Figure 1-22.** A *vibration cycle* is motion from a neutral position to the upper limit, from the upper limit to the lower limit, and from the lower limit back to the neutral position. Vibration frequency is the number of completed vibration cycles within a specified unit of time. Vibration frequency is most commonly expressed in cycles per minute (cpm). Rotating equipment is expressed in revolutions per minute (rpm), which corresponds to vibration frequency expressed in cpm. Vibration frequency can also be expressed in cycles per second (cps) or hertz (Hz). Vibration frequency at 1 Hz is equal to 60 cpm. *Vibration displacement* is the maximum range of motion from the upper limit to the lower limit of the vibration cycle. Vibration displacement is commonly expressed in mils.

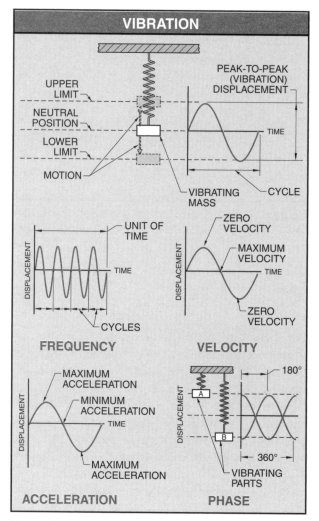

Figure 1-22. Vibration is analyzed for frequency, velocity, acceleration, and phase.

Vibration velocity is the speed of travel from an extreme limit of the vibration displacement to the neutral position. Vibration velocity is typically expressed in inches per second (ips). *Acceleration* is the change in velocity. Maximum acceleration occurs at the upper and lower limits of the cycle. *Vibration phase* is the relationship between the peak of a vibration and a moment in time. The phases of multiple vibrations can be compared by noting the time between their respective peaks.

Vibration analysis techniques include broadband, narrowband, and signature analysis. *Broadband analysis* is a vibration analysis that compares current condition readings with baseline condition readings to detect a wide variety of changes. The baseline readings are established by the manufacturer's specifications or recorded when the equipment is new. For example, when new bearings are installed the frequency is measured and recorded. As the bearings wear, the vibration frequency changes. At a certain frequency, the bearing is scheduled for replacement before failure occurs.

Narrowband analysis is a vibration analysis that focuses on specific vibration frequencies that correspond to equipment components or failure features. For example, in equipment with multiple causes of vibration, such as unbalance, misalignment, and worn bearings, the overall vibration signal is complex and difficult to interpret. **See Figure 1-23.** However, the vibration analysis reveals the narrow bands of frequencies that correspond to each problem.

Signature analysis is the visual comparison between two vibration frequency patterns (signatures) to detect differences. The comparison can be between different equipment or the same equipment at different times. For example, two centrifugal pumps installed at the same time should display similar vibration signatures. Likewise, equipment in good condition operating over a period of time should display similar vibration signatures.

Ultrasonic Analysis

Ultrasonic analysis is analysis that detects high-frequency vibrations to create an image or reading. Ultrasonic analysis is similar to vibration analysis but involves higher frequencies that cannot be heard. With amplification and proper sound isolation and analysis techniques, ultrasonic analysis can detect conditions that may lead to mechanical problems that cannot yet be detected by vibration or lubrication-oil analysis.

Ultrasonic sensors can also measure the flow of liquids and gases from outside the pipe or vessel by sensing the vibrations created by their motion. **See Figure 1-24.** Leaks can be detected by comparing the results with data collected on similar equipment or at different times.

⊕ SAFETY TIP

Predictive maintenance analysis tasks may require the removal of safety guards while equipment is running, which should be done with extreme caution.

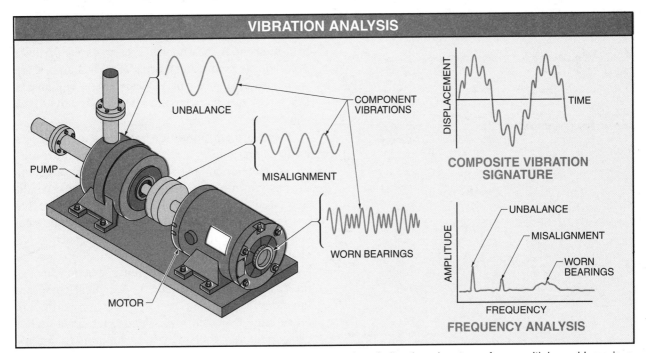

Figure 1-23. A complex vibration signature is usually a composite of vibration signatures from multiple problems in a system.

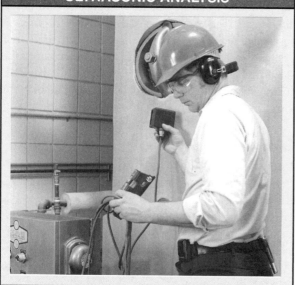

ULTRASONIC ANALYSIS

Figure 1-24. Ultrasonic analysis measures high-vibration frequencies that indicate liquid flow or leaks.

Internal flaws or cracks in metal can be located using ultrasonic testers that consist of a sending and receiving unit. Ultrasonic waves from the sender are passed through the metal being tested. Internal flaws or cracks distort the sound waves as they pass through the metals. These distortions are interpreted by the receiver, which generates a picture or readout of the flaw.

Datastream Systems, Inc.
Predictive maintenance data can be entered into a CMMS for analysis and recordkeeping.

Ultrasonic analysis is also used in leak detection. Different vibration frequencies emitted from pressure and vacuum leaks are recorded. This allows testing for leaks at the source of any gas without the limitations of sensing equipment designed for specific gases.

Oil Analysis

Oil analysis is a predictive maintenance technique that detects and analyzes the presence of acids, dirt, fuel, and wear particles in lubricating oil. Lubricating-oil analysis is performed on a scheduled basis. An oil sample is taken from a machine and sent to a company specializing in lubricating-oil analysis. **See Figure 1-25.** Abnormal results can indicate improper lubrication, worn parts, seal leaks, or other machine problems.

Measured physical properties of the lubricating oil include viscosity (resistance to flow), percentage of water, oxidation, nitration, sulfate, solvents, refrigerants, and insolubles. Trace elements are listed in parts per million (ppm). *Wear particle analysis* is the study of the size, frequency, shape, and composition of wear particles present in the lubricating oil. This provides further detail into equipment condition.

Normal wear occurs as equipment surfaces slide past each other. An increase in the frequency and size of wear particles in the lubricating oil indicates excessive wear, predicting possible failure. For example, lubricating oil samples having consistent wear particle readings over a period of time provide a baseline measurement. A subsequent increase in wear particles may indicate premature wearing of parts. Large, sharp wear particles indicate parts sheared in the equipment. Fractured wear particles indicate broken parts in the equipment.

The lubricant condition rating is identified as normal, marginal, or critical based on the results of the sample, comparison with previous data, and experience with similar equipment. A normal condition rating indicates the lubricant is within expected levels and requires no corrective action. A marginal condition rating indicates that critical physical properties and/or trace elements are outside expected levels and require minor maintenance action such as increased sampling frequency. A critical condition indicates that the majority of physical properties are outside the expected levels, and a lubricant and/or wear condition problem exists that requires definitive maintenance action.

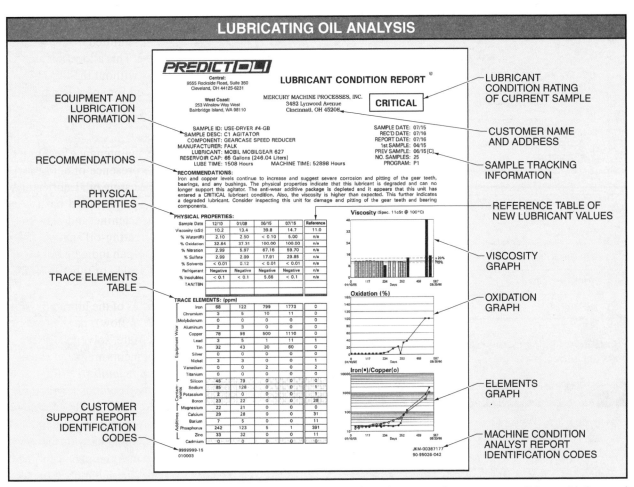

Figure 1-25. Oil analysis can predict potential equipment failures by detecting contaminants, wear particles, oxidation, and loss of viscosity in lubricating oil.

Thermometry

Thermometry is the use of temperature-indicating or measuring devices to quantify temperature or temperature changes. Thermometry is used in maintenance work to check equipment operating conditions. Temperature changes or extreme temperatures, either high or low, may indicate a condition that could cause problems in the future. Temperature is usually measured in degrees Fahrenheit (°F) or degrees Celsius (°C). Temperature-indicating devices commonly used in industrial maintenance can be either portable or stationary. **See Figure 1-26.**

Portable Temperature-Indicating Devices. Portable temperature-indicating devices include glass thermometers, temperature-indicating crayons, and electronic thermometers. Glass thermometers indicate temperature through the uniform expansion of a liquid in a sealed glass tube. Marks from temperature-indicating crayons

melt at a certain temperature. Crayons are available that melt at different temperatures.

Portable electronic thermometers include thermocouples, thermistors, and infrared-sensing devices. These devices are hand-held instruments that can be used at different worksites to measure the temperature of various components or environments. Thermocouples use two dissimilar conductors with different voltage-producing characteristics to measure temperature change. Thermistors measure temperature by the change in electrical resistance of a material. Infrared thermometers sense and measure infrared radiation from an object to determine temperature. All objects having a temperature above absolute zero emit infrared radiation. This allows temperature measurements to be taken from a distance without direct contact. Some infrared thermometers have a laser sight that allows precise targeting of the object to be measured. **See Figure 1-27.**

TEMPERATURE-INDICATING DEVICES

	DEVICE	TEMPERATURE RANGE (°F)*					
		−460　　　0　　　500　　　1000　　　1500　　　2000　　　2500					
PORTABLE	GLASS THERMOMETER THERMOMETER	▬ (approx. 0 to 500)					
	TEMPERATURE-INDICATING CRAYON	▬ (approx. 100 to 2500)					
	THERMOCOUPLE	▬ (approx. −300 to 2500)					
	THERMISTOR	▬ (approx. −100 to 300)					
	INFRARED	▬ (approx. 0 to 2500)					
STATIONARY	TEMPERATURE-INDICATING PAINTS AND LABELS	▬ (approx. 100 to 500)					
	BIMETAL	▬ (approx. 100 to 600)					
	THERMOCOUPLE	▬ (approx. −300 to 2500)					
	THERMISTOR	▬ (approx. −100 to 300)					
	INFRARED	▬ (approx. 0 to 2500)					

* with various probes

Figure 1-26. Temperature-indicating devices help identify changes in operating temperatures that predict potential equipment malfunction or failure.

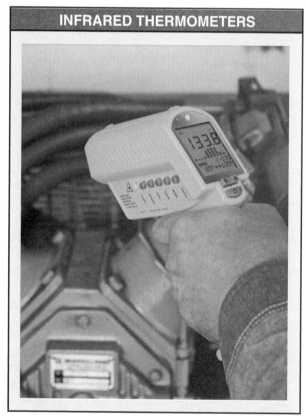

INFRARED THERMOMETERS

Fluke Corporation

Figure 1-27. Infrared thermometers measure the emission of infrared radiation to determine temperature without touching the object.

A thermal imager is an extension of single-point infrared thermometer technology that builds an infrared image of an area from many temperature measurements. These imagers work like digital cameras in that they display the image on a screen and can save it for later analysis. The electronic image files are transferred to a personal computer for storage, analysis, and inclusion into maintenance reports. Some imagers record both visible light and infrared images of the same field of view, which help identify specific components later. The temperature information is usually shown in false color to highlight the high and low temperatures. The use of thermal imagers is known as thermography.

Thermal imagers are an extremely effective tool for quickly inspecting a large area or many pieces of equipment and recording the results. Many maintenance problems, such as loose electrical connections, bad motor bearings, moisture penetration, insulation gaps, or leaky ductwork, are immediately apparent with a thermal imager. **See Figure 1-28.**

Infrared-based temperature-indicating devices, of any type, have some limitations. Temperature readings may be affected by the surface material and its condition, such as being dirty, painted, or insulated. Low-emissivity materials, such as polished metals, reflect their thermal surroundings, making readings confusing. The surrounding environment, such as water vapor, dust, and gases in the air, may also affect temperature readings.

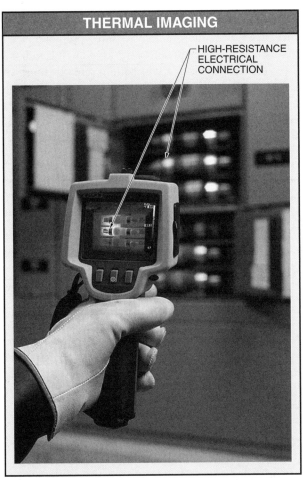

THERMAL IMAGING

HIGH-RESISTANCE ELECTRICAL CONNECTION

Fluke Corporation

Figure 1-28. Thermal imagers can scan a large area very quickly for temperature extremes and record the images for use in reports.

Stationary Temperature-Indicating Devices. Stationary temperature-indicating devices include temperature-indicating paint and labels, and permanently mounted temperature sensors. The sensors may use one of a variety of temperature-indicating technologies, such as bimetal strips, thermocouples, thermistors, or infrared sensors.

Temperature-indicating paint and labels are applied to surfaces and change color at one or more specific temperatures. Bimetal strips consist of two dissimilar metals bonded together in the shape of a coil. The different expansion and contraction rate for each metal causes the coil to expand or contract with temperature changes. This motion can be used to move a needle in reference to a calibrated scale. Thermocouples, thermistors, and infrared sensors provide equipment

temperature, which is then displayed on a gauge at the equipment location, on a panelboard, or on a computer screen at a remote location.

Electrical Analysis

Electrical analysis is a type of analysis that uses electrical monitoring devices and/or test instruments to evaluate the performance of equipment and the quality of electrical power delivered to the equipment. Electrical monitoring involves measuring minimum and maximum voltages, phase-to-phase voltage variation, loss of voltage, current levels, and other power quality characteristics. Power quality is particularly important to sensitive electronic equipment and can affect many types of equipment over time.

Monitoring devices can be permanently installed to provide continuous power-quality information. Alternatively, portable test instruments are used to measure power quality or equipment conditions on an as-needed basis at a specific location. **See Figure 1-29.** For example, the condition of an electric motor can be checked by performing an insulation spot test, which uses a megohmmeter to measure the resistance of motor winding insulation. Over time, the effects of heat, aging, and/or contamination can cause insulation breakdown. The values obtained from tests at scheduled time intervals are compared with benchmark acceptable readings to predict useful life and motor insulation failure.

TROUBLESHOOTING

Troubleshooting is the systematic investigation of the cause of system problems in order to determine the best solution. Troubleshooting methodically eliminates various system processes and components as causes until the malfunction is located and diagnosed. A *system* is a combination of components, units, or modules that are connected to perform work or meet a specific need. Almost all equipment operates as part of one or more systems. For example, a thermostat is part of the heating and cooling system. A *process* is a sequence of operations that accomplishes desired results. Maintenance personnel use troubleshooting skills on a regular basis. The troubleshooting process consists of investigating, isolating, remedying, and documenting problems. **See Figure 1-30.** As always, maintenance personnel must follow all applicable safety procedures while troubleshooting.

Fluke Corporation

Figure 1-29. Electrical monitoring equipment is used to evaluate the quality of incoming power and the performance of electrical equipment.

TROUBLESHOOTING CHECKLIST
1. Inspect equipment.
2. List safety concerns.
3. Question operator.
4. Gather resources.
5. Operate equipment (if possible).
6. List all symptoms.
7. List all possible causes.
8. Formulate testing sequence.
9. Determine cause of problem.
10. List possible repairs.
11. Formulate repair plan.
12. Perform repair.
13. Observe operation.
14. Document process.

Figure 1-30. A troubleshooting checklist guides maintenance personnel through diagnosing and remedying a problem.

TECH TIP

Predictive maintenance activities can lead to troubleshooting when the cause of newly discovered problems is unknown.

Investigating Problems

Investigation of the problem requires preparation by maintenance personnel before working on-site with the affected equipment. Certain actions greatly improve troubleshooting efficiency but may not be possible due to time constraints and circumstances. These actions are as follows:

- Review previous work orders, equipment age, and equipment operating history.
- Check plant procedures and equipment manufacturer's procedures for relevant information.
- Determine the effect of the equipment on overall plant production and operation.
- Consult with other maintenance personnel who have experience with the equipment.
- Gather required tools, test equipment, and safety equipment.

When arriving on-site, the first step is to communicate with the equipment operator or personnel familiar with the equipment and gather information about the problem. Determine when the problem started, how the equipment functions normally, if the equipment has a history of problems, and what actions were taken to remedy past problems. Verify that plant procedures were followed in the operation of the equipment and that no unauthorized changes were made.

Carefully inspect for problematic conditions such as leaks, broken parts, or unusual odors. If the problem occurs during operation, energize the circuit and follow specific plant procedures for startup. General start-up testing procedures include the following:

- Check fluid levels, guard positions, belt tightness, and loose parts before turning the machine ON.
- Turn ON and test one part of the equipment or system at a time.
- Test all manual operations first.
- Verify that all safety devices are working properly.
- Operate the equipment long enough to obtain normal operating fluid levels, pressures, belt tensions, and temperatures.
- List all symptoms of the problem. If similar equipment is available, compare operating characteristics.

Isolating Problems

Isolation of the problem begins with listing possible causes. Writing down the symptoms and the likely causes can help focus the process. Start with the simplest, most likely cause and establish a logical troubleshooting sequence to test each suspect part of the equipment. Consider all plausible causes, no matter how simple, such as checking if batteries are charged, tanks contain fuel, or equipment is plugged in. Consider other problems that have occurred in the past with similar maintenance tasks.

Consult with plant procedures and manufacturer's troubleshooting procedures, which may be in the form of a list of steps or a flow chart. **See Figure 1-31.** A list of steps starts with the simplest, most obvious cause. The order of steps eliminates the possibility of inadvertently skipping over the cause of the problem. When one component tests as "good," the next component in the sequence is tested. Troubleshooters must know what to expect as a normal result of the test or inspection and may organize their testing data with a troubleshooting test chart. **See Figure 1-32.**

> ### TECH TIP
> *Since downtime in production systems is very costly, troubleshooting is a central focus of maintenance. When solving problems, maintenance personnel use the best repair possible while considering how to prevent future problems and save energy.*

A *flow chart* is a diagram that shows a logical sequence of troubleshooting steps for a given set of conditions. Flow charts use symbols and interconnecting lines to help a troubleshooter follow a logical path when solving a problem. An ellipse indicates the beginning or end of a flow chart. A rectangle contains a set of instructions. A diamond contains a question that generates a yes or no answer, which determines the path to the next instruction. Arrows throughout the flow chart indicate the direction to follow.

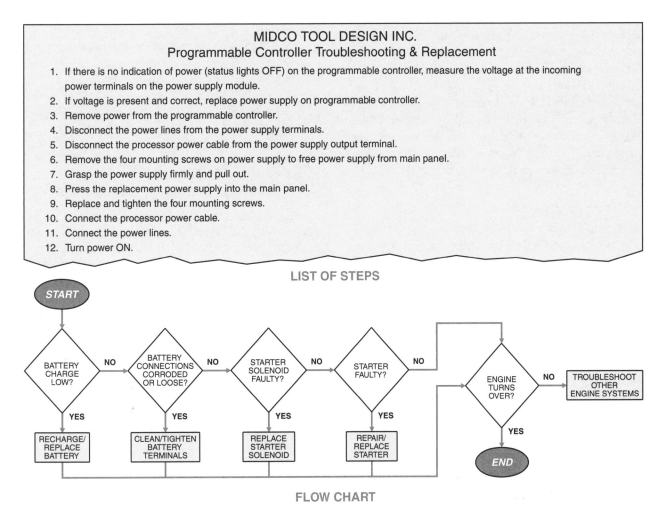

Figure 1-31. Plant troubleshooting procedures detail steps and procedures for investigating equipment problems.

TROUBLESHOOTING TEST CHART

COMPONENT TESTED	TEST EQUIPMENT	RESULTS IF GOOD	RESULTS IF BAD	ACTUAL RESULTS	NEXT ACTION
Incoming power	DMM voltmeter	about 120 VAC	high or low voltage	118 VAC	check power switch
Power switch	DMM voltmeter	120 VAC (open) 0 VAC (closed)	not within few volts	118 VAC (open) 0 VAC (closed)	inspect heater
Heating element	DMM ohmmeter	some resistance	OL	OL	replace heating element

Figure 1-32. A test chart helps organize a troubleshooting process and becomes a record of the troubleshooting strategy that can be analyzed later for improvements.

When troubleshooting, it is easy to become focused on only one possible cause of the problem. For example, if a maintenance engineer is convinced that a particular switch is the problem, other suspect devices may be overlooked. If a dead end is reached, reevaluate the problem and start over by listing all possible causes again. Look for potential secondary causes. For example, an air-conditioner motor stops working and the primary cause of failure is a blown fuse. However, the secondary cause is overheating of the motor caused by a clogged filter. When necessary, consult another maintenance engineer. Maintenance personnel must use all resources available to repair equipment quickly and economically.

Remedying Problems

After isolating the cause of the problem, list all the possible ways to remedy the issue. Determine which remedy is best by considering plant production, cost, and maintenance budget. Also consider safety, plant policy, and code and legal requirements. In some cases, it may be necessary to temporarily repair the problem until there is a better time to formulate a permanent repair.

The age of the equipment affects how to remedy the problem. Most equipment follows a typical life expectancy curve including the break-in period, useful life, and wear-out period. **See Figure 1-33.** The *break-in period* is the time just after installation when equipment achieves peak operating performance. The *useful life* is the period of time after the break-in period when most equipment operates as designed. Preventive maintenance is performed during the useful life of the equipment. The *wear-out period* is the period after the useful life of equipment when normal failures occur.

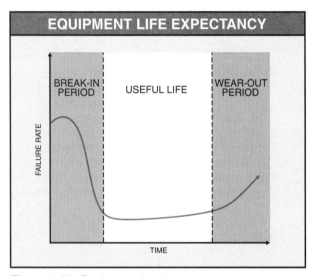

Figure 1-33. Equipment breakdowns are most common at the beginning and end of its life expectancy. The age of a piece of equipment within its life expectancy affects maintenance decisions.

If an equipment problem occurs during the break-in period, the cause could be improper installation or operating procedures. If the problem occurs during a machine's useful life, repair is typically the most economical choice. If the problem occurs during the wear-out period, the equipment may not be worth repairing.

Test equipment is used to verify that replaced or repaired parts operate properly. The equipment is monitored for proper function after the problem has been remedied. Sufficient time must be allowed for the equipment to reach normal operating levels and load conditions.

If a repair does not fix the original problem, then the troubleshooting process resumes by reevaluating the written lists of symptoms, causes, and troubleshooting test charts. It may be necessary to restart the process by gathering more information about the equipment. In the meantime, notify the appropriate personnel and lock out and tag out the equipment if the equipment should not be operated.

Documenting Problems

When a repair is successful and the equipment operation is restored, the troubleshooting process is recorded on a troubleshooting report. This includes symptoms, causes, the test conducted, the repairs chosen, and condition of the equipment after the repair. **See Figure 1-34.** The spare parts used in the repair are also recorded and replacement parts ordered as needed. In addition, suggested procedures that could prevent similar problems from occurring are noted. If the equipment requires additional work, the appropriate work orders are generated and scheduled.

Troubleshooting is an excellent opportunity for acquiring and improving skills. The key is to follow a logical troubleshooting sequence and to think through each step carefully while working on the problem. After the problem has been solved, filling out troubleshooting reports or work-order documentation is an opportunity to evaluate the troubleshooting process. It is recommended that maintenance personnel keep personal logs to record troubleshooting details, such as an evaluation of the troubleshooter's actions, what could be done better in the future, and what needs to be learned to be able to troubleshoot more effectively.

INTERPERSONAL SKILLS

Interpersonal skills are strategies and actions that allow an individual to communicate effectively with others in a variety of situations. Maintenance personnel often work with individuals that depend on equipment to meet production deadlines. Equipment failure is costly and frustrating and at times this frustration may be directed at the maintenance personnel.

TROUBLESHOOTING REPORT

Date _10/7_   Technician _Susan W._

Department _Shipping_   Equipment _Conveyor_

Problem _conveyor stopped operating_

Symptoms _motor operates, but conveyor not moving_

Cause(s) _Conveyor drive belt became loose due to drive pulley misalignment_

Repair _Realigned and tightened drive pulley_
Checked and reinstalled drive belt to proper tightness
Tested drive system under load

Preventive Maintenance Action _Check belt tightness weekly_

Notes _Build additional covers for drive system to keep it cleaner_

Figure 1-34. A troubleshooting report is a record to guide other troubleshooters if a similar problem occurs again.

Communication with an upset individual is best accomplished by remaining calm. Listening carefully to an explanation of the problem without interruption allows time for the individual to collect their thoughts and calm down. **See Figure 1-35.** Emphasize willingness to work with the individual as a team in finding a solution. Focus efforts on equipment problems. Take clear notes during the conversation as necessary. Avoid discussing related personnel issues. Identify the specific steps to be taken and the time frame required to return the equipment to service. Discuss the options and agree on the appropriate action to take.

Clear communication eliminates misunderstandings and reduces troubleshooting time. However, personnel issues sometimes prevent quick resolution of a maintenance problem. In this instance, communicate the equipment problem and related issues to a supervisor.

ADVANCEMENT IN MAINTENANCE TRADES

Successful maintenance personnel practice lifelong learning. Once basic skills are mastered, higher levels of skill mastery are required. Maintenance personnel make the most of each task by evaluating and refining strategies and skills. In addition, new materials, tools, and equipment are being developed regularly. The ability to adapt as the industry changes is crucial for success. Reading trade publications, attending classes and seminars, and participating in professional organizations are activities that provide valuable information on current topics and trends. New skills are required to remain current with advancing technology and grow in the maintenance trades.

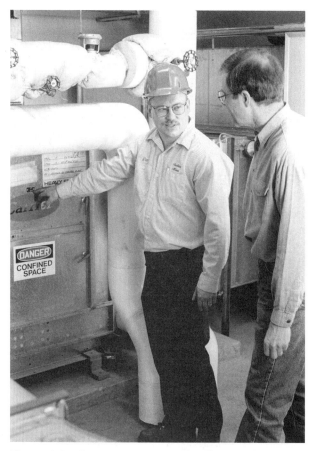

Figure 1-35. Clear communication between maintenance personnel and the equipment operator is necessary for efficient troubleshooting.

TECH TIP

Establishing a preventive/predictive maintenance program can boost productivity by reducing unplanned shutdowns and decreasing repair downtime.

Workplace Safety

Although facility safety is the responsibility of all workers, maintenance personnel are often the first to be aware of safety problems. Much of the work performed by maintenance personnel involves preventing hazardous conditions for all facility workers and the general public. This responsibility requires them to understand and follow safety regulations established by regulatory agencies and other specific workplace rules. Safety must be an integral part of any workplace activity, no matter how small or time-sensitive the task.

SAFETY PLANNING

Working safely is a habit established through training, development of technical skills, and work experience. A key to working safely is for maintenance personnel to focus attention on the work at hand and include safety considerations in all task planning. Even emergency troubleshooting tasks require planning. All work orders must contain a section for identifying safety concerns.

Safety planning begins with identifying the specific hazards of a machine or process. Then, personnel must formulate a plan for addressing each hazard based on established procedures and safety requirements. If there is no specific safety procedure for a certain task, maintenance personnel are responsible for applying general safety rules to the specific situation. This is done with a combination of manufacturer recommendations, common sense, knowledge of safety regulations, and work experience.

Maintenance personnel must also consider the effects of changing conditions on safety planning. They must anticipate the necessary safety equipment and procedures for weather, work areas, tasks, and facility conditions that may change quickly.

SAFETY CODES AND STANDARDS

Safety is a primary concern in all maintenance work. Maintenance personnel are potentially exposed to many different hazards and both the employer and the employee are responsible for ensuring that work is performed in a safe manner.

Many government agencies and independent organizations help ensure safe work environments by developing rules regarding safety equipment and procedures. Some rules are mandatory and some are voluntary. A *regulation* is a rule made mandatory by a federal, state, or local government. A *code* is a collection of regulations related to a particular trade or environment.

Units of government can develop their own regulations, but since this is a very time-consuming process, they often adopt existing standards as regulations.

A *standard* is a collection of voluntary rules developed through consensus and related to a particular trade, industry, or environment. Standards are developed and published by the members of standards organizations, often industry-specific, which periodically review and refine the rules in the standards. Standards themselves have no authority, unless they are adopted as regulations. Then they become mandatory and are enforceable by that adopting government unit.

Company procedures must comply with minimum federal, state, and local regulations. **See Figure 2-1.** A company may also adopt additional rules or standards that exceed the minimum regulation requirements. These are not enforceable by law, but compliance may be a condition of employment.

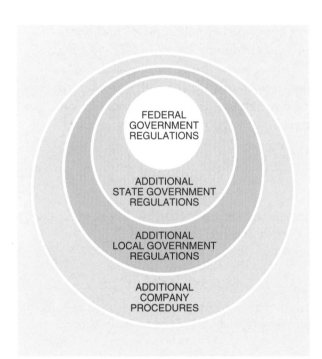

Figure 2-1. Company procedures must comply with the minimum federal regulations and any additional state and local regulations.

There are codes and standards that cover a wide variety of workers, industries, and situations. **See Figure 2-2.** However, no amount of rules can cover every possible hazardous situation that maintenance personnel might face. Personnel should understand the reasons for each rule and be able to apply this knowledge when formulating effective safety plans in situations that are not covered by an existing rule.

SELECTED SAFETY CODE AND STANDARD ORGANIZATIONS	
OSHA Occupational Safety and Health Administration www.osha.gov	Requires employers to provide a safe working environment
	Regulations are mandatory
	Enforces compliance through inspections
NFPA National Fire Protection Association www.nfpa.org	Provides guidance in assessing hazards of products of combustion
	Publishes the National Electrical Code®
	Develops hazardous materials information
ANSI American National Standards Institute www.ansi.org	Coordinates and encourages activities in national standards development
	Identifies industrial and public needs for standards
	Acts as national coordinator and clearinghouse for consensus standards
NEMA National Electrical Manufacturers Association www.nema.org	Assists with information and standards concerning proper selection, ratings, construction, testing, and performance of electrical equipment
NIOSH The National Institute for Occupational Safety and Health www.cdc.gov/niosh/	Acts in conjunction with OSHA to develop recommended exposure limits for hazardous substances or conditions located in the workplace
	Recommends preventive measures to reduce or eliminate adverse health and safety effects
UL Underwriters Laboratories Inc.® www.ul.com	Tests equipment and products to verify conformance to national codes and standards
CSA Canadian Standards Association www.csa.ca	Tests equipment and products to verify conformance to Canadian national standards

Figure 2-2. Standards organizations develop codes and standards to ensure worker safety.

Occupational Safety and Health Administration

In the United States, many agencies issue safety regulations, but the Occupational Safety and Health Administration is the most wide-reaching of these authorities. The *Occupational Safety and Health Administration (OSHA)* is a federal government regulatory agency that requires all employers to provide a safe working environment for their employees. Specific regulations have been created to cover more specific hazards or unhealthy situations. Employers must follow all safety and health regulations and ensure that their employees also follow regulations. For example, employees must wear the proper personal protective equipment (PPE) and immediately report any hazardous conditions, workplace injuries, or health problems to their supervisors. OSHA administers and enforces regulations through inspections by trained OSHA inspectors.

OSHA regulations cover most private sector employees. Under OSHA guidance, states may develop and administrate their own occupational safety and health plans, as long as they cover public-sector employees and meet or exceed the federal regulations. There are 26 states with their own occupational safety and health plans. **See Figure 2-3.**

The Office of the Federal Register publishes all adopted OSHA standards and required amendments, corrections, insertions, and deletions. Each year, all current OSHA standards are reproduced in the Code of Federal Regulations (CFR). OSHA standards are included in Title 29 of the CFR, Parts 1900-1999. These documents are available online and at many libraries.

National Fire Protection Association®

The *National Fire Protection Association® (NFPA®)* is a national standards organization that provides guidance in assessing fire-related hazards. The NFPA publishes the National Electrical Code® (NEC®), which is one of the most widely adopted and recognized standards in the world. The purpose of the NEC® is the practical safeguarding of individuals and property from hazards arising from the use of electricity. The NEC® is updated every three years. Many federal, state, and local agencies adopt the NEC® to set requirements for electrical installations.

The NFPA® also publishes other guides to electrical and fire safety. *NFPA 70E®, Standard for Electrical Safety in the Workplace,* outlines safe work practices and personal protective gear for electricians installing and maintaining electrical equipment. The *Fire Protection Guide to Hazardous Materials* provides information regarding properties of hazardous materials and required fire-fighting procedures.

American National Standards Institute

The *American National Standards Institute (ANSI)* is a national organization that helps identify industrial and public needs for national standards. ANSI does not develop standards, but acts as a national coordinator of standards activities and an approval organization and clearinghouse for standards. ANSI standards are developed by member technical societies, trade associations, and United States government departments. When approved, the standards become American National Standards and are copublished with ANSI. Standards are submitted and approved through democratic procedures. This ensures impartial consensus and elimination of conflict or duplication with other ANSI standards.

> **TECH TIP**
>
> *Designation as an American National Standard does not make a standard mandatory, but the more rigorous review and consensus may increase its adoption.*

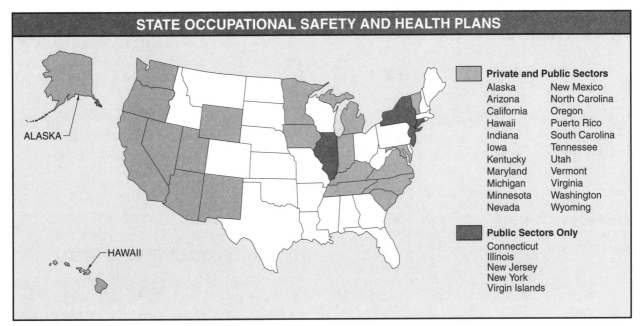

STATE OCCUPATIONAL SAFETY AND HEALTH PLANS

Private and Public Sectors

Alaska	New Mexico
Arizona	North Carolina
California	Oregon
Hawaii	Puerto Rico
Indiana	South Carolina
Iowa	Tennessee
Kentucky	Utah
Maryland	Vermont
Michigan	Virginia
Minnesota	Washington
Nevada	Wyoming

Public Sectors Only

Connecticut
Illinois
New Jersey
New York
Virgin Islands

Figure 2-3. States may administer their own occupational safety and health plans if plans meet or exceed minimum OSHA standards.

A *technical society* is an organization of technical personnel united by a professional interest. For example, the American Welding Society (AWS) is a nonprofit organization whose primary goal is advancing the science, technology, and application of welding and related joining disciplines. A *trade association* is an organization that represents the producers of specific products. For example, the American Gear Manufacturers Association (AGMA) is a national organization that assists with information and standards concerning the specification, design, and application of gears. Departments of the federal government are also often responsible for developing specifications, such as military standards (MIL-SPECs).

A standard does not need to be an American National Standard to be widely used or adopted. However, approval provides additional endorsement, which typically leads to even greater adoption. When approved, standards are identified by the ANSI acronym, sponsor organization acronym, and publication number of the organization. **See Figure 2-4.** For example, "ANSI/AWS F4.1" refers to the standard *Recommended Safe Practices for the Preparation for Welding and Cutting of Containers and Piping,* which was developed by the American Welding Society (AWS). The year of the last revision or reaffirmation is also often appended to the standard designation, for example "ANSI/ISEA Z89.1-2009."

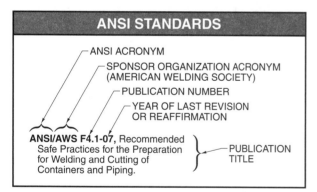

ANSI STANDARDS

ANSI ACRONYM
SPONSOR ORGANIZATION ACRONYM (AMERICAN WELDING SOCIETY)
PUBLICATION NUMBER
YEAR OF LAST REVISION OR REAFFIRMATION

ANSI/AWS F4.1-07, Recommended Safe Practices for the Preparation for Welding and Cutting of Containers and Piping.
PUBLICATION TITLE

Figure 2-4. Standards are designated by the sponsor organization, publication number, year of revision or reaffirmation, and title. American National Standards are copublished with ANSI.

ANSI also manages United States participation in international standards activities. The *International Organization for Standardization (ISO)* is a nongovernmental, international organization that provides a worldwide forum for the standards-developing process. The ISO is composed of national standards institutions from more than 90 countries.

National Electrical Manufacturers Association

The *National Electrical Manufacturers Association (NEMA®)* is a national trade association that provides information and develops standards concerning proper selection, rating, construction, testing, and performance of electrical equipment. NEMA standards are used as guidelines for the manufacture and use of electrical equipment.

National Institute for Occupational Safety and Health

The *National Institute for Occupational Safety and Health (NIOSH)* is a national organization that acts in conjunction with OSHA to develop and periodically revise recommended exposure limits for hazardous substances or conditions in the workplace. NIOSH also recommends preventive measures to reduce or eliminate the adverse health and safety effects of these hazards. NIOSH is primarily concerned with research, while OSHA is responsible for enforcement.

Underwriters Laboratories Inc.®

Underwriters Laboratories Inc.® (UL) is an independent organization that tests products to verify conformance to national codes and standards. Equipment tested and approved by UL carries the UL label. UL-approved equipment and products are listed in their annual publication.

Canadian Standards Association

The *Canadian Standards Association (CSA)* is a Canadian government agency that tests products to verify conformance to national codes and standards. The CSA is similar in function to UL.

PERSONAL PROTECTIVE EQUIPMENT

Personal protective equipment (PPE) is clothing and/or equipment worn by a worker to reduce the possibility of an injury. **See Figure 2-5.** All PPE and its use must meet requirements specified in applicable safety regulations. PPE regulations are listed in OSHA 29 CFR 1910

Subparts G and I, which also references various standards for each type. For example, appropriate protective helmets are worn in areas with overhead hazards, and safety shoes with steel toes are worn to provide protection from falling objects.

Figure 2-5. Personal protective equipment (PPE) is used to reduce the possibility and severity of an injury.

Common PPE includes protective clothing, head protection, eye protection, respiratory protection, foot protection, hearing protection, arm and hand protection, and fall protection. Special working conditions may require additional forms of protection.

Protective Clothing

Protective clothing is clothing that provides protection from contact with sharp objects, hot equipment, and harmful materials. Durable materials, such as denim, should be worn and the fit should be snug, yet allow ample movement. Clothing made of synthetic materials such as nylon, polyester, or rayon should not be worn because these materials are flammable and could melt to the skin. Some electrical work requires flame-resistant (FR) clothing or coveralls for additional protection.

Head Protection

Protective helmets (hard hats) are used to prevent injury from impact, falling and flying objects, and electrical shock in the workplace. Protective helmets protect workers by resisting penetration and absorbing the blow of an impact. The shell of the protective helmet is made of a durable, lightweight material. A shock-absorbing lining consisting of crown straps and a headband keeps the shell of the protective helmet away from the head and allows ventilation.

Standards for protective helmets are specified in ANSI/ISEA Z89.1, *Industrial Head Protection.* Protective helmets are identified by type and class for protection against specific hazardous conditions. **See Figure 2-6.** Type I helmets protect against impacts at the top and Type II helmets protect against top and side impacts. Type II helmet designs often include wide brims all around the helmet to help deflect side impacts.

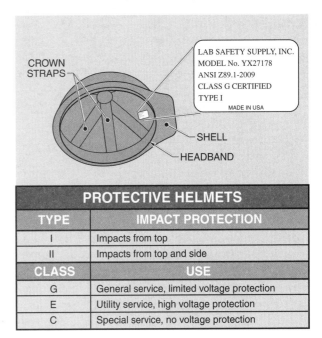

PROTECTIVE HELMETS	
TYPE	**IMPACT PROTECTION**
I	Impacts from top
II	Impacts from top and side
CLASS	**USE**
G	General service, limited voltage protection
E	Utility service, high voltage protection
C	Special service, no voltage protection

Figure 2-6. Protective helmets are identified by type and class for protection against specific hazards.

Electrical protection is classified as Class G, Class E, and Class C for helmets. Class G (general) protective helmets protect against impact hazards and contact with voltages up to 2200 V. These are the most common helmets used in manufacturing, mining, and construction. Class E (electrical) protective helmets protect against impact hazards and contact with high voltages up to 20,000 V. Class E protective helmets are used by electrical workers and maintenance personnel subject to electrical hazards. Class C (conductive) protective helmets are designed for impact protection only and may be used only when there is no danger from electrical hazards.

Hard hats must be maintained by the employees to keep them in good shape. Employees should inspect their hard hats for cracks or punctures and ensure that the headband is not stretched or worn. The shell can be washed in soap and water and rinsed clean. Hard hats should be stored in a clean and cool location.

Hard hats may be required to be worn at all times in a facility.

Eye and Face Protection

OSHA requires eye protection when exposed to flying particles, molten metal, chemicals, or radiant energy. Standards for eye protection are specified in ANSI/ISEA Z87.1, *Occupational and Educational Eye and Face Protection Devices.* There are several types of general eye and face protection, including safety glasses, goggles, and face shields. **See Figure 2-7.** Some tasks may require specialized eye protection due to especially dangerous hazards, such as the radiant energy from welding. All types of eye protection must protect the eyes from debris and minor impacts at the front and sides.

Safety glasses are the most common type of eye protection and are adequate for most maintenance tasks. Most industrial applications require the use of clear safety glasses, though tinted lenses may be permitted for outdoor work to protect against strong sunlight. Goggles may be required when chemical splash hazards are present, because they fit more snugly against the face. Full face shields are used around flying debris, such as when using a grinder. Usually, safety glasses are worn under face shields. Individuals wearing corrective lenses must wear face shields, goggles, or glasses that are made from impact-resistance material that include side shields.

Figure 2-7. Eye and face protection must be worn to prevent injuries caused by flying particles, chemical splashes, or radiant energy.

Eye protection must be maintained to provide protection and clear visibility. Pitted or scratched lenses should be replaced because they impair vision and are more likely to break. Eye protection should be cleaned and disinfected regularly. When not in use, they should be kept in cases that keep them clean and prevent scratching.

If debris or chemicals do get in the eyes or on the face, emergency eyewash and shower stations must be available to quickly flush the area with clean water. **See Figure 2-8.** Requirements for these safety devices are listed in ANSI/ISEA Z358.1, *Emergency Eyewash and Shower Equipment.*

EMERGENCY EYEWASH

Figure 2-8. Eyewash stations allow flushing of contaminants from the eyes or face.

TECH TIP

Per OSHA 29 CFR 1910.134 – Respiratory Protection, the employer must develop and implement a written respiratory protection program in any workplace where respirators are necessary to protect the health of the employee. The employer must provide respirators and applicable training. Employees should use respirators as directed to avoid respiratory hazards.

Respiratory Protection

A *respirator* is a device that protects the wearer from inhaling airborne contaminants. A respirator type is selected based on the type of material or chemical hazard an individual is being exposed to. **See Figure 2-9.** The most basic respirator type filters out dust or other particulates from the breathing air. Advanced respirators include cartridges with special filter materials that absorb chemicals in the air. Since chemical contaminants can vary widely, different cartridge materials are needed for each type. The cartridges are color-coded for the type of contaminant.

Respirators are broadly classified as air-purifying or supplied-air respirators, depending on the source of breathing air. Air-purifying respirators remove airborne contaminants before air is inhaled by the worker. Supplied-air respirators provide a separate, clean air supply to the worker. All respirators have limitations, and it is critical that users be fully trained on their use. The length of exposure, work effort, temperature, and ventilation are additional factors in determining the required respirator type. Respirator selection should be made according to NIOSH 87-108, *Respirator Decision Logic,* and ANSI/AIHA Z88.2, *Respiratory Protection.*

Warning: Respirators must be selected by qualified personnel. Improper use and/or selection of a respirator may result in serious injury or death.

A material safety data sheet (MSDS) or other documentation should be checked to determine the hazard type. Hazardous ingredients are listed with a threshold limit value and a permissible exposure limit. A *threshold limit value (TLV®)* is an estimate of the average airborne concentration of a substance to which workers may be exposed day after day without adverse effect. A TLV time-weighted average (TWA) value is based on exposures for 8 hr/day, 40 hr/week. A TLV short-term exposure limit is based on exposures of 15 min or less, no more than four times per day. A TLV ceiling is an absolute exposure limit that should never be exceeded. TLV guidelines are reviewed and published annually by the American Conference of Governmental Industrial Hygienists (ACGIH).

The *permissible exposure limit (PEL)* is the OSHA estimate of the average airborne concentration of a substance to which workers may be exposed day after day without adverse effect. PEL values may also include time-weighted, short-term, and ceiling limits. While similar in purpose, the TLV and PEL values for identical hazards may vary. The difference is that OSHA regulations have regulation status and are thus enforced by law. Exposure exceeding a PEL requires the use of respiratory protection, which is often listed by PEL. For example, 10xPEL means respiratory protection up to 10 times the permissible exposure limit. A supplied-air respirator is required if the contaminant concentration is above the level considered immediately dangerous to life or health (IDLH).

RESPIRATORS	
Air-Purifying	
Respirator	**Suggested Use**
 Disposable Particulates Mask	Low profile, lightweight, designed for limited use. Low-cost protection against dusts, mists, and fumes (not for mists containing gases, vapors, or nonabsorbed contaminants). Completely disposable. No cleaning or spare parts required.
 Reusable Half-Mask Respirator	Lightweight, easy to maintain, very little restriction of movement or vision. Uses replaceable cartridges and filters. Limited number of parts. Protects against chemical hazards such as dust, fumes, mists, and vapors.
 Reusable Full-Face Respirator	Offers greater eye and face protection than half-mask. Uses replaceable cartridges and filters. Easy to maintain (no intricate parts). Protects against chemical hazards such as dust, fumes, mists, and vapors.
 Powered Air-Purifying Respirator (PAPR)	Cooler, less exhausting for worker. Provides easier breathing for higher productivity. Uses cartridges or filters. Face- or belt-mounted with a battery for power. Includes air blower that pulls air through the cartridges and filters into the face piece.
Supplied-Air	
 Airline Respirator	Uses outside air source to keep worker cooler and offers greater protection than an air-purifying respirator. Available in two styles: constant flow and pressure demand. Uses Grade D air supply from ambient air pump, plant compressor, or bottled air. Not for use in IDLH situations or where the oxygen content is less than 19.5%.
 Self-Contained Breathing Apparatus (SCBA)	Provides greatest protection available. Pressurized bottle of air is carried on worker's back. For use in oxygen-deficient atmospheres, IDLH, and emergency situations. Available in two different types of cylinders: aluminum and composite. Provides good mobility with few restrictions because air source is carried on back.
 Emergency Escape Breathing Apparatus (EEBA)	For use in escape situations only. IDLH and oxygen deficiency. Service life depends on a 5 min to 10 min bottle of air. Not designed for rescue use.

COLOR CODE FOR CARTRIDGES/CANISTERS	
Contaminant	**Color Assigned**
Acid gases only	White
Organic vapors only	Black
Ammonia gas	Green
Acid gases and organic vapors	Yellow
Radioactive materials (except tritium and noble gases)	Purple
Dust, fumes, and mists (other than radioactive materials)	Orange
Other gases and vapors (not listed above)	Olive

Lab Safety Supply, Inc.

Figure 2-9. Respirators are selected for the specific contaminants and concentrations present in the hazard area.

Proper fit is critical for safe respirator operation. It must fit snugly so that it does not slip or allow contaminated air to enter around the edges of the respirator. It should not pinch the nose or interfere with the correct placement of safety glasses or goggles. The respirator should be adjusted for proper fit each time it is used. Facial hair may interfere with the fit of a respirator.

Two types of respirator fit tests are the positive pressure test and the negative pressure test. In the positive pressure test, the exhaust valve is closed and the wearer breathes out gently. The front of the respirator should bulge slightly but no air should leak around the edge of the respirator. For the negative pressure test, the inlet valve is closed and the wearer inhales gently. The respirator should flex in towards the face with no air leaking inside for ten seconds. Always follow the manufacturer's recommendations for proper testing and cleaning. It is important that respirators be cleaned after each use and stored in a clean, dry location. Respirators that are not cleaned properly can develop harmful mold and bacterial growth.

Maintenance personnel should always plan for the use of the correct respirator before starting a task. They should also be aware that if conditions change, the area should be evacuated until it is safe or the correct respirator can be used.

Foot Protection

Maintenance personnel perform many tasks that require handling objects that could injure a foot if dropped. Safety shoes with reinforced steel toes provide protection against injuries caused by compression and impact. Some safety shoes have protective metal insoles and metatarsal guards for protection from punctures from below, such as stepping on nails. Additional foot guards or toe guards may be used over existing work shoes.

Safety shoes should fit well and be worn with comfortable socks, as working on hard surfaces like concrete is a common cause of foot pain and fatigue. Specialized safety shoes are available for certain situations, such as metal-free shoes for electrical work or shoes with soles that prevent slips on wet or oily surfaces. Tall boots can protect against hot metal or other debris getting inside the boot. Rubber boots may be needed in wet or chemically corrosive situations.

Protective footwear was previously covered by ANSI Z41, *Personal Protection–Protective Footwear*. This standard was withdrawn in 2005 and replaced with ASTM F2413, *Standard Requirements for Protective Footwear*, though the requirements are similar.

⊕ SAFETY TIP

A common type of disposable particulate respirator is designated as N95. This respirator has a minimum filter efficiency of 95% and is used for non-oil related hazards only.

Hearing Protection

Many industrial settings have high noise levels, which can cause hearing loss or impairment. Hearing protection is required whenever workers are exposed to noise equal to or exceeding an 8 hr TWA of 85 decibels (dB), equivalent to the sound level of busy street traffic. However, because the requirement evaluates the effect of sound over time, workers may be permissibly exposed to higher average sound levels when it is for shorter periods of time. **See Figure 2-10.**

PERMISSIBLE NOISE EXPOSURES*		
Duration†	Sound Level‡	Approximate Equivalent
8.0	90	Heavy truck at 3′
6.0	92	Power lawnmower at 3′
4.0	95	Newspaper Press
3.0	97	Major road at 7′
2.0	100	Jackhammer at 7′
1.5	102	Jet flyover at 1000′
1.0	105	Bulldozer
0.5	110	Chainsaw at 3′
0.25 or less	115	Pneumatic chipper at 3′

* OSHA regulation 29 CFR 1926.52(D)(1)
† in hours per day
‡ in dBA slow response

Figure 2-10. Hearing protection must be used when noise levels exceed the time-weighted permissible exposures.

Hearing protection devices are broadly classified as earplugs or earmuffs. **See Figure 2-11.** An *earplug* is a compressible device inserted into the ear canal to reduce the level of noise reaching the eardrum. Earplugs are made of soft moldable rubber, foam, or plastic materials that expand to fill the ear canal. Most earplugs are inexpensive and disposable. An *earmuff* is a device worn over the ears to reduce the level of noise reaching the eardrum. A tight seal around the earmuff is required for proper protection. To protect electrical workers, some earmuffs have no metal parts. Electronic earmuffs can reduce certain sound frequencies while still allowing voices to be heard. Earmuffs must be washed and dried daily according to manufacturer's instructions.

HEARING PROTECTION

EARPLUGS

Fluke Corporation

EARMUFFS

Figure 2-11. Hearing protection consists of earplugs or earmuffs. Each type reduces the noise level reaching the ear by an amount given as a noise reduction rating (NRR).

Hearing protection devices are rated for noise reduction. A *noise reduction rating (NRR)* is the amount of the reduction of sound level (in decibels) provided by a hearing protection device. For example, an NRR of 27 means the noise level is reduced by 27 dB when tested at the factory. For field use, 7 is subtracted from the NRR, so the effective reduction is approximately 20 dB. Earplugs reduce noise level by up to 30 dB if fitted properly, and earmuffs reduce noise levels by up to 25 dB. In very noisy locations, workers might wear both earplugs and earmuffs.

Maintenance personnel often work in many areas of a facility. Thus it is important for them to carry hearing protection at all times and use it immediately upon entering a noisy area.

Arm and Hand Protection

Arm and hand protection is required to prevent injuries to arms and hands from burns, cuts, electrical shock, amputation, and chemical exposure. **See Figure 2-12.** A wide assortment of gloves, hand pads, sleeves, and wristlets are available to provide protection against various hazards.

Ansell Edmont Industrial, Inc.

Figure 2-12. Various types of protective gloves are available to guard against specific hazards, while allowing adequate dexterity.

The type of maintenance task determines the degree of dexterity required and the duration, frequency, and degree of exposure to potential hazards. Various types of gloves are used for different hazards. Glove selection is then based on required protection and glove test data, according to ANSI/ISEA 105, *Hand Protection Selection Criteria*. For example, protection against chemicals measures the ability of the gloves to restrict the passing of chemicals through to the skin. Gloves made from wire mesh, leather, or canvas provide protection from cuts and burns.

Gloves can be loose fitting or snug fitting. Loose-fitting gloves are easier to put on and take off, but if they are too large, they may pose a safety hazard when working around machinery. Snug-fitting gloves provide better sensitivity and control than loose-fitting gloves.

Certain maintenance tasks require specific gloves. For example, electrical work may require gloves made from rubber to provide maximum insulation from electrical shock hazards. Neoprene gloves provide protection against alcohol and a variety of chemicals and acids. Latex gloves provide maximum touch sensitivity.

✚ SAFETY TIP

Gloves that can protect against more than one hazard or double gloving may be required for workers who are exposed to multiple physical hazards or chemical mixtures.

Fall Protection

Fall protection equipment is required when working in elevated (over 6′ above the ground) or confined spaces. This includes work in areas such as elevated platforms, catwalks, scaffolds, tanks, manlifts, and vaults. Fall protection can be in the form of equipment worn by a worker to reduce the potential for injury from a fall or precautions taken in the work area to prevent a fall. Fall protection equipment and procedures must comply with OSHA 29 CFR 1926 Subparts E and M and ANSI/ASSE Z359.1, *Safety Requirements for Personal Fall Arrest Systems, Subsystems, and Components*.

A *full-body harness* is a fall protection device that evenly distributes fall-arresting forces throughout the body to prevent further injury. **See Figure 2-13.** The harness secures the worker around the legs, torso, and shoulders, and includes a metal D-ring at the back to attach a lanyard. The lanyard secures the harness to a beam, structural member, secured ladder, or lifeline. A shock absorber in the lanyard absorbs most of the force

of stopping the fall. Full-body harnesses are also used in confined spaces as a means to lift or pull an injured worker out of the hazardous area.

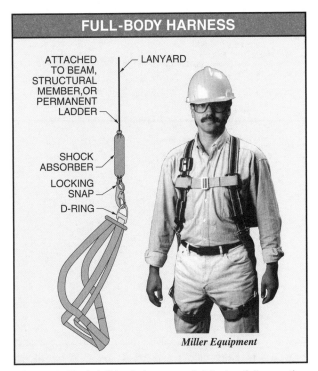

FULL-BODY HARNESS

ATTACHED TO BEAM, STRUCTURAL MEMBER, OR PERMANENT LADDER

LANYARD

SHOCK ABSORBER

LOCKING SNAP

D-RING

Miller Equipment

Figure 2-13. A full-body harness distributes fall-arresting forces throughout the body, which reduces the likelihood of injury.

Guardrail systems can also provide protection against falling. This type of fall protection consists of three railings that must be secured to the platform side of upright structural members. A top guardrail is required at 39″ to 45″ above the platform and a midrail exactly in between. A toeboard is a bottom railing in contact with the platform that keeps tools or materials from falling off the platform. **See Figure 2-14.** The railing system may be permanently affixed to the elevated platform or installed temporarily to work areas as required.

In large, flat areas where work will not be required within 6′ of the edge, a warning-line system may be used. A warning line is a high-visibility rope marking a controlled-access zone near the edge of the roof. Workers inside the warning line do not need additional fall protection, but workers between the warning line and the edge are required to have some other method of fall protection. A safety monitor watches for fall hazards, allows no workers without additional fall protection within the controlled access zone, and ensures that workers are aware of the warning line if they are working near it.

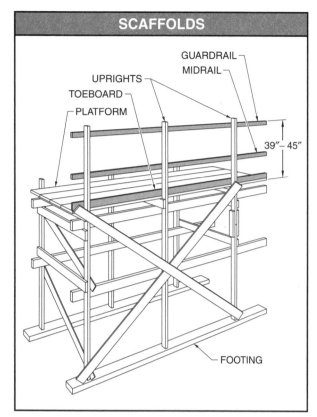

SCAFFOLDS

GUARDRAIL
MIDRAIL
UPRIGHTS
TOEBOARD
PLATFORM

39″– 45″

FOOTING

Figure 2-14. Fall protection can be in the form of guardrails on scaffolding or other types of elevated platforms. Guardrails must meet certain installation requirements.

Fire safety practices involve the proper storage and handling of flammable substances.

FIRE SAFETY

Fire safety procedures reduce or eliminate conditions that could cause a fire. Guidelines in assessing hazards of the products of combustion are provided by the NFPA. Maintenance personnel are responsible for preventing conditions that could result in a fire. This includes proper use and storage of lubricants, oily rags, and solvents, and immediate cleanup of combustible spills.

Warning: Always dispose of oily rags in a specially designated container that is emptied daily.

In the event of a fire, maintenance personnel must act quickly to minimize injury and damage from the fire. All facilities must have a fire safety plan. A fire safety plan establishes procedures that must be followed if a fire occurs. **See Figure 2-15.** The plan includes guidelines on fire response and assigns responsibilities to various employees. For example, one person checks that all employees have evacuated the facility. A floor plan shows the locations of key fire-related equipment, such as the main electrical breaker, fire main, exits, fire alarms, and fire extinguishers.

Fire Tetrahedron

Traditionally, a fire was believed to require only three elements: fuel, heat, and oxygen. Removing any single element extinguishes the fire. For example, removing additional fuel, such as clearing away nearby combustibles, allows the fire to burn itself out. Cooling a fire with water or carbon dioxide removes the heat required to continue burning. Smothering a fire with fire blankets or dry sand also extinguishes a fire by removing oxygen. These three elements are known as the fire triangle.

Further study of combustion and extinguishing technologies has revealed that a fourth critical element is a sustaining chain reaction. Some extinguishing agents act by specifically disrupting the chemical reactions that sustain a fire. By adding this fourth element, the fire triangle is now commonly known as the fire tetrahedron (triangular pyramid). **See Figure 2-16.**

Fire extinguishing agents remove one or (commonly) more of the four elements in order to extinguish a fire. For example, water and carbon dioxide both cool and smother a fire. However, depending on the type of fire, some agents may not be effective, or may even make the fire hazard worse. Fire extinguishers are selected based on the class of fire.

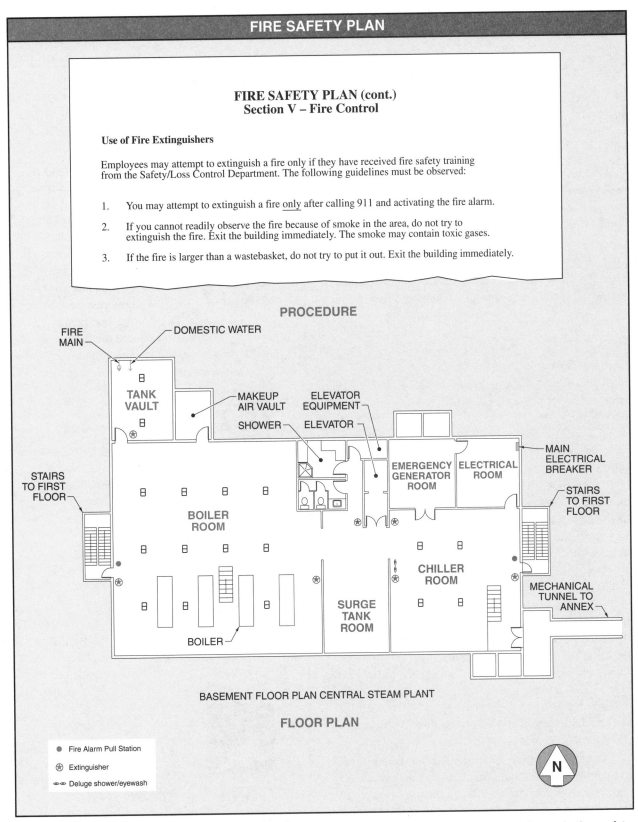

FIRE SAFETY PLAN

FIRE SAFETY PLAN (cont.)
Section V – Fire Control

Use of Fire Extinguishers

Employees may attempt to extinguish a fire only if they have received fire safety training from the Safety/Loss Control Department. The following guidelines must be observed:

1.   You may attempt to extinguish a fire <u>only</u> after calling 911 and activating the fire alarm.

2.   If you cannot readily observe the fire because of smoke in the area, do not try to extinguish the fire. Exit the building immediately. The smoke may contain toxic gases.

3.   If the fire is larger than a wastebasket, do not try to put it out. Exit the building immediately.

PROCEDURE

FIRE MAIN — DOMESTIC WATER

TANK VAULT

MAKEUP AIR VAULT — ELEVATOR EQUIPMENT

SHOWER — ELEVATOR

EMERGENCY GENERATOR ROOM | ELECTRICAL ROOM

MAIN ELECTRICAL BREAKER

STAIRS TO FIRST FLOOR

STAIRS TO FIRST FLOOR

BOILER ROOM

CHILLER ROOM

MECHANICAL TUNNEL TO ANNEX

SURGE TANK ROOM

BOILER

BASEMENT FLOOR PLAN CENTRAL STEAM PLANT

FLOOR PLAN

● Fire Alarm Pull Station

⊛ Extinguisher

⊂⊃ Deluge shower/eyewash

N

Figure 2-15. A fire safety plan details the locations of fire alarms, fire extinguishers, emergency exits, and other safety features, along with employee procedures to be followed if a fire occurs.

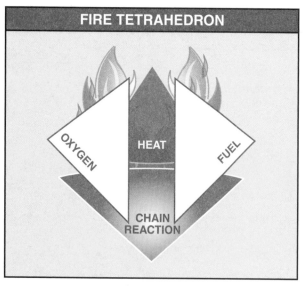

Figure 2-16. A fire requires four elements to be present to burn. Removing one or more extinguishes a fire.

Fire Classes

Fires are classified by their combustible material as either Class A, Class B, Class C, Class D, or Class K fires. **See Figure 2-17.** Class A fires are fires that burn wood, paper, textiles, and other ordinary combustible materials. These fires are easily extinguished with water or foam.

Class B fires are fires that burn oil, gas, grease, paint, and other flammable liquids. Water should never be used to extinguish this type of fire because it can cause the fuel to scatter, spreading the flames. Class B fires are extinguished by inhibiting the chemical reaction or by smothering the fire with carbon dioxide or foam.

Class C fires are fires caused by energized electrical equipment, such as motors or transformers. This sort of fire may be caused by short circuits or overloaded electrical conductors. Electrical fires can be extinguished in the same way as an ordinary combustible fire, but water, foam, and other conductive agents cannot to be used if there is a possibility that the electrical source is still energized. Extinguishing agents specifically rated for electrical fires include carbon dioxide and dry chemical powder extinguishers.

Class D fires involve certain combustible metals, such as titanium, magnesium, sodium, and potassium. Water and other common extinguishing agents can make the fire worse or cause the metals to explode. Dry chemical agents are recommended for cooling and smothering the fire.

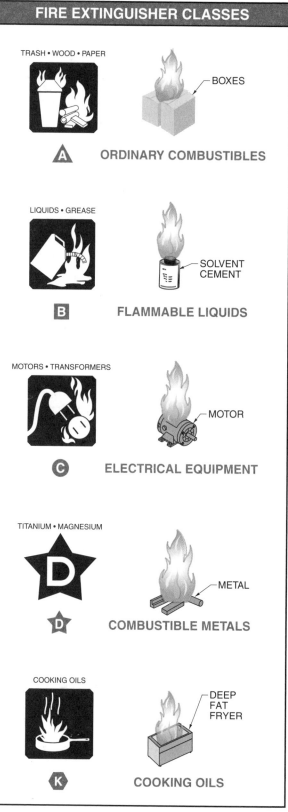

Figure 2-17. Fire extinguisher classes are based on the type of combustible material fueling the fire.

Class K fires include cooking oils and fats. Such fires are technically under the Class B category, but the unique characteristics of such fires justify a separate classification. Special extinguishing systems and agents are used in commercial kitchens to extinguish these types of fires.

Fire Extinguishers

Fire extinguishing equipment must be available throughout the facility, though this does not eliminate the need for plant fire protection personnel or the local fire department. Proper authorities must be notified in the event of a fire.

Warning: Always notify proper authorities as soon as possible in the event of a fire. Fires and smoke can spread quickly.

Personnel must know the locations of all fire extinguishing equipment in the facility. Fire extinguishing equipment, such as fire extinguishers, water hoses, and sand buckets, must be routinely checked according to plant procedures. The number and type of fire extinguishers required is determined by the authority having jurisdiction and is based on how quickly a fire may spread, potential heat intensity, and accessibility to the fire.

When a fire occurs, the fire alarm system must be immediately activated, manually if necessary. Employees then follow their assigned rolls as specified in the fire emergency plan. Maintenance personnel might be authorized to fight a small fire, after it has been reported and the alarm activated. The danger in fighting a fire is primarily inhalation of smoke, which can contain dangerous chemicals that can quickly overcome or even kill a person. If uncertain about the risks from fighting a fire, all personnel should evacuate the facility.

A small fire is fought using a portable fire extinguisher. All employees should know how to use extinguishers and which ones are safe to use on a particular type of fire. The general procedure for using a fire extinguisher is as follows:

1. Hold the extinguisher carefully. Pull the safety pin and aim the nozzle at the base of the fire at a distance of about 8′, if that is a safe distance.

2. Squeeze the nozzle and sweep it side to side while aiming at the base of the fire.

3. Always face the fire and make sure there is a clear exit to the rear. If the fire is not extinguished or is growing when the extinguisher is empty, evacuate the facility immediately.

Fire safety procedures are critical when welding, as sparks or hot slag can easily ignite nearby combustibles.

ELECTRICAL SAFETY

Electrical safety is promoted by the efforts of the NFPA, OSHA, and state safety laws. The NFPA sponsors the development of the NEC®, which is one of the most widely used and recognized standards in the world. The purpose of the NEC® is to protect people and property from hazards that arise from the use of electricity. Improper procedures when working with electricity can cause permanent injury or death. Maintenance personnel must minimize conditions that could lead to electrical shock.

The use of PPE is required whenever work may occur on or near energized and exposed electrical circuits. The requirements are specified in NFPA 70E, *Standard for Electrical Safety in the Workplace,* OSHA regulations, and other applicable safety mandates. All PPE and tools are selected for the task and must be rated for at least the operating voltage of the equipment or circuits involved.

Electrical Shock

Electrical shock occurs when an individual becomes part of an electrical circuit, such as when touching two conductors. Electrical shock causes muscle spasms that can cause a victim to fall or become unable to let go of the electrical source. Even a minor electrical shock can cause the heart and lungs to stop functioning correctly. Electricity can also damage internal organs and cause burns where it enters and leaves the body. The severity of the shock depends on voltage, current, duration, frequency, path through the body, and other factors. In general, currents over about 8 mA cause a painful and potentially harmful shock. **See Figure 2-18.**

EFFECT OF ELECTRICAL CURRENT	
Current*	Effect on Body
less than 8	Sensation of shock but probably not painful
8 to 15	Painful shock Removal from contact point by natural reflexes
16 to 20	Painful shock May be frozen or locked to point of electrical contact until circuit is de-energized
over 20	Causes severe muscular contractions, paralysis of breathing, heart convulsions

* in mA

Figure 2-18. The severity of electrical shock and the effect on the body increases with greater current flow.

The amount of current that passes through the body depends on the voltage and resistance of the resulting circuit. The resistance of a body to the flow of current varies. Sweaty hands have less resistance than dry hands. A wet floor has less resistance than a dry floor. As the current increases, the severity of the electrical shock increases. Insulation provided by rubber-soled shoes, insulated tools, and insulated work and floor surfaces increases resistance and reduces the possibility of electrical shock. **See Figure 2-19.**

Following basic electrical safety rules can reduce the possibility of electrical shock. The safety rules are as follows:
- Plan each job carefully.
- Unless unavoidable, de-energize, lock out, and tag out an electrical system prior to performing work.
- Assume a circuit is energized before working on it. Always use a test instrument to make sure the circuit is de-energized. Retest the circuit after leaving and returning to the circuit.
- Use proper lockout/tagout procedures before working on electrical equipment.
- Use one hand when working on an energized circuit. This prevents electric current from traveling across the chest and through the heart.
- Repair or replace tools that have cracked or missing insulation.
- Wear rubber-soled safety shoes or stand on rubber pads to avoid becoming part of the electrical pathway to the ground.
- Do not work alone.
- Use wood or fiberglass ladders around electrical circuits. A metal ladder can become part of the circuit.
- Never use electrical test equipment beyond its safe voltage and current limits.
- Remove rings, watches, key chains, jewelry, or large belt buckles that could conduct electricity into the body.
- Never use a green ground wire as a current-carrying wire.

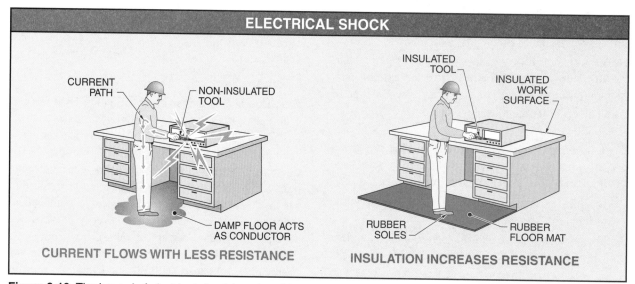

Figure 2-19. The hazard of electrical shock is reduced by insulating the body from the sources of electricity.

• Know cardiopulmonary resuscitation (CPR) and basic first aid procedures.

Warning: If an individual is receiving an electrical shock, power should be removed as quickly as possible. If the equipment circuit disconnect switch is nearby and can be operated safely, shut off the power. If power cannot be removed quickly, the victim's contact with the live parts must be broken. Insulated equipment such as a hot stick, rubber gloves, wood poles, or plastic pipes can be used to separate the victim from the energized circuit. Do not attempt to free the victim by grabbing without insulated equipment. When free from the electrical hazard, the victim should receive immediate medical attention.

Arc Flash and Arc Blast

Arcing is the discharge of electric current across an air gap. Arcs are caused by excessive voltage ionizing an air gap between two conductors, or by accidental contact between and then separation of the two conductors. When an arc occurs in certain situations, there is the possibility of arc flash or arc blast.

An *arc flash* is an extremely high-temperature discharge produced by an electrical fault in the air. Arc flashes can cause serious skin burns and damage eyesight. Arc flash temperatures reach 35,000°F, which can melt or vaporize metal and propel the droplets outward, causing serious burns or igniting clothing. Most electrical accidents requiring hospitalization are from arc flash burns, not electric shock. Arc flashes can be fatal up to about 10′ from the center of the flash.

An *arc blast* is an explosion that may accompany an arc-fault event. The intense heat of an arc fault causes the sudden expansion of air, resulting in a high-pressure wave with the power of several sticks of dynamite. Workers nearby can be thrown several feet away, be hit with flying debris, and suffer permanent hearing damage. Electrical systems of 480 V and higher have sufficient energy to produce an arc blast. Arc blasts are possible in lower voltage systems, but are less likely and not as destructive.

Arc flashes and arc blasts are possible hazards when working with electrical test instruments, so proper safety equipment must always be used. An arc flash can occur when using a test instrument to measure voltage at the same moment as a power-line transient, such as a lightning strike or power surge. Most test instruments indicate a CAT rating that specifies the tolerance limit for voltage transients. **See Figure 2-20.**

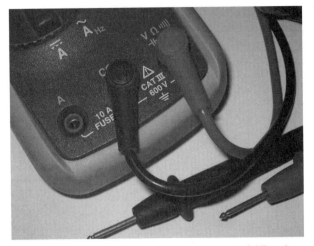

Figure 2-20. Test instruments specify their CAT rating, which is its degree to tolerance for voltage transients.

Also, a potential cause for arc flash and arc blast is improper test instrument use. For example, an arc blast can occur making an in-line current measurement on an energized circuit, which turns the meter into a low-resistance device. If a troubleshooter then switches to measuring voltage without changing test probe locations, the result is a short circuit that can cause an arc blast.

It is important to use a test instrument that is fused in the current-measuring mode to avoid this situation. Ensure that the test instrument has been tested by a reputable testing agency, such as UL, and never substitute off-brand fuses for the fuses specified. Only qualified personnel are allowed to work on energized circuits of 50 V or higher.

> **TECH TIP**
> A useful supplement to the NFPA 70E is the NFPA 70E Handbook for Electrical Safety in the Workplace, which clarifies each NFPA 70E code section with explanations and diagrams that are easier to understand.

Limits of Approach

In order to protect personnel from shock and arc hazards, the NFPA 70E defines several distances from live parts, each requiring increasing protection. **See Figure 2-21.** The *flash protection boundary* is the distance from live parts within which individuals could receive a second-degree burn due to an arc flash. PPE is required within the flash protection boundary. Per NFPA 70E, most systems of 600 V and less require a flash protection boundary of at least 4′.

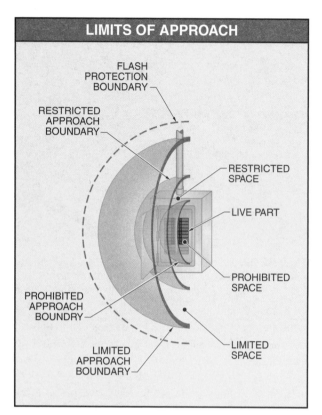

LIMITS OF APPROACH

FLASH PROTECTION BOUNDARY

RESTRICTED APPROACH BOUNDARY

RESTRICTED SPACE

LIVE PART

PROHIBITED SPACE

PROHIBITED APPROACH BOUNDRY

LIMITED SPACE

LIMITED APPROACH BOUNDARY

Figure 2-21. NFPA 70E specifies a flash protection boundary and a group of shock protection boundaries to define the PPE and procedures required based on the distance from live parts.

Shock protection boundaries are established to reduce the risk of electric shock. These are separate from the flash protection boundary in both purpose and requirements. The *limited approach boundary* is the distance from live parts within which a shock hazard exists. **See Figure 2-22.**

Personnel must be qualified to work within the limited approach boundary. An unqualified worker, such as a machine operator, must never cross the limited approach boundary unless trained in the hazards, wearing appropriate PPE, and constantly supervised by a qualified person. A *qualified person* is one with skills and knowledge of the relevant system and is trained to recognize and avoid related hazards. Depending on the circumstances, either the flash protection boundary or the limited approach boundary may be farther from the live parts.

The *restricted approach boundary* is the distance from live parts within which there is an increased risk of shock than at the limited approach boundary and further precautions are required. To work within this zone, a qualified worker must wear at least properly rated leather gloves over insulated rubber gloves, properly rated flame resistant clothing, safety glasses, and a hard hat with face shield and perhaps a balaclava shock hood. This PPE is required to prevent the accidental touching of energized conductors and electrical arcing to the body.

The *prohibited approach boundary* is the distance from live parts within which work is considered the same as making contact with the live part. It is considered forbidden for personnel to allow any part of their body within this space. For example, when testing a 480 V circuit, a troubleshooter's fingers may come no closer than 1″ to the energized conductors. Most high-quality test instrument probes are designed to maintain this safe distance by providing guards that prevent the fingers from sliding closer to the conductor. However, under extreme and justified circumstances, a highly trained worker may enter this space when wearing the required PPE and only after completing the necessary planning and risk analysis.

Nominal System Voltage, Phase To Phase*	Limited Approach Boundary		Restricted Approach Boundary	Prohibited Approach Boundary
	Exposed Movable Conductor	Exposed Fixed-Circuit Part		
0 to 50	N/A	N/A	N/A	N/A
51 to 300	10′-0″	3′-6″	Avoid contact	Avoid contact
301 to 750	10′-0″	3′-6″	1′-0″	0′-1″
751 to 15,000	10′-0″	5′-0″	2′-2″	0′-7″

SHOCK PROTECTION BOUNDARIES

* in V

Figure 2-22. Shock protection boundaries reduce the risk of electric shock by requiring certain levels of PPE within different boundaries. The boundary distances vary depending on the system voltage.

Even if the circuit being worked on is de-energized, the possibility exists that nearby circuits are still energized and within a limited approach boundary. Barriers such as insulation blankets, along with the proper PPE, must be used to protect against hazards from nearby energized circuits.

Electrical Work PPE

While it is safest to work on de-energized electrical circuits, this is not always possible, especially when troubleshooting. If work must be done on energized circuits, the NFPA 70E may require an energized electrical work permit, which requires personnel to carefully plan their movements and actions when working around electric circuits. They must inspect their PPE and test equipment, decide where to place their hands and arms to avoid accidentally touching any energized parts, and ensure that there are no obstacles or hazards that could hinder their movements. There are no "safe" electrical circuits and all electrical work requires prior training and careful planning.

Proper PPE must always be worn when working with energized electrical circuits. The level of PPE is determined by the type of activity and the voltage level of the circuits involved. For example, measuring voltage on a circuit lower than 240 V is a Hazard/Risk Category 1. This requires the use of rubber insulating gloves combined with leather over gloves and insulated and insulating tools. **See Figure 2-23.** Other PPE for this work includes FR clothing, hard hat, safety glasses, and hearing protection. Working on circuits higher than 240 V requires more PPE and higher-rated clothing, and at times arc-flash suits.

Protective clothing for electrical work is categorized by its arc rating. An *arc rating* is a PPE rating for the maximum amount of thermal energy the article can resist before exposing the wearer to second-degree burns. Arc ratings are quantified as energy per unit area, commonly measured in calories per cubic centimeter (cal/cm^2). A single layer of flame-resistant (FR) clothing should have a rating of 4 to 8 cal/cm^2. Two layers of FR clothing should provide 25 cal/cm^2 of protection. Certain tasks on or near equipment over 1000 V require a multilayer flash suit. **See Figure 2-24.** Flash suits are similar to fire fighter turnout gear and must satisfy at least 40 cal/cm^2 in thermal energy protection.

The requirements for boundaries and PPE can be confusing, but all maintenance personnel should carry and use the minimum required PPE and be vigilant regarding approach boundaries. When in doubt as to the level of hazard, always opt for the higher level of PPE.

PPE REQUIRED FOR ELECTRICAL WORK		
Hazard/Risk Category	Other Clothing	Other PPE
0	Non-melting fabrics	Safety glasses Hearing protection Leather gloves*
1	FR clothing with minimum 4 cal/cm² arc rating	Hard hat Safety glasses Hearing protection Leather gloves* Leather safety shoes
2	FR clothing with minimum 8 cal/cm² arc rating	Hard hat Safety glasses Hearing protection Leather gloves* Leather safety shoes
3	FR clothing and arc flash suit with minimum 25 cal/cm² arc rating	Hard hat FR hard hat liner Safety glasses Hearing protection Arc-rated gloves* Leather safety shoes
4	FR clothing and arc flash suit with minimum 40 cal/cm² arc rating	Hard hat FR hard hat liner Safety glasses Hearing protection Arc-rated gloves* Leather safety shoes

* Hand protection may consist of rated rubber insulating gloves and leather protectors, depending on voltage.

Figure 2-23. PPE required for electrical work is based largely on Hazard/Risk Categories, which are listed for various tasks in NFPA 70E.

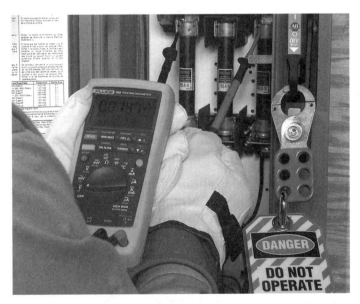

Safety procedures and proper PPE minimize the risk of electrical shock when testing electrical circuits.

ARC FLASH SUITS

Figure 2-24. Arc flash suits protect workers from the intense thermal energy of arc blasts.

LOCKOUTS AND TAGOUTS

Lockouts and tagouts are applied to power sources to prevent equipment operation during inspection, maintenance, and repair. Power sources include all electrical, pneumatic, and hydraulic power to the equipment. *Lockout* is the use of locks, chains, or other physical restraints to prevent the operation of specific equipment. **See Figure 2-25.** Lockout devices may employ lightweight enclosures to disable control devices such as valves, switches, and plugs. Colors are used to match standard hazard-level color codes. Locks used to lock out a device may be color-coded and individually keyed. A multiple lockout hasp is used when more than one worker must lock out a power source.

Tagout is the attachment of a danger tag to the source of power to indicate that the equipment may not be operated until the tag is removed. A tagout does not prevent the startup of equipment, but serves as a warning to operating and service personnel. A tagout is used when a lockout is not possible. Danger tags may include

warnings such as "Do Not Operate" and provide space to enter worker name, date, and reason for lockout/tagout. Tagouts are attached by hand and must be easy to read, durable, and resistant to accidental removal. Both lockouts and tagouts must be tough and able to resist damage from the working environment.

Employers are responsible for supplying lockout and tagout equipment, and training employees on lockout/tagout procedures. Written lockout/tagout procedures must be established for each piece of equipment in the facility. Before working on any equipment, maintenance personnel must locate sources of potentially hazardous energy that could be released, notify people working near the equipment that it is being shut down, and apply locks and tags to make the equipment safe before starting work. It is important to lock out equipment even if the work will only take a short time. If the lockout/tagout extends beyond a shift change, incoming workers must follow the established procedures of the lockout.

Warning: Lockout/tagout procedures must conform to OSHA 29 CFR 1910.147, *The Control of Hazardous Energy (Lockout/Tagout),* and company rules and procedures. A lockout/tagout shall not be removed by any individual other than the individual who installed it, except in an emergency. In an emergency, only supervisory personnel may remove a lockout/tagout, and only upon notification of the authorized individual.

CONFINED SPACES

A *confined space* is a space large enough that an employee can physically enter and perform assigned work, but has limited or restricted means for entry and exit, and is not designed for continuous employee occupancy. Confined spaces include storage tanks, process vessels, boilers, ventilation or exhaust ducts, sewers, underground utility vaults, pipelines, and open top spaces more than 4′ in depth such as pits and ditches.

TECH TIP

There must be at least one outside attendant whenever workers are inside a permit-required confined space. The attendant is responsible for periodically testing the atmospheric conditions and monitoring the workers inside. The attendant must never enter the confined space. This ensures that there is always a person able to call for help in the event that workers are injured or overcome by harmful gases.

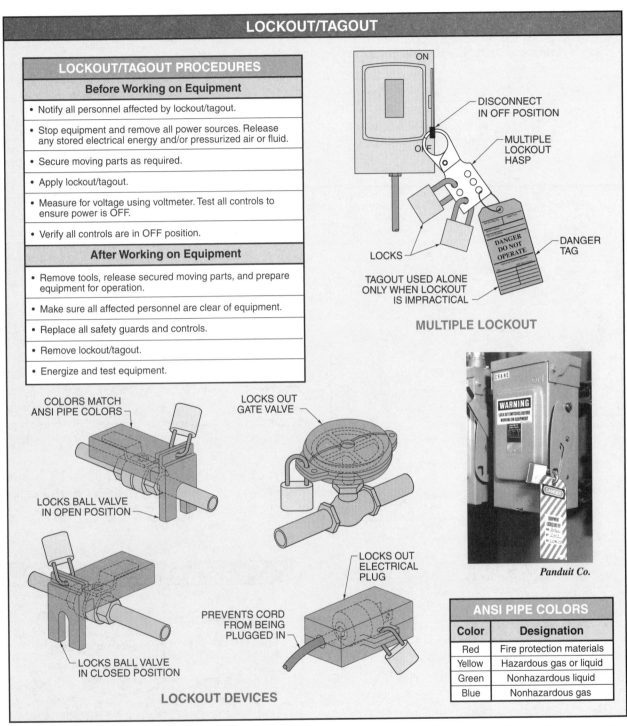

LOCKOUT/TAGOUT

LOCKOUT/TAGOUT PROCEDURES

Before Working on Equipment

- Notify all personnel affected by lockout/tagout.
- Stop equipment and remove all power sources. Release any stored electrical energy and/or pressurized air or fluid.
- Secure moving parts as required.
- Apply lockout/tagout.
- Measure for voltage using voltmeter. Test all controls to ensure power is OFF.
- Verify all controls are in OFF position.

After Working on Equipment

- Remove tools, release secured moving parts, and prepare equipment for operation.
- Make sure all affected personnel are clear of equipment.
- Replace all safety guards and controls.
- Remove lockout/tagout.
- Energize and test equipment.

ON

OFF

DISCONNECT IN OFF POSITION

MULTIPLE LOCKOUT HASP

LOCKS

DANGER TAG

TAGOUT USED ALONE ONLY WHEN LOCKOUT IS IMPRACTICAL

MULTIPLE LOCKOUT

Panduit Co.

COLORS MATCH ANSI PIPE COLORS

LOCKS OUT GATE VALVE

LOCKS BALL VALVE IN OPEN POSITION

LOCKS BALL VALVE IN CLOSED POSITION

PREVENTS CORD FROM BEING PLUGGED IN

LOCKS OUT ELECTRICAL PLUG

LOCKOUT DEVICES

ANSI PIPE COLORS

Color	Designation
Red	Fire protection materials
Yellow	Hazardous gas or liquid
Green	Nonhazardous liquid
Blue	Nonhazardous gas

Figure 2-25. Lockout and tagout procedures protect maintenance workers by preventing equipment operation and warning other personnel about potential dangers.

Confined Space Hazards

Confined spaces are particularly susceptible to containing oxygen-deficient, toxic, or explosive atmospheres. Oxygen deficiency is caused by the displacement of oxygen by leaking gases or vapors, combustion or oxidation processes, oxygen absorbed by the vessel or product stored, and/or consumption by bacteria. Oxygen-deficient air can result in injury or death. **See Figure 2-26.**

POTENTIAL EFFECTS OF OXYGEN-DEFICIENT ATMOSPHERES*

Oxygen Content†	Effects and Symptoms‡
19.5	Minimum permissible oxygen level
15 – 19.5	Decreased ability to work strenuously. May impair condition and induce early symptoms in persons with coronary, pulmonary, or circulatory problems
12 – 14	Respiration exertion and pulse Increase. Impaired coordination, perception, and judgment
10 – 11	Respiration further increases in rate and depth, poor judgement, lips turn blue
8 – 9	Mental failure, fainting, unconsciousness, ashen face, blue lips, nausea, and vomiting
6 – 7	Eight min, 100% fatal; 6 min, 50% fatal; 4 to 5 min, recovery with treatment
4 – 5	Coma in 40 seconds, convulsions, respiration ceases, death

* Values are approximate and vary with state of health and physical activities
† % by volume
‡ at atmospheric pressure

Figure 2-26. Oxygen-deficient atmospheres in confined spaces can be life-threatening.

Common toxic gases include hydrogen sulfide, natural gas, and carbon monoxide. Toxic gases are released by cleaning solvents, chemical reactions, heated materials, and other sources.

Combustible atmospheres are commonly caused by gases such as methane, carbon monoxide, and hydrogen sulfide. An increase in the oxygen level, above the normal 21%, further increases the explosive potential of combustible gases. Finely ground materials including carbon, grain, fibers, metals, and plastics can also cause explosive atmospheres.

The *explosive range* is the difference between the lower explosive limit and the upper explosive limit of combustible gases. **See Figure 2-27.** The *lower explosive limit (LEL)* is the lowest concentration (air-fuel mixture) at which a gas can ignite. Concentrations below this limit are too lean to burn. The *upper explosive limit (UEL)* is the highest concentration (air-fuel mixture) at which a gas can ignite. Concentrations above this limit are too rich to burn.

COMBUSTIBLE ATMOSPHERES

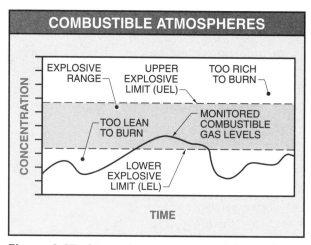

Figure 2-27. Atmospheres are potentially explosive when the concentration of a combustible gas is within a certain range.

However, combustible gases at any concentration are a concern. Lean mixtures can collect in areas and reach combustible levels. Also, lean mixtures may still be toxic. Rich mixtures can be diluted with air and become combustible. Detection instruments are set to sense the presence of combustible gases at levels that forewarn workers of potentially hazardous combustible atmospheres before the LEL is reached. **See Figure 2-28.**

Warning: Confined-space procedures vary in each facility. For maximum safety, always refer to specific facility procedures and applicable federal, state, and local regulations.

CONFINED SPACE ATMOSPHERE TESTING

Mine Safety Appliances Co.

Figure 2-28. Confined spaces must be tested for oxygen level and air contaminants before entering.

Confined-Space Permits

A *permit-required confined space* is a confined space that has specific health and safety hazards capable of causing death or serious physical harm. Permit-required confined spaces are spaces containing or having a potential to contain a hazardous atmosphere; containing a material that has the potential for engulfing an entrant; having an internal configuration such that an entrant could be trapped or asphyxiated by inwardly converging walls or a floor that slopes downward and tapers into a smaller cross-section; or containing any other recognized safety or health hazard. OSHA 29 CFR 1910.146, *Permit-Required Confined Spaces*, contains the requirements for practices and procedures to protect workers from the hazards of permit-required confined spaces. **See Figure 2-29.** Even if a confined space does not initially require a permit, personnel must be aware that these conditions can change with tasks such as welding, painting, or solvent use within the confined space.

Employers must evaluate the workplace to determine if spaces are permit-required confined spaces. If confined spaces exist in the workplace, the employer must inform personnel of their existence and location, and of the danger they pose. This is accomplished by posting danger signs or by other equally effective means. In addition, the employer must develop a written permit-required confined space program that specifies entry procedures, hazard identification, access restriction, hazard control, and monitoring of the space during entry.

Entry-Permit Procedures

An entry permit must be posted at confined-space entrances or otherwise made available to entrants before entering a permit-required confined space. The permit is signed by the entry supervisor and verifies that preentry preparations have been completed and that the space is safe to enter. **See Figure 2-30.** A permit-required confined space must be isolated before entry. This prevents hazardous energy and materials from entering the space. Plant procedures for lockout/tagout of permit-required confined spaces must be followed.

> **SAFETY TIP**
> *It is the responsibility of all facility personnel to look out for any potential safety hazards, ensure that they have the required safety gear and training, use the equipment properly, and avoid any procedures or situations that are beyond their training.*

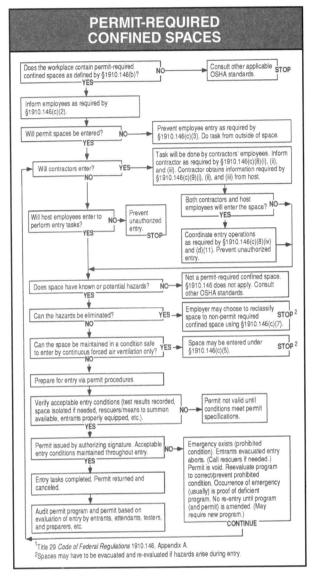

Figure 2-29. Procedures for entering a confined space must follow established OSHA safety standards.

Mine Safety Appliances Co.
Workers entering a confined space must have the proper safety equipment, such as a rescue harness and a portable gas detector.

ENTRY PERMIT

✔ CONFINED SPACE ✔ HAZARDOUS AREA

PERMIT VALID FOR 8 HOURS ONLY. ALL COPIES OF PERMIT WILL REMAIN AT JOB SITE UNTIL JOB IS COMPLETED

SITE LOCATION and DESCRIPTION _Bunker Fuel Oil Tank #2_

PURPOSE OF ENTRY _Routine Maintenance/Inspection_

SUPERVISOR(S) in charge of crews. Type of Crew Phone #
Michael Green   _Maintenance Shift II - X5924_

* BOLD DENOTES MINIMUM REQUIREMENTS TO BE COMPLETED AND REVIEWED PRIOR TO ENTRY*

REQUIREMENTS COMPLETED	DATE	TIME	REQUIREMENTS COMPLETED	DATE	TIME
Lock Out/De-energize/Try-out	10/2	09:00	**Full Body Harness w/"D" ring**	10/4	08:00
Line(s) Broken-Capped-Blanked	10/2	11:00	**Emergency Escape Retrieval Equip**	10/4	08:00
Purge-Flush and Vent	10/3	09:00	**Lifelines**	10/4	08:00
Ventilation	10/3	10:00	Fire Extinguishers	10/4	08:00
Secure Area (Post and Flag)	10/2	08:00	Lighting (Explosion Proof)	10/4	08:00
Breathing Apparatus	10/4	08:00	Protective Clothing	10/4	08:00
Resuscitator - Inhalator	10/4	08:00	Respirator(s) (Air Purifying)	10/4	08:00
Standby Safety Personnel	10/4	08:00	Burning and Welding Permit	N/A	N/A

Note: Items that do not apply enter N/A in the blank.

** RECORD CONTINUOUS MONITORING RESULTS EVERY 2 HOURS

CONTINUOUS MONITORING**   Permissible
TEST(S) TO BE TAKEN   Entry Level   10/4

	Permissible Entry Level						
PERCENT OF OXYGEN	19.5% to 23.5%		20.5	20.6	20.7	20.5	20.5
LOWER FLAMMABLE LIMIT	Under 10%		5	5	5	5	6
CARBON MONOXIDE	+35 PPM		0	0	0	0	0
Aromatic Hydrocarbon	+ 1 PPM	* 5PPM	2	1	2	1	1
Hydrogen Cyanide	(Skin)	* 4PPM	N/A				
Hydrogen Sulfide	+10 PPM	*15PPM	N/A				
Sulfur Dioxide	+ 2 PPM	* 5PPM	3	2	2	2	2
Ammonia		* 35PPM	N/A				

* Short-term exposure limit:Employee can work in the area up to 15 minutes.

+ 8 hr. Time Weighted Avg.:Employee can work in area 8 hrs (longer with appropriate respiratory protection).

REMARKS: _____

GAS TESTER NAME & CHECK #	INSTRUMENT(S) USED	MODEL &/OR TYPE	SERIAL &/OR UNIT #
Marty James	_Combination Gas Meter_	_Industrial Scientific_	_15 A_

SAFETY STANDBY PERSON IS REQUIRED FOR ALL CONFINED SPACE WORK

SAFETY STANDBY PERSON(S)	CHECK #	NAME OF SAFETY STANDBY PERSON(S)	CHECK #
Kate Washington	_3312_		
Tony Linder	_3318_		

SUPERVISOR AUTHORIZING ENTRY
ALL ABOVE CONDITIONS SATISFIED _Michael Green_   AMBULANCE 2800   FIRE 2900
Safety 4901   Gas Coordinator 4529/5387

Figure 2-30. A confined space entry permit documents preparations, procedures, periodic monitoring, and required equipment.

A plant may develop its own confined space permit, but OSHA requires that certain checklist items be included. These items include the permit date, time, expiration, supervisor name, results of atmospheric testing both prior to entry and periodically during work, the applicable safety equipment, and communication and rescue procedures. Standby personnel, commonly called a "hole watch," must remain just outside a confined space while workers are inside. Hole watches are responsible for periodically testing and recording the atmospheric conditions, monitoring the workers inside the space, and assisting rescue personnel in the event of an emergency.

The duration of entry permits must not exceed the time required to complete an assignment. The entry supervisor must terminate entry and cancel permits when an assignment has been completed or when new conditions exist. All canceled entry permits must be filed for at least one year.

Training is required for all personnel who are required to work in or around permit-required confined spaces. A certificate of training includes the worker's name, the signature or initials of the trainer, and the dates of training. The certificate must be available for inspection by authorized officials.

⊕ SAFETY TIP

Carbon monoxide is a colorless, odorless, and tasteless gas that is a common hazard in confined spaces, where it can easily collect.

HAZARDOUS MATERIALS

A *hazardous material* is a substance that could cause injury to individuals or damage to the environment. The characteristics of specific hazardous materials dictate their proper handling and disposal procedures. **See Figure 2-31.**

Justrite Mfg. Co.

Figure 2-31. Characteristics of specific hazardous materials may dictate certain handling and disposal procedures.

Worker information and education on proper handling procedures are most effective at preventing accidents involving hazardous materials. Employers must develop, implement, and maintain a written, comprehensive hazard communication program that includes provisions for container labeling, chemical inventory, collection, availability of material safety data sheets, and employee training. It must also contain a list of hazardous chemicals in each work area.

Container Labeling

Containers that contain hazardous materials must be labeled, tagged, or otherwise marked with the identity of the hazardous material and appropriate hazard warnings per OSHA 29 CFR 1910.1200(f), *Labels and Other Forms of Warning*. Container labels may vary in format, but must include basic right-to-know (RTK) information to convey hazards of the chemical according to federal and state standards. **See Figure 2-32.** Some of this information can be provided in symbol form by using the NFPA or HMIG hazardous material labeling systems. Using symbols or colors to represent hazards provides information at a glance, which speeds response in hazardous-material emergencies.

The NFPA Hazard Signal System uses a four-color diamond sign to display basic information about hazardous materials. **See Figure 2-33**. Colors and numbers identify potential health (blue), flammability (red), reactivity (yellow), and specific (no color) hazards. Numbers indicate the degree of severity, ranging from 0 (no hazard) to 4 (severe hazard).

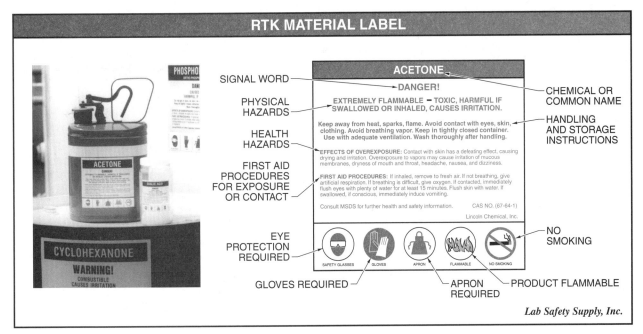

Figure 2-32. Hazardous material container labels must include basic right-to-know (RTK) information.

NFPA HAZARDOUS MATERIAL LABELING SYSTEM

HEALTH HAZARD (BLUE)
4 DEADLY
3 EXTREME DANGER
2 HAZARDOUS
1 SLIGHTLY HAZARDOUS
0 NORMAL MATERIAL

SPECIFIC HAZARD
OX   OXIDIZER
ACID   ACID
ALK   ALKALI
COR   CORROSIVE
W̶   USE **NO WATER**
☢   RADIATION HAZARD

CHEMICAL NAME:
ACETONE

FIRE HAZARD (RED)
FLASH POINTS
4 BELOW 73°F
3 BELOW 100°F
2 BELOW 200°F
1 ABOVE 200°F
0 WILL NOT BURN

REACTIVITY (YELLOW)
4 MAY DETONATE
3 SHOCK AND HEAT MAY DETONATE
2 VIOLENT CHEMICAL CHANGE
1 UNSTABLE IF HEATED
0 STABLE

Reprinted with permission from NFPA 704, Standard System for the Identification of the Fire Hazards of Materials, Copyright ©1990, National Fire Protection Association, Quincy, MA 02269. This reprinted material is not the complete and official position of the National Fire Protection Association on the referenced subject, which is represented only by the standard in its entirety.

Identification of Health Hazard Color Code: BLUE		Identification of Flammability Color Code: RED		Identification of Reactivity (Stability) Color Code: YELLOW	
Signal	Type of Possible Injury	Signal	Susceptibility of Materials to Burning	Signal	Susceptibility to Release of Energy
4	Materials that on very short exposure could cause death or major residual injury	4	Materials that will rapidly or completely vaporize at atmospheric pressure and normal ambient temperature, or that are readily dispersed in air and that will burn readily	4	Materials that in themselves are readily capable of detonation or of explosive decomposition or reaction at normal temperatures and pressures
3	Materials that on short exposure could cause serious temporary or residual injury	3	Liquids and solids that can be ignited under almost all ambient temperature conditions	3	Materials that in themselves are capable of detonation or explosive decomposition or reaction but require a strong initiating source or which must be heated under confinement before initiation or which react explosively with water
2	Materials that on intense or continued but not chronic exposure could cause temporary incapacitation or possible residual injury	2	Materials that must be moderately heated or exposed to relatively high ambient temperatures before ignition can occur	2	Materials that readily undergo violent chemical change at elevated temperatures and pressures or which react violently with water or which may form explosive mixtures with water
1	Materials that on exposure would cause irritation but only minor residual injury	1	Materials that must be preheated before ignition can occur	1	Materials that in themselves are normally stable, but which can become unstable at elevated temperatures and pressures
0	Materials that on exposure under fire conditions would offer no hazard beyond that of ordinary combustible material	0	Materials that will not burn	0	Materials that in themselves are normally stable, even under fire exposure conditions, and which are not reactive with water

Figure 2-33. Hazardous material labels using the NFPA system use color and numbers to indicate health, fire, reactivity, and other hazards.

A *health hazard* is the likelihood of a material to cause, either directly or indirectly, temporary or permanent injury or incapacitation due to an acute exposure by contact, inhalation, or ingestion. Health hazard is indicated on a blue background in the left diamond of the NFPA Hazard Signal System. The degree of health hazard is ranked by number according to the probable severity of the effects of exposure to the hazardous material. This determines the protective and respiratory equipment required by emergency response and fire-fighting teams.

A *flammability hazard* is the degree of susceptibility of materials to burning based on the form or condition of the material and its surrounding environment. Flammability hazard is indicated on a red background in the top diamond of the NFPA Hazard Signal System. The degree of flammability hazard is ranked by number according to the susceptibility of hazardous materials to burning.

A *reactivity hazard* is the degree of susceptibility of materials to explode or release energy by themselves or by exposure to certain conditions or substances. Reactivity hazard is indicated on a yellow background in the right diamond of the NFPA Hazard Signal System. The degree of reactivity is ranked by number according to the ease, rate, and quantity of energy released.

A *specific hazard* is a designation on the NFPA Hazard Signal System that specifies special properties and hazards associated with a particular material. This information is particularly useful for determining the appropriate emergency or fire-fighting response. For example, a letter W with a horizontal line through the center is used to indicate materials that demonstrate unusual reactivity with water. The letters OX indicate materials that possess oxidizing properties. Specific hazard symbols are shown in the bottom diamond of the NFPA Hazard Signal System or immediately above or below the entire symbol. No special color is specified.

The Hazardous Materials Identification System (HMIS) provides information using the same color code as the NFPA Hazard Signal System. The HMIG label also lists hazard ratings for health, flammability, reactivity, and protective equipment requirements using a number from 0 to 4. The protective equipment required is identified by letter. **See Figure 2-34.**

Material Safety Data Sheets

A *material safety data sheet (MSDS)* is a document containing hazard information about a certain chemical. This information is used to inform and train employees on the safe use of hazardous materials. **See Figure 2-35.** An MSDS must be in English, be easy to read, and include certain information required by OSHA 29 CFR 1910.1200, *Toxic and Hazardous Substances, Hazard Communication.*

All chemical products used in a facility must be inventoried and have an MSDS on hand. This includes chemicals used by maintenance personnel, custodial staff, building occupants, and contracted grounds maintenance and pest-control workers. MSDSs are available from chemical manufacturers, importers, or distributors. MSDS files must be kept up-to-date and well organized to allow quick access to the information in an emergency. Information may be filed according to product name, manufacturer, company-assigned number, or Chemical Abstracts Service (CAS) number. The CAS indexes information published in *Chemical Abstracts* by the American Chemical Society. For example, the CAS number for isopropyl alcohol is 67-63-0.

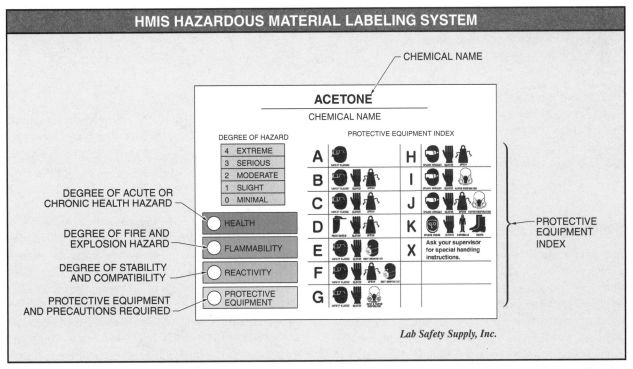

Figure 2-34. Hazardous-material labels using the HMIG system use symbols to indicate PPE required for handling the material.

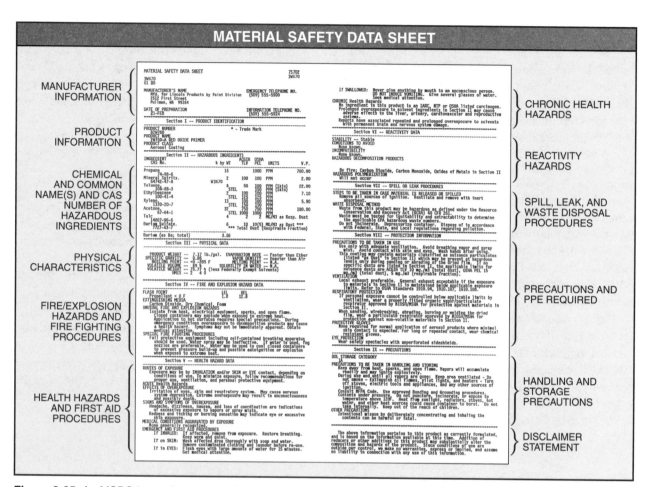

Figure 2-35. An MSDS is used to communicate chemical hazard information to all personnel using, handling, or working near the chemicals.

Workers must study the associated MSDS when using chemical products. The MSDS contains information about the hazards associated with the chemical, required PPE, storage and handling instructions, and response in case of personal exposure, fire, or a spill of the chemical.

INDUSTRIAL HYGIENE

Industrial hygiene is the science of anticipating, recognizing, evaluating, and controlling workplace conditions that may cause worker injury or illness. An industrial hygienist is an individual who analyzes, identifies, and measures workplace hazards or stresses that can cause sickness, impaired health, or significant discomfort in workers. Industrial hygienists then recommend corrective administrative, engineering, or PPE measures to eliminate or control these conditions.

Workplace Cleanliness

Many hazards identified by an industrial hygiene program are reduced by maintaining clean and organized workspaces. Many workers are injured because of slipping on wet or oily surfaces or tripping over uneven ground, objects on the floor, or loose rugs or carpeting.

Good housekeeping in the workplace is critical to preventing injuries. When a task is finished, the equipment and surrounding area should be completely cleaned and all parts and tools returned to their correct locations. Walkways should be clear of any tripping hazards. Extension cords should be routed and guarded to prevent someone from tripping. Floors should be kept clean and all spills cleaned quickly and completely. Signs and barriers should be placed around wet or slippery floors. Work areas should have adequate lighting so that potentially dangerous situations can be identified quickly.

Slip and trip accidents often occur when workers become distracted from their surroundings. It is important to plan every move and to remain aware of all obstacles that might cause problems. Extra caution is required when carrying any load on stairs. Workers should check their pathway up or down the stairs and ensure that the steps are not wet, slippery, or damaged.

Ergonomic Hazards

Ergonomics is the study of the effects of job-related tasks and work-area arrangement on the health of the worker. An *ergonomic hazard* is a physical task or body position that causes musculoskeletal stress or injury. Common ergonomic hazards include eyestrain, repetitive motions, and lifting. Ergonomic injuries commonly result from extreme exertion, awkward or immobile positions, repetition, vibrating tools, and poorly designed tasks and work areas. These injuries can be painful and debilitating, causing numbness, swelling, weakness, and reduced range of motion.

Employers can conduct ergonomic studies and adjust tasks or work areas to reduce hazards. Personnel can also protect themselves by using proper lifting techniques, avoiding awkward positions, and changing tasks as often as possible. Stretching before and after working in awkward or uncomfortable positions can help reduce stress and strains.

Lifting and Moving. Improper lifting and moving procedures can result in back injuries. Back injuries are extremely common, but preventable through a combination of safe lifting and moving practices. Before lifting, the weight of the object should be determined and the lifting and moving method should be planned based on the object's size, shape, weight, and path of travel. Assistance should be sought when necessary. The entire pathway should be cleared of obstructions or other hazards.

Back injuries are common when bending forward to pick up an object. The lower back supports the upper body's weight, cantilevered out from the waist. The weight of the object being lifted adds further to the force being exerted on the lower back. For proper lifting technique, the object should be close to the body and the worker should rise slowly using the strong leg muscles. When standing upright, the worker should not twist. This puts extreme stress on the back and legs. **See Figure 2-36.**

PROPER LIFTING TECHNIQUE

1 BEND KNEES AND GRASP OBJECT FIRMLY

KEEP BACK STRAIGHT

2 LIFT OBJECT BY STRAIGHTENING LEGS

3 MOVE FORWARD AFTER WHOLE BODY IS IN VERTICAL POSITION

Figure 2-36. Proper lifting technique uses the legs to lift an object.

A back-support belt provides additional support and direction during lifting. **See Figure 2-37.** Carts or hand trucks should be used when objects are too heavy or oddly shaped. The load must be balanced to avoid tipping or falling. When moving a load on a cart, the worker's body weight should be used while taking even, short steps.

BACK SUPPORT BELTS

Lab Safety Supply, Inc.

Figure 2-37. Lifting with proper technique and a back-support belt reduces the chance of a back injury.

Poor physical condition is a major contributor to all types of back problems. Being overweight significantly increases the pressure on the lower back. Therefore, as part of working safely, it is the worker's personal responsibility to maintain a healthy weight, good physical strength, and flexibility. Flexibility is required to avoid putting additional strain on the muscles and tendons that support the spinal column during every day work activities.

Physical Hazards

A *physical hazard* is a hazard caused by excessive levels of noise, vibration, illumination, temperature, or radiation. Noise can be reduced by installing equipment and systems that have been engineered, designed, and built to operate quietly. Enclosure of noisy equipment and proper maintenance and replacement of worn or unbalanced parts can also reduce noise. **See Figure 2-38.** Mufflers, baffles, and/or special mounts that reduce vibration provide additional sound deadening. Acoustical material on ceilings and walls can absorb and reduce reflected noise.

NOISE HAZARDS

Figure 2-38. Specially designed enclosures around noisy equipment reduce overall sound levels in the facility.

Vibration, illumination, and temperature hazards are controlled by environment and equipment design. Radiation hazards require special monitoring, shielding, and response procedures. (These hazards are uncommon in most industrial facilities.)

Ladder and Manlift Safety

Many injuries are caused by falls from ladders. Causes of falls generally involve improper maintenance or use of the ladder. Ladders should be kept clean, used in the manner intended, and stored properly. A ladder should be inspected before each use to ensure that all parts are clean, undamaged, and solidly in place. The ladder must be rated to support the weight of the worker and any necessary tools or supplies. The weight restrictions are found on the ladder's label.

The base of an extension ladder must be 1′ away from the wall for every 4′ of height. **See Figure 2-39.** The upper end of the ladder is placed against a secure surface capable of supporting the weight on the ladder. When accessing an elevated area, such as a rooftop, the ladder should extend 3′ above the upper support point. Shoes should be clean so they do not slip on the ladder rungs. The worker should never climb higher than the fourth rung from the top of an extension ladder. The body should remain in the center of the ladder by not reaching out to the left or right too far.

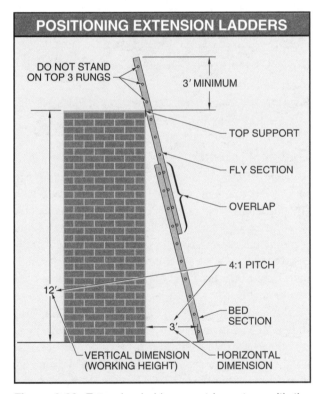

Figure 2-39. Extension ladders must be set up with the base 1′ from the wall for every 4′ of height leading to the top support.

Stepladders are self-supporting or fold-out ladders. The top of the ladder must be kept clear to prevent items from falling. Only one person may use a stepladder and the back support of the ladder must never be used for

climbing. Do not stand above the second rung from the top of a stepladder.

Manlifts are mechanized devices that lift a worker beyond the usual reach of a ladder. Manlifts can become unstable as the workers and tools rise, so stabilizing outriggers may be required. The manufacturer's instructions should include information on height, weight, and extension limits. Workers in the manlift bucket must wear fall protection equipment. When raising a manlift, particular attention must be paid to overhead obstructions. The required distances to overhead electrical wires must be maintained.

Fatigue

Fatigue is the cause of many safety problems and accidents, particularly during night shifts. Fatigue is caused by insufficient sleep, sustained mental concentration, physical work, and stress. Fatigue leads to a loss of concentration, which can cause lapses in safety procedures and failure to recognize hazardous situations. The results are less effective planning, slowed mental and physical reactions, increased risk-taking, and judgment errors.

While difficult working conditions can cause fatigue, poor sleep patterns are a major factor. Getting regular exercise and nutritious meals, avoiding caffeine, and establishing a regular sleep time with no disruptions are important steps to getting adequate sleep.

When fatigued, workers should take a break in a different location or eat a healthy snack. Sugary foods should be avoided because they only temporarily boost energy, followed by continued fatigue. Excessive caffeine use and consuming caffeine five hours before sleep should also be avoided. Many individuals are not aware of their own level of fatigue, so workers should be alert for fatigue in coworkers as well.

Also, many over-the-counter cold and congestion medications cause drowsiness and reduced alertness. Workers should be aware of the side effects of any medication and stay away from work if those medications will impact safety on the job. Those suspected of being under the influence of drugs or alcohol while on the job should be reported to a supervisor.

⊕ *SAFETY TIP*

The back vertebrae are constantly under stress from standing, bending, lifting, pushing, twisting, reaching, and even sitting. Proper posture and lifting techniques are vital to avoiding painful back injuries.

Air Contaminants

An *air contaminant* is an undesirable element in the air. The most common particulate contaminants include dusts, fumes, aerosols, and fibers. *Dust* is a collection of minute solid particles suspended in the air or settled on surfaces. Dust is created by handling, crushing, grinding, colliding, exploding, or heating materials such as rock, ore, metal, coal, wood, and grain. A *fume* is vapor from volatilized solids that condenses in cool air. An *aerosol* is a suspension of very fine solid particles or liquid droplets in air. A *fiber* is a solid particle whose length is several times greater than its diameter. Special worker certification and handling procedures are required when mitigating or working around known air contaminants. **See Figure 2-40.**

ASBESTOS HANDLING

Figure 2-40. Individuals working with air contaminants, such as asbestos fibers, require special training and must use specific handling procedures.

Gas contaminants in the air are typically measured in parts per million (PPM). Carbon monoxide is one of the most dangerous and common gases. **See Figure 2-41.** Carbon monoxide is invisible and odorless, so it gives no warning to its victims. Carbon monoxide is slightly lighter than air. In addition, carbon monoxide rises as it is warmed in the combustion process. However, during hot, humid weather, carbon monoxide may remain almost stationary, resulting in areas of high concentration. The primary source of carbon monoxide is incomplete combustion of carbon-based fuels such as gasoline, natural gas, oil, propane, coal, and wood.

POTENTIAL EFFECTS OF CARBON MONOXIDE EXPOSURE		
PPM*	Effects and Symptoms†	Time
50	Permissible exposure level	8 hr
200	Slight headache, discomfort	3 hr
400	Headache, discomfort	2 hr
600	Headache, discomfort	1 hr
1000 to 2000	Confusion, headache, nausea	2 hr
1000 to 2000	Tendency to stagger	1½ hr
1000 to 2000	Slight palpitation of the heart	30 min
2000 to 2500	Unconsciousness	30 min
4000	Fatal	Less than 1 hr

* parts per million values are approximate
† effects vary with state of health and physical activity

Figure 2-41. Carbon monoxide is a clear, odorless gas that can cause serious health hazards.

Chemical Hazards

A *chemical hazard* is a chemical in any form that has toxic effects when inhaled, absorbed, or ingested. Chemical hazards vary depending on the manner in which they are introduced into the body. For example, some chemicals are toxic through inhalation, while others are toxic by absorption through the skin or through ingestion. Some are toxic by all three routes and may be flammable as well.

The degree of risk from exposure to any given substance depends on the nature and potency of toxic effects and magnitude and duration of exposure. Chemical monitoring badges can be used to indicate the amount of exposure. **See Figure 2-42.** The type of PPE required is determined by the hazards of the chemical.

Information on the risk to workers from chemical hazards is listed on the MSDS. Chemical exposure may cause or contribute to many serious health effects, such as heart ailments, kidney and lung damage, cancer, sterility, burns, and rashes. Some chemicals may also have the potential to cause fires, explosions, and other serious accidents.

CHEMICAL HAZARD PRECAUTIONS

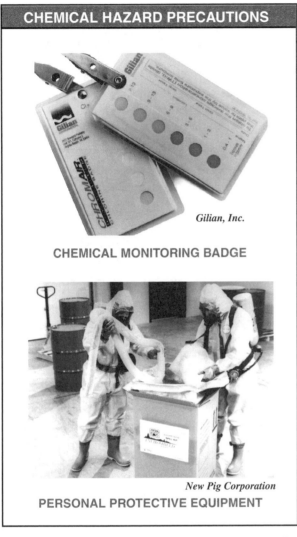

Gilian, Inc.

CHEMICAL MONITORING BADGE

New Pig Corporation

PERSONAL PROTECTIVE EQUIPMENT

Figure 2-42. Chemical monitoring badges indicate the amount of exposure to chemicals present. PPE is required when working with toxic chemicals.

Biological Hazards

A *biological hazard* is a bacterium, virus, fungus, or other microorganism that can cause acute and chronic infections by entering the body directly or through breaks in the skin. Tasks that require contact with food, plants, animals, animal products, or human bodily fluids may expose workers to biological hazards.

Where there is potential exposure to biological hazards, personnel must practice proper personal hygiene, such as thorough hand washing. Biological hazard waste is collected and transported in specially labeled containers. **See Figure 2-43.** Proper PPE, such as gloves and respirators, must be used when handling biological waste.

BIOLOGICAL HAZARD DISPOSAL

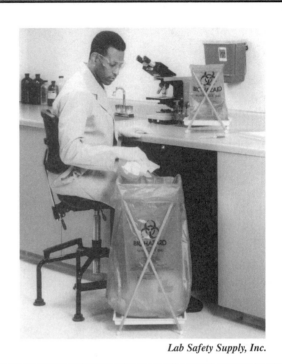

Lab Safety Supply, Inc.

Figure 2-43. Biological hazard waste requires special labeling and handling procedures.

Bloodborne Pathogens. Industrial personnel are at risk for biological hazards through bloodborne pathogens. A *bloodborne pathogen* is a pathogenic (disease-causing) microorganism present in human blood that can infect and cause disease in individuals through exposure. These pathogens can infect the human body through openings in the skin, such as cuts or abrasions. The two most significant bloodborne disease-causing viruses to industrial workers are the hepatitis B virus (HBV) and the human immunodeficiency virus (HIV).

HBV is the most common infectious bloodborne hazard faced by workers. Hepatitis means "inflammation of the liver." HBV-infected individuals experience flu-like symptoms that may require hospitalization. However, some individuals show no symptoms at all. Blood, saliva, and other bodily fluids may be infectious. The HBV virus is particularly dangerous since it can survive on dry surfaces at room temperature for at least a week without any visible signs. A vaccine is available to reduce or eliminate the risk of infection.

Mine Safety Appliances Co.
Chemical hazards are managed by properly storing hazardous chemicals and periodically inspecting containers for leaks.

HIV attacks body immune systems, causing acquired immune deficiency syndrome (AIDS). Currently, there is no vaccine to prevent this infection. An individual infected with HIV may carry the virus for several years without developing symptoms, then suffer from flu-like symptoms, fever, diarrhea, or fatigue. AIDS-related illnesses commonly include neurological problems, cancer, and other infections. Although HIV can be transmitted through contact with infected blood, it is not transmitted by touching or working around individuals who carry the disease.

Exposure Control. OSHA, along with state agencies, designs rules to protect workers from exposure to these serious diseases while performing lifesaving services. All workers must adhere to the policies, engineering controls, and work procedures used when there is an exposure risk. All workers share responsibility with their coworkers for the prevention of disease transmission.

Each worker must know how to recognize occupational exposure, and communicate changes in the exposure classification to their supervisor, before performing tasks that involve increased risk. A bloodborne pathogen exposure control plan dictates universal precautions to be followed for all activities involving contact with blood, tissue, bodily fluids, and equipment or materials that may be contaminated by these substances. **See Figure 2-44.** Precautions must be taken when providing first aid to reduce the risk of exposure to bloodborne pathogens.

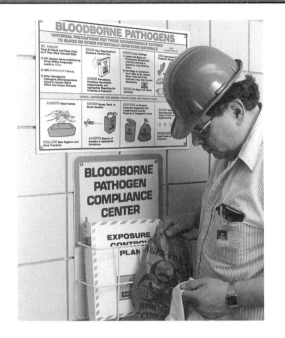

- Wash hands with antiseptic towelettes if there is any possibility of contact with blood, body fluids, or human tissue from an injured individual. Wash hands with soap and running water as soon as possible thereafter.

- Wear approved gloves when anticipating contact with blood, body fluid, tissues, mucous membranes, contaminated surfaces, or if breaks in the skin are present. Leather gloves are not recommended since leather absorbs liquid.

- Wear an impervious gown or apron if splattering of clothing is likely.

- Wear a mask if there is to be contact with an infectious disease spread by splatter droplets.

- Wear appropriate protective equipment at all times including a mask and eye protection if aerosolization or splattering is likely to occur when attending an injured individual.

- Use mouth pieces, resuscitation bags, and other ventilation devices available during emergency resuscitation.

- Handle sharp objects carefully.

- Dispose of all spills that contain or may contain biological contaminants in accordance with policies for hazardous waste disposal. Until cleanup is complete, the area should be restricted to other individuals.

- Post universal precautions signs in all areas designated for first aid and on emergency response boxes and first aid kits.

- Report all exposure of blood, body fluid, and needle sticks immediately to the supervisor.

- If exposed, go to the hospital emergency room for baseline HIV blood testing within 1 hour of exposure.

- Follow up test findings with authorized hospital personnel following established procedures.

Figure 2-44. A bloodborne pathogen exposure control plan dictates the precautions necessary to prevent the transmission of disease.

FIRST AID

First aid is emergency care or treatment given to an injured or ill individual before professional medical assistance is available. First aid stations provide basic medical supplies for treating minor injuries and are located for easy access in an emergency. **See Figure 2-45.**

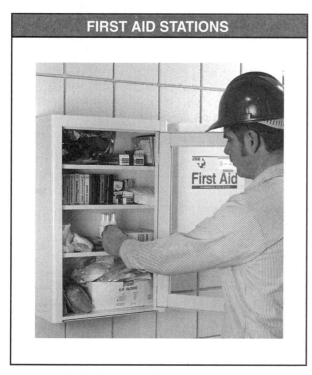

Figure 2-45. First aid stations contain basic medical supplies for treating minor injuries in the facility.

Immediately after an accident or sudden illness, and before medical help arrives, there is a critical period in which first aid techniques can mean the difference between life and death for the victim. Depending on state and local requirements, personnel may be required to be trained in first aid and recertified on a regular basis. First aid does not replace professional medical treatment, but attempts to stabilize the victim until medical personnel arrive. **See Figure 2-46.** Always send for medical personnel immediately in cases of serious injury. Victims with minor injuries should visit a physician as soon as possible.

When evaluating an emergency situation, the individual taking charge must obtain information from witnesses to the accident, assess the condition of the victim, and communicate with the victim. A primary

survey of the victim checks for obvious injuries. Next, the individual should attend to the immediately life-threatening conditions, such as respiratory arrest and severe bleeding. Once life-threatening conditions are under control, attention should be focused on the other obvious injuries. Open chest or abdominal wounds should be sealed, fractures immobilized, and burns and less serious bleeding wounds covered. The causes of the injuries should be identified in order to assess the extent of physical damage.

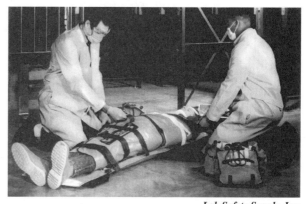
Lab Safety Supply, Inc.

Figure 2-46. First aid procedures stabilize a victim until medical personnel arrive.

A secondary survey should be performed after treating primary injuries to detect less obvious injuries. Individuals administering first aid should not assume that the obvious injuries are the only injuries present. A secondary survey is a head-to-toe examination in which an individual starts by examining the victim's head, neck, trunk, and extremities. Abnormalities such as swelling, discoloration, lumps, and tenderness might indicate a hidden injury.

Unnecessary movement of the victim should be avoided as it can cause further injury. A victim with a possible spinal injury or bone fracture must be handled with extreme care. Spinal injury can lead to paralysis or death. A closed fracture can become an open fracture if not properly immobilized.

✚ **SAFETY TIP**
Basic first aid training is available through employers or other organizations. Many facilities designate a certified first responder. This more advanced role involves 40 to 60 hrs of training in providing emergency medical care.

INCIDENT REPORTS

An *incident report* is a document that details facts about an injury-causing accident in a facility. **See Figure 2-47.** Regardless of the circumstances, work-related incidents that require medical attention beyond first aid must be recorded within seven days of the occurrence. Incidents that result in the death of an employee or in-patient hospitalization of three or more employees must be reported to OSHA within eight hours of the incident. There may also be reporting requirements for other organizations, such as a state occupational safety and health agency, corporate safety officers, and insurance companies. Incident reports become a permanent part of plant operation records.

Incident reports must contain certain information, including the name of the injured person, the date, time, and place of the accident, the name of the physician, the nature of the medical treatment given, and the circumstances surrounding the accident. A log must be kept of all work-related injuries and incidents, which is reported to OSHA at the end of each year. This log also includes the results of the incident, such as the number of days away from work or restricted duty for the injured employee. A summary of the information in the log must be posted for all employees to review. OSHA forms 301, 300, and 300A fulfill these record-keeping requirements.

EMERGENCY PLANS

Facilities must have an emergency plan. An *emergency plan* is a document that details procedures, exit routes, and assembly areas for facility personnel in the event of an emergency. **See Figure 2-48.** A designated official or safety committee is responsible for developing and implementing the emergency plan.

The emergency plan assigns specific responsibilities to facility personnel. Facility personnel do not take the place of emergency response personnel, but may be emergency coordinators. Emergency coordinators are responsible for performing additional duties during an emergency that can save lives and prevent serious damage. For example, maintenance personnel may be tasked with de-energizing equipment or utilities in an emergency, such as shutting off the natural gas supply to the building after an earthquake. Maintenance personnel may also have increased responsibilities regarding facility security because they are often the only people with access to all areas of the facility.

Evacuation, if required, must occur in a safe and orderly manner. All personnel must evacuate the building as quickly as possible, except emergency coordinators. Emergency coordinators stay only long enough to check that all other personnel have left and to perform vital emergency functions. In any emergency, the proper authorities must be notified immediately by calling 911 or other appropriate emergency numbers.

Employers and facility personnel must also be prepared for emergencies due to acts of terrorism. Some preparation, such as establishing an emergency plan and assembling emergency supplies, are the same for all emergencies. However, terrorist attacks present additional potential hazards from explosive, biological, chemical, nuclear, and radiological weapons. Each of these attacks requires specific actions as recommended by the Department of Homeland Security.

The Department of Homeland Security is a federal agency whose primary mission is to help prevent, protect against, and respond to acts of terrorism within the United States. The Department of Homeland Security provides a unifying core for the vast national network of organizations and institutions involved in efforts to secure the nation.

> **TECH TIP**
> *Natural disasters include a variety of weather-related phenomena such as hurricanes, tornadoes, floods, and droughts, as well as forest fires, plagues, volcanic eruptions, earthquakes, tsunamis, and landslides.*

Explosion and Fire

An *explosion* is a blast that is caused by a bomb or the ignition of existing flammable material. A fire usually results from an explosion. Other explosion hazards may include shock waves that knock over equipment or topple personnel, extreme brightness and sound that damage eyesight and hearing and cause disorientation, and flying debris tossed around by the force of the blast. Following an explosion, a possible long-term hazard is contamination by chemicals or other hazardous substances that were dispersed by the blast.

An explosion and the resulting hazards require immediate action. If an explosion is imminent, individuals should take shelter under a desk or a sturdy table and exit the building without using elevators as soon as possible. Facility fire and emergency procedures should be followed as required.

INJURY OR ILLNESS INCIDENT REPORT

Employee Status
☒ Full Time
☐ Part Time
☐ Other _____
_____

Accident Classification
☐ First Aid Only
☒ Medical Treatment

PLEASE COMPLETE **ALL** INFORMATION

1. Name of Employee _Jack Morgan_ _____ Case Number: _046-10-0625_

2. Address _7604 Garden Court_ _____ Home Phone # _555-5924_

3. Age _38_ Sex _M_ Occupation/Title _Maintenance Engineer_

4. Was employee engaged in regular course of his duties at time of accident? ☒ Yes ☐ No

5. If no, explain

6. Experience at this work activity _____5_____ years _____6_____ months _____ weeks

7. Total job experience for County _____10_____ years _____4_____ months _____ weeks

8. Department/District _Facilities Maintenance_ Date Reported to Supervisor _10/4_

9. Date of Accident _10/3_ Time _11:35_ am /(pm) Last Day Worked _10/3_

10. Location of Accident _Boiler room 1 - drop cord connection to overhead bus_

11. How many (days)/ hrs per week is employee employed? _____5_____ If seasonal employment give total weekly hours _____ Regular Days Off _Mon, Tue_

12. Shift Hours _4_ am /(pm) to _12_ am /(pm) Rotating Shift

13. Name of physician or other health care professional treating employee _Dr. Mark James_

14. If treated away from worksite, where was it given?

Facilty _Memorial Hospital_

Address _1175 Main Street_

Nature of Injury or Illness (cut, bruise, poisoning, etc.) _Burns - Fall_ Part of Body Affected _Hand/shoulder_ ☐ Left ☒ Right

15. Was employee treated in an emergency room? ☒ Yes ☐ No

16. Was employee hospitalized overnight as an in-patient? ☐ Yes ☒ No

17. **TYPE OF INCIDENT** (circle one **only**)

1. Slip and/or fall-same level
2. Slip and/or fall-different level
3. Struck by falling/flying object
4. Contact with tools/knives/power equip.
5. Contact with/by temperature extremes
(6.) Contact with/by electrical current
7. Contact with/by liquid/gas/vapor/etc.

8. Struck against
9. Caught in, under, between
10. Exposure-disease, parasite
11. Over exertion-lifting/pulling/pushing
12. Other_____

CAUSE OF INCIDENT (circle one **only**)

1. Inadequate guards or protection
2. Defective equip/tool/material substance
3. Congestion
4. Inadequate warning system
5. Fire and/or explosion
6. Substandard housekeeping
7. Hazardous Atmospheric conditions
8. Excessive noise
9. Radiation exposure
10. Occupational Illumination/ventilation
11. Poor layout, planning, design

12. Sharp/rough/frayed/cracked edges
13. Water/oil, etc. in walkway
14. Foreign object in walkway
15. Unexpected movement hazard
16. Ice/snow
17. Operating without authority
18. Errors of others
(19.) Error of employee
20. Improper procedures
21. Horseplay
22. Other_____

18. What was job assignment at time of incident? _Taking high voltage reading on bus._

19. What happened? (Explain in detail, use extra paper if needed) _Accidently contacted high voltage bus when reaching to take voltage reading._

20. What object or substance directly harmed the employee? _electricity_

21. If the employee died, when did death occur? _N/A_

22. Corrective action taken to prevent future incident of this kind _Review electrical safety procedures._

23 Witness(s) (Attach their written statements) _N/A_

24. Date Safety/Loss Control Manager contacted _10/4_ , if not contacted, explain why _____

NOTE: INCIDENT REPORTS MUST BE COMPLETELY FILLED OUT FOR EVERY INCIDENT AND DELIVERED TO SAFETY/LOSS CONTROL OFFICE WITHIN TWENTY-FOUR (24) HOURS.

Craig Linden
Immed. Supervisor's Signature
(Yellow Copy)

Bob Whitmore
Safety/Loss Control Signature
(White Copy)

PLEASE SEND ORIGINAL WHITE COPY TO SAFETY/LOSS CONTROL

Figure 2-47. An incident report details facts about an injury or illness in a facility.

EMERGENCY PLAN

Emergency Plan Procedures

1. When the fire alarm sounds, ALL EMPLOYEES AND GUESTS are to leave the building IMMEDIATELY by the nearest safe exit and report to their assembly area for roll call by emergency coordinators as follows:
    Assembly Area A
    Product Development
    Order and Marketing
    Assembly Area B
    Accounting
    Warehouse
    Production
    Area C
    Report Attendance
2. Emergency coordinators will report attendance to Area C when the safety supervisor or emergency response personnel arrive.
3. **DO NOT** reenter the building for any reason unless authorized by the safety supervisor.
4. Employees are responsible for their guests.
5. If smoke is present, crawl low under the smoke and feel doors with palm of hand. If door is hot, do not open door.

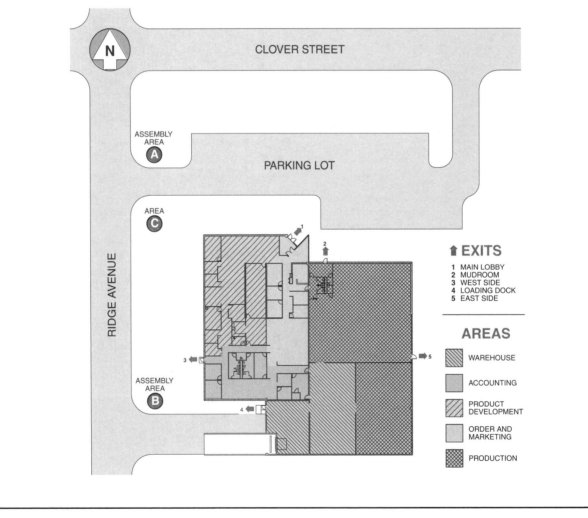

Figure 2-48. Emergency plans detail exit routes and assembly areas for facility personnel.

Biological Attack

A *biological attack* is the deliberate release of illness-causing microorganisms. Many biological agents can cause illness from being inhaled or ingested, or enter through a cut in the skin. Some biological agents, such as anthrax, do not cause contagious diseases. Others, like the smallpox virus, cause diseases that spread easily between people.

Unlike an explosion, a biological attack may or may not be immediately obvious. While it is possible that signs of a biological attack are evident, it is more likely that local health-care workers will report an unusual pattern of illnesses or that a sudden wave of infected individuals will seek emergency medical attention.

Chemical Attack

A *chemical attack* is the deliberate release of a toxic gas, liquid, or solid that can poison individuals and the environment. Signs of chemical attack include large groups of individuals suffering from watery eyes, twitching, choking, breathing trouble, or loss of coordination. Since birds, fish, and small animals are more likely to be sensitive to toxic chemicals, a large number of suddenly ill or dying animals is also cause for suspicion.

Nuclear Blast

A *nuclear blast* is an explosion with a damaging pressure wave, intense light and heat, and widespread radioactive material (fallout) that can contaminate the air, water, and ground surfaces. Shielding, distance, and time are effective at limiting the amount of radiation exposure. Shields are thick, dense materials that block much of the radiation from radioactive materials, reducing overall exposure. Greater distance from the blast and fallout lowers the exposure. Minimizing the time of exposure also reduces the effects.

Radiation Threat

A *radiation threat* is the use of conventional explosives to spread radioactive material over a targeted area. This type of attack is commonly referred to as a "dirty bomb" or radiological dispersion device (RDD). It is not a nuclear blast. The force of the explosion is localized but the radioactive contamination can potentially spread over a large area, depending on the weather conditions. While the blast is immediately obvious, the presence of radiation may not be known until trained personnel with specialized equipment are on the scene. As with any radiation hazard, the primary task is to limit exposure and avoid breathing radiological dust that may be in the air.

Emergency Supplies

When preparing for emergency situations, the basics of survival (fresh water, food, clean air, shelter, and warmth) are the priority. An emergency supply list details requirements for survival and other necessities, such as a battery-powered commercial radio and a National Oceanic and Atmospheric Administration (NOAA) weather radio with an alert function. A commercial radio is a good source for news and information from local authorities. An NOAA weather radio can provide weather emergency information or announcements from the Department of Homeland Security.

Service and Repair Principles

3

Maintenance personnel perform troubleshooting, service, and repair tasks using basic knowledge of equipment operating principles. Mechanical systems used on industrial equipment are comprised of simple machines. Equipment documentation provides information regarding system operation and equipment service and repair. Equipment service and repair requires a planned sequence to reduce cost and downtime. Maintenance personnel use a variety of tools to complete service and repair tasks.

EQUIPMENT OPERATION PRINCIPLES

All operating equipment uses energy. *Energy* is the capacity to do work. Two forms of energy are potential energy and kinetic energy. **See Figure 3-1.** *Potential energy* is stored energy a body has due to its position, chemical state, or condition. For example, water contained behind a dam has potential energy because of the position of the water. Gasoline has potential energy based on its chemical state. A compressed spring has potential energy because of its condition. *Kinetic energy* is energy of motion. Kinetic energy includes falling water, a speeding automobile, or a released spring. Kinetic energy is released potential energy. Energy is used to produce force.

Force

Force is anything that changes or tends to change the state of rest or motion of a body. A body, when referring to force, is anything with mass. **See Figure 3-2.**

For example, if an individual pushes against a vehicle trying to move it, a force is exerted on the vehicle. Force can be measured in pounds (lb). Forces can act on an object in several ways. Force acting on an object does not always result in motion. Force applied in different ways produces pressure, work, or torque. When pressure and area are known, force is found by applying the formula:

$$F = P \times A$$

where

F = force (in lb)

P = pressure (in psi)

A = area (in sq in.)

For example, what is the force on a flat head of a steam boiler drum with a pressure of 100 psi and an area of 1000 sq in.?

$$F = P \times A$$
$$F = 100 \times 1000$$
$$F = \mathbf{100,000\ lb}$$

69

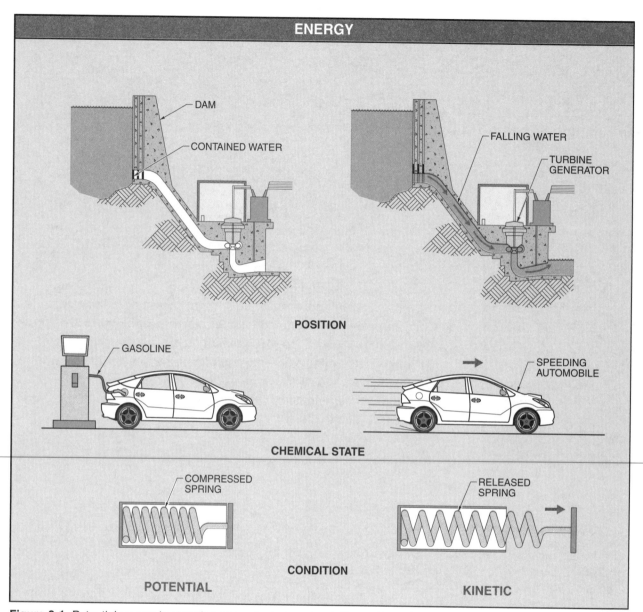

Figure 3-1. Potential energy is stored energy a body has due to its position, chemical state, or condition. Kinetic energy is energy of motion.

Pressure

Pressure is force per unit area. *Area* is the number of unit squares equal to the surface of an object.

In hydraulic systems, the pressure of the hydraulic fluid exerted on a piston and the area of the piston determine the force generated by the piston. If the hydraulic fluid is pressurized to 50 psi, it exerts 50 lb of pressure on each inch of the surface of the piston. If the piston is 10 sq in., the force generated is 500 lb (50 psi × 10 sq in. = 500 lb). If the pressure of the hydraulic fluid is increased to 100 psi, 1000 lb of force is generated

(100 psi × 10 sq in. = 1000 lb). **See Figure 3-3.** When force and area are known, pressure is found by applying the formula:

$$P = \frac{F}{A}$$

where
P = pressure (in psi)
F = force (in lb)
A = area (in sq in.)

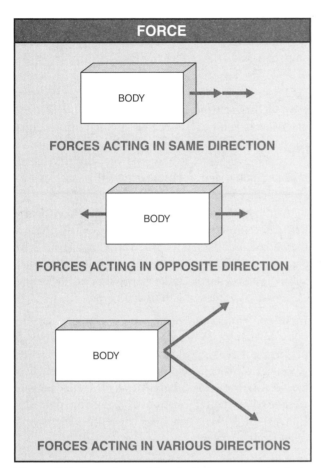

Figure 3-2. Force is anything that changes or tends to change the state of rest or motion of a body.

For example, what is the pressure exerted if a 60 lb force is applied to an area of 4 sq in.?

$$P = \frac{F}{A}$$

$$P = \frac{60}{4}$$

$$P = \textbf{15 psi}$$

Work

Work is the movement of an object by a force to a specific distance. Work occurs only when force results in motion. For example, work is done by lifting a weight from the ground against gravity. Work is not done if the weight is not lifted from the ground, even if force was applied. Work occurs when applied force overcomes resistance and results in motion. Work can be measured in foot-pounds (ft-lb). **See Figure 3-4.**

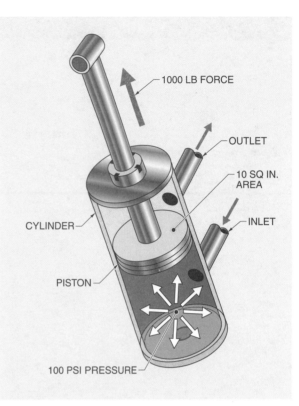

Figure 3-3. Pressure is the force per unit area.

Work requires only enough force to accomplish the desired task. For example, if a person lifts a 50 lb weight, only enough force is exerted to lift the weight. Additional force applied results in acceleration. When force and distance are known, work is found by applying the formula:

$$W = F \times D$$

where

W = work (in ft-lb)

F = force (in lb)

D = distance (in ft)

For example, what is the amount of work performed when lifting a 68 lb engine from the floor to the top of a 3′ high workbench?

$$W = F \times D$$

$$W = 68 \times 3$$

$$W = \textbf{204 ft-lb}$$

TECH TIP

The amount of useful work performed by a machine can never be greater than the work supplied to it. There will always be some loss due to friction or other dissipative forces.

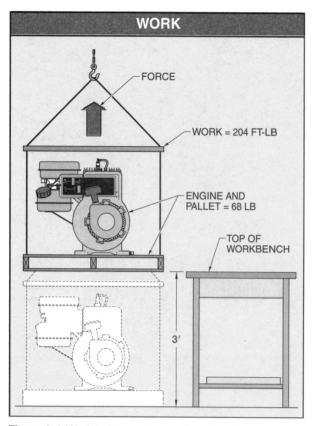

WORK

FORCE

WORK = 204 FT-LB

ENGINE AND PALLET = 68 LB

TOP OF WORKBENCH

3′

Figure 3-4. Work is the movement of an object by a constant force to a specific distance.

Simple Machines

A *simple machine* is any device that transmits the application of a force into useful work. Mechanical systems in industrial equipment are comprised of many simple machines. Simple machines are used to perform work and to change the size, direction, or speed of the applied force. Simple machines are used to produce linear, reciprocating, rotary, and elliptical motion. Simple machines include the lever, inclined plane, pulley, wheel and axle, wedge, and screw. **See Figure 3-5.**

Lever. A *lever* is a simple machine that consists of a rigid bar that pivots on a fulcrum with resistance and effort applied. A lever is used to obtain mechanical advantage to overcome a large resistance with less effort (force). The lever principle is commonly used on industrial equipment. Levers are classified as first, second, and third class levers depending on the location of the fulcrum in relation to the effort.

A *first class lever* is a lever that has the fulcrum located between the resistance and the effort. A first class lever reverses the direction of motion. The mechanical

advantage achieved depends on the distance of the effort and the resistance from the fulcrum. A pry bar, tin snip, and industrial vehicle control lever are examples of first class levers.

A *second class lever* is a lever that has the resistance located between the fulcrum and the effort. Second class levers multiply the effort applied but sacrifice the distance the object is moved. The object is moved in the same direction as the lever movement. A hand-operated pump and a wheelbarrow are examples of a second class lever.

A *third class lever* is a lever that has the effort located between the fulcrum and the resistance. The effort is placed at a point along the bar. Third class levers have a negative mechanical advantage, but multiply the distance the object moves. A damper control and the human elbow are examples of a third class lever.

Inclined Plane. An *inclined plane* is a simple machine that allows force to be applied over a long, horizontal distance to move heavy loads vertically. The mechanical advantage of an inclined plane is the ratio of the length of the sloped surface to the height it spans. Inclined planes were used by the Egyptians to construct the pyramids from heavy blocks. An inclined conveyor is an example of an inclined plane.

Pulley. A *pulley* is a simple machine consisting of a cylinder rotating freely on its axis, which uses a belt, rope, or chain to change the direction of an applied force or transmit rotational motion. Pulleys may be smooth, grooved, or studded with teeth to engage a chain. When a system with multiple pulleys is used, it creates a mechanical advantage in either linear or rotational systems of motion. A common industrial example of a pulley system is a cable hoist.

Wheel and Axle. A *wheel and axle* is a simple machine consisting of a wheel and an axle fixed together along the same axis. A wheel and axle is used to gain a mechanical advantage. The force on the axle is multiplied as the wheel is rotated. Many industrial-equipment applications use the wheel and axle as a drive system. A screwdriver is an example of a wheel and axle.

Wedge. A *wedge* is a simple machine that converts force applied to its blunt end into force that is perpendicular to its sloped surface. A wedge applies force through a long distance similar to the inclined plane. This provides mechanical advantage and changes the direction of the resulting force. Many cutting tools use the wedge to remove chips. A chisel is an example of a wedge.

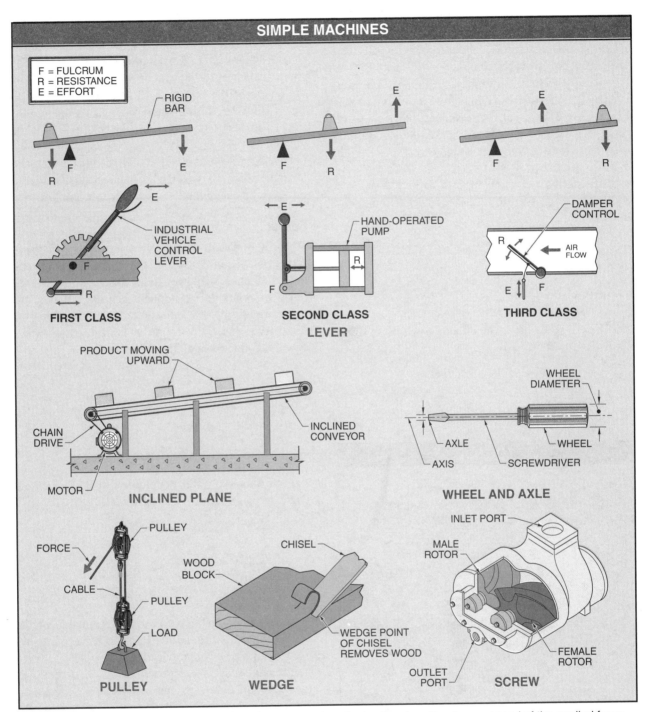

Figure 3-5. Simple machines are used to perform work and to change the size, direction, or speed of the applied force.

Screw. A *screw* is a simple machine consisting of a continuous spiral on a cylinder. A screw provides a large mechanical advantage. Screws convert between rotational force and linear force. A screw is used for fasteners, jacks, and pumps. For example, a rotary screw compressor changes rotary motion into linear motion.

TECH TIP

Simple machine principles are used in industrial machines to increase the efficiency of the operation. For example, a wheel and axle is used as the drive system on industrial milling machines to gain the mechanical advantage required for the efficient machining of metal.

Torque

Torque is rotational force. A linear force acting at a distance from a point of rotation results in torque. Torque is equal to the force multiplied by the perpendicular distance between the axis and the point of force. **See Figure 3-6.** Like linear force, torque may or may not result in motion. Units of torque may be distinguished from those for work by using pound-feet (lb-ft), though use of foot-pounds (ft-lb) for torque is also very common. The difference is only in the notation. When force and distance are known, torque is found by applying the formula:

$$T = F \times d$$

where

T = torque (in lb-ft)

F = force (in lb)

d = distance (in ft)

For example, what is the torque from a 60 lb force applied at the end of a 2′ arm?

$$T = F \times d$$

$$T = 60 \times 2$$

$$T = \textbf{120 lb-ft}$$

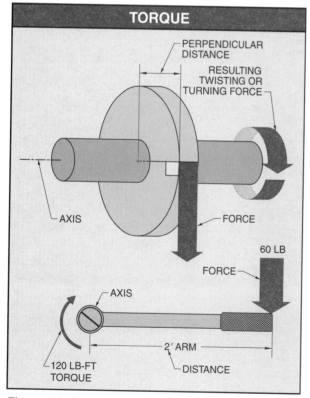

Figure 3-6. Torque is a force acting on a perpendicular distance from a point of rotation.

The length of the arm can be adjusted for the specific task requirements. For example, a 1′ wrench is used to tighten a bolt to 50 lb-ft by applying 50 lb of force. The same amount of torque can be obtained by increasing the length of the wrench to 2′ and applying a 25 lb force. However, with a length of 2′, the wrench must travel a greater distance to move the bolt head the same amount as a shorter wrench. The time required to move the wrench may also increase with the greater distance from the bolt head. Change in mechanical advantage is achieved by changing the wrench handle length.

Power

Power is the rate of doing work or using energy. Power adds a time factor. Therefore, power is work divided by time. Examples of power ratings include horsepower (HP) and watts (W). Although they are different numeric scales, both horsepower and watts measure how fast work is done. When force and distance are known, power is found by applying the formula:

$$P = \frac{W}{T}$$

where

P = power (in ft-lb/s)

W = work (in ft-lb/s)

T = time (in sec)

For example, what is the power output of an engine that performs 90,000 ft-lb of work in 120 sec?

$$P = \frac{W}{T}$$

$$P = \frac{90,000}{120}$$

$$P = \textbf{750 ft-lb/s}$$

Horsepower (HP) is a unit of power equal to 746 W or 550 ft-lb/s. **See Figure 3-7.** Horsepower is commonly used to rate the power produced by an engine based on a finite engine speed. Horsepower can also be determined by the amount of power consumed by an electric motor. Electrical power consumed (wattage) can be calculated by multiplying voltage and current. Horsepower is found by applying the formula:

$$HP = \frac{W}{T \times 550}$$

where

HP = horsepower (in HP)

W = work (in ft-lb)

T = time (in sec)

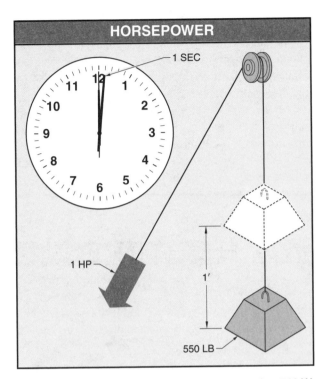

Figure 3-7. Horsepower is a unit of power equal to 746 W or 550 ft-lb/s.

For example, what is the horsepower rating of an engine that produces 412,500 ft-lb of work in 150 sec?

$$HP = \frac{W}{T \times 550}$$

$$HP = \frac{412,500}{150 \times 550}$$

$$HP = \frac{412,500}{82,500}$$

HP = **5 HP**

When calculating horsepower, efficiency may also be calculated. *Efficiency* is a measure of a device's useful output energy compared to its input energy. No energy-converting device is 100% efficient. Energy is lost in the form of friction, vibration and heat. This energy loss determines the efficiency of the device. Efficiency is found by applying the formula:

$$\eta = \frac{P_{out}}{P_{in}}$$

where

η = efficiency (percentage)

P_{out} = output power

P_{in} = input power

For example, what is the efficiency of an electric motor with a power consumption of 5 W and an output of 4.8 W?

$$\eta = \frac{P_{out}}{P_{in}}$$

$$\eta = \frac{4.8}{5}$$

η = **0.96** or **96%**

Horsepower is also used to quantify steam produced in a boiler. For example, a high pressure boiler is a boiler that has a maximum allowable working pressure above 15 psi and a power over 6 boiler horsepower (BHP). A *boiler horsepower (BHP)* is the power available from the evaporation of 34.5 lb of water per hour at a feedwater temperature of 212°F. An adequate supply of water must be available to evaporate or produce the required amount of steam. One boiler horsepower is equivalent to 33,472 Btu/hr. Boiler horsepower is found by applying the formula:

$$BHP = \frac{W_s \times FE}{34.5}$$

where

BHP = boiler horsepower (in BHP)

W_s = pounds of steam per hour (in lb/hr)

FE = factor of evaporation (in Btu/lb)

For example, what is the boiler horsepower of a boiler that generates 10,000 lb of steam per hour and has a factor of evaporation of 1.24?

$$BHP = \frac{W_S \times FE}{34.5}$$

$$BHP = \frac{10,000 \times 1.24}{34.5}$$

$$BHP = \frac{12,400}{34.5}$$

BHP = **359.42 BHP**

TECH TIP

The unit of horsepower was originally defined in the 18th century to compare the power output of draft horses to the power output of the steam engines gradually replacing them. Given the variation between different horses, the definition of one horsepower was in flux for many decades, though it still overestimates the power of the average horse. Ironically, James Watt, whose name went on to denote an alternate unit of power, coined the term "horsepower."

Fluid Power Transmission

A *fluid* is a substance that tends to flow or conform to the outline of its container. This includes liquids and gases. Fluids can be used to transmit power. All fluids respond in the same way when a pressure differential exists between two areas. Fluids flow from high-pressure areas to low-pressure areas. When pressure is applied to a fluid in a confined space, the pressure is the same throughout the entire fluid regardless of where the measurement is taken. **See Figure 3-8.**

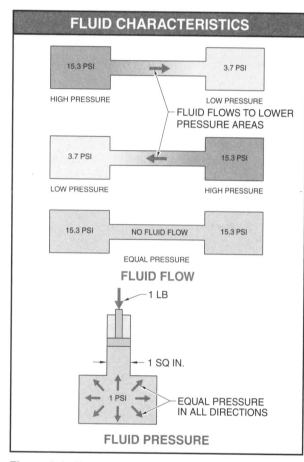

Figure 3-8. Fluids flow from high-pressure areas to low-pressure areas and exert equal pressure in all directions in a system.

Inertia

Inertia is the tendency of a physical body to persist in its state of rest or uniform motion until acted upon by an external force. For example, when operating a bench grinder, the grinding wheel continues to rotate for some time after the power is turned OFF. The grinding wheel is continuing its state of motion without being driven by the electric motor, though friction eventually slows it to a stop. Inertia is commonly used in industrial machinery to maximize energy efficiency. For example, on an air compressor, a heavy pulley acts as a flywheel to supply inertia for maintaining speed between pumping intervals. This reduces loss of pump speed when the piston works to compress air in the cylinder. Internal combustion engines also take advantage of flywheels with high inertias to minimize fluctuations in crankshaft speed. **See Figure 3-9.**

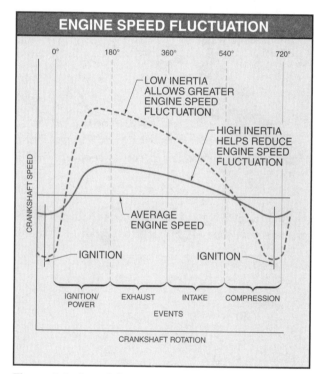

Figure 3-9. Speed fluctuation within the operating cycle of a single-cylinder engine is reduced by inertia from the engine flywheel and driven equipment.

When a motor with high inertial load is stopped, it takes extra power to get it moving. High current that rushes into the motor at startup supplies this extra power. Once the motor is running, less power is required to keep it moving, so the motor requires less current.

Heat

All matter is composed of atoms and molecules that are in a constant state of motion. *Heat* is a form of energy identified by a temperature difference or change of state.

Heat added to a substance causes an increase in molecular motion. This increases internal energy. Heat removed from a substance causes a decrease in molecular motion. This decreases internal energy.

A substance can be in a solid, liquid, or gas (vapor) state. For example, water may be solid (ice), liquid (water), or gas (steam). **See Figure 3-10.** The state of a substance depends on the intensity of the heat energy motion of the molecules. The change to or from one state to another requires the addition or removal of heat energy. When heat is added to ice, it changes to water. When heat is added to water, it eventually turns to steam. For example, a change in state occurs in a boiler as water is heated and steam is produced. As the steam is cooled, it condenses to water.

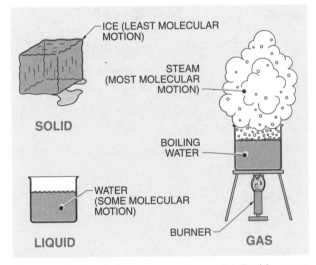

Figure 3-10. A substance can be in a solid, liquid, or gas (vapor) state.

Heat Transfer. Heat is transferred from one substance to another when a temperature difference exists between the two substances. Heat is always transferred from a substance with a higher temperature to a substance with a lower temperature. Heat transfer rates increase with the temperature difference between two substances. The three methods of heat transfer are conduction, convection, and radiation. **See Figure 3-11.**

Conduction is heat transfer that occurs when heat is passed from molecule to molecule through a material. For example, if one end of a metal rod is heated, heat is transferred by conduction to the other end. By heating one end of a metal rod, molecules are heated and move faster. The faster-moving molecules transfer energy to the slower-moving molecules in the metal rod. In a boiler, conduction occurs as heat increases on the fire side of the metal. The heat from the fire side of the boiler tube passes through the metal until it reaches the water side of the metal. The heat then moves to the water that contacts the metal.

Heat conduction occurs in a gearbox through lubricating oil. The lubricating oil comes in direct contact with gears that have a much higher temperature than the oil. When this occurs, the oil conducts the heat away from the gears and into the gearbox case. Heat conduction also occurs as the gearbox case conducts heat from the oil to the surrounding outside air.

Convection is heat transfer that occurs when currents circulate between warm and cool regions of a fluid. For example, as air is warmed by a fire, the warm air rises and is replaced by cool air. The movement of air creates a current that continues as long as heat is applied.

Heat transfer by convection occurs in an automobile engine radiator. Fins on the radiator channels increase the radiator's surface area in contact with passing air. This improves heat transfer efficiency from the coolant in the radiator. The warm coolant from the engine gives up its heat to the air as the coolant passes through the radiator. The cooled coolant is then drawn from the bottom and returned to the engine to repeat the cooling cycle.

In a boiler, heat transfer by convection occurs as the water touching the metal is heated and the hot water rises and cool water moves into its place. The cool water is also heated and moves upward. This starts a circular current that moves hot water to the top of the drum and cool water to the bottom of the boiler where it is heated.

Radiation is heat transfer from electromagnetic waves radiating outward from the source. The amount of heat transferred depends on the intensity of heat (temperature) of the substance. Radiant energy waves move through space without producing heat. Heat is produced when the radiant energy waves contact an opaque object. Heat produced on earth by light from the sun is a form of radiation.

Heat transfer by radiation occurs in all operating industrial equipment. Bearings, shafts, motors, chains, and other equipment components reject heat into the atmosphere. For example, an electric motor transfers a significant amount of heat by radiation while it is running. In a boiler, the radiant energy produced by the flames in the furnace section strikes the metal of the boiler tubes and causes the metal to heat.

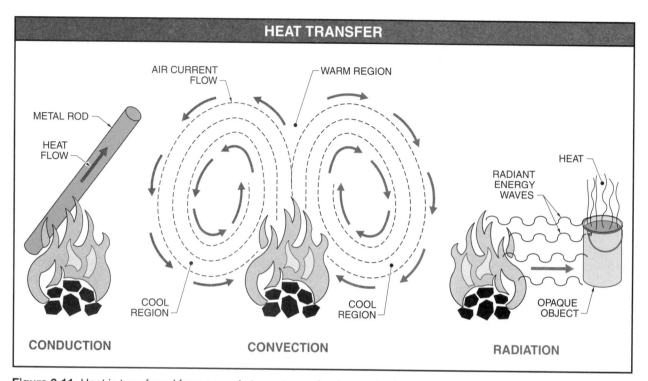

Figure 3-11. Heat is transferred from one substance to another by conduction, convection, or radiation.

Temperature. *Temperature* is the measurement of the intensity of heat. The temperature of matter is the degree of hotness or coldness on a certain scale. Temperature can be measured with a thermometer. There are many types of thermometers available for different scales, ranges, environments, and other factors.

A quantity of heat is the amount of energy needed to produce an accepted standard of physical change in the material. A common unit for quantity of heat measurement is the British thermal unit. A *British thermal unit (Btu)* is the quantity of heat required to raise the temperature of 1 lb of water 1°F.

Temperature Scales. Temperature is commonly expressed using the Fahrenheit or Celsius scales. On the Fahrenheit scale, 32°F is the freezing point and 212°F is the boiling point of water at normal atmospheric pressure (14.7 psi). On the Celsius scale, 0°C is the freezing point and 100°C is the boiling point of water at normal atmospheric pressure. It is sometimes necessary to convert temperature readings from one scale to the other. **See Figure 3-12.** To convert a Fahrenheit reading to Celsius, subtract 32 from the Fahrenheit reading and divide by 1.8. This is the same as applying the formula:

$$°C = \frac{°F - 32}{1.8}$$

where

°C = degrees Celsius

°F = degrees Fahrenheit

For example, what is 86°F converted to the Celsius scale?

$$°C = \frac{°F - 32}{1.8}$$

$$°C = \frac{86 - 32}{1.8}$$

$$°C = \frac{54}{1.8}$$

$$°C = \mathbf{30°C}$$

To convert a Celsius reading to Fahrenheit, multiply the Celsius reading by 1.8 and add 32. This is the same as applying the formula:

$$°F = (1.8 \times °C) + 32$$

where

°F = degrees Fahrenheit

°C = degrees Celsius

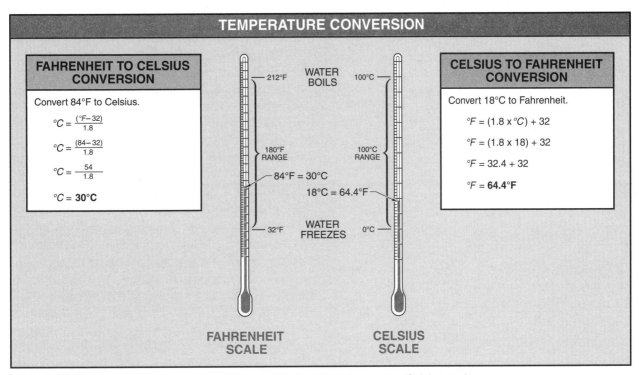

Figure 3-12. Temperature is commonly expressed by using the Fahrenheit or Celsius scales.

For example, what is 18°C converted to the Fahrenheit scale?

$$°F = (1.8 \times °C) + 32$$
$$°F = (1.8 \times 18) + 32$$
$$°F = 32.4 + 32$$
$$°F = \textbf{64.4°F}$$

SERVICE AND REPAIR DOCUMENTS

Equipment manuals and other documentation are critical to maintaining and troubleshooting industrial and facility equipment. All system documentation should be stored properly and must be used when troubleshooting and repairing equipment.

System Documentation

The system documentation most commonly used by maintenance personnel is the operator's manual. An *operator's manual* is a document that contains instructions for the safe and efficient installation, operation, troubleshooting, and repair of equipment. Operator's manuals are usually supplied with new equipment. An operator's manual should contain a space to record purchase information such as the date of purchase, serial and model numbers, and warranty and registration information. This information should be filled in even though it may be transferred to a computerized preventive maintenance system later.

Some operator's manuals contain important startup information that serves as baseline data from which to evaluate equipment operation. Some manuals contain detailed startup and installation instructions. This information is useful when troubleshooting, or when problems result from poor installation or when equipment must be removed and reinstalled. For example, the data on a refrigeration startup log can be used to compare the operation of the system to when it was new and presumably operating correctly. **See Figure 3-13.**

Safety concerns are noted throughout the operator's manual. Illustrations and explanations of operating controls are usually included. Alarm or warning signals are explained, such as when the air exiting a compressor is too hot. In addition, the normal operating temperatures for the equipment are indicated. This is critical information because temperature and humidity extremes can impact equipment operation. A maintenance schedule is also included. **See Figure 3-14.** The operator's manual includes specific information such as the frequency of the drive motor lubrication and that the periods between re-greasing can vary based on the severity of the service conditions.

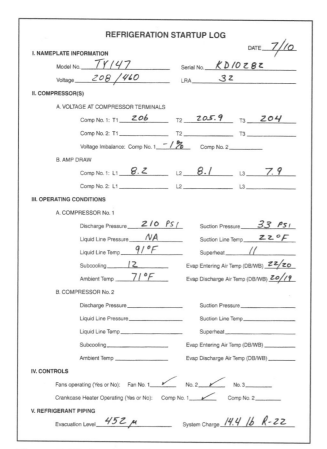

REFRIGERATION STARTUP LOG

DATE _7/10_

I. NAMEPLATE INFORMATION

Model No. _TY147_   Serial No. _KD10282_

Voltage _208/460_   LRA _32_

II. COMPRESSOR(S)

A. VOLTAGE AT COMPRESSOR TERMINALS

Comp No. 1: T1 _206_   T2 _205.9_   T3 _204_

Comp No. 2: T1 _____   T2 _____   T3 _____

Voltage Imbalance: Comp No. 1 _-1%_   Comp No. 2 _____

B. AMP DRAW

Comp No. 1: L1 _8.2_   L2 _8.1_   L3 _7.9_

Comp No. 2: L1 _____   L2 _____   L3 _____

III. OPERATING CONDITIONS

A. COMPRESSOR No. 1

Discharge Pressure _210 PSI_   Suction Pressure _33 PSI_

Liquid Line Pressure _NA_   Suction Line Temp _220°F_

Liquid Line Temp _91°F_   Superheat _11_

Subcooling _12_   Evap Entering Air Temp (DB/WB) _22/20_

Ambient Temp _71°F_   Evap Discharge Air Temp (DB/WB) _20/19_

B. COMPRESSOR No. 2

Discharge Pressure _____   Suction Pressure _____

Liquid Line Pressure _____   Suction Line Temp _____

Liquid Line Temp _____   Superheat _____

Subcooling _____   Evap Entering Air Temp (DB/WB) _____

Ambient Temp _____   Evap Discharge Air Temp (DB/WB) _____

IV. CONTROLS

Fans operating (Yes or No): Fan No. 1 _✓_   No. 2 _✓_   No. 3 _____

Crankcase Heater Operating (Yes or No): Comp No. 1 _✓_   Comp No. 2 _____

V. REFRIGERANT PIPING

Evacuation Level _452 μ_   System Charge _14.4 lb R-22_

Figure 3-13. A refrigeration startup log can be used to compare the operation of a system to when it was new.

Loctite Corporation

Anti-seize lubricants protect mating metal parts against friction, galling, and corrosion and reduce the torque required for assembly and disassembly of threaded connections.

The operator's manual also includes a troubleshooting chart. A written description of system operation and normal conditions is often included in the manual. The description is often coupled with schematic diagrams or illustrations to show equipment operation. Electrical schematics and mechanical plans should be consulted because they represent the directions most likely to ensure correct operation. Any change in system layout must be completed in consultation with the manufacturer to avoid causing severe operational problems. When a non-standard installation is encountered in a malfunctioning system, the repair can entail returning the system to its manufacturer-designed installation.

Traditionally, system documentation has been supplied as paper documents, but the information can be supplied on computer disks or accessed on-line. Information can be located through searching the electronic files and printed as needed, or viewed from a computer screen.

Printreading

Many maintenance tasks require consulting prints or technical drawings before beginning work. These drawings may show the original configuration of rooms and equipment, the location and routing of important utilities, or the feature dimensions required for new equipment installation.

Building Prints. Building prints are used to locate electrical, plumbing, or other fixtures found in or on a building. Building prints are usually in the form of floor plans and plot plans. A *floor plan* is a view of a building level looking directly down from a horizontal cutting plane 5′ from the floor. Floor plans show the layout of rooms and give information about built-in features such as windows, doors, and mechanical fixtures, and locations of major equipment. **See Figure 3-15.** Floor plans posted in public areas may also include emergency information such as escape routes and locations of fire extinguishers or eyewash stations.

A *plot plan* is a scaled drawing that shows the shape and size of a lot and the buildings on it. Plot plans show the features of the land, such as trees and drainage, but more importantly to maintenance personnel, also include information on buried utilities and their entry points into the buildings. Plot plans may also detail outdoor structures that require maintenance, including culverts, parking lots, lights, and fences.

MAINTENANCE SCHEDULE									
Action	Part or Item	Running Hours	Time Interval (Whichever Comes First)						
			Daily	1 Wk	1 Mo	3 Mo	6 Mo	Yearly	2 Yr
Inspect	Coolant level	8	X						
Inspect	Discharge temperature (air)	8	X						
Inspect	Separator element differential	8	X						
Inspect	Air filter delta P (at full load)	8	X						
Inspect	Oil filter delta P	8	X						
Replace	Coolant filter*	150		X†					
Inspect	V-belt/belt tension	500			X				
Check	Temperature sensor	1000			X				
Replace	Coolant filter*	2000					X‡		
Clean	Separator scavenge screen and orifice	4000						X	
Clean	Cooler cores§	4000					X		
Replace	Air filter*	4000						X	
Replace	Separator element*	*							
Replace	Shaft seal	8000							X
Replace	Ultra coolant*	8000							X
Replace	Food grade coolant (when used)	3000					X		
Inspect	Starter contactors	8000						X	

* In very clean operating environments and where inlet air filter is changed at the above prescribed intervals. In extremely dirty environments, change coolant, filters, and separator elements more frequently.

† initial change only

‡ subsequent changes

§ Clean cooler cores if discharge air temperature is excessive or if unit shutdown occurs on high air temperature.

Figure 3-14. A maintenance schedule specifies all recommended maintenance required to keep equipment in proper operating condition.

Utilities Prints. Maintenance personnel may also use detail prints of the locations and routing of utilities in their work. These types of systems typically run throughout the building and inside walls and ceilings, so accurate information is essential to minimize the impact that work on these systems has on personnel, equipment, and processes. Electrical, plumbing, and HVAC drawings are the most common types of detail prints used in maintenance. Depending on the worksite, additional prints for pneumatic lines, fire suppression systems, telephone, network, and security systems, and other systems may be necessary.

Installation Prints. An *installation print* is a print that outlines the general configuration and information needed to install a specific piece of equipment. Mounting dimensions, outline dimensions, clearance requirements, and feature information may be included. **See Figure 3-16.**

A *mounting dimension* is a dimension used to locate holes or threads for mounting equipment with screws, studs, brackets, or clips. This information is useful for preparing areas for equipment installation before the equipment arrives.

An *outline dimension* is a dimension that indicates the minimum space required to install the piece of equipment. They show the overall size and contour of the equipment and the surfaces related to the mounting dimensions.

A *clearance requirement* is a specification of the operating space required around a piece of equipment in addition to its outline dimensions. Clearances may be function- or service-related. Functional clearances cover all extremes of extended and retracted positions, angles of operation, or moving components. Service clearances are given for all areas such as brushes, covers, or panels that may require service.

Feature information is information about the relationships between equipment components and facility features. For instance, it may be essential for a component to be perpendicular to the floor following installation, within a certain tolerance requirement.

Abbreviations and Symbols. Standard abbreviations and symbols are used for consistency and to conserve space on drawings. Standards and industry organizations specify the abbreviations and symbols used for a particular type of system. For example, the Institute of Electrical and Electronics Engineers has developed a comprehensive set of standard symbols for electrical components. **See Figure 3-17.** Many other system elements can be represented by symbols developed by other organizations. **See Appendix.**

Symbols are preferred over abbreviations whenever possible because abbreviations are language-specific. Symbols are universal and do not require knowledge of a particular language to understand them.

Some types of lines can also be like symbols in that they convey additional information about the outlined object. Dashed lines are used for objects hidden from view, lines with alternating long and short dashes indicate ranges of moving equipment or previous positions, and thin parallel lines may indicate cutaway sections. These representations can be important to understanding all aspects of a maintenance situation. An *alphabet of lines* is a description of the various line types and their uses. Maintenance drawings should conform to this standard.

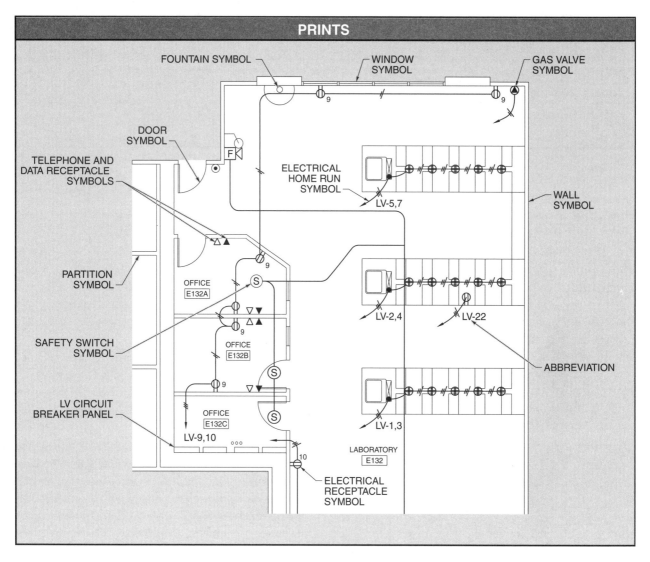

Figure 3-15. Building prints are used to locate electrical, plumbing, and/or other fixtures found in or on a building.

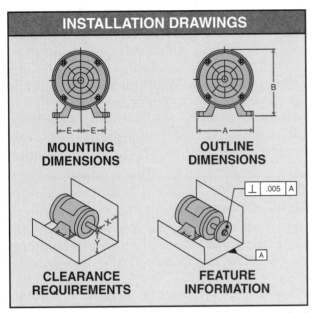

Figure 3-16. Installation drawings give information required to install a specific piece of equipment.

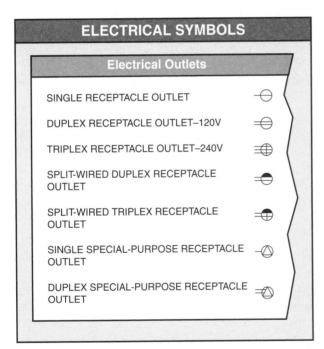

Figure 3-17. Symbols are standardized graphics developed to represent elements in a system.

SERVICE AND REPAIR PROCEDURES

Equipment service and repair requires a planned sequence to reduce cost and downtime. After the problem is isolated, service and/or repair tasks are performed. In some cases, it is more costly to repair a part than replace it. The overall repair cost must consider the part cost and labor required to repair the equipment. In addition, service and repair of the suspect component requires removal, cleaning, and possibly re-coating.

Parts Cleaning

Parts cleaning is commonly performed in a cleaning tank. A *cleaning tank* is a tank used for cleaning parts in flammable solvents with a lid that automatically closes to contain flames during a fire. **See Figure 3-18.** A low-temperature solder joint on the safety prop is quickly melted from the heat of a fire. When the prop breaks down, gravity closes the lid. Most cleaning solvents used for parts cleaning are considered hazardous materials.

Figure 3-18. A cleaning tank discharges cleaning solvent through a nozzle to flush grease and dirt off parts.

In the cleaning tank, cleaning solvent is pumped through a hose and nozzle to flush grease and dirt off parts. Like aerosol spray cleaners, the cleaning solvent breaks down and washes away grease, oil, and sludge. A pump directs the cleaning solvent through a filter. The filter collects fine particles, and the larger dirt particles collect at the tank bottom. Parts cleaning procedures must be performed in a well-ventilated area using proper eye and hand protection. Parts cleaning solvents are considered a hazardous material and must be disposed properly.

Parts can also be cleaned using rags, cleaning pads, or non-abrasive cloth commonly referred to as crocus cloth. Surface dirt can be removed using rags. Lint free rags should be used for parts cleaning. Otherwise, lint can be left behind on the part when it is replaced. The lint could then enter the equipment and disrupt lubrication. Dirt that adheres to the surface can be removed using cleaning pads or non-abrasive cloth. The dirt must be removed without removing or damaging surface material.

Loctite Corporation
Coatings are used to protect, seal, insulate, prevent corrosion, color code, and add gripping power to tools and equipment while providing excellent resistance to acids, alkalis, salt, and moisture.

Hazardous Material Disposal

Industrial plants and facilities commonly use and dispose hazardous materials such as oil, process chemicals, and cleaning solvents. When disposed, these materials become hazardous waste. In 1986, the Resource Conservation and Recovery Act (RCRA) regulations covering generators of small quantities of hazardous waste went into effect. This Act makes removal and proper disposal of hazardous waste materials the responsibility of the waste generator. Disposal options include transporting hazardous waste to an approved disposal site or contracting with a firm to pick up and dispose the hazardous material. A manifest lists the content and quantity of hazardous material transported.

Some hazardous waste is recycled or blended for safe burning to recover its heat value. For example, waste engine oil is typically processed into water (10%), light oil such as kerosene (12%), lubricating oil (65%), and asphalt used for roads (13%). If waste oil has been mixed with other hazardous materials, it is commonly burned above 2600°F in high-temperature cement kilns. These temperatures exceed the temperature in commercial incinerators and completely destroy hazardous materials.

Material Properties

Equipment components have specific designs and are made from certain materials for maximum efficiency. Any replacement parts or materials must meet the design requirements and properties of the original component. This is especially important with safety-related devices. In some cases, a replacement part is not available or is too costly. A fabricated replacement component must be made from materials having the similar mechanical, physical, and chemical properties as the original and must not adversely affect the designed operation of the equipment. In addition, service conditions should not be modified so they exceed the capability of the material, or a higher-strength material must be used.

Mechanical Properties. A *mechanical property* is the characteristic of a material that describes its behavior under applied load. A *mechanical load* is an external force applied to a body that causes the stress in the material. *Stress* is the internal effect of an external force applied to a solid material. Every machine part or structural member is designed to safely withstand a certain amount of stress. Stress is classified according to position and direction as tension, compression, shear, bending, and torsion. **See Figure 3-19.**

Tension is stress caused by forces acting along the same axial line to pull an object apart. Tension tends to stretch an object. *Compression* is stress caused by forces acting along the same axial line to squeeze an object. *Shear* is stress caused by parallel forces acting upon an object from opposite directions. Shear tends to cause one side of the object to slide in relation to the other side. *Bending* is stress caused by forces acting perpendicular to the horizontal axis of an object. Bending tends to bend an object and results in a combination of tension and compression. *Torsion* is stress caused by forces acting in opposite twisting motions. Torsion tends to twist an object.

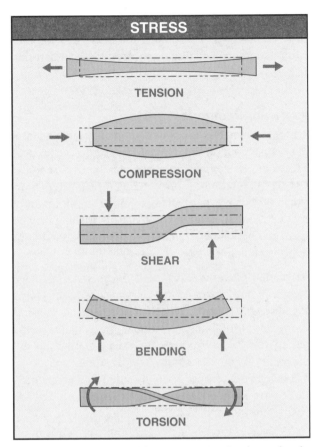

STRESS

TENSION

COMPRESSION

SHEAR

BENDING

TORSION

Figure 3-19. Every machine part or structural member is designed to safely withstand a certain amount of tension, compression, shear, bending, and torsion stress.

Stress on a material can cause strain. *Strain* is a material's deformation per unit length under stress. *Elastic deformation* is the ability of a stressed material to return to its original size and shape after being unloaded. *Plastic deformation* is the failure of a stressed material to return to its original size and shape after being unloaded. *Elastic limit* is the maximum stress with which a material can be deformed and still return to its original shape.

The reaction to stress must be considered when selecting materials for replacement parts. *Ductility* is the ability of a material to stretch, bend, or twist without breaking or cracking. High ductile metals such as copper resist stress from loads, deform easily, and fracture gradually. *Brittleness* is the lack of ductility in a material. Brittle materials fracture quickly in low-stress conditions. *Hardness* is the ability of a material to resist deformation by indentation. *Toughness* is a combination of resistance to stress and ductility. *Malleability* is the ability of a material to be deformed by compression without developing defects.

Physical Properties. Physical properties include thermal, electrical, and other properties of a material. *Melting point* is the temperature at which a solid changes to a liquid. *Boiling point* is the temperature at which a liquid changes to a gas. *Thermal conductivity* is the rate at which heat moves through a material. *Thermal expansion* is the change in volume of a material in relation to temperature. *Electrical conductivity* is the ability of a material to conduct the flow of electrons. This is the opposite of electrical resistivity. Many other types of physical properties, such as magnetic, optical, acoustic, and radioactive, are defined for metals and other materials. These can be found in material reference books or online.

Chemical Properties. Chemical properties are related to the molecular composition and chemical reactivity of materials. In regard to metals, chemical reactivity includes corrosion, oxidation, and reduction of a material. *Corrosion* is the combining of metals with elements in the surrounding environment that leads to the deterioration of a material. *Oxidation* is the combination of metal and oxygen to form metal oxides. Rust is the oxidation of steel or iron. When making a repair that is subject to moisture, stainless steel components are used to prevent damage from rusting. *Reduction* is the loss or removal of oxygen from a material. Reduction commonly occurs during the welding process.

> **TECH TIP**
> *A material's mechanical behavior is illustrated in a plot of stress versus strain, which shows elastic deformation, plastic deformation, and ultimate strength*

Material Identification

During service or repair of equipment, maintenance personnel may not have information regarding the material of the equipment requiring service or repair. Before welding, brazing, or gluing materials, the materials must be correctly identified. For example, a welding rod used to join aluminum cannot be used to successfully join steel.

Certain adhesives and glues are designed for specific surfaces. For example, an adhesive made to bond porous surfaces such as foam insulation may not successfully bond smooth plastic surfaces. In addition, some surfaces require special preparation to adhere properly. Using a structural adhesive on a smooth plastic surface requires that the surface be cleaned with a specific chemical and roughened using sandpaper.

Quick identification of the material reduces the time required to join broken parts. Not all methods of identification work on all materials. However, some quick observations can simplify the identification process. For example, if a motor guard appears rusted, it contains iron and can be successfully welded. Identification methods used on metals include the magnetic, color, spark, chemical, file, torch, chip, and fracture tests.

Magnetic Test. The *magnetic test* is a metal identification test that uses a magnet to check for the presence of magnetic iron in a metal.

Color Test. The *color test* is a metal identification test that identifies metals by their color. The color of the metal is examined and compared to known material colors. For example, gray usually indicates carbon steel. Zinc and lead are bluish white, tin is silvery white, and so on.

Spark Test. The *spark test* is a metal identification test that identifies metals by the shape, length, and color of spark emitted from contact with a grinding wheel. For example, a small red spark emitted from a metal sample indicates white or gray cast iron.

Chemical Test. The *chemical test* is a metal identification test that uses chemicals that react when placed on certain metals. Chemical tests can also be used on metal shavings immersed in chemicals, which change color to indicate composition.

File Test. The *file test* is a metal identification test in which a file is used to indicate the hardness of a metal compared with that of the file. For example, if steel can be marked with a file only after great effort, it is likely to be high-carbon steel or harder. If the file marks the sample easily, it is likely to be mild steel.

Torch Test. The *torch test* is a metal identification test that uses the application of heat to identify a metal by its color change, melting point, and behavior in the molten state. For example, if there is no color change before melting occurs and a great deal of heat is required to obtain a puddle, the metal is probably aluminum.

Chip Test. The *chip test* is a metal identification test that identifies metal by the shape of its chips. A chisel is used to remove small chips of the sample for examination. Long and curled chips are taken from mild steel and soft metals such as aluminum. Short, broken chips are taken from cast steel. Sample chips are very difficult to obtain from high-carbon steel.

Fracture Test. The *fracture test* is a metal identification test that breaks the metal sample to check for ductility and grain size. For example, mild steel deforms before fracturing.

Fastening Methods

Fastening methods used by equipment manufacturers vary depending on function and production efficiency. For example, components that must be disassembled for service commonly use threaded fasteners. Components that are permanently joined are commonly riveted, welded, or adhesive bonded. Parts that may be damaged by heat are commonly joined by riveting or adhesive bonding.

Threaded Fasteners. A *threaded fastener* is a device such as a bolt, screw, or nut that joins or fastens parts together with threads. Threaded fasteners have several advantages for joining parts, including availability in a variety of sizes, styles, strengths, and materials and the capability of joining similar and dissimilar materials. Threaded fasteners are easily installed by the manufacturer or in the field with power or hand tools and are easily removed and replaced.

Threaded fasteners are based on the principle of an inclined plane wrapped around a cylinder. A *screw thread* is a ridge of uniform section in the form of a helix on an internal or external surface of a cylinder or cone. A *helix* is the curve traced on a cylinder or cone by a spiral. Threads may be right- or left-handed. **See Figure 3-20.**

Threaded fasteners commonly used include bolts, screws, and nuts. Bolts, screws, and nuts are purchased parts and are listed in documentation and prints with notes giving specific information. Threaded fasteners replaced during service and repair must match the characteristics of the original threaded fastener. The Society of Automotive Engineers (SAE®) has established standards for classifying bolts and screws. Bolts and screws are classified into grades based on their tensile strength and yield strength. The SAE has designated markings to put on bolt and screw heads to indicate the grade. The markings consist of radial slashes. Threaded fasteners are used with anchors to provide maximum structural strength.

Washers. A *washer* is a device used with threaded fasteners to distribute load, affect friction, prevent leakage, and/or ensure tightness. Washers are classified as flat, spring lock, and tooth lock washers. **See Figure 3-21.** Flat washers are round and flat, and are also called plain washers. Flat washers are used under the head of a screw or bolt or under a nut to spread a load over a greater area than the head of the screw or bolt. Flat washers also prevent damage to the part from the rotation of the fastener.

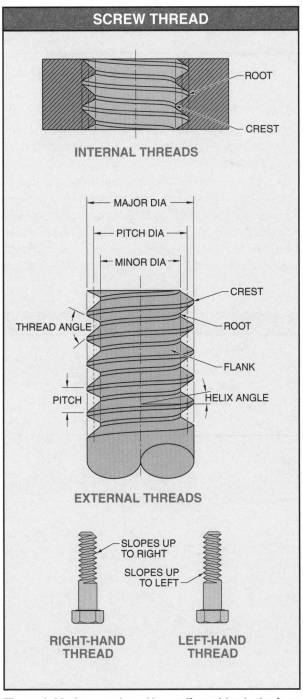

Figure 3-20. A screw thread is a uniform ridge in the form of a helix on the internal or external surface of a cylinder.

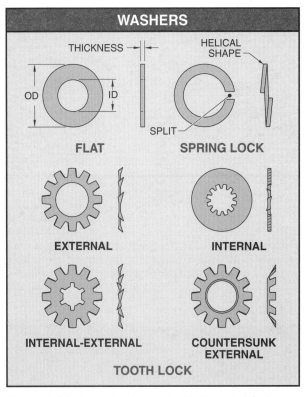

Figure 3-21. A washer is used with threaded fasteners to distribute load, reduce friction, prevent leakage, and/or ensure tightness.

Spring lock washers are split on one side and helical in shape. Spring lock washers provide pressure to compensate for looseness between joined parts and increase the frictional resistance between the bearing surface of the fastener and part. Tooth lock washers are washers that have hardened, twisted teeth to prevent the loosening of the fastener from vibration. Tooth lock washers can be external, internal, internal-external, or countersunk external. Washers are sized for use with standard threaded fastener sizes.

Thread Lock Coatings. A *thread lock coating* is a liquid coating applied to a threaded fastener that prevents the loosening of assembled parts from vibration, shock, and/or chemical leakage. **See Figure 3-22.** Thread lock coatings can sometimes be used in place of lock washers. The thread lock coating selected is determined by the fastener and application. Thread lock coatings can be applied before or after assembly. A wicking thread lock coating has a low viscosity and is used to penetrate between bolts and nuts after assembly. Special high-temperature and chemical-resistant thread lock coatings are available. Fasteners can also be purchased with threads pre-coated.

I.T.W. Devcon

Figure 3-22. A thread lock coating is applied to a threaded fastener to prevent loosening of assembled parts from vibration, shock, and/or chemical leakage.

TECH TIP

Lock wire is another method of preventing fasteners from loosening. It is also called safety wire because it prevents critical components from unfastening.

Nonthreaded Fasteners. A *nonthreaded fastener* is a device that permanently joins or fastens parts together without threads. Nonthreaded fasteners include rivets, pins, and keys. The most common nonthreaded fastener is the rivet. **See Figure 3-23.**

A *rivet* is a cylindrical metal pin with one preformed head that is deformed on the opposite side in order to hold parts together. The rivet shank is inserted through a hole in the parts and expanded by pressure applied by a hammer or special tool to join the parts. The size of the rivet required is determined by the thickness of the parts being joined. Rivets are considered permanent fasteners. They can be removed, but the process destroys the rivet and may damage the parts being joined. To remove a rivet, the center is drilled out with the appropriate size drill bit. The remaining parts of the rivet then fall apart.

A *pin* is a cylindrical nonthreaded fastener that is placed into a hole to secure the position of two or more parts. **See Figure 3-24.** A wide variety of pin types, sizes, and materials are commercially available. Standard pins include straight, dowel, taper, clevis, cotter, slotted spring, spirally coiled, and grooved.

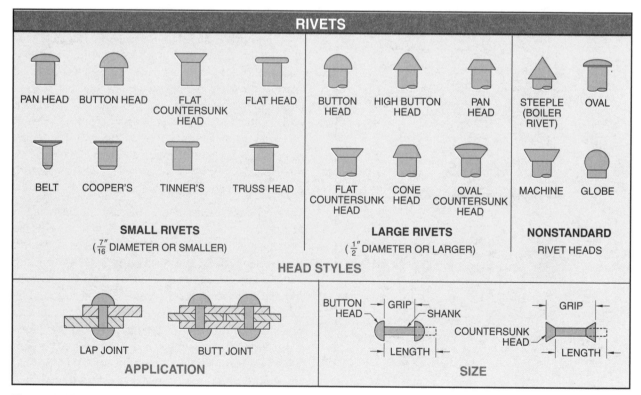

Figure 3-23. A rivet is a cylindrical metal pin with a preformed head and a cylindrical shank used to permanently join parts together.

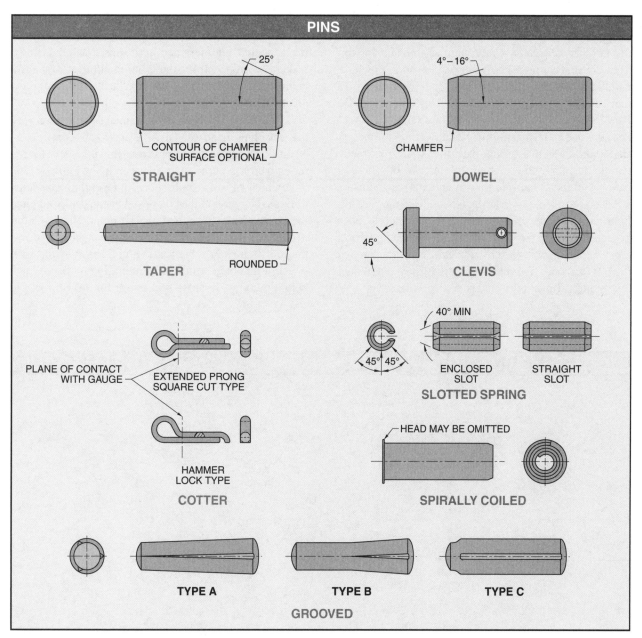

Figure 3-24. A pin is a cylindrical nonthreaded fastener that secures the position of two or more parts.

Straight pins are usually fabricated from bar stock. The ends are either square or chamfered. Straight pins are often used to transmit torque in round shafts. Dowel pins are fabricated from bar stock. Hardened dowel pins are bullet-nosed on the entry end. Soft dowel pins are chamfered on both ends. Dowel pins are used in machine and tool fabrication. Taper pins are used to transmit small torques or to position parts. Clevis pins are used to attach clevises to rod ends and levers and to serve as bearings. Cotter pins are used with clevis pins to prevent the clevis pin from becoming disengaged.

The entry ends are opened after insertion to keep the cotter pin in place.

Slotted spring pins are tubular with one longitudinal slot. The ends are chamfered and the pins are self-locking. Spirally coiled pins are rolled from spring stock. The spring stock compresses upon entry and expands to hold firmly. Grooved pins are solid with three parallel, equally spaced grooves. The grooves provide a tight fit and a locking feature. Grooved pins are used for semi-permanent fastening of levers, collars, gears, cams, and other components, to shafts.

A *key* is a removable part that provides a positive engagement between a shaft and a hub when mounted in a keyseat. A *keyseat* is a rectangular groove along the axis of a shaft or hub. The basic shapes of keys include parallel, Woodruff, and taper. Keys are designated by their name, type, and dimensions. **See Figure 3-25.** Parallel keys are square or rectangular in shape. Parallel keys are used for transmitting unidirectional torques in shafts and hubs that do not have heavy starting loads. Parallel keys may be easily withdrawn. Woodruff keys are shaped like a half-circle with either a full radius or a flat bottom. Woodruff keys are used for transmitting light torques or locating parts on tapered shafts. Taper keys may be plain taper, alternate plain taper, or gib head taper. Taper keys are used for transmitting heavy torques in shafts and hubs that are reversed frequently and subject to vibration. Taper keys may be easily withdrawn.

Threaded Fastener Repair. Threaded fasteners may break and threaded holes may become stripped during repeated maintenance activities. Fasteners may become corroded from the environment, or damaged from improper tool use, leaving heads that tools are unable to grip.

The simplest way to remove a damaged bolt or screw is to break off the head using a chisel and remove the threaded portion using an extractor. An *extractor* is a tool for removing broken bolts, studs, or screws. **See Figure 3-26.** An extractor is used by first drilling a certain size hole through the center of the bolt or screw and gently tapping the extractor into the hole with a hammer. The extractor is then turned counterclockwise with a wrench to draw the tool tightly into the remaining portion of the bolt or screw and then remove it. Penetrating lubricant may help loosen the threaded section for removal.

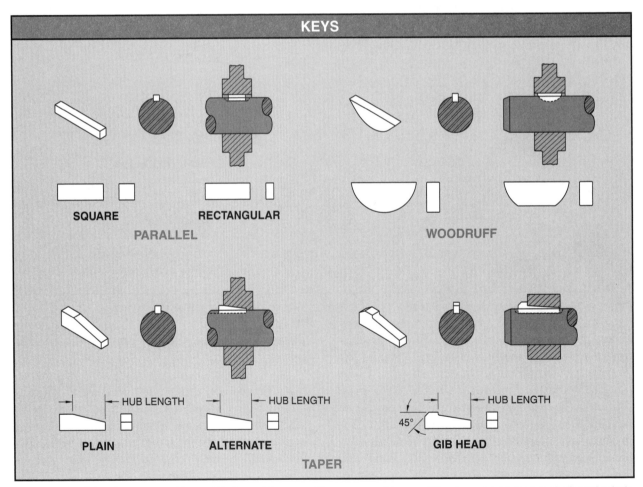

Figure 3-25. A key is a removable part that provides a positive means of transmitting torque between a shaft and a hub when mounted in a keyseat.

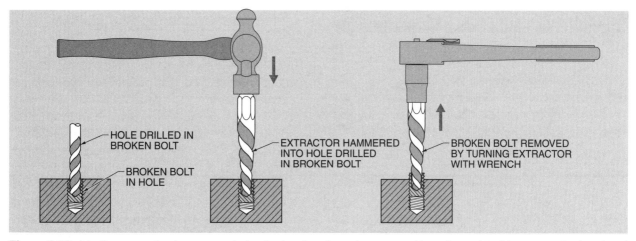

Figure 3-26. A bolt or screw having a corroded or broken head may be removed by using a chisel to remove the head and an extractor to remove the bolt or screw.

Nuts, studs, or bolts with severely damaged threads should always be replaced with new ones. Damaged internal threads can be repaired by creating new ones with a tap or by using a thread insert. A *tap* is a tool for cutting internal threads. A *thread insert* is a small coil with outside threads to hold the coil in a tapped hole and inside threads to fit a standard fastener. Thread inserts may be used where an internal thread is damaged or stripped.

A thread insert is installed by enlarging a hole to a specified oversize, tapping it with a special tap designed for the thread insert, and screwing the insert into the hole. **See Figure 3-27.** The insert stays in the hole as a permanent repair. Thread inserts enable the use of the original bolt because the internal threads are the same size as the original. When correctly installed, a thread insert is often stronger than the original threads.

Welding. *Welding* is a joining process that fuses materials by heating them to melting temperature. A filler metal can be used to reinforce the weld joint. Welding processes commonly used in maintenance work are oxyfuel welding (OFW) and shielded metal arc welding (SMAW). Other welding processes may be required for certain metals or specialized welding tasks.

Oxyfuel welding (OFW) is a welding process that produces heat from the combustion of a mixture of oxygen and a fuel gas. Oxyfuel welding is commonly used for maintenance work because it does not require electricity. Fuels commonly used with oxygen include acetylene, MAPP (methylacetylene-propadiene) gas, natural gas, and propane. *Oxyacetylene welding (OAW)* is an oxyfuel welding process that uses oxygen mixed with acetylene. **See Figure 3-28.**

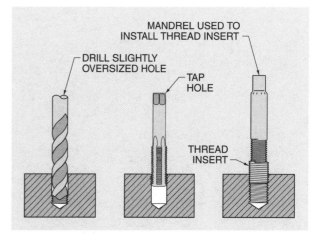

Figure 3-27. A thread insert is installed by drilling a hole to a specified oversize, tapping it with a special tap, and screwing the insert into the hole.

The Lincoln Electric Company
The shielded metal arc welding process is commonly used in industry to maintain and repair various plant equipment.

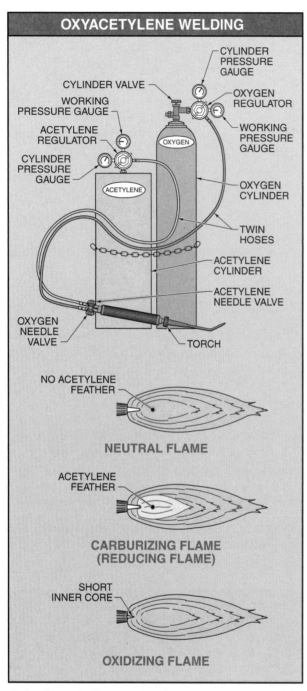

OXYACETYLENE WELDING

CYLINDER PRESSURE GAUGE

CYLINDER VALVE

WORKING PRESSURE GAUGE

OXYGEN REGULATOR

ACETYLENE REGULATOR

WORKING PRESSURE GAUGE

CYLINDER PRESSURE GAUGE

OXYGEN

OXYGEN CYLINDER

ACETYLENE

TWIN HOSES

ACETYLENE CYLINDER

ACETYLENE NEEDLE VALVE

OXYGEN NEEDLE VALVE

TORCH

NO ACETYLENE FEATHER

NEUTRAL FLAME

ACETYLENE FEATHER

CARBURIZING FLAME (REDUCING FLAME)

SHORT INNER CORE

OXIDIZING FLAME

Figure 3-28. Oxyfuel welding uses oxygen combined with a fuel to sustain a flame that generates the heat necessary for welding.

Oxygen and acetylene can also be used for cutting operations. Metal to be cut is first preheated, then pressurized oxygen is blown onto the heated metal, which rapidly oxidizes and burns it. The burned metal is also blown away by the oxygen, leaving a clean cut.

Shielded metal arc welding (SMAW) is an arc welding process in which the arc is shielded by the gases from the decomposition of a consumable electrode covering. The SMAW process is commonly used in maintenance work because of its versatility, low cost, and minimal equipment required. The SMAW process uses electricity to generate the heat necessary for melting the workpieces. Relatively high currents (55 A to 200 A) and low voltage (18 V to 35 V) are used. In the welding process, heat is generated from the resistance of electricity traveling through air and melts the base metal. **See Figure 3-29.** The electrode directs the flow of current and melts during the welding process to provide the filler metal required for weld strength. The type of electrode required is determined by the base metal, current, and weld requirements.

Gas metal arc welding (GMAW) is an arc welding process that uses a separate shielding gas and a continuous wire electrode. The wire electrode maintains the arc and is fed continuously by a wire feeder as it is consumed in the weld. The weld area is protected during the welding process by an inert gas shield. An *inert gas* is a gas that does not readily react with other elements.

Gas tungsten arc welding (GTAW) is an arc welding process in which a shielding gas protects the arc between a tungsten electrode and the weld area. The electrode directs the arc and is not consumed during the welding process. Filler metal must be added separately. Like GMAW, flux is not required because an inert gas shields the weld area from possible contamination. The GMAW and GTAW processes are commonly used for repair of food processing or other sanitary equipment.

Parts that must be joined without melting the base metal commonly use the brazing (B) or soldering (S) processes. *Brazing* is a joining process that joins parts by heating the filler metal to temperatures greater than 840°F, but less than the melting point of the base metal. *Soldering* is a joining process that joins parts by heating the filler metal to temperatures up to 840°F, but less than the melting point of the base metal. Brazing and soldering produce joints that are weaker than welded joints. The filler metal is bonded to the weld parts by adhesion in brazing and soldering compared to fusion in welding. This requires preparation of the joints for maximum surface contact with the filler metal by cutting, grinding, or filing. Brazing and soldering are used when the high heat required for welding may damage the parts repaired.

application of surfacing material that has hardness properties to reduce damage from wear. Worn parts that can be surfaced instead of replaced offer significant savings. The application of a hard, wear-resistant layer can extend the life of the part. Surfacing material is commonly applied in overlapping beads. **See Figure 3-30.** If necessary, the part is then machined or ground to specifications.

The Lincoln Electric Company

Figure 3-30. Surfacing material is applied in overlapping beads until the desired dimensions are obtained.

In maintenance repair tasks, surfacing material is commonly applied using the shielded metal arc welding process. The layer of material applied must have the desired properties for the application. For example, the type of wear, such as impact, abrasion, or corrosion, determines the type of surfacing material required. Consult the electrode manufacturer for the recommended electrode required for the application.

Adhesive Bonding. *Adhesive bonding* is the joining of parts with an adhesive placed between mating surfaces. Adhesive bonding is useful for joining dissimilar metals, plastics, and composites in manufacturing and repair operations. Adhesive bonding can be used to reduce the numbers of fasteners required and strengthen joints prone to failure from vibration. Adhesive bonding products should be applied with proper eye and hand protection to reduce the possibility of injury.

Thin parts subject to heat distortion can be joined with adhesives. For example, sheet metal panels joined with adhesives do not have distortions caused by resistance welding heat. Also, joint member dimensions do not affect bonding strength. Thin parts can be joined with thick parts. Adhesives fill voids between parts without breaking surface contours. The flexibility of many adhesives also allows distortion without failure. Adhesive bonding joints require large contact areas for adhesion similar to brazing and soldering.

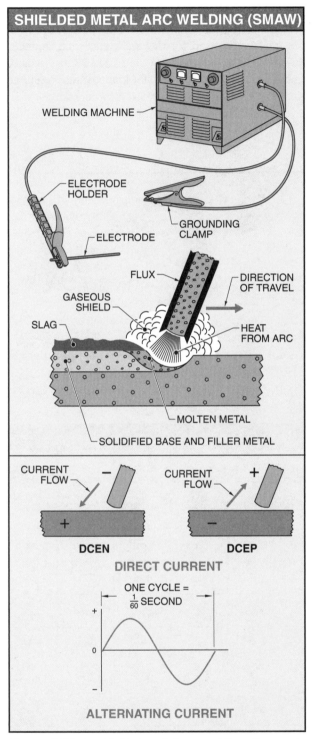

Figure 3-29. In shielded metal arc welding, the arc is shielded by the decomposition of the electrode covering.

Surfacing. *Surfacing* is the process of applying a layer of material by welding or brazing to obtain the desired dimensions or surface properties. *Hardfacing* is the

Adhesive bonding requires proper surface preparation, application, and curing procedures. The mating surfaces must be clean and free of foreign matter. Adhesives are cured by chemical action using catalyst cure, evaporation, ultraviolet (UV) light, heat, pressure, or both heat and pressure. Equipment required for adhesive bonding varies depending on the application and curing methods.

Adhesives are available in various viscosities. *Viscosity* is the measure of the resistance of a fluid to flow. Low-viscosity adhesives are liquid in form and flow readily into small spaces. High-viscosity adhesives range from gel to plastic-like forms. High-viscosity adhesives are good for filling small voids and gaps between mating surfaces. Adhesives are classified by chemical content or base as acrylic, anaerobic, cyanoacrylate, epoxy, hot melt, polyurethane, polysulfide, silicone, solvent-base, or water-base adhesives. **See Figure 3-31.**

An *acrylic adhesive* is a one-part UV- or heat-cure or a two-part adhesive that can be used on a variety of materials. Acrylic adhesive has a fast cure time and good flexibility. An *anaerobic adhesive* is a one-part adhesive or sealant that cures by the absence of air displaced between mated parts. Low-viscosity anaerobic adhesives are commonly used for locking metal parts together such as screws, nuts, and other fasteners. High-viscosity anaerobic adhesives are used for joining parts with large gaps between mating surfaces.

Loctite Corporation
Gasket material is used on pipe flanges to prevent gas or liquid leakage.

ADHESIVE BONDING										
Adhesive	Components	Cure Time	Viscosity	Void-Filling	Flexibility	Heat Resistance	Cold Resistance	Thermal Resistance	Water Resistance	Metal Bonding
Acrylic	two-part one-part (UV or heat cure)	medium to fast	medium	good	good	good	good	good	good	good
Anaerobic	one-part	medium	low	poor to fair	good	good	good	good	good	fair
Cyanoacrylate	one-part	fast	low	poor to fair	poor to fair	fair	fair	good	fair	good
Epoxy	two-part	slow to medium	medium to high	excellent	fair	good	fair	good	good	good
Hot Melt	one-part	fast	high	excellent	fair to good	poor to fair	fair	fair	good	fair
Polyurethane	one-part two-part	medium	medium	good	good	fair	good	good	fair	good
Polysulfide	one-part two-part	medium	high	excellent	good	good	good	excellent	good	good
Silicone	one-part two-part	medium	high	excellent	excellent	excellent	excellent	excellent	excellent	fair
Solvent-Base	one-part	medium	low to medium	poor to fair	good	good	good	good	good	good
Water-Base	one-part	medium	low to medium	poor to fair	poor to fair	fair	fair	poor	poor	poor to fair

Figure 3-31. Adhesive selection is determined by the material and application of the parts joined.

A *cyanoacrylate adhesive* is a one-part adhesive that cures instantly by reacting to trace surface moisture to bond mated parts. Cyanoacrylate adhesives are commonly referred to as super glue and have a fair resistance to high temperatures, moisture, vibration, and shock. An *epoxy adhesive* is a two-part adhesive that cures when resin and hardener are combined. Some epoxy adhesives are heat-cured.

A *hot melt adhesive* is a thermoplastic material applied in a molten state that cures to a solid state when cooled. Hot melt adhesives are not as strong as epoxy but are very fast-curing. A *polyurethane adhesive* is a one- or two-part adhesive with excellent flexibility that cures by evaporation, catalyst, or heat. A *polysulfide adhesive* is a one- or two-part adhesive or sealant that cures by evaporation or catalyst. Polysulfide adhesives are commonly used in the aerospace and building materials industry. *Silicone* is an adhesive or sealant that has excellent flexibility, resilience, and strength over a wide temperature range. A *solvent-base adhesive* is a one-part adhesive with a rubber or plastic base that cures by solvent evaporation. Solvent-base adhesives are commonly used as contact cement for bonding large surface areas and lamination applications. A *water-base adhesive* is a one-part adhesive that cures by water evaporation. Water-base adhesives have fair flexibility and are primarily used for wood and paper products.

Sealing Methods

Sealing methods for equipment components are specified by the manufacturer. The sealing method used must consider heat, expansion, and chemical resistance. If specifications are not available from the equipment manufacturer, a reputable gasket or sealant supplier should be contacted.

A *gasket* is a rigid or semirigid pliable material placed between mating surfaces to prevent gas or liquid leakage. Gaskets can be purchased as precut parts for a specific equipment application or as sheets of uncut material for custom applications. Gaskets are used to seal close-fitting pieces of metal such as parts of an engine or pipe flanges. An oil-resistant gasket is used to join two pipe flanges in piping that carries fuel oil in a boiler plant.

A *sealant* is a product used to seal, fill voids, and waterproof parts. Like a gasket, a sealant is selected based on the material and application of the parts joined. Sealants can be used as gasket material. For example, a sealant can be used as a gasket when a valve cover is replaced on an engine.

Puncture Sealing. Puncture sealing requires the repair of a breach in a sealed container or pipe. Puncture sealing usually requires shutdown, removal, and repair/replacement of the container or pipe. In some cases, the repair is temporary, and a permanent repair or replacement is made later when there is a minimum impact on production. If the container repair requires welding, brazing, or soldering, the container must be properly cleaned and vented to prevent fire hazards or pressure buildup. Contents previously held in the container must be identified. If necessary, the MSDS should be referenced. All pressure should be released, contents removed, and attached components potentially affected by heat disconnected.

Warning: Before cleaning a container, the container contents must be identified to determine the proper cleaning operation. As a precaution, the container should be partially filled with water and adequately vented before welding.

When making temporary repairs, patching products provide an alternative to welding, brazing, soldering, or replacement. **See Figure 3-32.** Patching products are available as liquids, putty, or patches and allow quick repairs without the problems associated with draining the container, cleaning, and heat from welding. Manufacturers provide specific products for container materials and contents. Patching products should be applied with proper eye and hand protection to prevent possible injury.

Patches can be permanent or temporary. A small hole in an aluminum evaporator of a refrigerator can often be patched using a patching compound made for that purpose. Diaphragm or pneumatic actuators can be patched using tape or rubber patches if a new diaphragm is not readily available. A hole or worn spot in the refractory brickwork found in a boiler furnace can be patched using mortar that withstands high temperatures. The patching material must suit the application because an improper patch or adhesive material could further damage the material being repaired.

TECH TIP

For optimal adhesive bonding, remove all oil, grease, residue, and paint. Scuff the surfaces to be bonded with an abrasive to increase surface area and friction. Apply clamp pressure and allow the maximum cure time possible.

Coatings

Equipment is typically finished with a protective coating from the manufacturer. Wear and repair handling may remove the protective coating. After repair, the equipment should be cleaned and recoated with a protective coating suitable for the application and operating environment. Coating manufacturers can be used as a resource for coating recommendations. In addition, specific colors may be selected for functional and aesthetic purposes. The composition of the previously applied coating must be determined to assure proper adherence and product performance.

For example, the paints used in a desalination plant, which converts salt water to fresh water, help prevent corrosion of the underlying metal. Paint used on slippery stairs may have a fine, clean sand added to produce a rough surface to prevent falls and slips. Exposed metal parts of lathes and other precision equipment are coated with a grease-like sealant to prevent corrosion of the metal before the equipment is shipped. When metal becomes rusted, the rust should be removed down to bare metal and immediately coated to help prevent further corrosion.

In some cases, a coating is used to provide a smoother, more efficient surface in the equipment. For example, a low-friction surface can be applied to a pump housing to reduce wear and increase overall efficiency. **See Figure 3-33.**

Mechanical Repairs

Many repairs of defective equipment are accomplished by simply replacing a defective component. For example, a broken ON/OFF switch or dirty hydraulic filter is replaced. While this may bring the equipment back into service, the repair is not complete until the cause of the problem has been located. For example, if the switch was broken because it was located in a position where it could be hit by moving equipment, the switch may break again. A complete repair would be to move the switch or install a guard to protect it.

A hydraulic filter might become dirty in the normal course of its operating life or it could become dirty very quickly because a filler cap has been left off the oil reservoir, allowing dirt to enter. While changing the filter, maintenance personnel should look for signs of excessive dirt buildup or a way that dirt might be entering the system.

When making a mechanical repair, the condition(s) that caused the failure should be determined. Usually, there is physical evidence in the pieces of the damaged equipment or in the position of the equipment when the failure occurred. Once the cause of the failure is determined, the mechanical repair should be made sufficiently strong enough to resist the condition(s) that caused the failure.

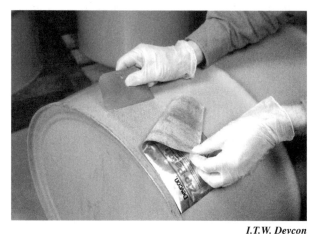

I.T.W. Devcon

Figure 3-32. Patching products are available to allow quick repairs without the problems associated with draining the container, cleaning, and heat from the repair.

Loctite Corporation

Figure 3-33. A coating may be used to build up a worn part or provide a more efficient surface in the equipment.

All mechanical equipment is subjected to various stresses that could damage the structural material of the equipment. If the stress is great enough, it can deform the material. *Deformation* is a change in the shape of a material caused by stress. If the deformation is too great, the material breaks or bends beyond the point at which it can return to its normal shape. For example, the metal brackets of storage shelving bend and break if the weight on the shelves creates excessive shearing or bending stress.

When making a repair to mechanical equipment, the stresses acting on the equipment must be considered. For example, when repairing the collapsed shelving, stronger brackets would withstand the weight of the items stored on the shelves. Try to visualize the stresses acting on equipment or structures by careful observation of similar equipment while operating. Other conditions to consider when making a mechanical repair are the amount and types of vibration, impacts, and extremes of heat and cold that can cause expansion and contraction of the material.

Always ensure that the mechanical structure will resist the stresses likely to act on it. If the safety of the structure cannot be ensured, outside engineering help is needed to ensure that repairs are safe and equipment will not break under the stresses that caused the original damage.

Service and Repair Cost Justification

Maintenance personnel must often determine the cost and benefit of specific service and repair to equipment. The cost of parts is often a small portion of the total repair cost. Downtime, labor costs, and parts acquisition costs must also be considered when determining the best course of action. Replacement parts can be purchased from the manufacturer, an authorized parts supplier, or an aftermarket parts vendor.

The best replacement part is not always the most or least expensive. The manufacturer is the best source for parts because the part should precisely match equipment design requirements. If the part is no longer available from the manufacturer, an authorized parts supplier or an aftermarket parts vendor should be contacted. In all service and repair tasks, equipment warranty stipulations must be considered.

Some equipment service and repair tasks may require estimates from contractors outside the plant. These estimates may or may not include warranty agreements on parts and service. Whenever possible, it is best to solicit advice from appropriate plant personnel or other professionals. Some equipment, such as boiler safety valves and welding regulators, must be serviced by authorized personnel at the manufacturer or certified repair agency.

The choice to repair a machine using plant personnel or seek outside help varies from plant to plant. Insurance companies or local regulations may require that equipment such as elevators or gas and electric meters are serviced by outside personnel with special training. In some cases, it is simply less expensive to have work done by outside specialists. For example, many plants do not repair electronic components. It is cheaper to have these components repaired at a specialty repair facility or by the manufacturer. In some cases, maintenance personnel have no choice in repair options. The manufacturer requires that their technicians make repairs to maintain a warranty or as part of the purchase contract. When new equipment is purchased, manufacturer's requirements must be followed to maintain the warranty.

After the warranty period has ended, the equipment is repaired in the most efficient manner possible. In most cases, skilled maintenance personnel in the plant repair the equipment. They know the equipment and learn its common issues. This leads to faster response and faster, more effective repairs.

In some cases, it is more cost effective to have outside contractors do specialized work. For example, in a plant with only a few refrigeration and air conditioning systems, plant personnel performs most maintenance. If a serious problem requires that refrigerant be removed from the system, an outside contractor is called. The cost of refrigerant recovery equipment is too large for the plant given the frequency that the equipment is needed. It is cheaper for outside contractors to do highly specialized work requiring expensive equipment. In a plant with many refrigeration systems, the cost of owning refrigerant recovery equipment may be justified because it is likely to be used more frequently. Such decisions are made on a plant-by-plant basis.

TECH TIP

Repairs to parts under significant stress or involving safety systems often need to be engineered to ensure adequate strength. A professional mechanical or structural engineer studies the forces involved and determines the necessary material and repair method.

TOOLS

Maintenance personnel use a variety of tools to complete service and repair tasks. Tools are commonly classified by function as measuring, fastening, cutting, striking/driving, and specialty tools. Tools required for specific tasks vary, but some basic tools are frequently used for common service and repair tasks. Most tools may be purchased from a variety of tool manufacturers or retailers. For specific tasks on certain equipment, specialty tools are purchased from the equipment manufacturer.

Large plants have equipment and tools in-house to perform some machine parts repair and/or replacement. Small plants may contract specialty machine work to an outside company.

Tools can be classified as hand tools and power tools. General tool safety rules that apply to hand and power tools include:

- Always wear proper eye protection, hearing protection, and protective clothing.
- Secure hair and loose clothing.
- Work in areas with good lighting.
- Ensure all personnel are at a safe distance before using a tool.
- Keep tools free of oil, grease, and foreign matter.
- Use the tool for its designed use.
- Secure small work in a clamp or vise.
- Repair or replace damaged tools.
- Report any injuries to the supervisor.

Hand Tools

Hand tools are tools that are used manually. Hand tools are usually acquired and maintained by maintenance personnel. Hand tools used for measuring include rules, power return steel tapes, and folding rules. Hand tools used for fastening include wrenches, screwdrivers, and pliers. Hand tools used for cutting include wire strippers, hacksaws, files, and chisels. Hand tools used for striking/driving include hammers, mallets, and punches.

Hand Tool Safety. Accidents with hand tools can be reduced by following basic hand tool safety rules. Hand tool safety rules include:

- Keep tools sharp and in proper working order. Look for wear that could cause an injury such as a pitted hammer face, damaged insulation on pliers, or splintered handle.
- Point cutting tools away from the body during use.

- Grind off excess metal from mushroomed chisels.
- Organize tools to protect and conceal sharp cutting surfaces.
- Never use a hammer on another hammer. The impact of the hardened surfaces may cause the heads to shatter.
- Do not carry tools in a pocket. Transport sharp tools in a holder or with the blade pointed down.
- Remove fasteners by pulling the tool toward the body or pushing the tool away from the face.

Power Tools

Power tools are tools that are electrically, pneumatically, or hydraulically powered. Power tools allow greater efficiency when performing service and repair tasks. Power tools present in a facility vary with each company. Power tools can be portable or stationary. Portable power tools can be transported with the operator. Stationary power tools cannot be transported and are commonly installed in a fixed location. Common power tools include grinders, drill presses, air chisels, and air impact wrenches. Power tools can cause serious injury if proper safety procedures are not followed.

Power Tool Safety. Safety risks when using power tools can be reduced by following basic power tool safety rules. Power tool safety rules include:

- Follow manufacturer's recommended operating instructions.
- Use UL or CSA approved power tools that are installed in compliance with the NEC®.
- Do not use an electric tool on a wet or damp floor.
- Ensure that power tools are double-insulated or have a third conductor grounding terminal to provide a path for fault current.
- Ensure the power switch is in the OFF position before connecting the tool to the power source.
- Ensure all safety guards are in place before starting.
- Arrange cords and hoses to prevent accidental tripping.
- Stand to one side when starting and using a grinder.
- Stand clear of operating power tools. Keep hands and arms away from moving parts.
- When using pneumatic tools, use tool components specifically designed for compressed air service.
- Shut off, lockout, and tagout disconnect switches of power tools requiring service.

ELECTRICAL TEST INSTRUMENTS

Maintenance personnel must often determine the electrical condition of a component within a system. A *test instrument* is an electrical measurement tool used to test the condition or operation of an equipment component or system. Maintenance personnel must determine if the component is receiving an electric current when required, if the conductors are able to carry an electric current, and/or if an electric component operates when electricity is present. These questions may be answered by using a variety of test instruments, depending on the complexity of the problem.

Continuity Testers

A *continuity tester* is a test instrument that indicates if a circuit is open or closed. An *open circuit* is a circuit having an incomplete path, which prevents current flow. A break in a conductor or an open switch causes an open circuit, preventing the flow of electricity. A *closed circuit* is a circuit that has a complete path for current flow. Continuity is generally determined using a continuity tester or a multimeter. A continuity tester supplies its own voltage and current by the use of batteries. A continuity tester must be used only in a circuit with the power OFF. **See Figure 3-34.**

A continuity tester is used on an electric component to check a wire to determine if it is continuous or broken. The circuit is complete and the continuity tester lights when the tester leads are touching both ends of a continuous (unbroken) wire. The wire may be broken if the tester does not light.

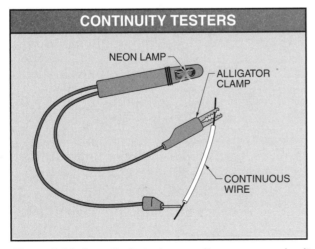

Figure 3-34. A continuity tester uses its own power circuit to determine if a wire is continuous or broken.

Voltage Indicators

A *voltage indicator* is a test instrument that indicates the presence of voltage when the test tip touches, or is near, an energized conductor or metal part. The tip glows and/or the device emits a sound when voltage is present at the test point. Voltage indicators are used to test receptacles and fuses/circuit breakers, to detect breaks in cables, and for other applications in which the presence of voltage must be detected. **See Figure 3-35.**

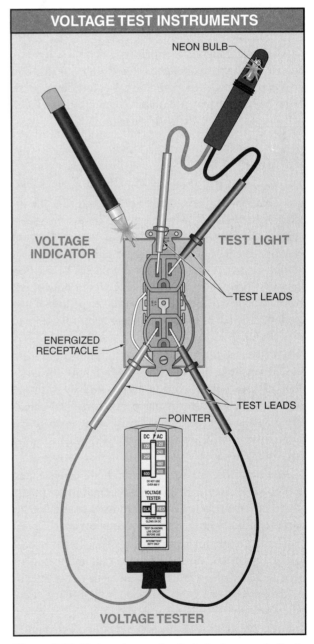

Figure 3-35. Various types of voltage testers can indicate the presence, type, or approximate level of voltage.

Voltage indicators are available in various voltage ranges, from a few volts to hundreds of volts, and in the different voltage types (AC, DC, AC/DC) for testing various types of circuits.

Test Lights

A *test light* is a test instrument with a bulb that is connected to two test leads to give a visual indication when voltage is present in a circuit. Test lights look similar to continuity testers, but they do not supply voltage. Voltage must be present in the tested circuit for the bulb to light.

The most common test light is a neon test light. A neon test light has a bulb that is filled with neon gas and uses two electrodes to ionize the gas (excite the atoms). Neon bulbs have long lifetimes compared to other bulbs because they draw very little current.

Voltage Testers

A *voltage tester* is a test instrument that indicates approximate voltage level and type (AC or DC). The indication can be by the movement of a pointer on a scale or by neon lights. AC voltage will also trigger vibration on some models. Basic voltage testers do not have a digital display. A voltage tester should not be used on low-power circuits because a voltage tester draws high current and no voltage indication is given on low-power circuits. The range for most voltage testers is between 90 V and 600 V.

Voltage testers are designed for intermittent use and should not be connected to a power supply for more than 15 sec. Simple voltage testers that are compact and rugged enough for maintenance personnel generally have the capability of being a voltage tester and a continuity tester. Caution must be taken when testing any voltage over 24 V.

Warning: Ensure no part of the body contacts any part of a live circuit, including the metal contact points at the tip of a tester. Always assume that circuits are energized (hot) until they have been checked.

The procedure for using a voltage tester is to connect one test probe to one side of the circuit and then connect the other test probe to the other side of the circuit. A reading is taken from the pointer. The white wire, when present, should be checked first to determine that it is the neutral conductor. A *neutral conductor* is a wire that carries current from one side of the load to the grounded neutral bar in the circuit breaker panel.

Neutral conductors are connected directly to loads and not through fuses, circuit breakers, or switches. Connecting a voltage tester between the neutral wire and a grounded surface should give no voltage indication. All other hot wires are then checked to ground or neutral and should show a voltage reading (voltage present) on the voltage tester.

Receptacle Testers

Testing receptacles is also possible with a receptacle tester. A *receptacle tester* is a test instrument that is plugged into a standard receptacle to determine if the receptacle is properly wired and energized. **See Figure 3-36.** Some receptacle tester models include a ground fault circuit interrupter or ground fault interrupter test button that allows the receptacle tester to be used on GFCI or GFI receptacles.

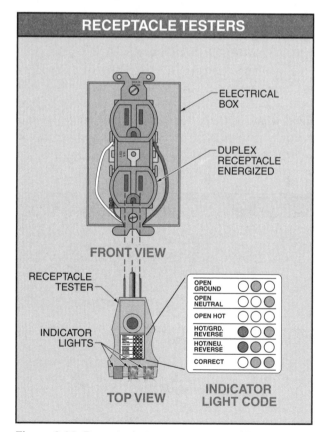

Figure 3-36. Receptacle testers use a set of lights to indicate wiring configurations.

Ammeters

An *ammeter* is a test instrument that measures the amount of current in an electrical circuit. *Current* is the flow of electrons through a conductor. Current is measured in amperes. An ampere relates to the number of electrons passing a given point per second. Large amounts of current are measured in amperes (A). Small amounts of current are measured in milliamperes (mA) or microamperes (μA). Current flows through a closed circuit when a power source is connected to a load that uses electricity.

Current measurements are used to determine the amount of circuit loading or the condition of an electrical component (load). Every load (light bulb, motor, heating element, speaker) that converts electrical energy into some other form of energy (light, rotating motion, heat, sound) uses current. The more electrical energy required, the higher the current usage. Every time a load is switched ON or a new load is added to a circuit, the power source must provide more current through the circuit for the new loads.

Ammeters measure current with noncontact jaws or when connected in line with the circuit. **See Figure 3-37.** Ammeters with open or clamp-on jaws measure current by detecting the strength of the magnetic field around a single conductor. These ammeters do not need to touch the conductor to measure current. Small amounts of current can be measured using test leads that connect the ammeter in line with the circuit. In-line current measurements are more accurate, but must be limited to circuits that can be easily opened and known to have currents less than 10 A. Ammeters are available for measuring AC only, or for measuring both AC and DC.

Typically, the first choice for measuring current in a circuit is a clamp-on ammeter or a multimeter with a clamp-on current probe accessory. In-line current measurements present greater opportunity for electrical shock. When a clamp-on ammeter cannot be used, or an in-line measurement is required, steps must be taken to prevent possible injury and/or damage to the electrician, test instrument, and equipment.

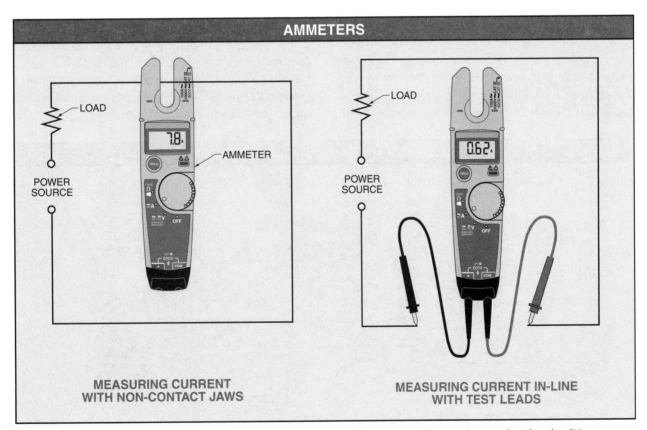

Figure 3-37. Current measurements can be taken by ammeters with noncontact jaws or by opening the circuit to use ammeter test leads.

Ohmmeters

An *ohmmeter* is a test instrument that measures resistance. *Resistance* is the opposition to the flow of electrons in a circuit. Resistance is measured in ohms (Ω). Prefixes are used to simplify the measurement displayed. Resistance measurements on ohmmeters are displayed as Ω, kΩ (kilohms), and MΩ (megohms).

Ohmmeters measure the amount of resistance in a de-energized circuit or component. Circuits or components are not required to be energized because the ohmmeter has an internal battery that is used to supply voltage to the test leads and the component being tested. In fact, resistance measurements must be taken with the circuit de-energized and the component removed from the circuit. When a circuit includes a capacitor, the capacitor must be discharged before taking any resistance readings of the circuit.

Ohmmeters may be either analog or digital. Analog ohmmeters have multiple scales, and a function switch is used to select the appropriate resistance range. Digital ohmmeters are usually a feature of multimeters. To ensure accuracy, analog ohmmeters must be zeroed before taking a measurement.

Resistance measurements are normally taken to indicate the condition of a circuit or component. The higher the resistance in a circuit, the lower the current flow through the circuit. Likewise, the lower the resistance in a circuit, the higher the current flow through the circuit.

Components designed to insulate, such as rubber or plastic, have very high resistance values. When insulators are damaged by moisture or overheating, resistance decreases. Megohmmeters are a type of ohmmeter that is used to measure the very high resistances of insulating components. **See Figure 3-38.**

Components designed to conduct, such as conductors or component contacts, have very low resistance values. When conductors are damaged by burning or corrosion, resistance increases. Other components such as heating elements and resistors have a fixed resistance value. Any significant change in the fixed resistance value typically indicates a problem.

Circuit Analyzers

Many loads found in industrial and commercial sites are connected to receptacles (outlets) for power. Various testers are used to verify that receptacles are wired correctly and operating at the proper voltage level. Circuit analyzers can be used to test multiple circuit operating functions and potential problems at once. A *circuit analyzer* is a receptacle plug and test instrument that determines circuit wiring faults (reverse polarity or open ground), tests for proper operation of ground fault circuit interrupters (GFCIs) and arc fault breakers (AFCIs), and displays important circuit measurements (hot/neutral/ground voltage, impedance, and line frequency). **See Figure 3-39.**

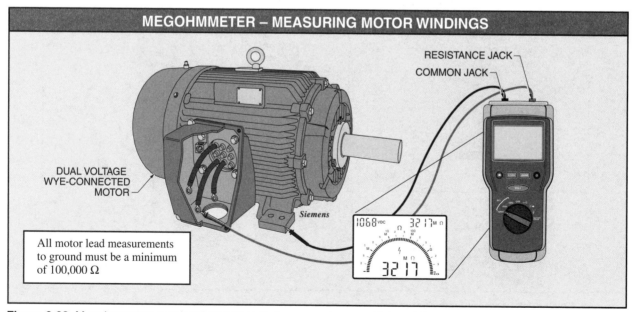

Figure 3-38. Megohmmeters test insulation integrity by measuring very high resistances.

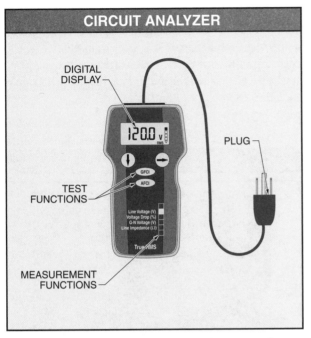

Figure 3-39. Several types of receptacle tests and measurements can be taken with a circuit analyzer.

Multimeters

A *multimeter* is a test instrument capable of measuring two or more electrical quantities. A function switch controls which electrical quantity is being measured. **See Figure 3-40.** Nearly all multimeters can measure AC and DC current, voltage, and resistance. Some can also measure quantities such as frequency, temperature, continuity, and capacitance. To take a measurement, the function switch is set to the correct quantity, the test leads properly connected, and the correct scale used for accuracy. Setting the scale might involve a range switch.

Multimeter batteries must be in good condition, and the meter should be switched OFF after taking measurements in order to conserve battery life. Multimeters are available as digital multimeters or analog multimeters.

A digital multimeter (DMM) has a digital display, which helps minimize human error when taking readings by displaying the exact values measured. DMMs are available with a wide assortment of functions and features. DMM types can be broadly classified as general-purpose, standard, or advanced DMMs. **See Figure 3-41.**

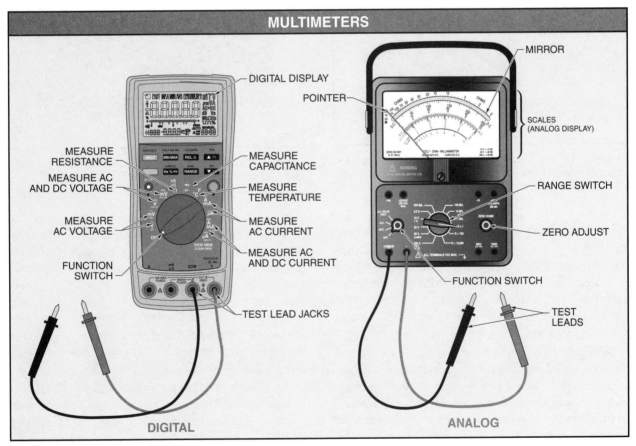

Figure 3-40. A multimeter is a test tool commonly used for electrical troubleshooting tasks.

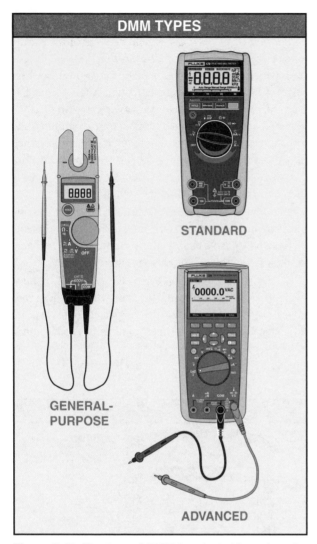

Figure 3-41. The type of DMM is selected based on the specific measurement and troubleshooting requirements.

equivalent to the measured quantity in its own units. For example, a pressure-measuring attachment may output 1 mV/psi. **See Figure 3-42.** The multimeter is set to measure millivolts and the attachment is plugged into the test-lead jacks. If the multimeter displays 61.05 mV, then the reading is 61.05 psi.

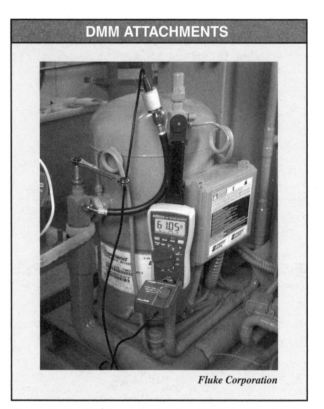

Fluke Corporation

Figure 3-42. Attachments plug into DMM test-lead jacks and add measuring capabilities to the meter.

A general-purpose DMM is the least expensive DMM type and is typically used for taking basic voltage, resistance, and current measurements when testing or troubleshooting. A standard DMM includes more functions and features. For example, a standard DMM includes specialized functions for diode testing, capacitance, or frequency.

Attachments can be used with general-purpose and standard DMMs to extend their measurement capabilities to current (with a clamp), temperature, pressure, light intensity, gas concentration, and other quantities. Typically, these attachments output a small voltage or current in proportion to the measured quantity. The DMM displays the voltage and current value, which is

Most DDMs auto range, meaning that the meter automatically chooses the most appropriate range in which to take the reading. Many DMMs incorporate a digital bar graph, usually at the bottom of the display. The bar graph reacts more quickly to changes in voltage and current, so it captures rapid fluctuations more effectively than the numbers on the digital display. This can be used to illustrate faults, such as a relay that opens and closes rapidly sending voltage pulses into a circuit.

An advanced DMM can be used for standard DMM applications, troubleshooting complex problems, and identifying and recording operating conditions. Monitoring with an advanced DMM is a predictive maintenance technique that can reveal equipment problems. An advanced

DMM is convenient when documentation and printed data are required as part of the maintenance task.

An analog multimeter indicates readings by the mechanical movement of a pointer across a scale. Analog meters typically require that the range be manually selected. A mirror on the scale is used to help minimize parallax error. *Parallax error* is a measurement error caused by an improper viewing angle of the pointer slightly above the scale. When viewing the scale properly, from directly above, the pointer hides its reflection in the mirror. If both the pointer and its reflection are visible, then the improper viewing angle may cause the user to misinterpret the reading.

Analog multimeters have been largely replaced by DMMs. However, although a DMM offers greater precision with numerical values displayed, many find that the movement of the pointer on an analog multimeter more effectively indicates trends.

Recording Modes. Most DMMs include special modes that are particularly useful for noting changes in measured values. The HOLD mode records the currently displayed reading. This is used when technicians cannot see the display while taking a measurement. The peak mode switches the displayed value to the peak value of a sine wave, rather than the default RMS (root-mean-square) value. The MIN/MAX mode records the minimum, maximum, and average values since the beginning of a measurement. The relative mode establishes a baseline value and displays the current measurement as a positive or negative difference from that baseline.

All of the modes can be used for any of the measured electrical quantities, and for measurements from attachments. Also, modes can be used in combination. For example, using both the MIN/MAX and relative modes records the maximum difference from a baseline value over time.

Data Logging. DMMs are invaluable as portable and temporary data logging devices. *Data logging* is the recording of measurements at specified intervals over a period of time. Recording measurements over time identifies stable and unstable values, which provide valuable information for troubleshooting or equipment operation analysis. A DMM with a data logging or trending function records measurements at specified intervals, defined in seconds, minutes, or hours. The recorded measurements can then be displayed as a plot on the DMM or downloaded to a computer for analysis, documentation, and distribution. **See Figure 3-43.**

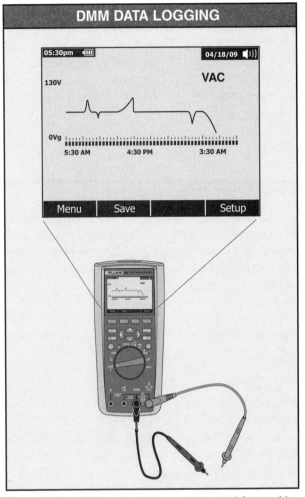

Figure 3-43. DMM data logging can be used for troubleshooting, or as part of a comprehensive predictive maintenance or energy auditing program.

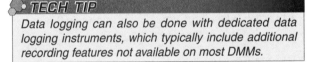

Data logging can also be done with dedicated data logging instruments, which typically include additional recording features not available on most DMMs.

Logging electrical measurements over time is common when troubleshooting a circuit problem. For example, repeated equipment failure at a certain time of day may be caused by voltage sags. Logging voltage measurements could confirm the time and help identify the cause of the voltage sag, such as loads energized simultaneously. Certain low transient voltages from switching motors and equipment can also be identified with DMM data logging. Although most low transient voltages do not cause immediate equipment damage, some may result in data loss and processing errors.

In addition to troubleshooting, DMM data logging can also be conducted as part of a comprehensive predictive maintenance or energy-auditing program. Reports utilizing baseline measurements, test measurement values, and charts illustrate trends and simplify data interpretation. In many instances, documenting a problem and providing information to supervisors are critical for planning, funding, and implementing these programs.

INVENTORY MANAGEMENT

Common replacement parts must be readily available to maintenance personnel. Parts used from the plant inventory require a procedure for reordering. If replacement parts are not inventoried in the plant, order records from suppliers must be maintained in order to place an order promptly. A preventive-maintenance system can be used to organize purchasing records. Purchase dates, reorder dates, warranty information, and other pertinent data should be filed for easy access. Inventory management is a common part of a CMMS. For example, parts may be automatically reordered as the part replaced is recorded on the troubleshooting report.

Managing inventory helps to ensure effective repairs. If an essential replacement part is not available, the machine remains inoperative until the part is available. In a perfect inventory control system, there would never be a situation when a needed part was not available. There would be just enough parts to repair the equipment. In addition, there would be no supplies that are never used. Excessive inventory costs money to purchase, store, and track. Stocking too many unused parts also makes finding necessary parts difficult.

Part Numbers

All spare parts should be numbered so their location and quantity can be tracked. A part number can be any number that clearly identifies the individual part, such as a manufacturer's number, a supplier's number, or a unique number specific to the plant. A complete description should be included with the number in the inventory system.

Using unique identifying numbers helps prevent confusion when selecting parts. For example, labeling a part as "3 A fuse" does not clearly identify the specific type of fuse. Two 3 A fuses may look alike and be the same size, yet have different operating characteristics.

For example, a non-time delay fuse is a fast-acting fuse that provides overcurrent protection. A time delay fuse is used with equipment-like motors that are subject to temporary startup or surge currents. A time delay fuse may damage equipment requiring a non-time delay fuse. Likewise, a non-time delay fuse used in place of a time delay fuse may cause needless disruption as it blows when a motor starts normally.

Inventory Control

When a part is removed from inventory, it should be noted in the inventory record on either a computerized or paper-based filing system. When the inventory of a certain part reaches a predetermined minimum number needed to keep the plant equipment running, a sufficient number should be reordered to bring the inventory back to the desired level. Some computerized systems automatically generate orders that can be faxed or emailed to the manufacturer. Computerized inventory systems can also identify how frequently parts are being used so that their minimum stock numbers can be increased or reduced.

Some parts are located together in a central storage area and others are kept close to their associated equipment. Each type of part should be stored in separate and clearly labeled boxes. It is the responsibility of maintenance personnel to ensure that the inventory is organized and accurately recorded. When searching for parts, the maintenance personnel should have the old part nearby or have a written copy of the part number to help ensure that the correct part is selected.

Ordering Parts

The cost of ordering parts includes both the cost of the parts and the time required for placing the order and sorting the parts when they arrive. Comparing prices helps locate the best price, and careful ordering saves time and money. Parts are most often ordered over the phone. Efficient part ordering requires preparation and organization. Clarity and efficiency lead to speedy, cost-efficient parts handling, which can be critical when a part is needed to repair malfunctioning equipment. The following steps should be taken to ensure efficient ordering:

1. Write down the complete part number. Have the part nearby to refer to when ordering if possible.
2. Locate the part in the parts supplier's catalog or manufacturer's documentation. Keep the pages marked and have them ready for easy access.

3. Locate the correct phone number and the name of a contact person if possible. Keep this information in case the line is busy.

4. Dial the number, identify yourself and your organization, and ask for the contact person. When speaking to the contact person, identify yourself again and clearly state the part number. Be prepared to refer to the documentation or answer any questions regarding the part.

5. Write down the part's cost, availability, and delivery date and method. If it is to be picked up, make sure the directions to the supplier are clear.

6. Log the parts order in a parts order book. **See Figure 3-44.** When the parts arrive, ensure that they are what was ordered and that the price and quantity are correct. Log the order as received. If a partial order is received, note the items still expected and when they should arrive.

Always be polite, even when encountering difficulties. Knowledgeable parts suppliers can be of great assistance when encountering problems in locating parts.

Cutler-Hammer

Efficient inventory management and control enables the optimum number of replacement parts to be on hand based on the maintenance requirements of plant equipment.

TECH TIP

Inventory management and careful recordkeeping of parts orders is a critical part of an effective preventive maintenance program.

Bill To:	Crete Machine Inc. 8601 E. Crawford Ave. Chicago, IL 60655 Country: USA Account: 12-345-6789		Purchase Order #	M23-172-341235
			Project/Job #	64364
			Date	06/26

	Item #	Description	Quantity	Price
Item(s)	8A4763	General purpose relay	3	$ 9.37
	7X6394	Time delay relay	2	$ 49.06

| Ship To: | Crete Machine Inc. 8601 E. Crawford Ave. Chicago, IL 60655 Country: USA ATTN: Maintenance Engineer Phone: 312-555-1100 |

Special Instructions

Deliver to building H.

Shipping Method | UPS Ground

Subtotal	$126.23
Shipping	$6.50
Total	$132.73

Figure 3-44. The record of a parts order should be kept until the parts arrive and are placed in inventory.

4

Electrical Systems

Most facility systems operate from electrical power, including lighting, manufacturing processes, and heating and cooling. Equipment in electrical systems is arranged in electrical circuits. Troubleshooting electrical circuits requires the use of test instruments to locate voltage source problems, incomplete or damaged electrical paths, and open or shorted loads. Preventive maintenance of electrical equipment reduces machine downtime, repair costs, and troubleshooting time.

ELECTRICAL PRINCIPLES

Electricity is the energy of the flow of electrons in a conductor. A *conductor* is a material that has little resistance and permits electrons to move through it easily. The flow of electrons cannot be seen because it occurs at the atomic level. An *atom* is the smallest building block of matter that cannot be divided into a smaller unit without changing its basic character. Atoms consist of protons, neutrons, and electrons.

A *proton* is a subatomic particle that has a positive electrical charge of one unit. A *neutron* is a subatomic particle that has no electrical charge. An *electron* is a subatomic particle that has a negative electrical charge of one unit. The *nucleus* is the dense center of an atom, consisting of protons and neutrons. The nucleus has a positive electrical charge. Electrons whirl around the nucleus in orbits (shells). Each shell can hold a specific number of electrons. **See Figure 4-1.**

For example, the nucleus of a copper atom contains 29 protons and 35 neutrons. A copper atom also contains 29 electrons located in four shells. The attraction between the positively charged protons and negatively charged electrons keeps the atom together. In a copper atom, there is one valence electron. A *valence electron* is an electron in the outermost shell of an atom. Valence electrons are capable of jumping into or out of the outer shell. A positive "hole" is left in the outer shell when an electron jumps from the outer shell. This positive hole may attract nearby electrons. For example, in the cord of a lamp that is OFF (de-energized), valence electrons jump around at random. When the lamp is turned ON (energized), the valence electrons jump in the same direction (from negative to positive) when the lamp is turned ON (energized) and electricity flows through the cord.

Current is the flow of electrons through a conductor. *Direct current (DC)* is current that flows in one direction only. *Alternating current (AC)* is current that reverses its direction of flow at regular intervals. **See Figure 4-2.** Current is measured in amperes. An ampere is related to the number of electrons passing a given point per second. The flow of electrons in a circuit is similar to the amount of water flow through a pipe.

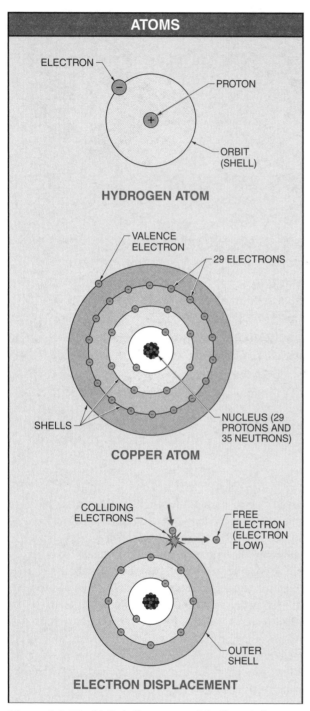

Figure 4-1. Electricity is the energy of the flow of electrons from atom to atom in a conductor.

Voltage is the amount of electrical pressure (electromotive force) that causes electrons to move in a circuit. Voltage is measured in volts (V). A difference in electrical potential (polarity) is required for electricity to flow in a circuit. *Polarity* is the positive (+) or negative (–) state of an object. The voltage required to make electrons move in a circuit is similar to the water pressure required to make water flow in a pipe. AC voltage reverses its polarity and alternates at regular intervals. DC voltage does not change its polarity.

Resistance is the opposition to the flow of electrons. Resistance is measured in ohms (Ω). A resistance of 100 ohms is written R = 100 Ω. Resistance limits the flow of current in an electrical circuit. The higher the resistance, the lower the current flow. Likewise, the lower the resistance, the higher the current flow.

Ohm's Law

Ohm's Law is the relationship between voltage, current, and resistance in a circuit. Ohm's Law states that current in a circuit is proportional to the voltage and inversely proportional to the resistance. For example, an increase in voltage increases current, if resistance is held constant. Likewise, a reduction in voltage reduces current, if resistance is held constant. Any value in this relationship can be found when the other two are known. The relationship between voltage, current, and resistance may be visualized by presenting Ohm's Law in pie chart form. **See Figure 4-3.**

Ohm's Law also states that an increase in resistance decreases current, if voltage is held constant. Likewise, a reduction in resistance increases current, if voltage is held constant. Voltage is used to force the electrons to move through the resistance offered by a load or conductor. Increasing or decreasing the resistance of the conductor changes the amount of current flow.

Power Formula

The *power formula* is the relationship between power, voltage, and current in a circuit. *Power* is the rate of doing work or using energy. Electrical power is measured in watts (W). The power formula states that the power from a circuit changes if current and/or voltage change. Usually, voltage remains relatively constant while current varies as loads change. Any value in this relationship may be found when the other two are known. The relationship between power, voltage, and current can be visualized by presenting the power formula in pie chart form. **See Figure 4-4.**

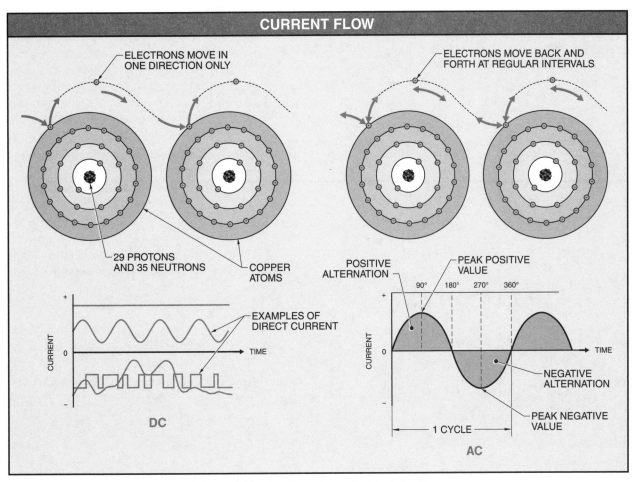

Figure 4-2. Current is the amount of electrons flowing through an electrical circuit.

Electrical Prefixes

Numbers are used in electrical systems when taking measurements, specifying equipment, and troubleshooting circuit components. Numbers used in electrical systems can range from very large numbers (5,000,000 Ω) to very small numbers (0.000002 A). Metric prefixes are used to simplify large and small numbers. Prefixes can be used with any electrical quantity, such as volts (1000 V = 1 kV), amperes (1000 A = 1 kA), or ohms (1000 Ω = 1 kΩ). For example, the rated output of a generator can be expressed as 4000 W or 4 kW. The most commonly used prefixes in the electrical and electronic fields are mega (M), kilo (k), milli (m), micro (μ), and pico (p).

Converting Units

To convert a base unit number that does not include a prefix to a unit that includes a prefix, the decimal point in the base unit number is moved to the left or right and

a prefix is added. For example, a voltage of 345,000 V is used in power transmission lines. To convert the base number to kilovolts (kV), the decimal point is moved three places to the left and the prefix "kilo" is added (345,000 V = 345 kV).

ELECTRICAL CIRCUITS

An *electrical circuit* is the interconnection of conductors and electrical components through which current flows. Electrical circuits commonly consist of a voltage source, conductors for current flow, and one or more loads. **See Figure 4-5.** There are a number of possible voltage sources, such as generators, transformers, batteries, fuel cells, or photovoltaic cells. The conductors, which are normally copper wire, connect the different parts of the circuit. Loads are devices in a circuit that use electricity, such as a lamp, heating element, or motor, to produce a useful output.

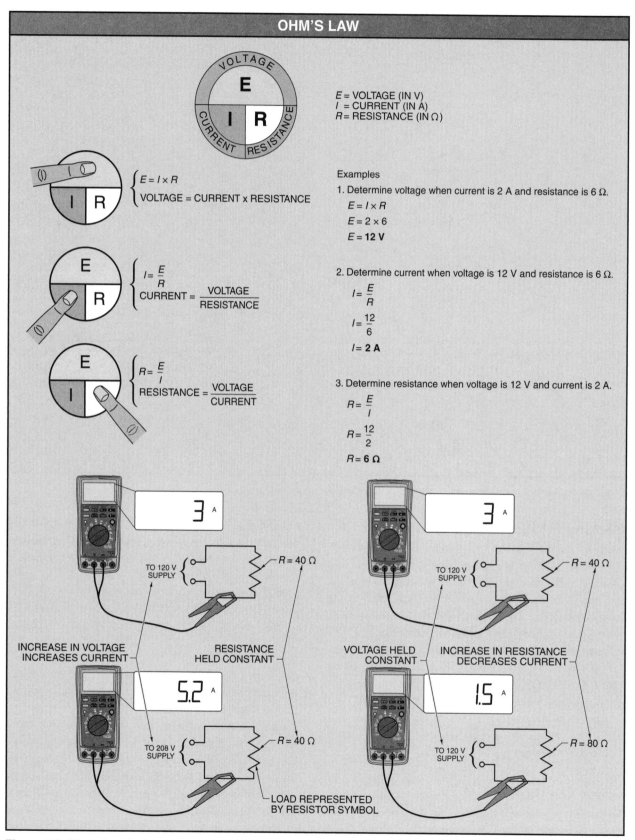

Figure 4-3. Ohm's Law is the relationship between voltage, current, and resistance in a circuit.

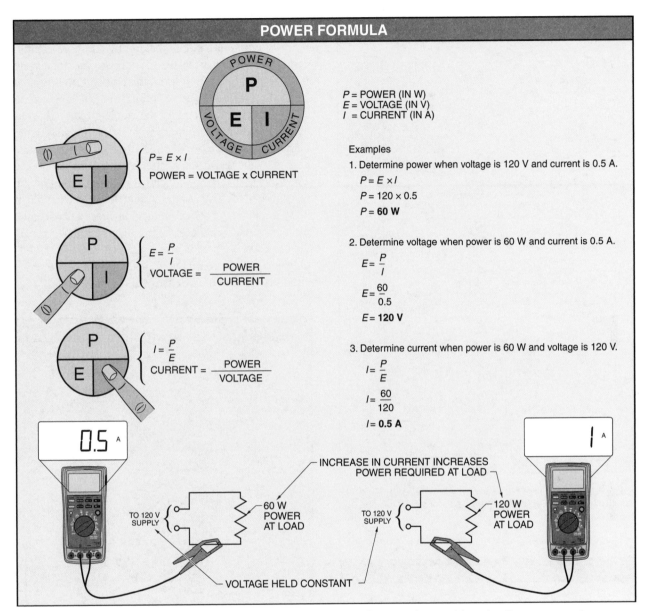

Figure 4-4. The power formula is the relationship between power, voltage, and current in a circuit.

Circuits are either open or closed. An *open circuit* is a circuit having an incomplete path, which prevents current flow. An open circuit is caused when a conductor is disconnected or the current path is otherwise disrupted. The load is de-energized and no electrical work is done in an open circuit. A common maintenance problem is an undesired open circuit that de-energizes all or part of a circuit. A *closed circuit* is a circuit having a complete path for current flow. The current flow is converted into heat, light, sound, or motion when the load is energized in a closed circuit.

Short Circuits

A *short circuit* is an undesirable, low-resistance path for current to leave the normal current-carrying path through a load. **See Figure 4-6.** For example, uninsulated and crossed wires create a low-resistance path around a load. This low resistance allows a high current flow. The high current flow may exceed the current rating for the circuit, causing overheating, damage, and possibly a fire. Fuses and circuit breakers are overcurrent protection devices that protect the source, path, and load from excessive current flow.

ELECTRICAL CIRCUITS

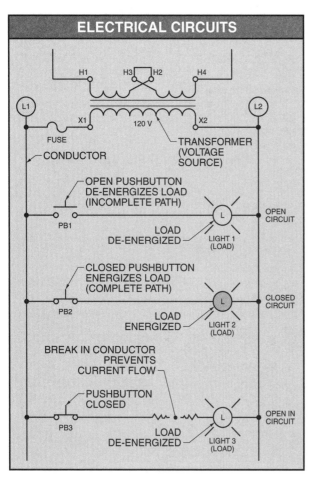

Figure 4-5. An electrical circuit is the interconnection of conductors and electrical elements through which current flows.

SHORT CIRCUITS

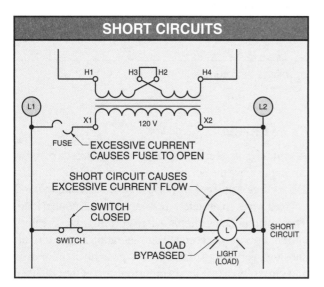

Figure 4-6. A short circuit creates a low-resistance path around a load, which allows high current flow.

Series Circuits

A *series circuit* is a circuit that has two or more components connected such that there is only one path for current flow. **See Figure 4-7.** Total circuit current flows through each component in a series circuit. An open anywhere in a series circuit de-energizes the whole circuit. All voltage applied to a series circuit is shared by all loads in the circuit. *Voltage drop* is the reduction of circuit voltage caused by the resistance of circuit conductors and loads. The voltage drop across each load is determined by the resistance of the load and the current flow. For example, if 24 V is applied to a series circuit with three lights of equal resistance, each light uses approximately 8 V. Current is the same everywhere in a series circuit because there is only one path for current flow. Total resistance in a series circuit is the sum of all individual resistances.

SERIES CIRCUITS

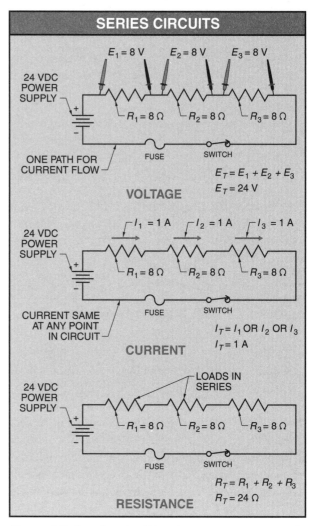

Figure 4-7. A series circuit has two or more components connected such that there is only one path for current flow.

Switches are often connected in series to control an entire circuit and develop circuit logic. A *switch* is any component that is designed to start, stop, or redirect the flow of current in a circuit. All lights in a series circuit are de-energized when the switch is turned OFF (opened). All series-connected switches in a circuit must be closed before any current flows. Opening one or more of the switches stops all current flow in the circuit.

Switches may be manual, mechanical, or automatic. A *manual switch* is a switch that is operated by a person. For example, ignition switches and light switches are manual switches. A *mechanical switch* is a switch that is operated by the movement of an object. For example, a safety switch installed under the seat of a lawn tractor is a mechanical switch. The switch is closed when the operator is seated and is open when the operator is off the seat. The switch can be used to prevent engagement of the blade without the operator seated, which is a safety feature. An *automatic switch* is a switch that responds automatically to changes in a system. Automatic switches include pressure, temperature, level, and flow switches. For example, on some engine low oil pressure warning systems, an automatic switch senses low oil pressure and de-energizes the ignition circuit to shut down the engine.

Parallel Circuits

A *parallel circuit* is a circuit that has two or more components connected such that there are multiple paths (branches) for current flow. An open in a branch of a parallel circuit de-energizes that branch only. Voltage is the same throughout the circuit. Current in each branch is controlled by the resistance of the branch. Total circuit current is the sum of current in each branch. **See Figure 4-8.** Adding loads to a parallel circuit reduces total circuit resistance and increases total current. Excessive current opens the fuse or circuit breaker to prevent damage to the circuit. Most industrial electrical circuits are parallel circuits. Switches are often connected in parallel to develop circuit logic. One or more switches connected in parallel must be closed before current flows. All closed switches connected in parallel must be opened to stop current flow.

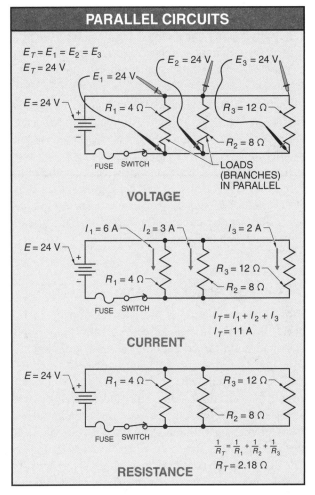

Figure 4-8. A parallel circuit has two or more paths (branches) for current flow.

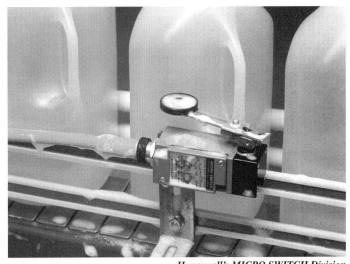

Honeywell's MICRO SWITCH Division
Mechanical switches use motion produced by physical contact with a moving object to control an electrical circuit.

Series/Parallel Circuits

Series circuits may be combined with parallel circuits to form series/parallel circuits. A *series/parallel circuit* is a circuit that contains a combination of components connected in series and parallel. **See Figure 4-9.**

Switches are commonly connected into series/parallel circuits to form circuit logic. All switches connected in series and one or more of the switches connected in parallel must be closed before current flows. Any one switch connected in series, or all switches connected in parallel must be opened to stop current flow. Each part of the circuit operates using the same rules as separate series and parallel circuits. Current flow in the circuit follows the path of least resistance.

DRAWINGS AND DIAGRAMS

Drawings and diagrams are used to convey facts, ideas, information, and directions, and to provide information on the operation of devices and circuits. Drawings and diagrams are commonly used when installing new equipment, troubleshooting, and making circuit modifications. Drawings and diagrams show how individual components, equipment, and systems are interconnected. Some show the control logic of a system. They are also used when troubleshooting to help determine the correct placement of test instruments and the expected measurements, and when making circuit modifications to determine circuit limitations and proper connections.

Standard practices for drawing electrical and electronic diagrams are detailed in ANSI/IEEE 315, *Electrical and Electronics Diagrams*. This standard provides definitions, general information, and recommended drafting practices.

Drawings

A *drawing* is a graphic representation of an object or idea. Standard drawings commonly used in industry include pictorial, sectional view, application, and architectural drawings. **See Figure 4-10.**

Pictorial Drawings. A *pictorial drawing* is a drawing that shows the length, height, and depth of an object in one view. True views of the surfaces are not obtained. Each surface is skewed, allowing three major surfaces to be seen. Pictorial drawings are used as working drawings to show the general location, function, and appearance of parts and/or assemblies.

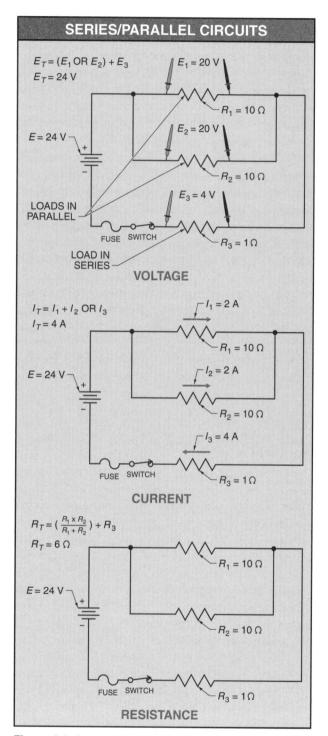

Figure 4-9. A series/parallel circuit contains a combination of components connected in series and parallel.

TECH TIP

An arrow placed across a symbol indicates that the component is variable (variable resistor, variable capacitor, variable flow control valve, etc.).

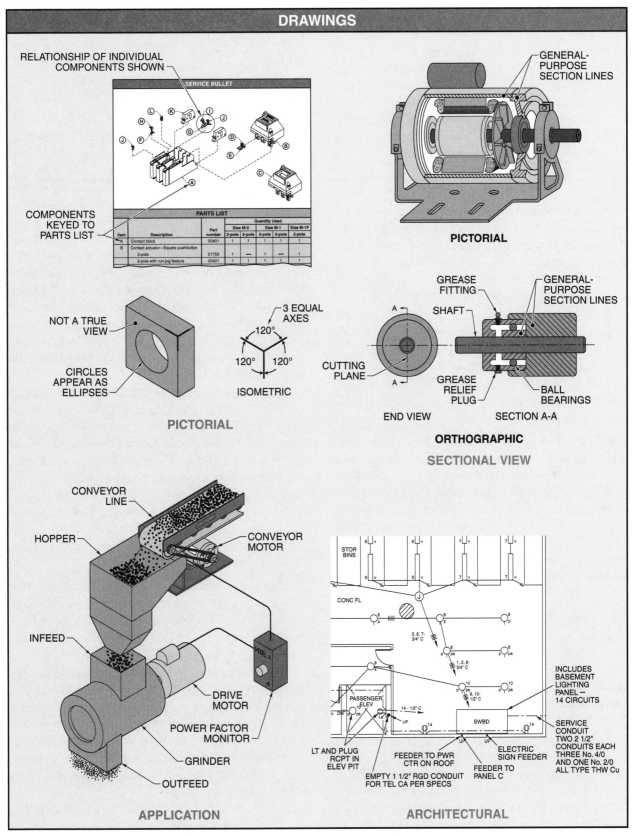

Figure 4-10. A drawing is a graphic representation of an object or idea.

An *isometric drawing* is a pictorial drawing with the three axes 120° apart. Isometric drawings show a 3D view of a component. Isometric drawings are used in promotional literature showing and describing products for sale. Isometric drawings are also used to show steps of assembly. Exploded isometric drawings are often used in service bulletins to show the relationships of individual components. Dashed lines connect the various components to show how they are assembled. The components are keyed to a parts list, which provides statistical information.

Sectional View Drawings. A *sectional view drawing* is a drawing that shows the internal features of an object. An imaginary cutting plane is passed through the object perpendicular to the line of sight. The portion of the object between the cutting plane and the observer is removed, revealing the internal features.

Section lines are drawn on all surfaces cut by the cutting plane. General-purpose section lines are commonly used. Whenever adjacent parts are shown in sectional views, the direction of the lines is changed for clarity. Specific section line patterns representing various materials, such as metals, soil, or insulation, may also be used.

Application Drawings. An *application drawing* is a drawing that shows the use of a particular piece of equipment or product in an application. Application drawings show product use and are not intended to show component connection or wiring. These drawings present ideas on the use of a product in problem solving and are used by manufacturers to promote products. Application drawings are also used in troubleshooting to show new or different components that can replace older or troublesome components.

Architectural Drawings. An *architectural drawing* is a drawing that contains building information in the form of floor plans, elevation views, section drawings, and detail drawings. Architectural drawings show the layout of a facility in a scale drawing. Floor plans, electrical requirements, sections, and electrical fixtures are detailed.

Architectural drawings are used to locate electrical devices such as outlets or lighting fixtures. The circuit breakers controlling all electrical devices are indicated on the drawings. Architectural drawings are consulted and the correct circuit breaker is opened when a circuit must be de-energized prior to maintenance or troubleshooting.

Diagrams

A *diagram* is a graphic that shows the arrangement and relationship of objects, areas, or parts. Diagrams commonly used in industrial facilities include line, wiring, and schematic diagrams. **See Figure 4-11.**

Line Diagrams. A *line diagram* is a diagram that uses lines and graphic symbols to show the logic of an electrical circuit. Line diagrams show the electrical connection of major equipment, protective relays, meters, and instruments. Line diagrams show the relationship between circuits and their components but do not show the actual location of the components. Line diagrams describe how the circuit works, how it is energized, and the sequence of operations. The arrangement of a line diagram should be clear. Graphic symbols, abbreviations, and device designations are drawn per industry and company standards. **See Appendix.** Connecting lines may be shown by the dot or no dot method. The main circuit is shown in the most direct path and logical sequence. Lines between symbols should be horizontal or vertical.

Line diagrams resemble a ladder, so they are also known as ladder diagrams. Line diagrams are read from left to right. For example, the common industrial circuit that contains a stop pushbutton, start pushbutton, motor starter, and overload is read as: coil M is energized when the start pushbutton is pressed, closing the normally open contacts M on line 2. This provides a holding circuit that keeps the circuit energized after the start pushbutton is released. The circuit is de-energized when the stop pushbutton is pressed, the overload (OL) opens, or when power to the circuit is lost.

Wiring Diagrams. A *wiring diagram* is a diagram that shows the electrical connections of a circuit. A wiring diagram shows, as closely as possible, the actual location of each component in a circuit, wire connections, wire numbers, and wire color.

Wiring diagrams are used in troubleshooting because they show the actual component layout and connections. Wiring diagrams are limited in quickly showing circuit logic because the connections are often hard to follow. A clear understanding of both the circuit wiring and operation is given when a wiring diagram is used with its line diagram. Symbols used with wiring diagrams may be the symbols used with line diagrams or may be simple rectangles and circles. No attempt is made to show exact sizes of component devices or parts.

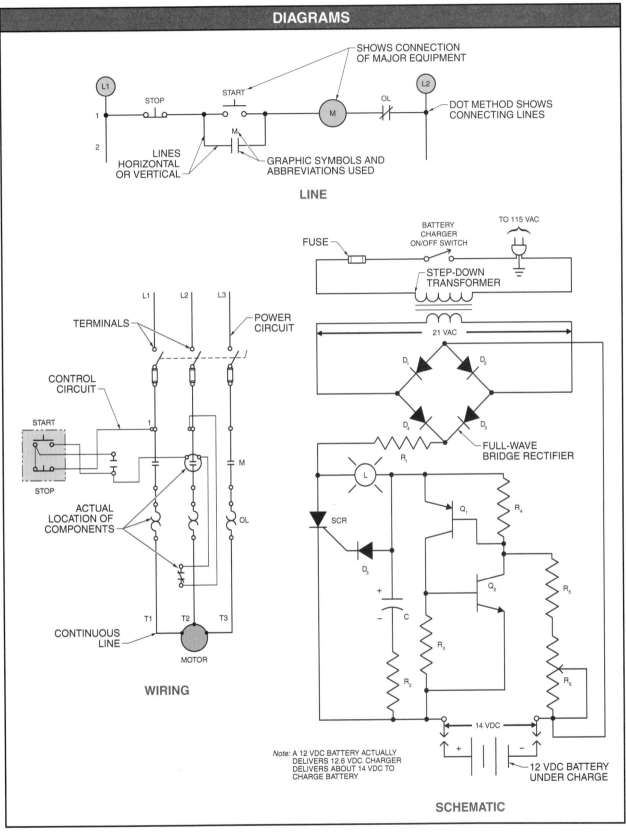

Figure 4-11. A diagram is a graphic design that shows the arrangement and relationship of objects, areas, or parts.

Schematic Diagrams. A *schematic diagram* is a diagram that shows the electrical connections and functions of a specific circuit arrangement with graphic symbols. A schematic diagram is similar to a wiring diagram, but does not show the physical relationship of the components in a circuit.

A schematic diagram is used to trace a circuit and its functions without regard to the actual size, shape, or location of the component device or parts. Schematic diagrams are also used to determine the function of each individual component within the circuit.

POWER DISTRIBUTION

Power distribution is the delivery of electrical power to where it is needed. **See Figure 4-12.** AC voltage is the most common voltage used in industry because AC permits efficient transmission of electrical power over long distances.

Electricity is commonly generated at a regional power generating station. The power generating station may be located hundreds or thousands of miles from the industrial facility. A power generating station and its network of transmission lines feeding customers form a regional electrical grid. Multiple grids are connected together to form the national power grid. This interconnection provides some measure of redundancy if one generating station goes off-line. Electricity may also be generated by a cogeneration plant, which generates electrical energy and process steam or heat simultaneously.

Outside the power generating station, voltage is increased by step-up transformers and routed across the country using high-voltage transmission lines. A *transformer* is an electric device that steps up or steps down alternating current. High-voltage electricity is transmitted from transmission substations to heavy industry and distribution substations. Distribution substations transmit electricity to industrial, commercial, and residential users. Step-down transformers at distribution substations reduce the voltage near the point of use. The voltage is sent to individual facilities, where it is reduced further and distributed to main switchboards. A *switchboard* is a piece of equipment in which incoming electrical power is broken down into smaller units for distribution throughout a building.

From the main switchboard, busways are used to distribute electricity for feeder circuits. A *busway* is a metal-enclosed distribution system of busbars available in prefabricated sections. Individual circuits are taken off the busway through fused, plug-in bus plugs. Wires from busways feed lighting and fusible panelboards. A *panelboard* is a wall-mounted distribution cabinet containing a group of overcurrent and short-circuit protection devices for branch circuits. Panelboards contain fuses, circuit breakers, disconnect switches, and other circuit protection devices for specific sections of a facility.

A *generator* is an electric device that converts mechanical energy into electrical energy by means of electromagnetic induction. Generators are driven by steam, wind, water turbines, or diesel engines. In a generator, the rotating coils create a strong rotating magnetic field that passes through stationary windings built in the housing of the generator. The magnetic field produced by the rotating coils induces voltage in the stationary windings. The induced voltage generates current flow through the wires leading from the stationary windings to the electrical system.

A generator produces AC voltage in the form of a sine wave. **See Figure 4-13.** A *sine wave* is a periodic, symmetrical waveform that varies over time according to the trigonometric sine function. The wave reaches its peak positive value at 90°, returns to zero at 180°, increases to its peak negative value at 270°, and returns to zero at 360°. Power generators create three-phase (3ϕ) voltage. Small auxiliary generators generate single-phase (1ϕ) voltage.

A *cycle* is one complete positive and negative alternation of a waveform over 360°. An AC sine wave has one positive alternation and one negative alternation per cycle. *Frequency* is the number of waveform cycles per second. Frequency is measured in cycles per second, or hertz (Hz). For example, a frequency of 5 cycles per second is equal to 5 Hz. The United States, Canada, Mexico, and most of the Caribbean distribute AC voltage at a frequency of 60 Hz. In most of the rest of the world, including Europe, Asia, and much of South America, AC voltage at a frequency of 50 Hz is commonly used for residential power.

+ SAFETY TIP

It is dangerous and illegal to improperly dispose of transformer oil containing poly-chlorinated biphenyls (PCBs). Check local and state codes for proper transformer oil handling.

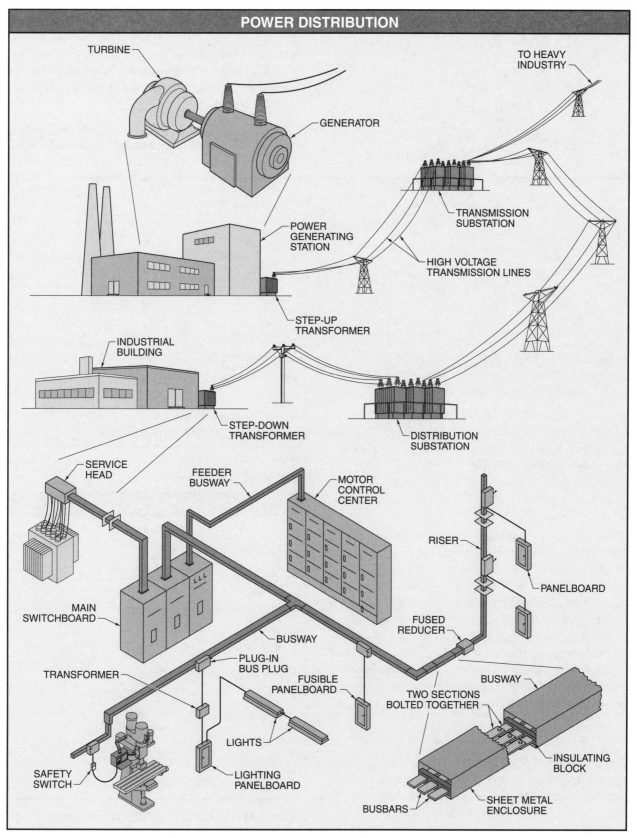

POWER DISTRIBUTION

Figure 4-12. Power distribution is the process of delivering electrical power to where it is needed.

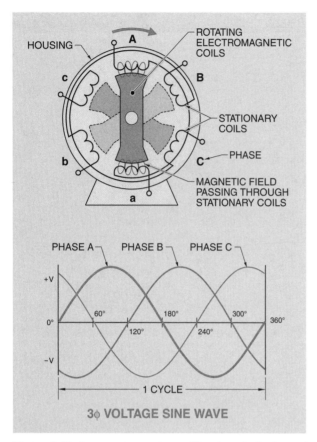

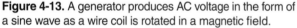

Figure 4-13. A generator produces AC voltage in the form of a sine wave as a wire coil is rotated in a magnetic field.

Single-Phase AC Voltage

Single-phase AC voltage contains one alternating voltage waveform. **See Figure 4-14.** Single-phase AC voltages range from 6 V to 277 V. Low-voltage 1ϕ power (120 V) is produced with one hot wire and a grounded neutral (white wire). A grounded neutral creates a potential difference between the hot and neutral wire, which causes electricity to flow. High-voltage 1ϕ power (240 V) is produced by using two hot wires. High-voltage circuits do not require a neutral wire to create a potential difference. Potential difference exists between the hot wires, causing electricity to flow in the circuits.

Three-Phase AC Voltage

Three-phase AC voltage is a combination of three alternating voltage waveforms, each displaced 120 electrical degrees apart. **See Figure 4-15.** Three-phase AC voltage is produced by the three coils in commercial generators. The coils are out-of-phase with each other. Three-phase AC voltage is commonly required

at industrial facilities. Common 3ϕ AC voltages are 208 V, 240 V, and 480 V. Three-phase circuits do not require a neutral wire as in residential 1ϕ circuits. The potential difference between each phase causes current to flow in each phase. The voltage between each phase should be within 2%.

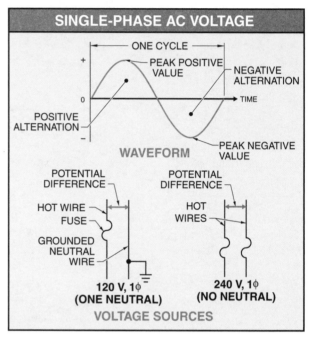

Figure 4-14. Single-phase AC voltage controls only one alternating current waveform.

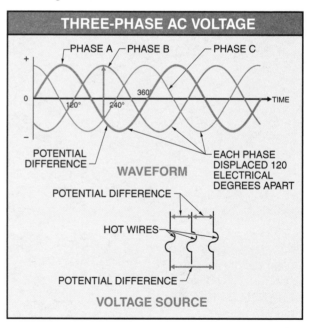

Figure 4-15. Three-phase AC voltage is a combination of three alternating voltage waveforms spaced 120 electrical degrees apart.

DC Voltage

DC voltage is voltage that flows in one direction only. **See Figure 4-16.** DC voltage powers motors, lamps, heaters, and other electrical devices in the same way as AC voltage. All DC voltage sources have a positive and negative terminal. Positive and negative terminals establish polarity in a circuit. Polarity is the positive (+) and negative (–) state of an object. All points in a DC circuit have polarity. A common source of DC voltage is a battery. In addition to obtaining DC directly from batteries, DC is also obtained by passing AC through a rectifier. A *rectifier* is an electrical component that converts AC to DC by allowing voltage and current to flow in only one direction. DC voltage obtained from a rectified AC voltage supply varies from almost pure DC voltage to half-wave DC voltage.

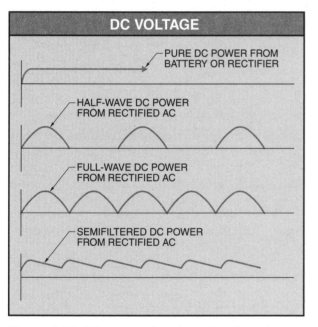

Figure 4-16. DC voltage is voltage that flows in one direction only.

Ground Circuits

A *ground circuit* is a conducting connection between electrical equipment and the earth. **See Figure 4-17.** A ground circuit is for safety and is not required for circuit operation. A circuit ground wire has green insulation or is bare and is connected to a metal grounding electrode buried in the ground close to the building. Ground circuits have 0 V potential. Current flows through the ground circuit to the ground if a hot wire touches the grounded metal of the equipment. This prevents the

operator from becoming part of the circuit and receiving an electrical shock. Current always follows the path of least resistance. Therefore, current flows in the ground circuit rather than through higher resistance of the operator. All industrial circuits are grounded for safety.

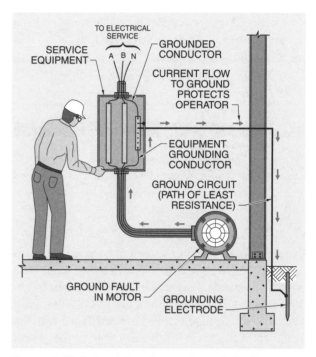

Figure 4-17. Industrial equipment ground circuits protect the operator from electrocution by providing a low-resistance path to ground.

Emergency Generators

An emergency generator is used to supply enough voltage to prevent critical systems from failing due to a loss of voltage. When sensors connected to incoming power lines sense a loss of voltage, a transfer switch opens, disconnecting the building from outside power. The emergency generator starts and generates voltage within a few seconds. Then, the switch connects the emergency generator to the building power distribution system. Emergency generators do not provide enough power for the whole facility, only enough for critical systems.

Larger emergency generators are usually powered by diesel engines and are required in hospitals, high-rise buildings, hotels, and wherever critical electrical systems are operated. Lubricating oil must be kept warm and circulating throughout the diesel engine when not in operation because diesel emergency generators must start and deliver voltage quickly.

Diesel emergency generator startup is tested weekly to ensure that it will start within the required time. Diesel emergency generator output is tested annually by connecting an electrical load equivalent to the output. The startup and shutdown sequence of the generators must be carefully planned so as not to overload the generator or cause damage to electrical systems in the facility. A generator that is unable to deliver the required power is repaired or replaced, or the load is reduced. The testing and maintenance of emergency generators is commonly regulated by health and safety authorities.

> **TECH TIP**
>
> *It is better to oversize a wire than undersize a wire. A larger wire (smaller AWG number) has less resistance than a smaller wire (larger AWG number).*

Wire Size

Every part of an electrical circuit must have the capacity to safely carry the current drawn by the load. Wire or switch contacts that are rated lower than the current drawn by the load can overheat and cause a fire. Wire size and switches are selected in accordance with the National Electrical Code® after determining the current load for the circuit.

The amount of current a wire can safely carry is determined by its size (diameter). Wires are sized using the American Wire Gauge (AWG) numbering system. The smaller the AWG number, the larger the diameter of the wire. Generally, the larger the diameter of the wire, the less resistance and more current it can safely carry. **See Figure 4-18.**

The three most common industrial wire sizes are No. 14, No. 12, and No. 10. Current ratings vary for insulation, type, application, and ambient temperature. Under typical conditions, No. 14 Cu (copper) wire is rated up to 20 A and must have a circuit breaker rated at 15 A or less. No. 12 Cu wire is rated up to 25 A and must have a circuit breaker rated at 20 A or less. No. 10 Cu wire is rated up to 35 A and must have a circuit breaker rated at 30 A or less.

Wires are commonly covered with an insulating material. An *insulator* is a material that has a high resistance. Insulators are commonly made from rubber, plastic, or paper. Excessive current in a wire can result in overheating, which can melt the insulation on the wire. The loss of insulation can lead to bare wires

contacting another conductor and causing a short circuit. A short circuit is current that leaves the normal current-carrying path by going around the load and back to the power source or ground. A short circuit offers little or no resistance.

WIRE SIZES				
AWG	**Solid**		**Stranded**	
	Diameter*	Resistance†	Diameter*	Resistance†
18	0.040	7.77	0.046	7.95
16	0.051	4.89	0.058	4.99
14	0.064	3.07	0.073	3.14
12	0.081	1.93	0.092	1.98
10	0.102	1.21	0.116	1.24
8	0.128	0.764	0.146	0.778
6	—	—	0.184	0.491
4	—	—	0.232	0.308
3	—	—	0.260	0.245
2	—	—	0.292	0.194
1	—	—	0.332	0.154
1/0	—	—	0.372	0.122
2/0	—	—	0.418	0.0967
3/0	—	—	0.470	0.0766
4/0	—	—	0.528	0.0608

* in in.
† in Ω/kft at 75°C (167°F)

Figure 4-18. The amount of resistance in a wire is determined by its size (diameter).

ELECTRICAL TEST EQUIPMENT

In most electrical circuits, measurements are taken to determine the presence and level of a measurable quantity, such as voltage, current, or resistance. The most common electrical test instrument is the digital multimeter. A digital multimeter (DMM) has a digital display and is used to measure two or more electrical values. DMMs accurately measure voltage, current, and resistance. This accuracy is critical when testing electrical and electronic circuits for correct levels of voltage. Some DMMs automatically select the most accurate range for displaying the voltage, current, or resistance measurements. In addition, accessories can be used to measure other quantities, such as temperature, pressure, or vacuum.

Voltage Measurement

Voltage measurements are taken in parallel with the component or section of the circuit being tested. The circuit must be energized to be tested. The two basic voltage measurements are source voltage and voltage drop. **See Figure 4-19.** *Source voltage* is potential electrical energy available to do work. For example, a measurement of 481 V at a disconnect means that there is 481 V of potential electrical energy available to do work when the switch is closed. When troubleshooting electrical systems, verifying correct voltage at the source is typically one of the first tests.

Voltage drop is the reduction of circuit voltage caused by the resistance of circuit conductors and loads. Voltage drop indicates the amount of electrical energy used by a load. As the disconnect switch is closed, there is a drop of about 476 V as the electrical power does work in the motor. Voltage drop through an intended load is expected.

Voltage also drops through unintended circuit resistances, such as conductors and connectors. Some voltage is needed to get the current through the circuit to the load. However, if too much voltage is dropped in the conductors and connections, then electrical loads will not operate at peak efficiency. Excessive voltage drop is caused by unintended sources of high resistance in the circuit, including damaged or loose connections, undersized conductors, and circuits that are too long. The NEC® includes recommendations for maximum voltage drops for different types of circuits, and manufacturers might specify an allowable voltage drop for circuits feeding their equipment. In general, a branch circuit should have a voltage drop of no more than 3% of the source voltage.

To measure the voltage drop of a circuit, the voltage drop across a load is subtracted from the source voltage reading. For example, if a 481 V circuit feeds a motor that uses 476 V, then the voltage drop is 5 V. If there are multiple loads in a series circuit, then the load voltage drop is the total of the individual voltage drops across each load. To express voltage drop as a percentage, divide by the source voltage. For example, 5 V dropped in a 481 V circuit is 1% of the voltage (5 V ÷ 481 V = 0.01 or 1%).

If the voltage drop is too great, then the causes of the voltage drop must be identified and corrected. Any loose connectors and damaged conductors should be identified. If necessary, the length of the circuit run should be measured and replaced with larger conductors that have less resistance.

Current Measurement

Current is measured with a clamp-on ammeter, a DMM connected in series (in-line ammeter), or a DMM with a clamp-on current probe accessory. **See Figure 4-20.** A *clamp-on ammeter* is a meter that measures current in a circuit by measuring the strength of the magnetic field around a single conductor. A DMM may be connected in series when taking current measurements less than 10 A. A DMM is normally connected in series in very low current electronic circuits of less than 250 mA. Current flows through the meter as it becomes part of the circuit. The current rating of the DMM must not be exceeded. Most industrial circuits have current levels too high to measure with a DMM connected in series. High current levels are measured using a clamp-on ammeter or a DMM with a clamp-on current probe accessory.

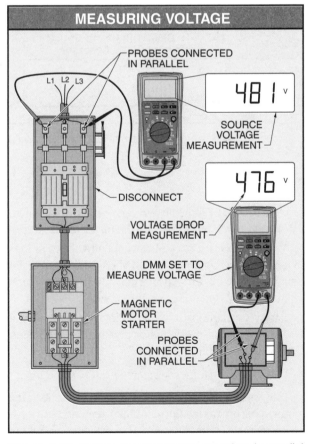

Figure 4-19. Voltage measurements are taken in parallel around the component or section of the circuit being tested.

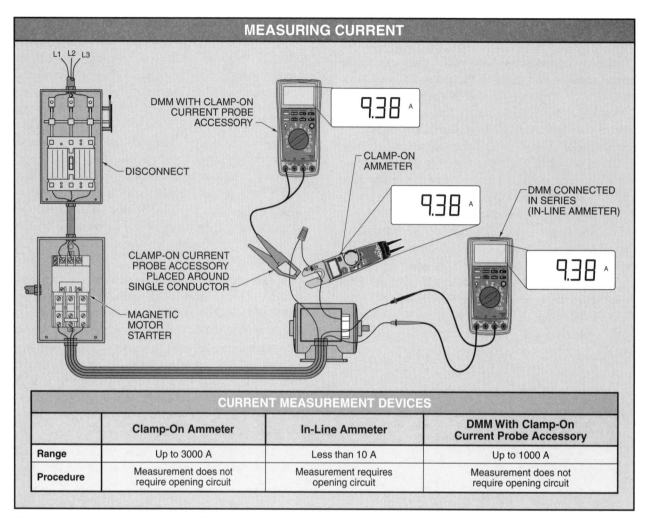

Figure 4-20. Current is measured with a clamp-on ammeter, a DMM connected in series (in-line ammeter), or a DMM with a clamp-on current probe accessory.

Resistance Measurement

Resistance measurements are taken with the circuit de-energized and the component removed from the circuit. The component being tested is checked to ensure that it is not connected to any other component that could alter the resistance. Resistance is displayed on the meter as ohms (Ω), kilohms (kΩ), or megohms (MΩ). The meter leads are moved to various pairs of wires on the component to test their resistance. **See Figure 4-21.**

A DMM set to measure resistance can also be used to test continuity. *Continuity* is the presence of a complete path for current flow. A DMM emits an audible beep for little or resistance, indicating continuity. For example, a short length of copper wire should read approximately 0 Ω. Copper wire reading infinite resistance

or overload (OL) indicates an open in the wire which causes too much resistance for the meter to measure.

Warning: Never exceed current and voltage ratings of a DMM. Never connect a DMM set to measure resistance on a live circuit. Always consult the meter user's manual before operating the equipment.

Safety precautions must be followed when using a DMM to prevent injury and/or damage to equipment. DMM safety precautions include:

• Use a meter that meets accepted safety standards.

• Use a meter with fused current jacks and check the fuses before taking current measurements.

• Inspect test leads for physical damage before taking a measurement.

• Use the meter to check the condition of the test leads.

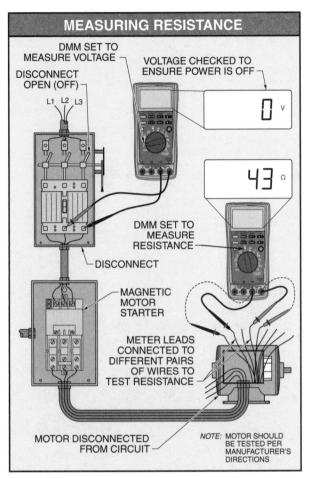

MEASURING RESISTANCE

DMM SET TO MEASURE VOLTAGE

DISCONNECT OPEN (OFF)

L1 L2 L3

VOLTAGE CHECKED TO ENSURE POWER IS OFF

0 V

43 Ω

DMM SET TO MEASURE RESISTANCE

DISCONNECT

MAGNETIC MOTOR STARTER

METER LEADS CONNECTED TO DIFFERENT PAIRS OF WIRES TO TEST RESISTANCE

MOTOR DISCONNECTED FROM CIRCUIT

NOTE: MOTOR SHOULD BE TESTED PER MANUFACTURER'S DIRECTIONS

Figure 4-21. Resistance measurements are taken with the component removed from the circuit and the circuit de-energized.

- Use only test leads that have shrouded connectors and finger guards.
- Use only meters with recessed jacks.
- Select the proper function and range for the measurement.
- Ensure the meter is in good operating condition.
- Follow all equipment safety procedures.
- Always disconnect the hot (red) test lead first.
- Never work alone.
- Use a meter that has overload protection on the resistance function.
- Turn the power OFF before connecting a meter in a circuit when measuring current without a clamp-on current probe accessory.
- Use the appropriate equipment such as high-voltage probes and high-current clamp-on current probe accessories in high current and high voltage situations.

POWER QUALITY

An *electrical system* is a system that produces, transmits, distributes, and delivers electrical power to satisfactorily operate electrical loads. When electrical power is properly supplied to a load, the load should operate for years without a problem. Damage to electrical equipment occurs when electrical power is not supplied at the proper level (voltage), amount (current), type (single-phase or three-phase, AC or DC), or condition (purity). Before a load is connected or when a load is serviced, the quality of the incoming power must be tested to ensure proper system and load operation. Power quality must also be tested as part of a preventive maintenance program.

Voltage Changes

Power line frequency is constant (60 Hz in the U.S.) and current changes as loads are added and removed. Voltage in a power distribution system should be within the acceptable range of +5% to –10% of nominal. **See Figure 4-22.** Large voltage fluctuations, however, are caused by storms, electrical faults, transformers that are overloaded or unbalanced, or inadequate wiring. Small voltage fluctuations typically do not affect equipment performance, but voltage fluctuations outside the normal range can cause circuit and load problems. Common voltage problems are low voltages, voltage surges, and power interruptions.

Fluke Corporation
Electrical equipment must be locked out and tagged out before maintenance or servicing is performed.

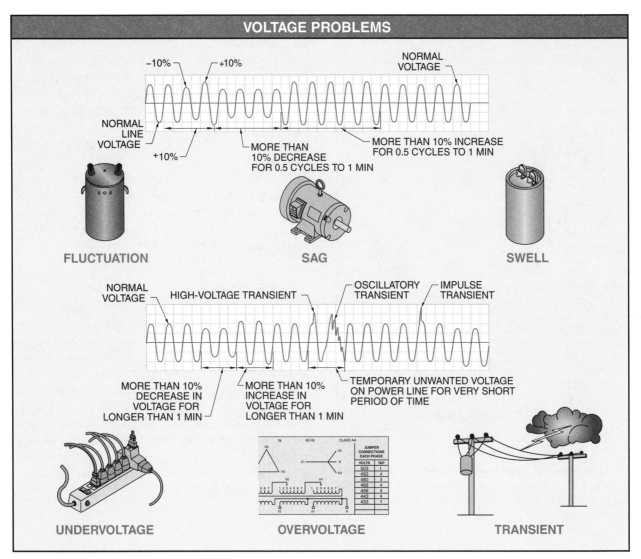

VOLTAGE PROBLEMS

Figure 4-22. Voltage problems can be difficult to detect because some occur intermittently.

Low Voltages. A *voltage sag* is a drop in voltage of more than 10% (but not to 0 V) below the normal rated line voltage that lasts from 0.5 cycles up to 1 min. Voltage sags commonly occur when high-current loads are turned ON and the voltage on the power line drops below the normal voltage fluctuation (–10%) for a short period of time. Voltage sags are often followed by voltage swells as voltage regulators overcompensate for the voltage sag.

Undervoltage is a drop in voltage of more than 10% (but not to 0 V) below the normal rated line voltage that lasts longer than 1 min. Undervoltages are commonly caused by overloaded transformers, undersized conductors, conductor runs that are too long, too many loads on a circuit, or peak power usage periods

(brownouts). A *brownout* is a drop in voltage by the power company to conserve power during times of peak usage or excessive loading of the power distribution system.

Voltage Surges. A *voltage swell* is an increase in voltage of more than 10% above the normal rated line voltage that lasts from 0.5 cycles up to 1 min. Voltage swells commonly occur when large loads are turned OFF and voltage on the power line increases above the normal voltage fluctuation for a short period of time. Voltage swells are not as common as voltage sags. However, voltage swells can be more destructive than voltage sags because voltage swells damage electrical equipment in very short periods of time.

Overvoltage is an increase in voltage of more than 10% above the normal rated line voltage that lasts longer than 1 min. Depending on the cause, voltage increases above the normal voltage fluctuation of +5% can occur for long periods of time. Overvoltages are caused when loads are near the beginning of a power distribution system or when taps on a transformer are not wired correctly.

Transients. A *transient voltage* is a temporary, undesirable voltage spike, ranging from a few volts to several thousand volts and lasting from a few microseconds up to a few milliseconds. Transient voltages are caused by the sudden release of stored energy due to lightning strikes, unfiltered electrical equipment, contact bounce, arcing, and generators being switched ON and OFF. High-voltage transients can permanently damage circuits or electrical equipment.

Transient voltages differ from low voltages and voltage surges by being larger in magnitude and shorter in duration, having a steep (short) rise time, and being erratic. Lightning is the most common source of transient voltages on utility power distribution systems. In most cases, the transient voltage is dissipated by the utility company grounding and protection systems after the lightning has traveled a certain distance. However, unprotected equipment is severely damaged when lightning strikes close to equipment that is in use, or when the system is not properly protected.

Voltage Measurements. To yield usable information about voltage changes, voltage measurements must be taken and recorded over time, at various times of the day and at various places within the facility. **See Figure 4-23.** Voltmeters (or multimeters) with a MIN MAX Recording mode are able to capture and display low-voltage conditions, though they do not indicate when or for how long a low-voltage condition existed. Typically, a power analyzer meter with recording mode is the best meter to use for detecting voltage problems. When voltage fluctuations are found to be more than ±8%, a voltage regulator must be added to the system. A *voltage regulator* is a device that provides precise voltage control to protect equipment from voltage fluctuations.

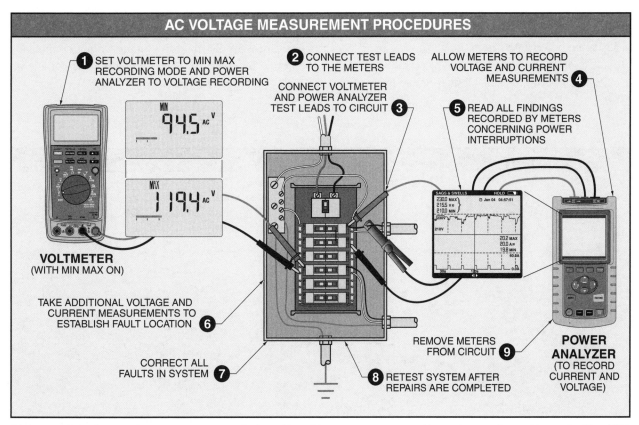

Figure 4-23. Voltage variations are more easily found by taking measurements with a voltmeter (or multimeter) with a MIN MAX Recording mode, or with a power analyzer.

When transient voltages are identified as a problem within a facility, transient voltage surge suppressors must be used. A *surge suppressor* is a device that limits the intensity of voltage transients that occur on a power distribution system.

Power Interruptions

Power interruptions are classified into standard industry categories as momentary, temporary, and sustained interruptions. **See Figure 4-24.** A *momentary power interruption* is a decrease in voltage to 0 V that lasts from 0.5 cycles up to 3 sec. All power distribution systems encounter momentary power interruptions during normal operation. Momentary power interruptions are caused by lightning strikes (nearby strikes) or some types of switching operations.

A *temporary power interruption* is a decrease in voltage to 0 V that lasts from 3 sec to 1 min. A temporary power interruption can be caused by the gap in time between a power interruption and when a circuit breaker resets or a backup power supply (generator) takes over, or if someone accidentally opens a circuit by switching the wrong circuit breaker or switch gear. When power is not restored, a temporary power interruption becomes a sustained power interruption.

A *sustained power interruption* is a decrease in voltage to 0 V on all power lines that lasts for more than 1 min.

Even the most reliable power distribution systems have a complete loss of power at some time. Sustained power interruptions (outages) are commonly the result of storms or circuit breakers tripping due to damaged equipment.

Several different test instruments are used to detect power interruptions. Voltmeters with MIN MAX Recording functions are used to record power interruptions over time, but cannot indicate when or for how long an interruption took place. Recording meters are used to show exactly when power was interrupted and for what length of time.

Voltage Unbalance

Voltage unbalance is the unbalance that occurs when the voltages of a 3ϕ power supply or the terminals of a 3ϕ load are not equal. When voltage to a 3ϕ motor is unbalanced, one or two windings overheat, causing thermal deterioration of the windings. Voltage unbalance also results in a current unbalance. Line (L1, L2, and L3) voltages must be tested for voltage unbalance periodically and during all service calls. **See Figure 4-25.** Voltage unbalance should not be more than 1%. The primary cause of voltage unbalances of less than 2% is too many single-phase loads on one phase of a 3ϕ distribution system. Voltage unbalance is found by measuring and comparing the voltage on each of the power lines.

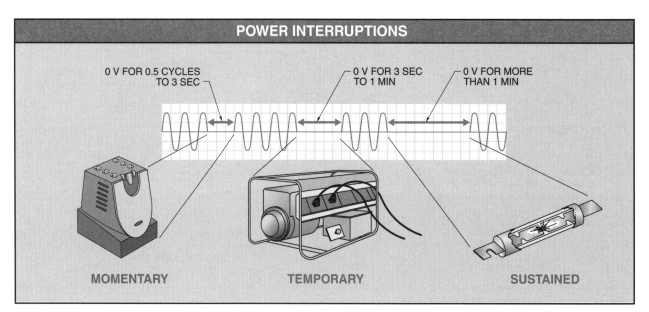

Figure 4-24. Power interruptions to an electrical system can be momentary, temporary, or sustained.

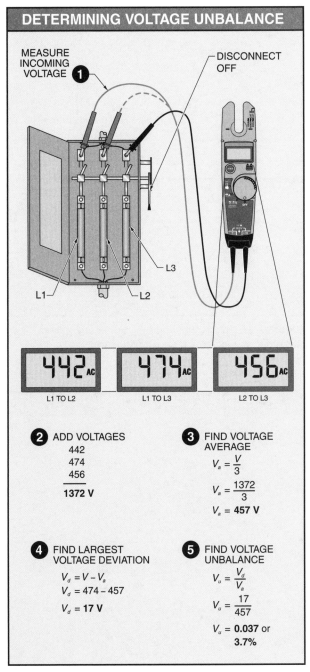

DETERMINING VOLTAGE UNBALANCE

MEASURE INCOMING VOLTAGE **1**

DISCONNECT OFF

L3

L1   L2

442 AC   474 AC   456 AC

L1 TO L2   L1 TO L3   L2 TO L3

2 ADD VOLTAGES
442
474
456
‾‾‾‾‾
1372 V

3 FIND VOLTAGE AVERAGE

$V_a = \dfrac{V}{3}$

$V_a = \dfrac{1372}{3}$

$V_a = \textbf{457 V}$

4 FIND LARGEST VOLTAGE DEVIATION

$V_d = V - V_a$

$V_d = 474 - 457$

$V_d = \textbf{17 V}$

5 FIND VOLTAGE UNBALANCE

$V_u = \dfrac{V_d}{V_a}$

$V_u = \dfrac{17}{457}$

$V_u = \textbf{0.037}$ or **3.7%**

Figure 4-25. Voltage unbalance should not be more than 1%.

Single phasing is the complete loss of one phase on a 3ϕ power supply. Single phasing is the maximum condition of voltage unbalance. Common causes of single phasing include blown fuses, mechanical failure within switching equipment, or a lightning strike on one of the power lines. Single phasing can go undetected on

systems because most 3ϕ motors continue to operate on two phases, though they will not start on two phases. When a motor is not properly protected, single phasing will cause the motor to eventually burn out because the motor draws all required current from the two remaining powered lines.

Measuring voltage at a motor does not always detect a single phasing condition because the open winding in the motor acts like a transformer and generates a voltage that is almost equal to the phase voltage that is lost. To test for a motor or load that is single phasing, both voltage and current measurements must be taken.

Current Unbalance

Current unbalance is the unbalance that occurs when current is not equal on each of the three power lines of a 3ϕ power supply or a 3ϕ load. Small voltage unbalances cause high current unbalances. High current unbalances cause excessive heat, resulting in insulation breakdown. Typically, for every 1% of voltage unbalance, current unbalance is 4% to 8%.

Current unbalances should never exceed 10%. Any time a current unbalance exceeds 10%, electricians must test for voltage unbalance. Likewise, any time there is a voltage unbalance of more than 1%, electricians must test for current unbalance. Current unbalance is found the same way as voltage unbalance is found, except current measurements are taken. **See Figure 4-26.**

Phase Unbalance

When 3ϕ power is generated and distributed, the three power lines are 120° out of phase with each other. *Phase unbalance* is the unbalance that occurs when 3ϕ power lines are more or less than 120° out of phase. **See Figure 4-27.** Phase unbalance of a 3ϕ power system occurs when single-phase loads are applied, causing one or two of the lines to carry more or less of the load. Electricians must balance the loads on 3ϕ power systems during installation.

Improper Phase Sequence

Improper phase sequence is the changing of the sequence of any two phases in a 3ϕ system or circuit. **See Figure 4-28.** Phase sequence determines the direction a motor shaft will rotate, so improper phase sequence reverses motor rotation. Reversing motor rotation can damage driven machinery or injure personnel.

DETERMINING CURRENT UNBALANCE

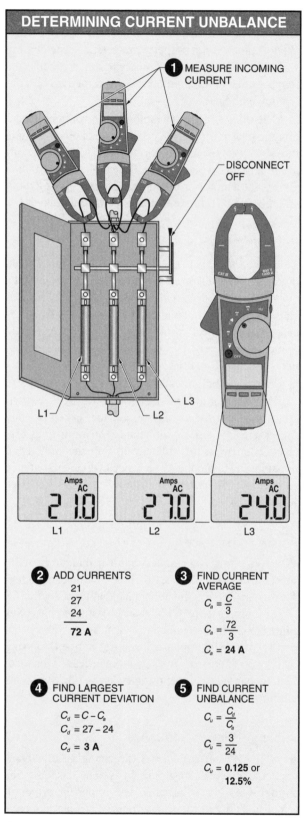

❶ MEASURE INCOMING CURRENT

DISCONNECT OFF

L1
L2
L3

Amps AC
21.0
L1

Amps AC
27.0
L2

Amps AC
24.0
L3

❷ ADD CURRENTS
21
27
24
‾‾‾‾‾
72 A

❸ FIND CURRENT AVERAGE
$C_a = \dfrac{C}{3}$
$C_a = \dfrac{72}{3}$
$C_a = \mathbf{24\ A}$

❹ FIND LARGEST CURRENT DEVIATION
$C_d = C - C_a$
$C_d = 27 - 24$
$C_d = \mathbf{3\ A}$

❺ FIND CURRENT UNBALANCE
$C_u = \dfrac{C_d}{C_a}$
$C_u = \dfrac{3}{24}$
$C_u = \mathbf{0.125}$ or **12.5%**

Figure 4-26. Small voltage unbalances cause high current unbalances.

PHASE UNBALANCE

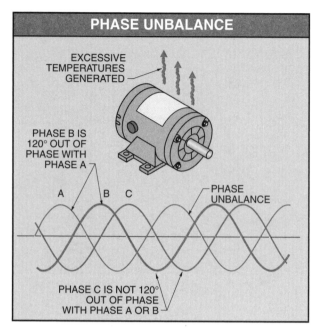

EXCESSIVE TEMPERATURES GENERATED

PHASE B IS 120° OUT OF PHASE WITH PHASE A

A   B   C

PHASE UNBALANCE

PHASE C IS NOT 120° OUT OF PHASE WITH PHASE A OR B

Figure 4-27. Phase unbalance occurs when 3φ power lines are more or less than 120° out of phase.

IMPROPER PHASE SEQUENCE

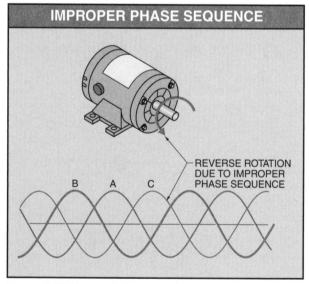

REVERSE ROTATION DUE TO IMPROPER PHASE SEQUENCE

B   A   C

Figure 4-28. Reversal of any two of the phases of a 3φ power supply will cause 3φ motors to reverse their rotation.

TECH TIP

Motors are rated for operation at a specific frequency and their performance is affected when the frequency varies. A motor operates satisfactorily with a frequency variation of up to ±5% from the frequency rating listed on the motor nameplate.

Phase reversals typically occur when modifications are made to power distribution systems or when maintenance is performed on electrical conductors or switching equipment. Phase reversal protection is required on all personnel transportation equipment such as moving walkways, escalators, and ski lifts, or other equipment damaged by accidental phase reversal.

A *phase sequence tester* is a test instrument used to determine which of the 3ϕ power lines are powered and which power line is phase A, which is phase B, and which is phase C. Phase sequence testers are available in two types—phase sequence tester only, or phase sequence tester and motor rotation tester combined.

TROUBLESHOOTING ELECTRICAL SYSTEMS

Troubleshooting is the systematic elimination of the various parts of a system, circuit, or process to locate a malfunctioning part. Troubleshooting electrical systems requires an organized, sequenced approach. Common electrical problems include no source voltage, open circuits, short circuits, low voltage, excessive current, abnormally high resistance, and incorrect wiring. Troubleshooting requires the use of electrical test equipment, drawings and diagrams, and manufacturer's specifications. Testing charts may be used to quickly and easily determine system problems.

Testing Charts

A *testing chart* is a chart that lists possible causes of a problem along with testing procedures. Testing charts simplify the troubleshooting process by listing causes from most likely to least likely. **See Figure 4-29.** The component and test method are listed next to each cause. Before the test is performed, the expected results are written on the chart. The component is good if it is operating within manufacturer's recommended specifications. The component is bad if it is not operating within manufacturer's recommended specifications.

Each test is performed and results are recorded on the testing chart. The component is inspected or tested further if it is bad. The next component to be tested is based on the results of the previous testing. A testing chart provides a useful record of the troubleshooting process required to solve complex problems. A testing chart can be used with troubleshooting charts and manufacturer's service procedures.

Tie-Down Testing Method

The tie-down testing method is commonly used when troubleshooting with a DMM. The *tie-down testing method* is a testing method in which one test instrument probe is connected to a point on a circuit (L1 or L2) and the other probe is moved along the circuit to test

TESTING CHART

1. Determine possibly faulty component(s) to test.
2. Determine test method: visual inspection, manual operation, or voltage, resistance, current reading, etc.
3. Determine expected results if component is good or bad.
4. Perform test and record results.

Problem: Light Does Not Turn ON				
Cause	**Component Tested**	**Test Method**	**Expected Results**	**Test Results**
Most Likely ↑ ↓ Least Likely	Bulb	Remove and replace	New light turns ON if old one was bad	New bulb did not light
	Voltage	Take voltage reading at bulb receptacle	Good voltage should be 120 V	No voltage
	Circuit Breaker	Inspect and reset circuit breaker	Breaker should not be tripped. If it is reset it and the light should turn ON	Breaker tripped

Figure 4-29. A testing chart simplifies the troubleshooting process by listing possible causes of a problem and testing procedures from most likely to least likely.

sections sequentially. **See Figure 4-30.** The tie-down method allows maintenance personnel to work quickly on a familiar circuit that is small enough for the probes to reach across. Generally, the tie-down method is a more advanced approach and the expected readings must be known to be successful.

The tie-down testing method allows a troubleshooter to work quickly on a familiar circuit that is small enough for the test probes to reach across the test points.

Testing Source Voltage

Source voltage provides the power to operate components in a circuit. De-energized circuits are tested in sequence beginning with the source voltage, circuit path, and load. Source voltage for a circuit is commonly produced by a transformer. **See Figure 4-31.** A transformer consists of two or more coils of insulated wire wound on a laminated steel core. Transformers operate on the principle of one coil inducing a voltage into another coil. A strong, pulsating magnetic field is produced when AC is applied to the primary coil. The *primary coil* is the transformer coil to which the incoming voltage is supplied. The steel core routes the pulsating magnetic field to the secondary. The pulsating magnetic field induces a voltage into the secondary coil. The *secondary coil* is the transformer coil that supplies output voltage to the circuit. It is the coil in which the voltage is induced.

A *step-up transformer* is a transformer in which the secondary coil has more turns of wire than the primary coil. A step-up transformer produces a higher voltage on the secondary coil than the voltage applied to the primary. A *step-down transformer* is a transformer in which the secondary coil has fewer turns of wire than

the primary coil. A step-down transformer produces a lower voltage on the secondary coil than the voltage applied to the primary.

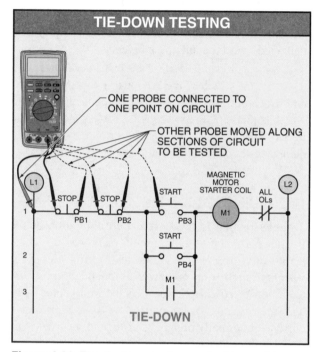

Figure 4-30. The tie-down testing method is used to systematically eliminate the various parts of a circuit.

The transformer supplying source voltage to a circuit is checked first when testing source voltage. The fully loaded voltage from the secondary side of the transformer should be within ±5% of the rating listed on the transformer frame. The primary voltage is checked if there is no voltage or low voltage in the secondary. The transformer is defective if the primary voltage is correct and the secondary voltage is incorrect. If the secondary voltage is approximately half of its rating, the transformer may be connected incorrectly. The transformer connections should be checked against the wiring schematic located on the transformer frame.

Overheating can cause transformer coils to melt open, short to the frame, or short together. Defective coils cannot be repaired and must be replaced. A transformer removed from power is tested with a DMM set to measure resistance. **See Figure 4-32.** A good primary or secondary coil offers some resistance. A bad coil indicates infinite resistance, overload (OL), or near zero resistance. There should be no continuity between the iron core and either coil and no continuity between the primary or secondary.

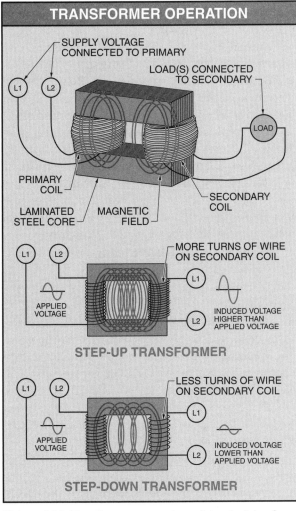

Figure 4-31. Transformers operate on the principle of one coil inducing a voltage into another coil.

The panelboard circuit breaker is checked if there is no primary voltage to the circuit transformer. **See Figure 4-33.** If the circuit breaker is tripped, the reason why it tripped must be determined before it is reset. The power coming into the circuit breaker panelboard is checked if there is no power coming into the circuit breaker. Loss of power is usually caused by a blown fuse or tripped circuit breaker upstream of the panelboard. The power company must be contacted if there is loss of voltage or low voltage into the facility.

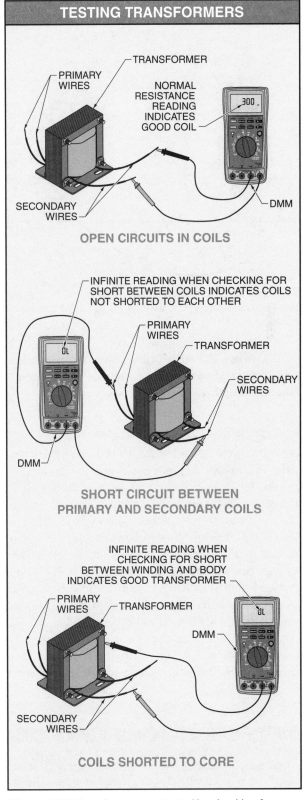

Figure 4-32. Transformers are tested by checking for open circuits in the coils, short circuits between the primary and secondary coils, and coils shorted to the core.

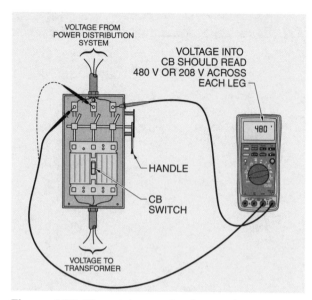

Figure 4-33. The power coming into a circuit breaker panelboard is checked if there is no power at the circuit breaker.

Warning: Do not work on high voltage unless trained and authorized to do so. Special test equipment is required for high voltage circuits.

Overcurrent Protection Devices

An overcurrent protection device is used in an electrical circuit to provide protection from short circuits and overloads. An *overcurrent protection device (OCPD)* is a device that automatically opens a circuit to prevent damage from a high-current condition. Fuses and CBs are OCPDs designed to automatically stop the flow of current in a circuit that has a short circuit or is overloaded. **See Figure 4-34.** A *fuse* is an OCPD with a fusible link that melts and opens the circuit when an overload condition or short circuit occurs. A *circuit breaker* is an OCPD with a resettable mechanism that automatically opens the circuit when an overload condition or short circuit occurs. Fuses and CBs are widely used in power distribution systems and individual pieces of equipment.

> **TECH TIP**
> For OCPD contacts that must interrupt high currents, refractory materials with antiwelding characteristics such as molybdenum and tungsten are used. These materials are essential for contacts that must close in on short-circuit currents and open at a later time under normal operating force.

In a normal circuit, the amount of current flow is determined by the size of the load. However, if there is a problem such as a short circuit or overloaded circuit, higher-than-normal current flows, resulting in an overcurrent condition. An *overcurrent condition* is a condition that occurs when the amount of current flowing in a circuit exceeds the design limit of the circuit. An overcurrent condition is dangerous and can cause damage to the equipment.

Fuses are connected in series with the circuit conductors. Fuses contain a metal strip that quickly melts and opens the circuit when the current flowing through the strip is greater than the rating of the fuse. Fuses are non-time delay or time delay. A *non-time delay fuse* is a fuse that opens the circuit almost instantly. A *time delay fuse* is a fuse that opens a short circuit almost instantly, but allows small overcurrents to exist for short periods of time. The metal strip melts at a specific temperature caused by current exceeding the rated limit of the fuse. In a time-delay fuse, a resistor slows the heating of the metal strip, allowing for temporary overcurrents.

Warning: Never replace a blown fuse with one of a higher rating.

CBs contain a spring-loaded electrical contact that opens the circuit. The spring is used to open and close the contacts with a fast snap-action. A handle is added to the contact assembly so the contacts may be manually opened and closed. The contacts are automatically opened on an overcurrent by a bimetal strip and/or an electromagnetic tripping device. The CB has one stationary contact and one movable contact. The movable contact is attached to a spring that provides a fast snap-action when tripped.

The bimetal strip is made of two dissimilar metals that expand at different rates when heated. The strip bends when heated and opens the contacts. The bimetal strip is connected in series with the circuit and is heated by the current flowing through it. Like the bimetal strip, the electromagnetic device is connected in series with the circuit. As current passes through the coil, a magnetic field is produced. The higher the current, the stronger the magnetic field. The magnetic field opens the contacts when it becomes strong enough.

A good fuse has a closed internal metal strip. A bad (blown) fuse has an open internal metal strip. A DMM set to measure resistance is used to test fuses. A good non-time delay fuse has no resistance. A good time delay fuse has a small resistance. A blown fuse of either type indicates infinite resistance or overload (OL).

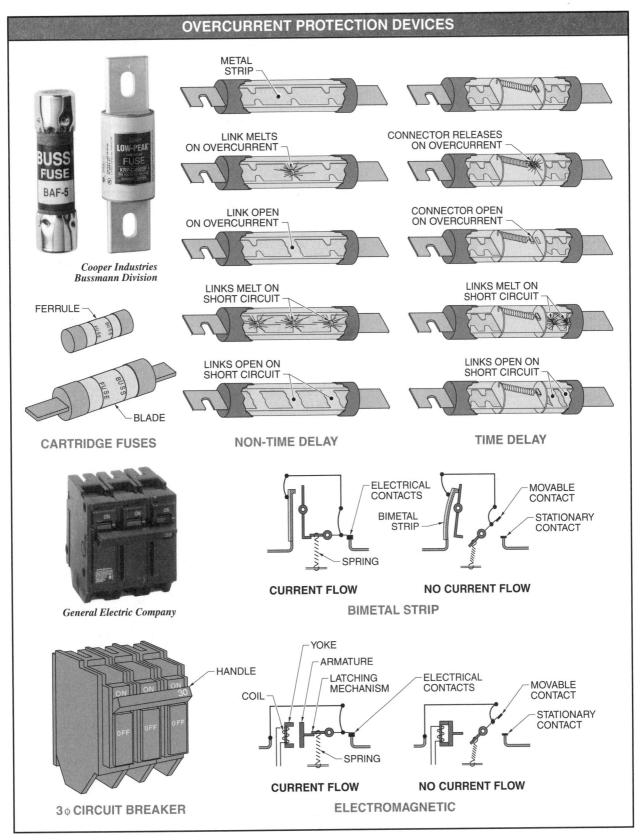

OVERCURRENT PROTECTION DEVICES

Cooper Industries Bussmann Division

FERRULE

BLADE

CARTRIDGE FUSES

METAL STRIP

LINK MELTS ON OVERCURRENT

LINK OPEN ON OVERCURRENT

LINKS MELT ON SHORT CIRCUIT

LINKS OPEN ON SHORT CIRCUIT

NON-TIME DELAY

CONNECTOR RELEASES ON OVERCURRENT

CONNECTOR OPEN ON OVERCURRENT

LINKS MELT ON SHORT CIRCUIT

LINKS OPEN ON SHORT CIRCUIT

TIME DELAY

General Electric Company

ELECTRICAL CONTACTS

BIMETAL STRIP

SPRING

CURRENT FLOW

MOVABLE CONTACT

STATIONARY CONTACT

NO CURRENT FLOW

BIMETAL STRIP

HANDLE

3 ϕ CIRCUIT BREAKER

YOKE

ARMATURE

LATCHING MECHANISM

COIL

SPRING

CURRENT FLOW

ELECTRICAL CONTACTS

MOVABLE CONTACT

STATIONARY CONTACT

NO CURRENT FLOW

ELECTROMAGNETIC

Figure 4-34. Fuses and circuit breakers open a circuit when an overcurrent condition or short circuit occurs.

Fuses connected to a voltage supply may also be tested using a DMM set to measure voltage. **See Figure 4-35.** A DMM placed across a good fuse indicates 0 V. A DMM placed across a bad fuse indicates source voltage. One of three fuses in a 3ϕ circuit is tested by taking a phase-to-phase reading across two fuses. The fuse being tested is the one with the probe on the load side. A good fuse indicates the phase-to-phase voltage. A bad fuse indicates a voltage other than the source voltage. (The level of this voltage reading varies depending on the level of the source voltage being tested.) This non-source voltage reading can cause confusion but can also be used to find multiple opens that are in series in electrical circuits. Analog meters and less sensitive DMMs may not display this type of voltage.

CBs are tested by placing a DMM set to measure voltage across the CB contacts. A good CB indicates 0 V when the contacts are closed and source voltage when open. A CB suspected of being bad is removed from the circuit and tested using a DMM set to measure resistance. A good CB indicates nearly 0 Ω when the contacts are closed and infinite resistance or overload (OL) when open.

A CB may be in the ON, TRIPPED, or OFF position. If it appears undamaged and is cool, reset a tripped CB by moving the handle to OFF and then to ON. CBs are designed so they cannot be held in the ON position if an overload or short is present. Check the voltage of the reset CB if resetting the CB does not restore power. Replace all faulty CBs because they are important safety devices that help prevent circuit overheating and fires.

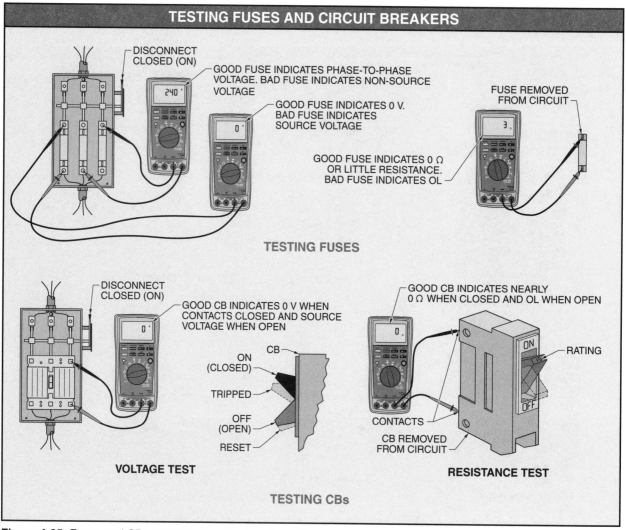

Figure 4-35. Fuses and CBs are tested by checking the voltage or resistance across them.

Open Circuits

An open circuit has an incomplete path that prevents current flow. An open in a series circuit deenergizes that circuit. For example, a circuit is completely de-energized if switch contacts are open. Switches are tested by toggling the switch to check if the contacts close.

The switch is then tested with a DMM set to measure voltage. **See Figure 4-36.** A good switch indicates source voltage when open and 0 V when closed. A bad switch indicates source voltage when open and closed.

A switch is tested with a DMM set to measure resistance or a jumper wire if the results of the voltage test indicate the switch is bad. A jumper wire is placed in parallel around the bad switch and the circuit is energized. The jumper wire closes the circuit, energizing the load.

An open in an individual branch of a circuit de-energizes only that branch. That branch of the circuit is tested with a DMM set to measure voltage. For example, the DMM might be placed across a section of wire to determine if there is a break in that part of the wire. This test is often taken across wire connection points. The meter indicates 0 V if the wire has no break. The meter indicates source voltage if the wire is broken and there is no other open in the branch. If there is an open in the branch, such as a switch, the meter reads non-source voltage across the second open. The meter reads non-source voltage when bridging all opens.

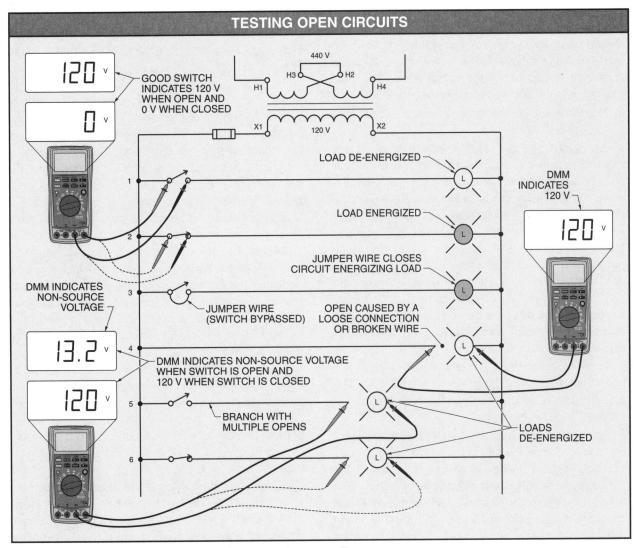

Figure 4-36. An open circuit does not provide a path for current flow.

Warning: Jumper wires can cause equipment to start unexpectedly and must be removed from the circuit when no longer needed for testing.

Short Circuits

A short circuit causes current to leave the normal current-carrying path by going around the load and back to the power source or to ground. Short circuits may be caused by incorrect wiring, equipment damage, or conductive debris that gets inside the equipment.

Short circuits may be dead shorts or operational shorts. A *dead short* is a short circuit that opens the OCPD as soon as the circuit is energized. An *operational short* is a short circuit that causes the circuit to malfunction and may not open the OCPD.

In a dead short, the CB is reset or the fuse replaced. The circuit is inspected for the location of the short if the CB trips or fuse blows again. The location of the short is indicated by signs of overheating, such as burn marks or discolored insulation. A DMM set to measure resistance is used to test the circuit if the dead short location is not found visually. **See Figure 4-37.**

Warning: Ensure the circuit is de-energized when measuring resistance.

All open contacts should be closed so the meter reads total circuit resistance. A dead short reads near 0 Ω. Isolate one branch of the circuit at a time by disconnecting a wire on the branch circuit. Reconnect the branch if the near 0 Ω total resistance does not change after isolating the branch. Continue by isolating the next branch. The branch that causes the DMM reading to jump from near 0 Ω to a higher resistance contains the dead short. Inspect that branch for signs of overheating and crossed, frayed, or loose wires. Verify the short by finding the location of the near 0 Ω resistance (usually in a load). A normal resistance reading indicates the dead short is repaired.

Large, complex circuits are tested one section at a time to determine which section contains the short. The individual branches of the section are then tested to find the exact location of the short.

An operational short causes the circuit to malfunction but does not draw enough current to open the circuit overcurrent protection device. For example, if two lights are energized when only one should be energized, two wires of the circuit are touching. Operational shorts are located by studying circuit drawings and diagrams. Probable causes and locations are checked for incorrect wiring or crossed wires.

Circuit Wiring

Incorrect wiring causes a circuit to malfunction with loads being energized at the wrong time or not at all. Circuit diagrams can be used to identify the problem in a circuit. All connections must be verified against circuit diagrams starting from simple to complex. For example, incorrectly numbered wires or swapped wires, such as number six for number nine, are checked against the circuit diagram and specifications.

Before a circuit can be wired, repaired, or checked, maintenance personnel must understand the operation of the circuit, the devices used to energize and de-energize it, the sequence of events, and any timing or counting functions. A line diagram details the electrical operation of a circuit. **See Figure 4-38.** A wiring diagram details circuit wiring but not its electrical operation. For example, a standard start-stop station used to control a 3φ motor is one of the most common motor control circuits.

The circuit line diagram indicates that pressing the start pushbutton energizes the magnetic motor starter contact coil (M). The magnetic motor starter contact coil closes the magnetic motor starter contacts (M), which energizes the motor. At the same instant, the M contacts on rung 2 of the line diagram close, providing a holding circuit for the start pushbutton. All rungs of the ladder diagram are numbered. The M contacts on rung 2 provide an alternate path for current flow (the holding circuit) that keeps the M coil energized when the start pushbutton is released.

The magnetic motor starter controls the opening and closing of any contacts attached to that motor starter. All normally open (NO) M contacts close when the magnetic motor starter is energized and open when the magnetic motor starter is de-energized. The contact reference numbers on the right side of the line diagram indicate which contacts on the other rungs are controlled by the load on that rung. Numbers that are underlined reference normally closed (NC) contacts and numbers that are not underlined reference NO contacts. The magnetic motor starter remains energized until the stop pushbutton is pressed or the overloads (all OLs) open due to excessive current flow to the motor. When either of these events occurs, the magnetic motor starter is de-energized, opening all M contacts and de-energizing the motor. At the same instant, the M contacts on rung 2 open, removing the holding circuit. The circuit does not re-energize unless the start pushbutton is pressed again. Loss of power also de-energizes the complete circuit and it does not re-energize when power is restored until the start pushbutton is pressed again.

Motor control equipment is wired using a standard numbering system. Each wire group in the line diagram is given a number. A *wire group* is any set of wires that are connected directly without being broken by a device such as a pushbutton, contact, or coil. Numbers are assigned from left to right and from top to bottom of the diagram. Numbers are not usually assigned to factory-installed wiring inside equipment. Wire groups are referenced when troubleshooting. A DMM should read continuity between the ends of a wire group.

After a line diagram is numbered by wire group, the individual wires are installed starting with number 1.

Most magnetic motor starters and start/stop pushbuttons are numbered so wire number 1 is connected between the incoming power line and 1 on the start/stop pushbutton. A correctly installed wire is checked off with a check mark or X on the line diagram. Wire 2 is connected between 2 on the stop pushbutton and 2 on the magnetic motor starter holding circuit contacts. Each wire is installed in succession until the wiring is completed. The same method is used to check for faulty wiring or to work through a circuit when troubleshooting. Each wire is traced and inspected in succession until the problem is located and corrected.

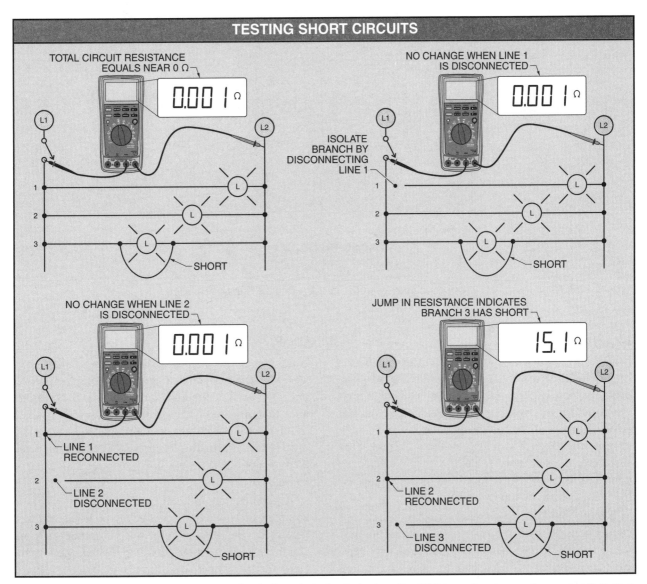

Figure 4-37. The branch containing the dead short is indicated when the DMM reading jumps from near 0 Ω to a higher resistance.

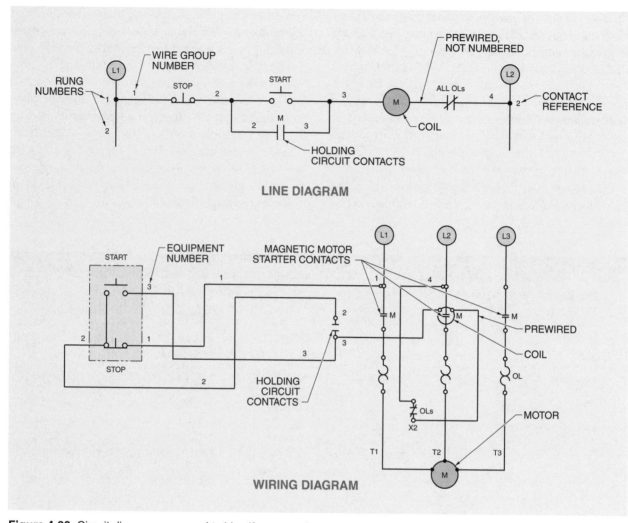

Figure 4-38. Circuit diagrams are used to identify proper circuit wiring and circuit operation.

Low Voltage at Load

Low voltage at a load causes the load to operate erratically or not at all. The voltage drop across the problem load is checked if the source voltage is at the correct level. **See Figure 4-39.** The circuit path is checked for unwanted resistance if the voltage drop is more than 5% below the circuit rating. This resistance is usually caused by dirty contacts, loose connections, wire that is too small, frayed wires, or loads mistakenly wired in series.

There should be no voltage drop across contacts or connections. A voltage drop occurs at the location of the resistance. This resistance can cause excessive heat, which is indicated by dark spots on wires or overheated components.

Warning: Use caution when testing energized circuits.

Contacts

A *contact* is the conducting part of a switch that operates with another conducting part to complete (close) or break (open) a circuit. Contacts are made of a silver alloy that offers little resistance. Contacts are classified by their position before operation, such as normally open (NO) or normally closed (NC). Contacts fail when they do not close, do not open, or offer resistance to current flow.

An NO contact in good condition indicates source voltage when open (not actuated) and 0 V when closed (actuated). **See Figure 4-40.** An NC contact in good condition indicates source voltage when open (actuated) and 0 V when closed (not actuated).

When two (or more) NO contacts are located in a circuit, a DMM set to measure voltage indicates non-source voltage across one contact when both (all) contacts are open. The DMM indicates source voltage across the open contact when the other contact(s) is closed.

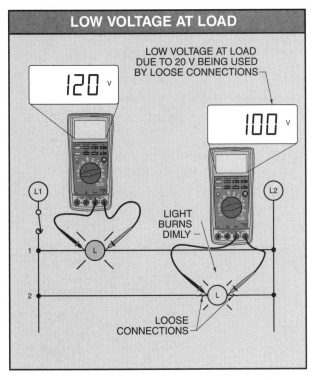

LOW VOLTAGE AT LOAD

LOW VOLTAGE AT LOAD DUE TO 20 V BEING USED BY LOOSE CONNECTIONS

120 V

100 V

L1

L2

LIGHT BURNS DIMLY

1

2

LOOSE CONNECTIONS

Figure 4-39. Low voltage at a load is caused by resistance in the circuit path.

When two (or more) NC contacts are located in a circuit, a DMM set to measure voltage indicates 0 V across one contact when closed regardless of the condition (open or closed) of the other contact. The DMM indicates non-source voltage across one contact when both NC contacts are open.

Changes in measured voltage are the best indication that contacts are actually opening or closing. Contacts may not open or close even though the contact operator works properly. An *operator* is the device that is pressed, pulled, or rotated by the individual operating the circuit. Contacts can be forced out of position from physical damage. Contacts fuse or melt away when they are unable to dissipate heat properly. The overheated contacts can fuse, and when cooled, are unable to open. In some cases, contacts melt away, leaving no contact metal to open or close.

Overheating of contacts can be caused by contacts not being large enough to dissipate the heat produced by the current in the circuit. Undersized contacts should be replaced with contacts rated for the proper amount of current. Also, the maximum amount of current a conductor can safely carry before overheating is lowered when contacts become pitted. *Pitting* is localized corrosion that has the appearance of cavities (pits). Contacts that open and close often develop pits from arcing when the contacts pull apart. *Arcing* is the discharge of an electric current across an air gap.

Pitted contacts can be detected before they become dangerous. Contacts that indicate a voltage drop or some resistance when closed may have dirty or pitted contact surfaces. The larger the voltage drop, the greater the damage. Badly pitted contacts should be replaced immediately. Filing contacts to renew the surface is not recommended as it removes metal and lowers the amount of current the contacts can safely carry. Contacts may appear dirty but offer no resistance. Dirty contacts can be cleaned with contact cleaner, cardboard, or contact cleaning pads to remove dirt without damaging the metal surfaces. Resistance readings of clean contacts are near 0 Ω when closed and infinity when open.

Fluke Corporation

Special test instrument features may include detachable displays that can be placed in a more convenient location for viewing measurements.

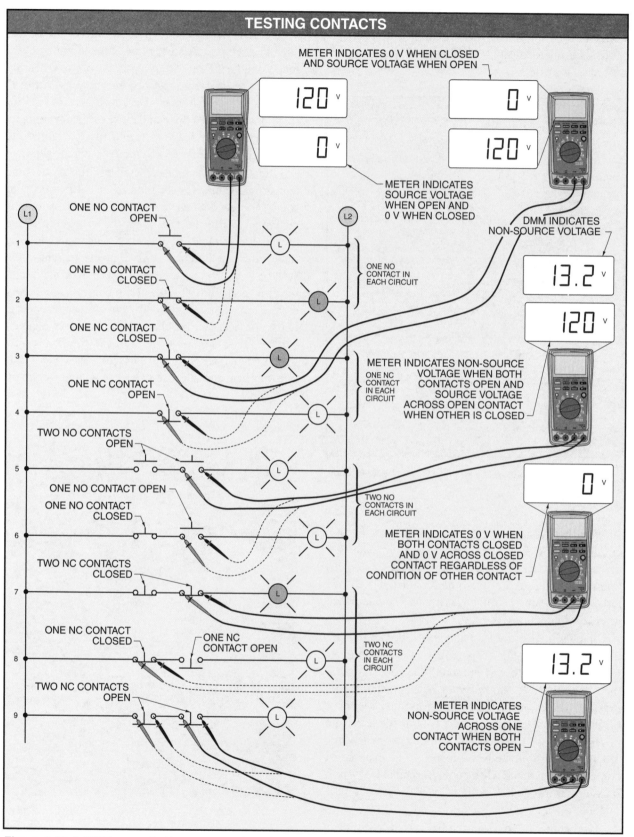

Figure 4-40. A DMM connected to measure voltage is used to locate defective contacts in a circuit.

Defective Loads

Most loads in electrical circuits include coils. A *coil* is a winding of insulated conductors arranged to produce a magnetic field. Coils produce a magnetic field similar to a transformer. Current flowing in a coil creates a magnetic field that moves an iron armature. The moving armature opens or closes relay or solenoid contacts. A *relay* is an electrical switch that is actuated by a separate circuit. A *solenoid* is an electric output device that converts electrical energy into a linear mechanical force. The resistance or voltage of a coil may be tested using a DMM.

A DMM set to measure resistance is used to test a coil when the coil is isolated from the circuit. **See Figure 4-41.** A good coil shows resistance. A near 0 Ω resistance reading indicates the coil is shorted. Shorting in a coil commonly occurs from wires that have overheated and melted the insulating varnish around each wire in the coil. An infinite resistance or overload (OL) reading indicates the coil is open with a break in the wire. The coil should be electrically isolated from its metal frame.

A voltage test of a good coil while connected in a circuit indicates near 0 V before it is energized and a voltage drop when it is energized. An open coil indicates non-source voltage or source voltage before it is energized.

Overheating in a coil is prevented by ensuring that the voltage at the coil is within ±5% of its rated voltage. Voltage above and below the rated voltage can overheat a coil. *Duty cycle* is the percentage of time a load or circuit is ON. A coil receives a momentary current overload that heats the coil each time it is turned ON. The coil is replaced with a coil that has a higher duty cycle rating if excessive duty cycles cause overheating and the number of duty cycles cannot be reduced. Coils should be checked routinely for signs of overheating. A supply of replacement coils reduces downtime when a coil must be replaced.

Overheating is also caused by an armature that is unable to move. An *armature* is the moving part of a coil. Instead of receiving the normal temporary overload at startup, the current level in the coil remains high because the armature cannot move. The coil receives continuous high startup current and burns out within a few minutes.

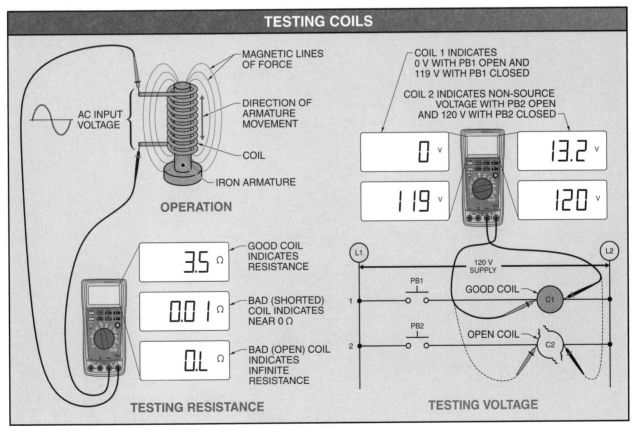

Figure 4-41. The resistance or voltage of a coil may be tested using a DMM.

Never apply power to a coil when it is removed from its armature. The iron core on which the coil is fitted helps to prevent high current. The coil is subject to excessive current flow and overheats rapidly if separated from the iron core. Identify the causes of overheating to prevent repeated coil failures. A bad coil is replaced, not repaired.

Relays

A relay controls one electrical circuit by opening and closing contacts in another circuit. A relay allows a low-voltage control circuit to open or close contacts in a high-voltage load circuit. The load in the control circuit is the relay coil. The coil becomes an electromagnet, causing the armature to move when the control circuit is energized. This closes the relay contacts to energize high-voltage loads such as motors, lights, or heating elements.

Relays are commonly used to control heating elements. **See Figure 4-42.** The control circuit and heating element (load) circuit are separate circuits. The control circuit operates at 24 V, 0.5 A. The heating element circuit operates at 460 V, 10 A. As temperature falls in a room, the NO contacts in the thermostat close the control circuit to energize the relay coil. Current flowing through the relay coil produces a magnetic field that closes the NO contacts in the high voltage load circuit. The closed contacts complete the circuit and power flows to the heating elements.

A common problem of defective relays is that the loads do not energize or do not de-energize. Troubleshooting is started by manually operating the relay to determine if the control circuit, relay, or load is the problem. The problem is in the control circuit if the load works properly when the relay is operated manually.

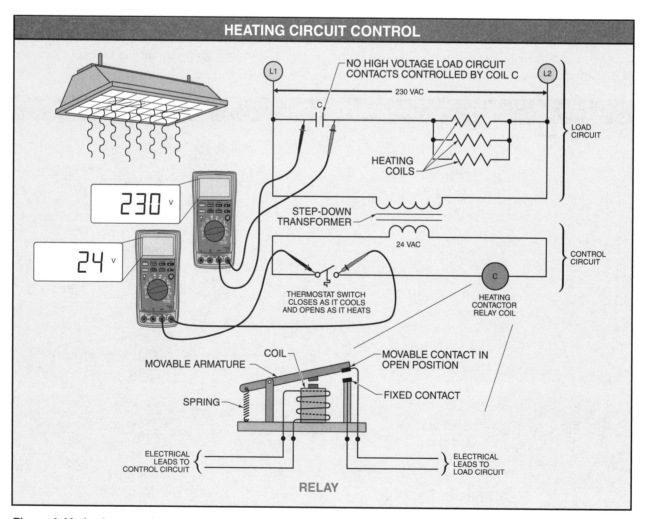

Figure 4-42. A relay controls one electrical circuit by opening and closing contacts in another circuit.

Warning: Use caution when manually operating a relay because loads may start or stop without warning.

The problem is in the load circuit if the load does not operate. Common problems in the load circuit include loose connections, open contacts, welded contacts, dirty contacts, or improperly installed contacts.

Solenoids

A solenoid converts electrical energy into a linear mechanical force. Solenoids are used to activate a switch or valve using a movable iron core (plunger). **See Figure 4-43.** Solenoids consist of a doughnut-shaped coil surrounding a movable plunger. Current in the coil produces an electromagnetic field that moves the plunger. The plunger is attached to a valve or other movable device.

Solenoids are either energized or de-energized. The electrical part of the solenoid is the coil, which is tested using a DMM set to measure resistance or voltage. The mechanical action of the solenoid is checked by listening, observing its operation, or feeling it to detect the motion of the plunger. A solenoid is tested in the same manner as a coil. Solenoid coil testing is one of the most common coil tests.

Heating Elements

A *heating element* is a conductor with an intentionally high resistance for producing heat when connected to an electrical power supply. Electric current flows through the conductor, making the heating element hot. Heating elements are used in electric ovens, dryers, and forced-air heating systems.

A good heating element has some resistance when tested. **See Figure 4-44.** An open is created in the element when heating elements overheat. A defective heating element with an open indicates an infinite resistance or overload (OL) when connected to a DMM set to measure resistance. A short in a heating element indicates $0\ \Omega$ resistance. An open heating element connected to a circuit indicates some voltage before it is energized and source voltage after it is energized. A good heating element indicates near 0 V before it is energized and a voltage drop after it is energized.

> **TECH TIP**
> *Ensure the switch for a circuit is open or disconnected before removing a fuse from a circuit. Always use an approved puller and break contact on the hot side of the circuit first.*

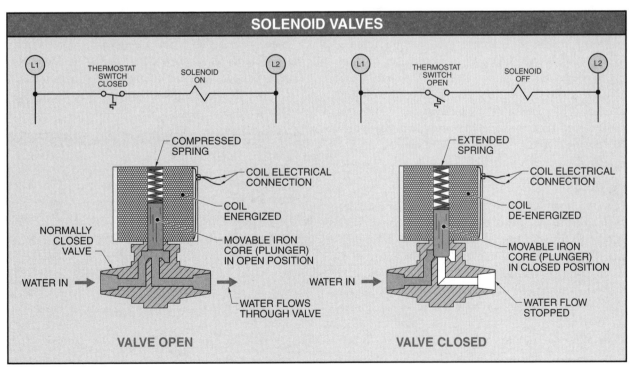

Figure 4-43. Solenoids activate a switch or valve using a movable iron core (plunger).

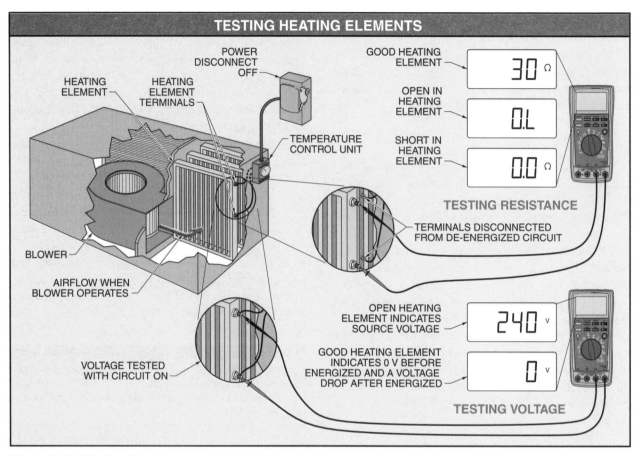

Figure 4-44. Heating elements are tested for resistance when disconnected from a circuit or voltage when connected to a circuit.

Opens in heating elements are commonly caused by overheating, high voltage, physical damage, or shorting to nearby metal. Forced-air heating elements need constant airflow to prevent overheating. The element overheats and opens or shorts rapidly if the airflow stops. Most forced-air heaters include safety switches to open the heating element circuit if air flow stops.

Lighting

Lighting commonly used in industry includes incandescent, fluorescent, and high-intensity discharge (HID) lamps. Each lamp has advantages and disadvantages when used in different applications. For example, an incandescent lamp has low initial cost and is dimmed inexpensively, but has a relatively short life. An *incandescent lamp* is an electric lamp that produces light by the flow of current through a tungsten filament inside a gas-filled, sealed glass bulb. A filament that has continuity is good and lights. A good incandescent lamp indicates some resistance when tested with a DMM set to measure resistance. **See Figure 4-45.** The lamp is bad when there is an open in the filament. A bad incandescent lamp indicates an infinite resistance or overload (OL). Incandescent lamps are commonly replaced when they fail.

> **TECH TIP**
> *Lamp poles should be spaced at a maximum of 2 times their mounting height from the edge of an area to be lit and 4 times their mounting height from any adjacent lamp pole.*

A *fluorescent lamp* is an electric lamp that produces light as electricity passes through a gas. Fluorescent lamps have small heating elements at each end of the lamp tube. The lamp tube has a phosphor coating on the inside and is filled with mercury and argon vapor. The heating elements at the ends of the tube heat up when energized, lowering the resistance of the gas. High voltage is applied to the tube ends and current starts to flow through the tube with the gas as the conductor.

The high voltage is then removed and the lamp operates on normal line voltage. Each fluorescent fixture uses a ballast to supply the high starting voltage. **See Figure 4-46.** A *ballast* is a transformer or solid-state circuit that limits current flow and supplies the high starting voltage for fluorescent and HID lamps. After starting, regular line voltage is used to keep the lamp lit. Lamp holders hold the lamp or lamps securely in place.

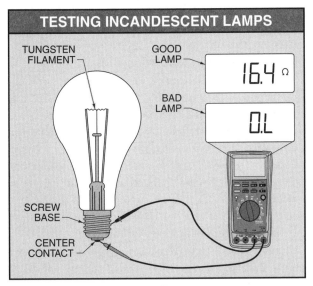

Figure 4-45. Incandescent lamps indicate some resistance when good and OL (infinite resistance) when bad.

Fluorescent lamps wear out and their ends discolor as they age. All fluorescent lamps in an area should be replaced at the same time to avoid uneven lighting. Check that the lamp bases are firmly fitted into the lamp holders if a lamp does not light or flickers. Replace the lamp if the lamp base is providing a solid mount for the lamp. The lamp ballast is checked according to manufacturer's recommendations if the problem persists after lamp replacement.

A bad ballast is replaced with a new ballast of the same type. Flickering and lamps that do not light can also be caused by low ambient temperatures and low supply voltages. Ballasts must be matched to specific fluorescent lamps. Most ballasts come with an installation wiring diagram. The lamps required are listed, as are any temperature considerations and operating voltage. Voltage supplied to the lamp and ballast should be within ±7% of the ballast rating to avoid shortening the life of the lamp and ballast. Reduced voltages can also cause flashing during startup and reduced light output.

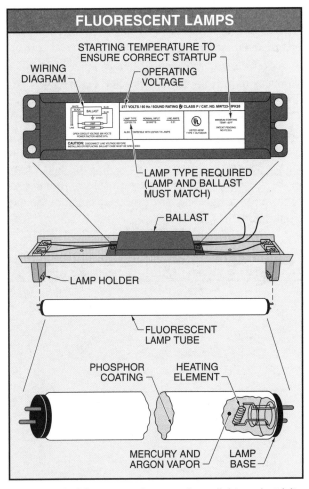

Figure 4-46. A fluorescent lamp produces light as electricity passes through mercury and argon vapor.

Fluorescent lamp bases must be matched to the lamp holder. For example, a single-pin lamp base cannot be used in a double-contact lamp holder. The lamp base must seat correctly into the lamp holder to light properly and to prevent the lamp from falling out of the lamp holder. The base mark of the lamp must be aligned with the center of the lamp holder. Both ends of the lamp must be secured in the lamp holder, which may be difficult with long lamps.

Care must be taken when changing fluorescent lamps because high voltages are present in the lamp bases. Change lamps with the circuits de-energized and locked out if possible. Extremely careful movement with and careful placement of the lamps helps to prevent electric shock and lamp breakage when it is not possible to de-energize the lighting circuits during maintenance. Broken lamps are extremely dangerous due to the sharp glass shards and the chemicals inside the lamps. Lamp replacement is usually accomplished with ladders or hydraulic lifts so additional care must be practiced when working at a height.

A *high-intensity discharge (HID) lamp* is a lamp that produces light from an arc tube. **See Figure 4-47.** An *arc tube* is the light-producing element of an HID lamp. The arc tube contains a gas that conducts the arc between the electrodes. HID lamps include low-pressure sodium, mercury-vapor, metal-halide, and high-pressure sodium lamps. The main difference is the pressure and vapor in the arc tube. HID lamps require several minutes to begin emitting light and most produce a humming sound from their ballasts.

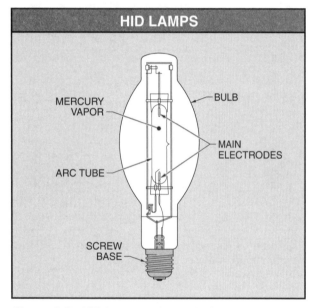

HID LAMPS

MERCURY VAPOR

BULB

MAIN ELECTRODES

ARC TUBE

SCREW BASE

Figure 4-47. HID lamps produce light from an arc tube.

Ballasts. A ballast is a transformer or solid-state circuit that limits current flow and supplies the starting voltage for fluorescent and HID lamps. Ballasts must operate within their normal operating temperatures. Temperatures that are too low can create problems in starting and can cause lamp flickering. Enclosing the complete fixture is one method for raising ballast temperature. However, this may lead to overheating during summer months, which can shorten ballast life. A common problem is poor installation that does not allow for correct heat dissipation. Fixture vents must be kept open and clean to allow heat to leave the fixture. Per the NEC®, a 3″ minimum clearance is required between a fluorescent lamp and thermal insulation. High operating temperatures can also be caused by

using the wrong number of lamps or an incorrect lamp. This often occurs when the wrong lamps are purchased and are used rather than returned. Matching lamp and ballast is critical for correct operation. Always follow manufacturer's recommendations regarding ballast and lamp matching and maintaining the correct operating temperatures. Lamps are labeled with the wattage of the bulb and other characteristics such as the light color.

All ballasts produce some noise. Low-noise ballasts may be installed in locations where excessive noise is distracting. Ballasts are sound rated with quiet units recommended for quiet office areas and loud units recommended for heavy production areas.

HID lamp ballasts can be damaged by lamps nearing the end of their life cycle. For example, flickering lamps can damage the ballast circuits through constant restarting. Energy-efficient ballasts are available for fluorescent and HID lamps, but many require the use of specific lamps. When upgrading to energy-efficient ballasts, all ballasts are changed and the correct lamps stocked to avoid installing incompatible lamps that could damage the new ballasts. Energy savings can be considerable from upgrading lighting fixtures.

Ballasts can also produce radio interference. Radio interference is caused by the direct radiation from a fluorescent lamp. Radio interference can be eliminated by separating the radio and lamp by 10′ or more, providing a good ground for the radio or sound system, adding an external filtering circuit, or connecting the radio and fluorescent lamp to separate circuits.

Relamping. Lamps may be replaced individually or in groups. Most manufacturers recommend group relamping. *Group relamping* is the replacement of all lamps in a given area when they reach 60% to 80% of their rated average life. HID and fluorescent lamps produce reduced light levels as they age while still consuming the same amount of electrical energy. For this reason, relamping before large numbers of lamp failures can save energy costs and reduce labor costs while maintaining safe lighting levels. Group relamping is managed by keeping records of lamp operating hours. For example, if a group of lamps is rated for 10,000 hours of life, they would be changed at about 8000 hours of operation. Hours of operation can be calculated or the lamps can be placed on an hours of operation meter that keeps track of their actual operating time. Lamp fixtures must be cleaned when lamps are changed. The combination of dirty walls and dirty lamp fixtures can reduce light output by up to 50%.

Handle lamps carefully when replacing old lamps. Certain HID lamps can explode if damaged and some emit potentially dangerous levels of ultraviolet radiation if the outer skin of the lamp is scratched or punctured. Lamps should be replaced with the power OFF and all safety precautions must be followed regarding elevated work. Follow all manufacturer's recommendations regarding lamp handling. Broken lamp pieces must be collected using recommended safety gear such as gloves and safety glasses. Lamps that present an explosion hazard must be protected by shields that stop falling glass from hitting those below the lamp.

Warning: Use extreme caution when working with the high voltages found in industrial lighting circuits.

Different lamps produce different colors of light. The color of a lamp distorts the true color of objects viewed in the light. *Color rendering* is the appearance of a color when illuminated by a particular light source. Specific lamps are selected in areas where authentic color representation is required. For example, low-pressure sodium lamps are not used in a printing press area because printed colors are distorted by the yellow-orange light produced. Mercury-vapor lamps have poor rendering of reds and blues. Metal-halide lamps produce good overall color rendering. Fluorescent lamps can also be selected in various color renderings depending on the requirements of the location. Specially-coated lamps can improve color rendering.

Lighting is often ignored in industrial settings. A lighting plan detailing lighting levels, color, lamps, ballasts, and a relamping schedule is an effective method of controlling lighting costs while maintaining safe and effective lighting. Such a plan can serve as a guide when troubleshooting and developing maintenance activities. Most major manufacturers offer advice and assistance in establishing a lighting plan.

Three-Phase Motors

The most common industrial motor is the 3ϕ induction motor. An *induction motor* is a motor that rotates due to the interaction between the magnetic fields of the stator and rotor. A 3ϕ motor consists of three coils of wire wound into a stator. A *stator* is the stationary part of an AC motor. A stator is a slotted circular iron frame. The stator windings do not move and surround the rotor, which rotates. A *rotor* is the rotating part of an AC motor. The rotor is mounted on the motor shaft, which is attached to the load the motor is powering.

Each coil in a 3ϕ motor is wound so that one half of the coil is opposite the other half of the coil in the stator. Each stator coil receives a single phase of electric current. **See Figure 4-48.** Each phase lags or leads the other two phases to create a pulsating magnetic field that creates a rotating magnetic field in the stator. The rotating magnetic field passes through the aluminum or copper bars of the squirrel cage built into the rotor. This induces current in the squirrel cage bars, producing a magnetic field around each rotor bar. The rotating magnetic field of the stator pushes on the magnetic field around each rotor bar, causing the rotor to rotate because magnetic fields cannot cross or pass through each other. The torque produced by the interaction of the three magnetic fields starts the motor and keeps it rotating.

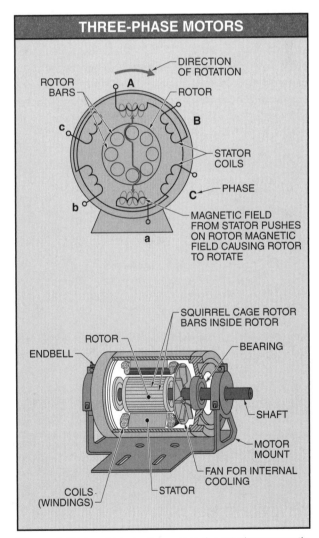

Figure 4-48. In a 3ϕ induction motor, the rotating magnetic field of the stator pushes on the magnetic field around each rotor bar, causing the rotor to rotate.

Overheating is the most common cause of motor failures. Overheating melts the insulation on the stator coil wires. The uninsulated wires touch, shorting the coils and destroying the motor. Overheating may be caused by overloading the motor. Overloading a motor causes excessive current flow in the stator coils, which become overheated. As a motor starts, it receives high levels of current (high inrush current) until the rotor begins to rotate. Normally, the high inrush current lasts a few seconds at startup. The motor overheats and fails quickly due to overheating if the motor does not start or is completely stalled by a heavy load. A motor that is started and stopped frequently or a motor that has its direction of rotation reversed frequently can overheat. Low voltage or high voltage can also cause overheating. Motor voltage should be within ±10% of the motor required voltage.

A motor that is excessively hot should be tested with a clamp-on ammeter or a DMM with a clamp-on current probe accessory. **See Figure 4-49.** The reading is compared to the motor nameplate data. Motor nameplates contain information concerning motor operation. A motor is normally loaded if the motor current is between 95% to 105% of the rated current. Current above 105% of rated current causes overloading if continued for excessive periods of time.

Most motors can handle brief overloads without damage. Input voltage is tested if motor current is high. The voltage source and path are checked if the voltage is not within ±10% of the rated voltage. The motor bearings are lubricated if input voltage is correct. If current falls to an acceptable level after the motor bearings are lubricated, they are most likely damaged and should be replaced. The bearings in the load are lubricated if current remains high after lubricating the motor bearings. The load bearings are defective if motor current becomes normal after load bearing lubrication. The motor is disconnected from its load and operated if current remains high after all bearings are lubricated. If current returns to normal after the motor is disconnected from the load, the problem is in the load, which is inspected for mechanical problems. The motor is replaced if current remains high with the load disconnected because the motor coils may be damaged. A motor that is overloaded with a properly functioning load should be replaced with a motor of greater horsepower.

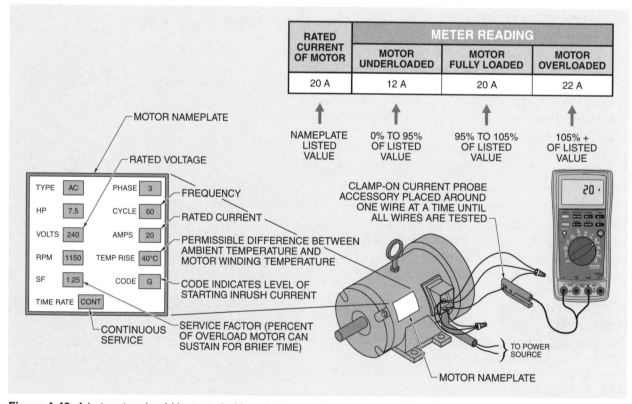

Figure 4-49. A hot motor should be tested with a clamp-on ammeter and/or DMM with clamp-on current probe accessory and the reading compared to the motor nameplate data.

Magnetic Motor Starters. A *magnetic motor starter* is a specialized relay used to energize and de-energize a motor and includes motor overload protection. Magnetic motor starters are used to control 3ϕ motors. When the magnetic motor starter coil (control circuit) is energized, its magnetic field closes three motor circuit contacts to allow current to flow to the motor. **See Figure 4-50.** The magnetic motor starter also contains auxiliary (holding) contacts that keep the coil energized after the start pushbutton or other starting device is released.

Each wire leading to the motor contains an overload contact. The overload contacts are operated by a heater unit. A *heater unit* is a heating element used to open overload contacts for overcurrent conditions. When one or more wires that carry excessive current for more than a few minutes, the overload heater to trip (open) the NC overload contact in the control circuit of the magnetic motor starter coil. The coil is de-energized and the load circuit contacts open, stopping the current supplied to the motor.

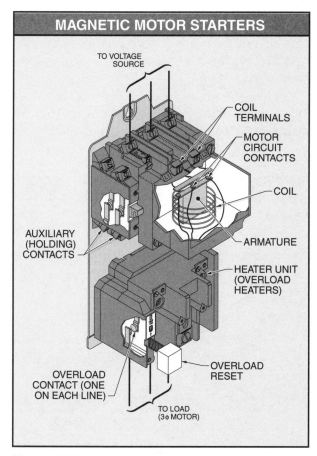

MAGNETIC MOTOR STARTERS

TO VOLTAGE
SOURCE

COIL
TERMINALS

MOTOR
CIRCUIT
CONTACTS

COIL

AUXILIARY
(HOLDING)
CONTACTS

ARMATURE

HEATER UNIT
(OVERLOAD
HEATERS)

OVERLOAD
CONTACT (ONE
ON EACH LINE)

OVERLOAD
RESET

TO LOAD
(3ϕ MOTOR)

Figure 4-50. A magnetic motor starter is a specialized relay in which the coil is energized and de-energized by the devices that control the motor.

Tripped overload contacts are reset by pressing the overload reset, which clicks them back into the closed position. Some overload contacts also use a trip indicator that shows if the overload contacts have tripped.

Overload contacts trip if a motor cannot start. A motor that cannot start allows high inrush current to continue beyond its normal, temporary duration at startup. The same problem occurs if a motor stalls or slows due to overloading. High current also results from sustained overloading of the motor, frequent stopping and starting or reversing, or low or high voltage applied to the motor.

When overload contacts are found tripped, motor current should be measured after the motor is restarted to determine if the motor is mechanically overloaded. Mechanical overloads should be corrected or the motor replaced with a larger motor. Another cause of overload contact tripping is heaters that are too small. Overload heaters must be correctly matched to motor current. Overload heaters that are too small open the overload contacts before normal current levels are reached. Overload heaters that are too large allow excessive current into the motor, causing motor overheating. Follow the manufacturer's recommendations when sizing and checking magnetic motor starter overload heaters.

A magnetic motor starter is operated manually when a motor is not working and the overload contacts are not tripped. This involves closing the power contacts in the starter, which connects the motor to its voltage source. This is accomplished by different means in different magnetic starters. Closing the power contacts should only be done if the circuit is properly fused and the auxiliary contacts are not used to control any other loads.

Warning: Always use proper safety equipment and follow manufacturer's recommended safety procedures when working on or near energized magnetic motor starters.

Motor source voltage is tested if the motor does not start with the magnetic starter power circuit contacts closed. The motor is replaced or tested if source voltage is at the correct level and is present at the motor when the magnetic starter is manually operated.

A 3ϕ motor is tested using a DMM set to measure resistance. **See Figure 4-51.** The three individual motor coils should have the same resistance. They should not have continuity with each other. If they do, the insulation around some of the wires has melted and bare wires from separate coils are touching. In addition, overheated motor windings are likely to be blackened and have a distinct burnt insulation smell. The three coils that are connected electrically in the center should have a different resistance reading from the three individual coils. There should be no continuity between the connected coils and the individual coils. The coil configuration varies for different motors. For this reason, manufacturer's literature must be consulted before conducting motor resistance tests. The motor is damaged and must be rewound or replaced if one or more coils indicates OL (infinite resistance) or the individual coils have continuity with the motor frame. Rewinding a motor involves the complete replacement of all the copper wire in the stator. It is sometimes economically feasible to rewind large motors; however, it is usually less expensive to replace small motors.

The reason for damage must be determined when a motor is found with an open or shorted coil. Check the size of the overload heaters, supply voltage, load on the motor, and the mechanical operation of the overload contacts. Excessive current overheats the motor windings if the overload contacts do not open when needed.

Single-Phase Motors

A *single-phase motor* is a motor designed to run on 1φ power. Single-phase motors are seldom larger than 1 HP and are used for light loads. Most 1φ motors are induction motors that operate on the same principle as 3φ induction motors. **See Figure 4-52.** A *split-phase motor* is a 1φ AC motor that includes a starting and running winding in the stator. A *starting winding* is a coil that is energized temporarily at startup to create the torque required to start a 1φ motor rotating. The starting winding is wound out-of-phase with the running winding of the motor. A *running winding* is a coil that continues to operate after the motor is started.

A split-phase motor is primarily a rotor, a stator consisting of the running and starting windings, and a centrifugal switch that is located on the motor shaft. A *centrifugal switch* is a switch that opens when a rotor reaches a certain preset speed and reconnects when the speed falls below a preset value.

At startup, both windings are energized and create two out-of-phase magnetic fields that produce the torque to start the motor rotating. *Motor torque* is rotational force that is generated in a split-phase motor by the strength of and phase angle difference between the magnetic fields in the stator. The centrifugal switch opens, disconnecting the starting winding from the circuit when the motor reaches approximately 75% of full speed. This allows the motor to operate on the running winding only.

A capacitor-start motor is selected if a heavy load must be operated. A *capacitor-start motor* is a 1φ motor that has a capacitor in the starting winding. A *capacitor*

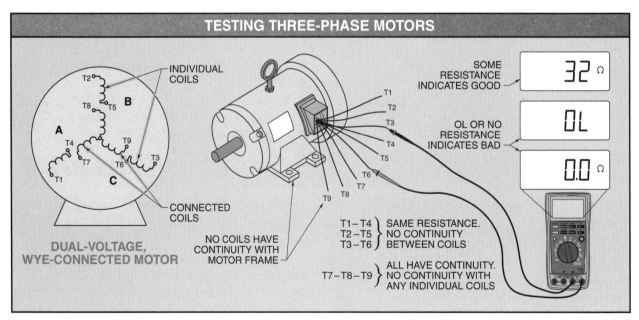

Figure 4-51. A 3φ motor is tested using a DMM set to measure resistance.

is a device that stores electrical energy in an electrostatic field. A capacitor consists of two metal plates separated by a dielectric (insulator). The capacitor causes the starting winding current to be nearly 90° out-of-phase with the running winding current. **See Figure 4-53.** Starting torque is increased by adding a capacitor in series with the starting winding.

A *capacitor start-and-run motor* is a 1ϕ motor that has capacitors in the starting and running windings. The capacitors create additional starting and running torque. The motor may not start and could overheat the starting windings if the starting capacitor is damaged and not producing its rated capacitance. Capacitors are checked by using a capacitor tester that compares the actual capacitance produced by the capacitor against its rated value. The instructions for each capacitor tester must be followed exactly. The capacitance of a capacitor should be within ±10% of its rated value. Capacitors are replaced when damaged.

The rotor of a split-phase or capacitor-start motor contains a centrifugal switch mounted on the motor shaft. The centrifugal switch is closed and the starting winding is energized when the rotor rotates below 40% of full speed. At approximately 75% of full speed, the centrifugal switch opens, disconnecting the starting winding. **See Figure 4-54.** The starting winding is not designed to carry current for long periods of time. The starting winding burns out quickly if the centrifugal switch does not open. After the centrifugal switch opens, the motor continues to run on the running winding.

Single-phase motors are usually operated by a switch and the motor is protected by a time-delay fuse and/or overload contacts. The overload contact is reset if a 1ϕ motor does not start. The motor is replaced if the motor still does not start and the voltage is at the correct level. Single-phase motors are generally inexpensive and are usually replaced without extensive testing.

Variable Frequency Drives

The frequency of a power supply determines the speed of AC motors. Since normal power line frequency is a constant 60 Hz (in the U.S.), it must be changed to increase or decrease the speed of the motor. A *variable frequency drive* is a motor controller that is used to change the speed of AC motors by changing the frequency of the supply voltage. In addition to controlling motor speed, variable frequency drives can control motor acceleration time, deceleration time, motor torque, and motor braking.

A drive changes the frequency of the voltage applied to a motor by converting the incoming AC voltage to a DC voltage, and inverting it back to an AC voltage that simulates the desired fundamental frequency. *Fundamental frequency* is the desired voltage frequency simulated by the varying ON/OFF pulses at a higher carrier frequency. *Carrier frequency* is the frequency of the short voltage pulses of varying length that simulate a lower fundamental frequency. The higher the carrier frequency, the more individual pulses there are to reproduce the fundamental frequency and the closer the output sine wave is to a pure fundamental frequency sine wave. **See Figure 4-55.**

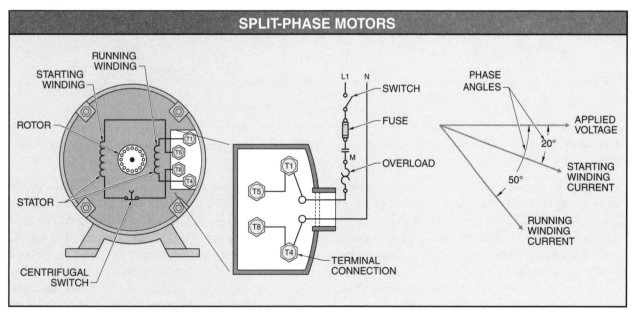

Figure 4-52. Single-phase motors have starting and running windings in the stator.

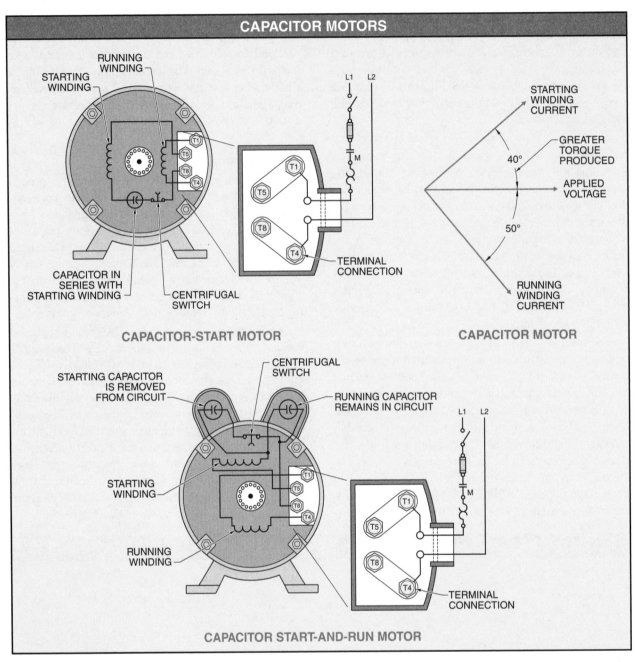

Figure 4-53. Capacitor motors have capacitors in the starting and/or running windings, which increase the starting and/or running torque.

The carrier frequency of most variable frequency drives ranges from 1 kHz to about 20 kHz. The carrier frequency of a variable frequency drive can be changed to meet particular load requirements. Higher carrier frequencies reduce heat in the motor because motors operate cooler the more closely the voltage simulates a pure sine wave. However, the fast switching of the inverter section of a variable frequency drive produces large voltage spikes that can damage motor insulation.

Increasing the carrier frequency also increases the noise produced by the motor. Noise is noticeable in the 1 kHz to 2 kHz range because it is within the range of human hearing and is amplified by the motor. Raising the carrier frequency of a variable frequency drive beyond the range of human hearing solves the noise problem, but can cause greater power losses (thermal losses) in the variable frequency drive because of the solid-state switches in the inverter section of the drive.

An oscilloscope can be used to look at the carrier frequency and the fundamental frequency of a variable frequency drive.

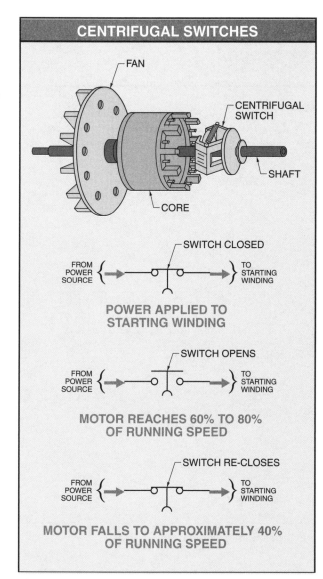

Figure 4-54. At approximately 75% of full speed, the centrifugal switch opens, disconnecting the starting winding.

PREVENTIVE MAINTENANCE

Preventive maintenance keeps equipment in peak operating condition. Electrical equipment must be kept clean, dry, and cool. The inside and outside of electrical equipment is kept clean to prevent dirt from creating paths for short circuits. Moisture lowers resistance

and damages insulation, causing short circuits. The number one cause of electrical equipment failure is overheating. Electrical equipment must have proper ventilation and be kept clean for proper heat dissipation. Electrical equipment that is kept clean and dry with its connections tight and its supply voltage within manufacturer's recommendations provides years of efficient and reliable service.

A thermal imager is used to locate overheating electrical equipment. A *thermal imager* is an infrared-measuring device that displays an image based on temperature. **See Figure 4-56.** Hot surfaces register brighter than surrounding surfaces. The brighter the image, the hotter the surface. Thermal imagers produce a still or video image, or display a digital temperature readout. The image is analyzed and further testing conducted when excessive heat is detected. The cause of the overheating is found and remedied before a dangerous situation develops. Special training may be required to operate thermal imagers and interpret the results.

TECH TIP

An oscilloscope is an instrument that displays the shape of a voltage waveform and can measure its voltage level, frequency, and phase. This makes them particularly useful for troubleshooting variable frequency drives.

Cutler-Hammer
Preventive maintenance requires the periodic checking of equipment operating characteristics to ensure the equipment is within operating parameters.

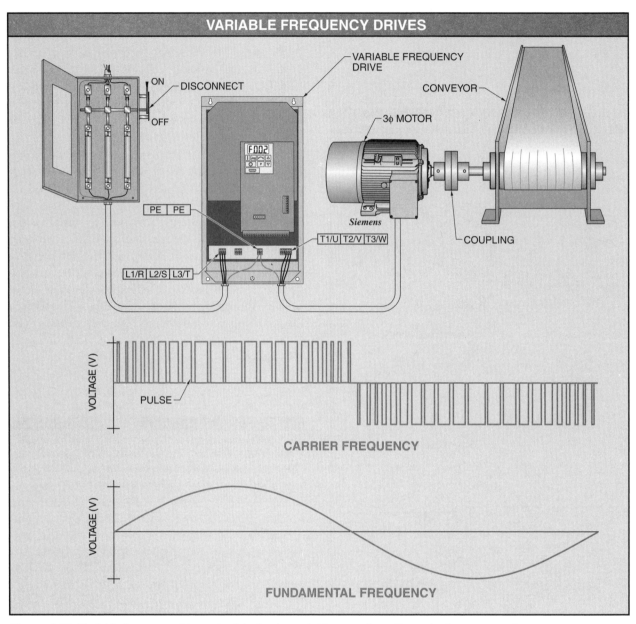

Figure 4-55. Variable frequency drives simulate fundamental frequencies with carrier frequency pulses.

Source Voltage

Preventive maintenance techniques are used to monitor the quality of the source voltage into a plant and at various points throughout the plant power distribution system. Electronic power monitors can constantly monitor the level of voltage and current at the plant service entrance. Power monitors de-energize the circuit to prevent damage to plant equipment if voltage is lost or becomes too low. Warnings are sounded or circuits are de-energized before serious damage occurs if voltage or current becomes dangerously unbalanced. In 3ϕ voltage, the phases rise and fall in a specific sequence. For example, the common phase sequence for 3ϕ circuits is phase A, phase B, and phase C. This keeps motors operating in the correct direction of rotation. Phase sequence reversal monitoring equipment can be installed to de-energize circuits if the phase sequence is reversed by the utility company. Motors operate in reverse rotation if the phase sequence is altered, causing serious damage to plant equipment.

Electrical Path

Loose connections creating opens in an electrical path are the most common electrical problem. The tightness of each electrical connection should be checked with a screwdriver at least annually. Connections can become loose from heating, cooling, and vibration. Loose connections cause high resistance, leading to dangerous overheating and open circuits. Busway connections and other electrical equipment should be checked according to manufacturer's instructions.

Warning: Always check connection tightness with the circuit de-energized.

VISIBLE LIGHT IMAGE

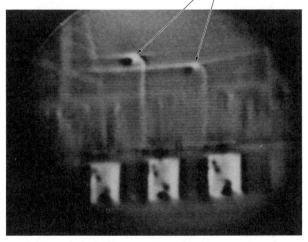

OVERHEATING ELECTRICAL CONDUCTORS

INFRARED IMAGE

John Snell & Associates

Figure 4-56. A thermal imager locates potential maintenance problems by sensing heat buildup.

Wire insulation should be checked annually for cracking, darkening, brittleness, or wearing away of the insulation on moving equipment. Chemicals, oil, and dirt cause insulation to deteriorate quickly. The correct insulation must be selected for use in the various operating environments. For example, high-temperature rated insulation is used in hot areas. The size of the wire and its insulation quality must comply with provisions of the NEC®. A clamp-on ammeter can be used to verify that wires are not carrying excess current.

Voltage drop across contacts should be checked annually for correct operation and to ensure that high resistance is not developing at the contacts. Equipment used infrequently, such as disconnects, overloads, and circuit breakers, should be operated at least annually to prevent them from becoming stuck in one position. The mechanical operation of relays, solenoids, and magnetic starters should be checked quarterly. All springs, interlocks, and mechanical stops should be in place. Humming or chattering is an indication of mechanical problems interfering with the closing of contacts. Overheating insulation has a distinct odor and indicates a serious problem. Changes in operating sounds, loudness, and/or frequency can indicate developing electrical problems. Equipment should be checked for excessive vibration and high temperatures. All equipment enclosures must be checked annually for cleanliness, physical and moisture damage, and solid mounting.

Motors

Electronic monitors are used on large or critical motors to monitor phase unbalance or single phasing. Phase unbalance occurs when one phase voltage or current is significantly higher or lower than other phases. Single phasing is the operation of a motor that is designed to operate on three phases operating on two phases because one phase is lost. Monitors stop a motor when voltage is lost in one or more phases to prevent damage from overheating. A single phasing motor does not start and can burn out rapidly. A running motor that loses a phase might continue to operate until it gradually fails due to overheating.

⬡ TECH TIP

At high altitudes (over 3000′), low-density air reduces a motor's cooling capacity. This is corrected by increasing the HP or service factor rating of the motor. A continuous-duty motor should be used if the motor must operate at full load for 1 hr or more in any 24-hour period.

Vibration analysis may be used to monitor changes in motor bearing operation. As changes in the vibration produced by a bearing indicate a problem, the bearing can be changed before serious problems develop. Special training is required to interpret the results of vibration analysis.

Motor current should be checked quarterly. Motor current that increases over time indicates problems in the motor or the motor load. Motor current at or exceeding its maximum rating could indicate an overloaded motor. Disconnect the motor from its load and take a current reading. The overloading problem is in the load or the connection to the motor if the current returns to normal with the load removed. The problem is in the motor if the current remains high with the load removed. The motor bearings are lubricated and current is measured. The bearings are damaged and must be replaced if current falls after motor bearing lubrication. The motor could also have damaged motor windings or low or high source voltage, which could be causing these problems.

All motor mounts and load mounts must be tightened at least annually. Motors are lubricated according to manufacturer's specifications. All connections to loads should be inspected for secure mounting and correct alignment of motor and load.

Motors must be kept clean to allow for adequate heat dissipation. Scheduled motor maintenance should include periodic inspection and cleaning of the motor windings. Dirt or debris inside the motor can cause high-resistance shorts if it becomes a path to ground, causing the insulation on the windings to overheat. If the insulation overheats, it melts, and current leaks through to the motor housing. To prevent this, low-pressure compressed air (below 15 psi) should be used to carefully remove any dirt or foreign matter. Excessive air pressure or contact with winding insulation can cause damage.

Damage to insulation can also occur from motor operating conditions. For example, if the oil in a refrigeration compressor becomes acidic and deteriorates the motor insulation, a high-resistance current path may flow through the acidic oil to the grounded metal of the compressor housing.

Even though this situation results in a short circuit, the circuit fuse may not blow because the high-resistance pathway limits the current flow. The short may go unnoticed. However, a high-resistance short to ground is still an electrical shock hazard and can cause overheating and overloading of the circuit. Ground-fault monitoring equipment can be used to sense current flow in the ground and automatically de-energize the suspected circuits. Otherwise, a DMM is used to periodically measure ground current with the in-line method.

Small, noncritical motors are often run until failure occurs. Large or critical motors require scheduled inspections and testing using a megohmmeter. A *megohmmeter* is a type of ohmmeter that is used to measure the very high resistances of insulating components. A megohmmeter uses a high DC voltage to detect insulation deterioration by measuring a decline in insulation resistance over time. **See Figure 4-57.**

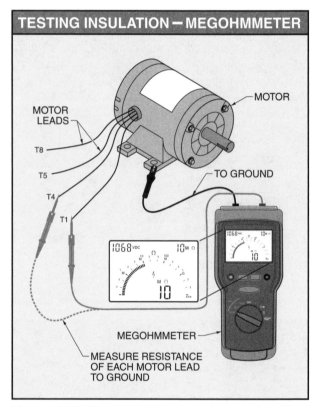

Figure 4-57. Motors can be tested regularly with a megohmmeter to determine the quality of the insulation on the wires of its coils.

TECH TIP

Typical multimeters cannot accurately measure very high or very low resistances. Very high resistance measurements, such as for motor winding insulation, require a megohmmeter. Similarly, very accurate low resistance measurements, such as for precision electronics, require a microhmmeter.

When using a megohmmeter, the motor must be disconnected from its circuit for testing. The megohmmeter is connected between each motor lead and the motor frame (ground). The amount of resistance should be in the hundreds of thousands or millions of ohms. Records should be kept of the resistance readings and compared to manufacturer's recommendations. If the resistance measurement is below manufacturer's recommendations, the insulation is deteriorating. The cause of the deteriorating insulation, such as overheating or excessive moisture, can be determined, or the motor can be serviced or replaced before it fails.

Warning: Always check manufacturer's recommendations and warnings before conducting megohmmeter tests. Never conduct a megohmmeter test on any circuit that is connected to a solid-state device or computer.

When testing insulation, the temperature of the conductor must be above the dew point of the ambient air. If the temperature is below the dew point, then moisture forming on the insulation surface, or absorbed by the material, will affect the readings. The surface of the conductor must be free of hydrocarbons and other foreign matter that can become conductive in humid conditions. Since insulation resistance is inversely proportional to insulation temperature (resistance goes down as temperature goes up), the recorded readings are altered by changes in the temperature of the insulating material. Three common types of insulation tests are the insulation spot test, dielectric absorption test, and insulation step voltage test.

An *insulation spot test* is a simple insulation resistance measurement. A megohmmeter is connected to each motor winding lead (in turn) and to ground, and the resistance is measured after 60 sec. A spot test should be performed at regular intervals (for example, every six months) to track the changes in insulation integrity. The test should also be performed after a motor is serviced. If a reading does not meet the minimum acceptable resistance, the motor should be serviced.

A *dielectric absorption test* is an insulation resistance test that checks the absorption characteristics of wet or contaminated insulation. A megohmmeter is connected to each motor winding lead (in turn) and to ground. Readings are recorded every 10 sec for the first minute and every minute thereafter for 10 min. The results are plotted on a graph. **See Figure 4-58.** The slope of the curve indicates the condition of the insulation. Good insulation gradually increases in resistance. Moist or cracked insulation has a relatively constant resistance.

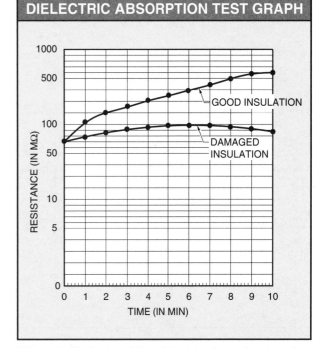

Figure 4-58. A plot of the dielectric absorption test data shows how insulation resistance changes over the 10 min test, which indicates the insulation integrity.

TECH TIP

Megohmmeters work by applying a high voltage to the insulation and measuring the resulting leakage current. Most common handheld models can conduct tests at up to 1000 V. Larger models may test insulation at up to 5000 V.

An *insulation step voltage test* is an insulation resistance test that puts electrical stress on internal insulation to reveal aging or damage not found during other motor insulation tests. The insulation step voltage test is performed only after an insulation spot test. The megohmmeter is set to 500 V output and connected to each motor winding lead (in turn) and to ground. The resistance is measured after 60 sec. The megohmmeter test leads are returned to the winding with the lowest resistance. The voltage setting is increased to 1000 V and another resistance is measured after 60 sec. Additional measurements are taken at 500 V intervals until the high voltage limit is reached for the motor. The results are plotted on a graph. **See Figure 4-59.** The resistance of good insulation remains approximately the same at different voltage levels. The resistance of deteriorated insulation decreases substantially at different voltage levels.

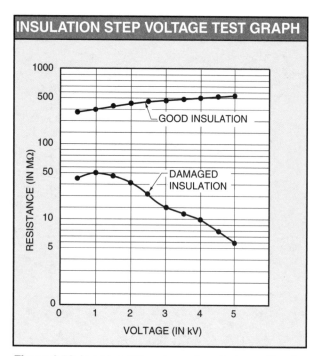

INSULATION STEP VOLTAGE TEST GRAPH

Figure 4-59. In an insulation step voltage test, good insulation has a relatively constant resistance while the resistance of damaged insulation decreases.

Lighting

Group relamping and cleaning is an important part of lighting maintenance. Replacement near the end of life expectancy can save labor cost and time. The shade and reflector should be wiped clean whenever a lamp is changed. The light fixture should be inspected for signs of overheating or other damage. Check all ventilation openings for accumulated dirt and clean as required. Most fluorescent and HID lamps produce humming from the transformers in their ballasts. The tightness of all mounts and fittings should be checked if excessive humming becomes a problem.

TECH TIP

Maintenance personnel usually find it convenient to keep their most commonly used electrical maintenance and troubleshooting tools and parts organized in a tool bag. Common electrical tools include screwdrivers, wire cutters, wire strippers, needle-nose pliers, cable ties, electrical tape, labeling stickers, twist-on wire connectors, and a digital multimeter.

5

Electronics and Programmable Logic Controllers

Electronic devices are very reliable and require less maintenance than electrical devices. All electronic devices are extremely sensitive to overvoltage and must be protected from overvoltage by stabilizers and surge and noise suppressors. Programmable controllers are used to automatically control many industrial processes. Programmable controllers are ruggedly constructed and allow easy circuit modification.

ELECTRONICS

Electronic circuits control electron flow through the use of solid-state devices. A *solid-state device* is an electronic component that switches or controls the flow of current in a circuit with no moving parts. Solid-state devices are extremely reliable and fast, and require less maintenance than electromagnetic relays and contactors.

Most electronic components operate through the movement of electrons at the atomic level. For this reason, electronic component actions cannot be seen, heard, or felt. Electronic circuits are usually assembled by robotic machines and can only be repaired by experts in electronic troubleshooting and specialized soldering techniques.

Maintenance personnel correct most common electronic circuit problems by simply replacing the printed circuit (PC) board. The inputs into and outputs from a PC board are measured to determine if the PC board is functioning correctly. The function of the PC board must be understood, but not necessarily the exact inner workings.

Printed Circuit Boards

A *printed circuit (PC) board* is a thin plate of insulating material, such as fiberglass or phenolic, with conducting paths laminated to one or both sides. Electronic circuits consist of electronic components mounted and tightly spaced on PC boards. **See Figure 5-1.** The components are connected by traces. A *trace* is a conducting path used to connect components on a PC board. A PC board must be handled with extreme care. Once installed, PC boards work reliably for decades if not subjected to overvoltage, excessive heat, or physical damage such as twisting or vibration.

Voltage Variations

Electronic devices are extremely sensitive to voltage variations. An electromechanical relay can tolerate as much as 10% overvoltage. The same percentage of overvoltage could easily destroy an electronic device. All electronic devices must be protected from overvoltage by voltage regulators and surge and noise suppressors installed according to manufacturer's recommendations.

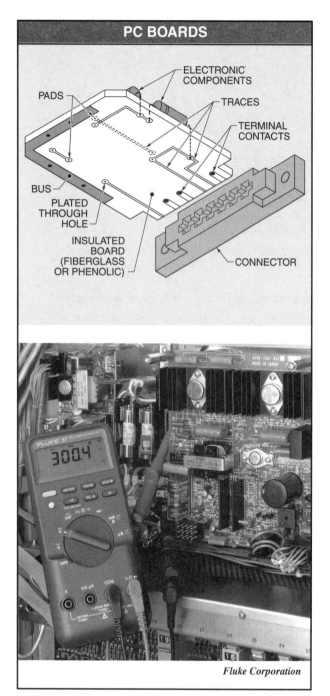

Figure 5-1. Electronic circuits consist of electronic components mounted and tightly spaced on PC boards.

TECH TIP

PC board components and traces are easily damaged. PC boards should be handled by the edges only and carefully inserted or removed from connectors when replacement is necessary.

A *voltage regulator* is a device that provides precise voltage control to protect equipment from voltage fluctuations. A voltage sag is a voltage drop of more than 10% (but not 0 V) below the normal rated line voltage that lasts from 0.5 cycles up to 1 min. A voltage swell is a voltage increase of more than 10% above the normal rated line voltage lasting from 0.5 cycles up to 1 min. **See Figure 5-2.** Voltage sags and swells are common in industrial settings. Voltage sags and swells are sometimes visible in the form of dimming or brightening of lights.

Caution: Ensure that voltage stays within manufacturer's recommended specifications.

A *transient voltage* is temporary, undesirable voltage spike, ranging from a few volts to several thousands volts and lasting a few microseconds up to several milliseconds. Transient voltages create no visible symptoms because they last for such a short time, but are very destructive. Transient voltages are caused by lightning strikes, equipment being energized and de-energized, and by other electronically controlled equipment. *Electrical noise* is a disturbance that distorts the voltage and current sine wave. Transients and electrical noise can be limited by proper electrical system design or corrected by surge and noise suppressors installed on power and data lines and as part of electronic equipment.

A *surge suppressor* is a device that limits the intensity of voltage transients that occur on a power distribution system. Surge suppressors limit voltage surges by discharging or bypassing the damaging voltage. Surge suppressors may be installed separately or may be included with a noise suppressor. A *noise suppressor* is a device that reduces the random electrical signals (noise) on power lines. Voltage regulators and surge and noise suppressors must be installed on power lines and data lines such as telephone wires and printer cables because high voltage can enter electronic devices on any of these pathways.

A high level of electrical noise is possible when many pieces of equipment share a common ground circuit. Electrical noise is further limited by using isolated grounded circuits. An *isolated grounded circuit* is a circuit that minimizes electrical noise by providing a separate grounding path. **See Figure 5-3.** Only electronic equipment should be used on isolated grounded circuits. Isolated grounded receptacles are marked with an orange face or an orange triangle on the receptacle.

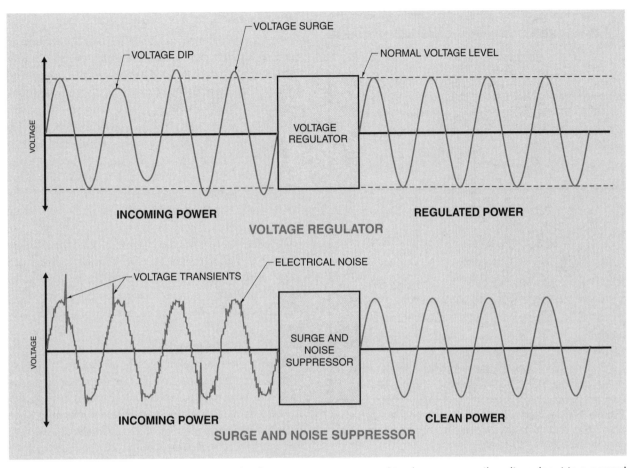

Figure 5-2. Voltage regulators and surge and noise suppressors are used to change an erratic voltage input to a normal voltage output for electronic equipment.

PC Board Handling

Handling a PC board can cause damage if static electricity is allowed to transfer from hands or clothing to the PC board. A discharge of static electricity can destroy a PC board immediately or cause future problems. Future failures might not be attributed to static electricity discharged during PC board handling.

A large percentage of electronic device failures are caused by static electricity discharges. The damage caused by a static electricity discharge is not visible because it occurs at the microscopic level. It is possible for charges of less than 3000 V to discharge from a person's fingertip without being felt. Grounded wrist straps, shielding bags, or grounded static-free floor mats must be used before handling electronic equipment. **See Figure 5-4.**

Warning: Exercise extreme caution when using any grounding equipment because the risk of electrocution is increased. When possible, grounding straps should be equipped with fuses that open if the person wearing the strap accidentally contacts high current.

Static electricity discharges can be caused by vacuuming or blowing dirt from electronic equipment. Static electricity discharge is a major cause of electronic equipment failure.

Caution: A PC board must be handled under static-free conditions only.

Removing, inserting, or handling a PC board can cause damage if the board is distorted, twisted, or compressed during handling. Gently rock a PC board when removing it. Install a PC board gently so it fits snugly into its connector.

A PC board must be stored in specially manufactured static-shielding bags. Maintaining the correct relative humidity helps to prevent static electricity discharges. Low relative humidity (dry air) causes the generation of stronger and more numerous static electricity discharges.

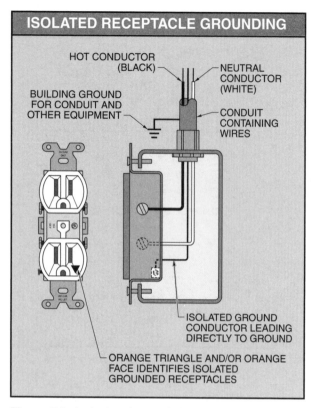

ISOLATED RECEPTACLE GROUNDING

HOT CONDUCTOR (BLACK)

NEUTRAL CONDUCTOR (WHITE)

BUILDING GROUND FOR CONDUIT AND OTHER EQUIPMENT

CONDUIT CONTAINING WIRES

ISOLATED GROUND CONDUCTOR LEADING DIRECTLY TO GROUND

ORANGE TRIANGLE AND/OR ORANGE FACE IDENTIFIES ISOLATED GROUNDED RECEPTACLES

Figure 5-3. An isolated grounded circuit uses a separate ground conductor leading from the receptacle or other connection device directly to ground.

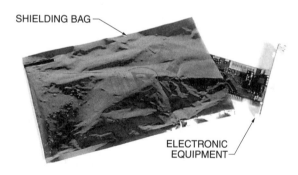

SHIELDING BAG

ELECTRONIC EQUIPMENT

Figure 5-4. Shielding bags are used to prevent damage to electronic equipment from static electricity discharges.

TECH TIP

Electronic components rarely fail on their own. Always check for additional circuit or system problems when replacing a bad electronic component. Common causes on electronic problems include high voltages, static discharges, and short circuits due to dust or debris.

Troubleshooting Electronic Circuits

In the PC board replacement method of troubleshooting, each PC board is viewed as a self-contained unit for testing and replacement. Even if a damaged component was found, it would be difficult to repair without damaging the rest of the PC board.

Testing procedure only measures the voltage and current inputs into the board and voltage and current outputs from the board. **See Figure 5-5.** For example, an input signal to a PC board should be 5 VDC when a pushbutton is closed. Output from the PC board should be 24 VDC. The PC board is replaced if the input to the PC board is correct and the output is not 24 VDC. The troubleshooter follows the circuit to locate the problem if the PC board output is 24 VDC. Actual readings may vary from circuit to circuit depending on the devices used. The voltage drop across the output of the solid-state relay is 5 V because 5 V are being used to force the electrons through the relay. This means that 115 V are left for use by the motor.

Identifying and replacing the defective PC board (PC board replacement method) is the usual repair for an electronic problem, unless a technician has specialized electronics training. This method works only when required input and output voltage and current levels are known. This information is normally obtained from the manufacturer's literature.

Caution: Obtain all system specifications before troubleshooting any electronic equipment.

Even when a technician has the skills to repair PC boards, the board schematics may not be available. Many manufacturers do not supply schematics for PC boards because their design is legally the private property of the manufacturer.

In addition, most PC boards are too complicated to be tested using a digital multimeter (DMM). Specialized equipment is required for PC board repairs. The DMM probes can cause short circuits across small traces. Voltage output from a DMM set to measure resistance can damage PC board components. Removing or replacing soldered components can generate enough heat to damage the PC board and its components.

Caution: Do not test individual PC board components or solder or desolder components on a PC board unless trained and authorized to do so.

Maintenance shops may employ specialists in electronic board repair, contract with outside electronic repair shops, or rely on manufacturers to repair damaged PC boards.

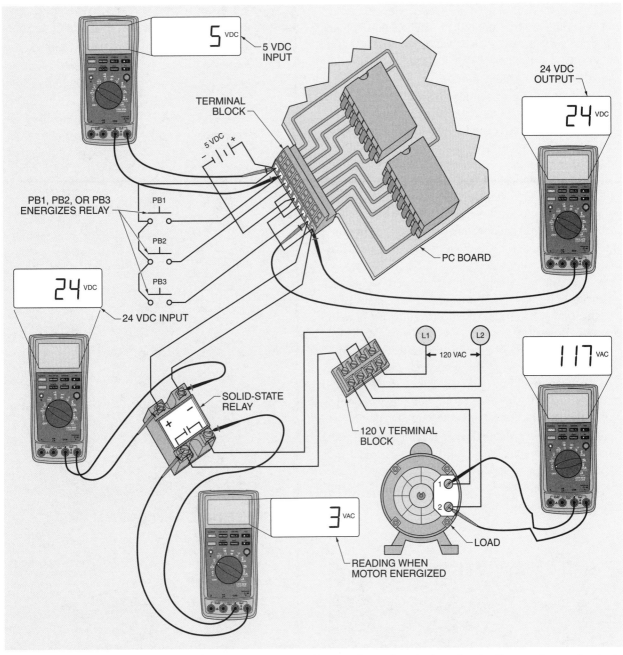

Figure 5-5. In the PC board replacement method, voltage and current inputs into the board and voltage and current outputs from the board are measured.

SOLID-STATE SEMICONDUCTOR DEVICES

A *conductor* is a material that has little resistance and permits electrons to move through it easily. An *insulator* is a material that has a high resistance. A *semiconductor* is a material with an electrical conductivity between that of a conductor (high conductivity) and an insulator (low conductivity). A semiconductor becomes a conductor when it contains certain chemicals and is exposed to specific electrical conditions. The most common semiconductors contain germanium and silicon, which have four electrons in the outer shell of their atomic structure. To make pure silicon or germanium into a useful semiconductor, tiny amounts of certain chemicals are added in the manufacturing process.

Doping is the altering of pure semiconductor material by adding small amounts of other elements. Doping alters the structure of semiconductor material to make the P-type material and N-type material used in solid-state devices. **See Figure 5-6.**

P-type material is semiconductor material with empty spaces (holes) in its crystalline structure. P-type material is made when chemicals such as indium or gallium, which have three electrons in their outer shell, are added to pure semiconductor material, which has four electrons in its outer shell. The impurity creates regular holes in the semiconductor crystalline structure. A *hole* is an electrically positive space created by the missing electron in P-type semiconductor material. The hole attracts free, negatively charged electrons when electrical conditions are correct.

N-type material is semiconductor material with free electrons in its crystalline structure. N-type material is created when chemicals having five electrons in their outer shell, such as arsenic and antimony, are added to pure semiconductor material. This creates regularly spaced free electrons in the semiconductor structure. The free electrons are available to move into the holes in P-type material when electrical conditions are correct. When this occurs, current flows through the device.

P- and N-type semiconductors are used to make all solid-state components. The difference between the various solid-state components is the number of layers, thickness, and manner in which the P-type and N-type materials are joined.

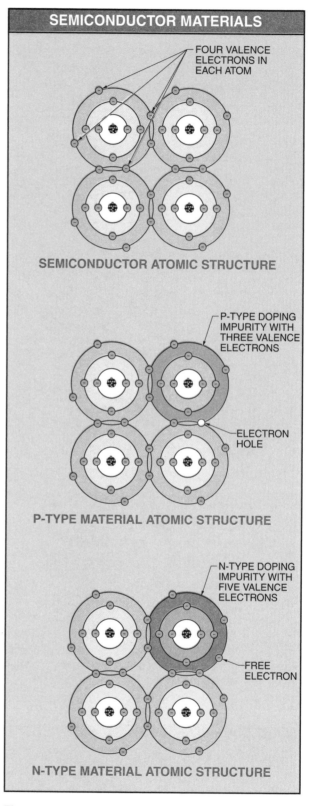

SEMICONDUCTOR MATERIALS

FOUR VALENCE ELECTRONS IN EACH ATOM

SEMICONDUCTOR ATOMIC STRUCTURE

P-TYPE DOPING IMPURITY WITH THREE VALENCE ELECTRONS

ELECTRON HOLE

P-TYPE MATERIAL ATOMIC STRUCTURE

N-TYPE DOPING IMPURITY WITH FIVE VALENCE ELECTRONS

FREE ELECTRON

N-TYPE MATERIAL ATOMIC STRUCTURE

Figure 5-6. Doping alters the structure of semiconductor material to make the P-type and N-type material used in solid-state devices.

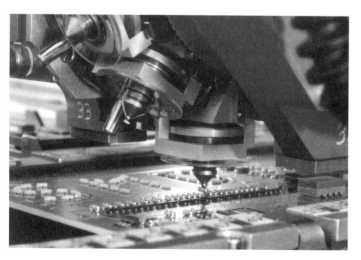

Siemens Corporation
A modular placement system can place thousands of components per hour at an accuracy of ±50 μm for the automatic manufacturing of PC boards.

Diodes

A *diode* is an electronic component that allows current to pass in only one direction. **See Figure 5-7.** A diode consists of one layer of N-type material joined to a layer of P-type material. The depletion region is located at the junction of the N- and P-type materials.

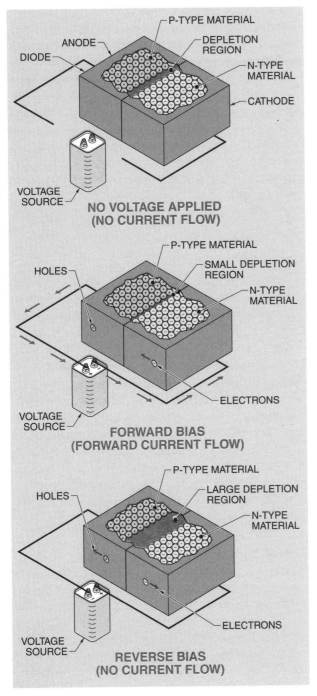

Figure 5-7. A diode allows current to pass in only one direction.

The negative terminal is referred to as the cathode and the positive terminal is referred to as the anode. A diode allows current to flow only when a positive voltage is applied to the anode (forward bias). The depletion region at the junction of the N- and P-type materials is flooded with electrons when a diode is forward-biased. Forward bias occurs when the negative terminal of a DC voltage source is connected to the negative terminal of an electronic device and the positive terminal of a DC voltage source is connected to the positive terminal of an electronic device. This causes a small depletion region, which offers less resistance to electron flow and allows current to pass through the diode. When a diode is reverse-biased, a large depletion region having a high resistance is produced. This prevents electrons from flowing over the region and current from flowing across the diode. In reverse-bias, the diode is receiving negative voltage at its anode and positive voltage at its cathode.

A diode may be destroyed if voltage or current exceeds the rating of the diode. The depletion region breaks down and the diode blows open or short circuits due to excessive heating of the diode.

Caution: Never exceed the voltage and current limitations of any electronic equipment. Like all electronic equipment, diodes can be destroyed by heat caused by excessive voltage or current.

Diodes are marked or shaped to indicate their anode (positive lead) and cathode (negative lead). **See Figure 5-8.** Diodes that are soldered onto PC boards are generally not removed or tested by maintenance personnel. Screw-in diodes may be removed from a circuit and tested. Diodes that are removed from a circuit must be returned with the anode and cathode correctly positioned.

Heidelberg Harris, Inc.

Electronic components and PLCs are an integral part of most automated industrial control systems.

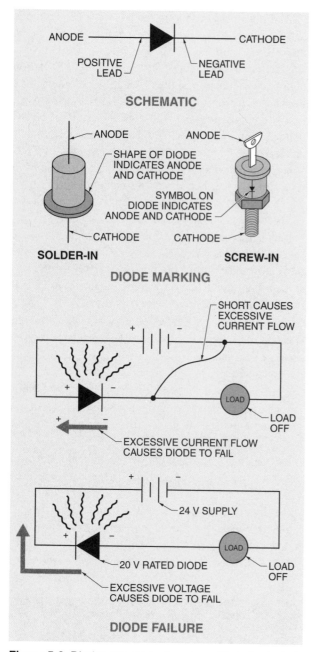

Figure 5-8. Diodes are marked or shaped to indicate their anode and cathode.

Testing Diodes. Although most electronic circuits are too complicated to allow testing of individual components, some components such as screw-in diodes are designed to be removed and tested. A high-quality DMM is used to conduct a basic diode test. **See Figure 5-9.** A diode is tested by applying the procedure:

1. Ensure that all power in the circuit is OFF. Carefully remove the diode from its connections.

2. Set the DMM function switch to the diode test mode.

3. Connect the test leads to the diode in forward bias. A tone sounds and a measurement is displayed by the meter if the diode is good. The voltage displayed is the voltage drop of the diode. The voltage drop should be within the specifications for the diode.

4. Reverse the test leads. No tone sounds and the meter reads OL if the diode is good.

A diode that does not meet these conditions is defective and should be replaced. Bad diodes are either open or shorted.

The tone produced by the DMM indicates low resistance (forward bias). When no tone is produced, the diode is offering high resistance (reverse bias). The digital readout on a DMM in diode test mode is the voltage drop offered by the diode. The exact voltage drop offered by a diode must be determined based on the manufacturer's literature. To operate correctly, the voltage drop must be within manufacturer's specifications. A typical voltage drop for a diode is 0.6 V.

Light Emitting Diodes

A *light emitting diode (LED)* is a diode that produces light when current flows through it. **See Figure 5-10.** An LED operates like a regular diode except that it produces light as electrons flow through the depletion region when forward-biased. The LED depletion region is wider and creates a higher voltage drop than in a regular diode.

Approximately 1.5 V are used to force electrons across the depletion region of an LED. Some of the voltage drop energy is converted into light. The cathode of an LED is indicated by a notch or flat side next to one of the leads. An LED cannot be tested using a regular diode tester because of the high voltage drop. An LED must be tested according to manufacturer's specifications. A failed LED is usually replaced. An LED may be used as a status indicator.

Power Supplies

A *power supply* is a device that provides the voltage required for the internal operation of a device. Power supplies are the most common electronic circuits. Power supplies contain an input section, rectifier, filter, and voltage regulator. **See Figure 5-11.** The input section often contains a transformer to change the level of AC. The rectifier converts the AC to DC. The filter and voltage regulator smooth the pulsating DC, converting it to a constant level of DC voltage. A constant level of DC voltage leaves the power supply and is supplied to the load.

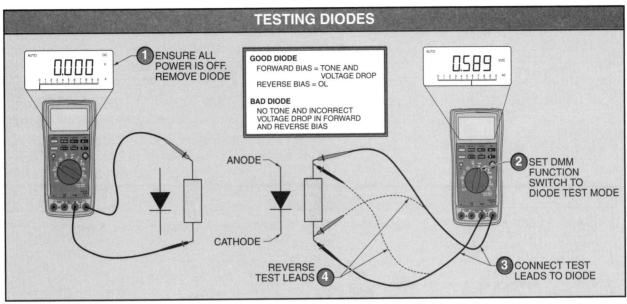

Figure 5-9. A good diode is indicated by a tone and voltage drop when forward-biased and high resistance (OL) when reverse-biased.

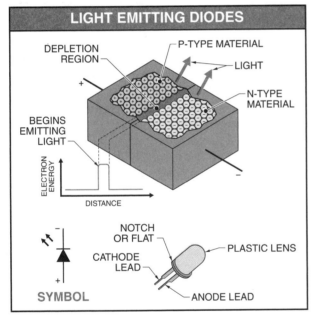

Figure 5-10. An LED produces light as electrons flow through the depletion region when forward-biased.

A *rectifier* is an electrical component that converts AC to DC by allowing voltage and current to move in only one direction. Alternating current is more efficiently generated and transmitted than DC, so AC is delivered to industrial facilities. However, AC must be changed to DC because most electronic equipment uses DC. Rectifiers include half-wave, full-wave, and full-wave bridge rectifiers. **See Figure 5-12.**

Half-Wave Rectifiers. A *half-wave rectifier* is a circuit containing a diode that permits only the positive half-cycles of an AC sine wave to pass. A half-wave rectifier acts like an electronic switch that opens and closes so rapidly that only the positive half of the AC sine wave passes through the diode. This occurs because during one half of the sine wave, the diode is forward-biased in relation to the applied AC voltage. During the other half of the sine wave, the diode is reverse-biased because the AC has changed polarity. The diode does not conduct in reverse bias, so only the positive half of the AC passes.

Full-Wave Rectifiers. A *full-wave rectifier* is a circuit containing multiple diodes that permit the positive and negative halves of an AC sine wave to pass. A full-wave rectifier converts the positive and negative halves of the AC sine wave to pulsating (unfiltered) DC. Pulsating DC voltage varies too much for most electronic applications. For this reason, filter and regulator circuits are included in power supplies to remove the pulsations and create the constant level of DC voltage required by electronic devices.

> **TECH TIP**
> *The heat generated from soldering when assembling PC boards must be precisely controlled to prevent damage to sensitive electronic components.*

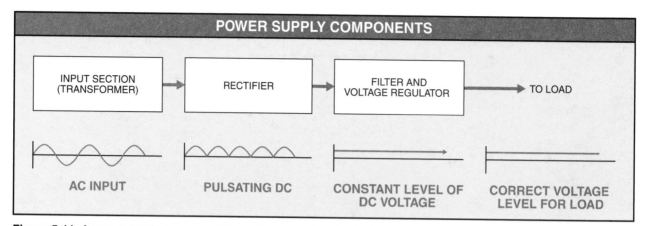

Figure 5-11. A power supply provides all the voltages required for the internal operation of a device or system.

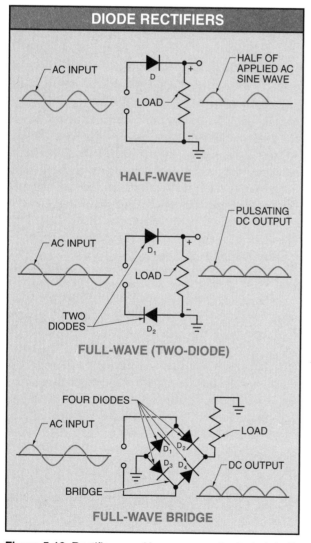

Figure 5-12. Rectifiers used in power supplies include half-wave, full-wave, and full-wave bridge rectifiers.

A two-diode full-wave rectifier uses two diodes placed back-to-back. One diode is forward-biased while the other is reverse-biased. When the incoming AC changes polarity, this changes the bias of the diodes. The positive half of the AC sine wave passes through one diode and the negative half passes through the other diode.

A *bridge rectifier* is a full-wave rectifier containing four diodes. A bridge rectifier produces the same full-wave DC output as a two-diode full-wave rectifier. Bridge rectifiers are more efficient than two-diode full-wave rectifiers because each diode blocks only half as much reverse voltage for the same output voltage. Bridge rectifier output is pulsating DC and must be filtered before it can be used in most electronic devices.

Testing Power Supplies. A power supply is tested by checking the input and output voltage of the power supply. This requires knowledge of the manufacturer's specified input and output voltages. **See Figure 5-13.** A power supply is tested by applying the procedure:

1. Measure the power supply output. A voltage output within manufacturer's specifications indicates a good power supply. A voltage output that varies greatly from manufacturer's specifications indicates a possible faulty power supply.

2. Measure the transformer primary and secondary voltage. Correct primary voltage indicates a bad transformer if the secondary voltage is incorrect. An incorrect primary voltage indicates a bad fuse or voltage source. Correct secondary voltage indicates a good transformer, fuse, and voltage source. Replace the power supply if bad. An incorrect secondary voltage indicates a possible bad transformer, fuse, or voltage source.

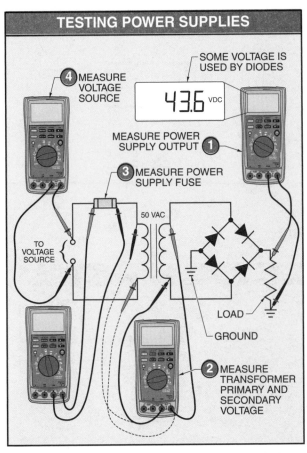

TESTING POWER SUPPLIES

SOME VOLTAGE IS USED BY DIODES

4 MEASURE VOLTAGE SOURCE

43.6 VDC

MEASURE POWER SUPPLY OUTPUT **1**

3 MEASURE POWER SUPPLY FUSE

50 VAC

TO VOLTAGE SOURCE

LOAD

GROUND

2 MEASURE TRANSFORMER PRIMARY AND SECONDARY VOLTAGE

Figure 5-13. A power supply is tested by checking the input and output voltage of the power supply.

3. Measure the power supply fuse. The fuse is replaced if it is blown and the cause of the failure is determined. The power supply is replaced if the voltage source and fuse are good and the power supply output is incorrect.
4. Measure the voltage source if the input voltage of the power supply is not correct. Spare power supplies must be available for replacement.

A voltage drop is produced across each diode in a bridge rectifier circuit. For example, a bridge rectifier with a 50 VAC input produces a 43.6 VDC output. The difference in input and output voltage results from the voltage drop across each diode in the bridge rectifier circuit.

TECH TIP
Many electronic components are analogous to components in a plumbing system. Diodes, SCRs, and transistors are like different types of valves.

Silicon Controlled Rectifiers

A *silicon controlled rectifier (SCR)* is a solid-state rectifier with the ability to rapidly switch heavy currents. The three terminals of an SCR are the anode, cathode, and gate. **See Figure 5-14.** An SCR acts like a diode except that it conducts only when triggered by a pulse of control voltage to the gate.

An SCR converts AC to DC. A single SCR controlling the speed of a DC motor allows more or less of the AC sine wave to pass to control the speed of the motor. The larger the portion of the AC sine wave that reaches the motor, the greater the overall voltage, and the faster the motor rotates.

When DC is used to trigger the gate, the SCR is switched ON (conducts) only when the gate is triggered with the correct level of voltage. **See Figure 5-15.** The SCR continues to conduct until the current flowing through the SCR drops below its holding current level. *Holding current* is the minimum current required to keep an SCR conducting.

Once the current flow in the SCR drops below its holding current level, the SCR does not conduct until the gate is triggered again. An SCR can control the level of voltage applied to a load by varying the timing of the gate trigger voltage. The gate control circuit contains high-speed switching and timing circuits that can trigger the gate at any time during the AC sine wave to pass only part of the sine wave. The result is that a greater or lesser amount of voltage reaches the load.

Fluke Corporation
A DMM may be used to test the electronic components of a motor control circuit.

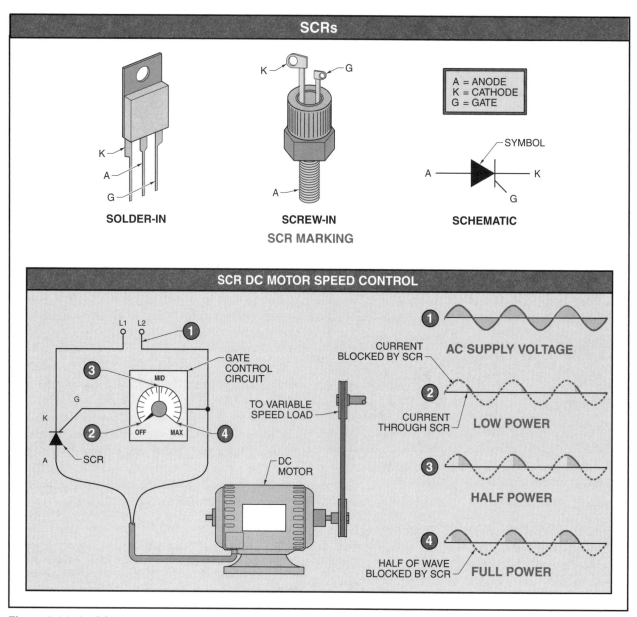

Figure 5-14. An SCR is a solid-state rectifier with the ability to rapidly switch heavy currents.

When AC is used to trigger the gate, the SCR stops conducting with each reversal of the sine wave cycle. This occurs because current falls below the holding current level with each sine wave reversal of the AC. One SCR can transmit only one half of an AC sine wave. Two SCRs installed back-to-back are needed to conduct both halves of the AC sine wave. Two SCRs can be replaced by a triac that can control both halves of the AC sine wave.

A *triac* is an AC semiconductor switch that is triggered into conduction in either direction. A triac operates like two back-to-back SCRs with one triggering gate.

See Figure 5-16. The three terminals of a triac are main terminal 1, main terminal 2, and the gate. A triac is commonly used to control the level of voltage that reaches a load such as an electric heater. The heating element becomes hotter when more of the sine wave reaches the heating element.

A triac or two SCRs can transmit all or part of an AC sine wave. They can be triggered at any time during the sine wave, acting as if they contained a switch that turns ON and OFF so quickly that it passes only a portion of an AC cycle. The result is that more or less of the sine wave is applied to the load.

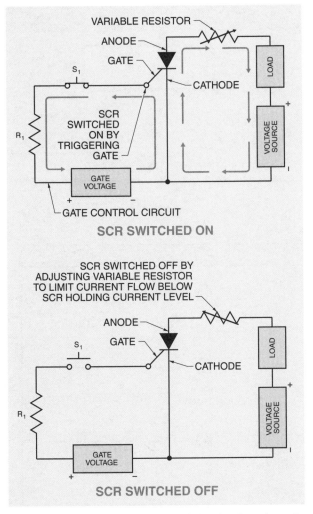

VARIABLE RESISTOR

ANODE

GATE

S₁

CATHODE

LOAD

SCR
SWITCHED
ON BY
TRIGGERING
GATE

R₁

VOLTAGE
SOURCE

+

GATE
VOLTAGE

+   −

GATE CONTROL CIRCUIT

SCR SWITCHED ON

SCR SWITCHED OFF BY
ADJUSTING VARIABLE RESISTOR
TO LIMIT CURRENT FLOW BELOW
SCR HOLDING CURRENT LEVEL

ANODE

GATE

S₁

CATHODE

LOAD

R₁

VOLTAGE
SOURCE

+

GATE
VOLTAGE

+   −

SCR SWITCHED OFF

Figure 5-15. An SCR is switched ON only when the gate is triggered with the correct level of voltage.

Testing SCRs. High-power SCRs should be tested using a test circuit. Low-power and some high-power SCRs may be tested using a DMM set to measure resistance. Screw-in SCRs can be removed from a circuit and tested using a DMM. **See Figure 5-17.** An SCR is tested using a DMM by applying the procedure:

1. Ensure all power is OFF to the SCR.
2. Set the function switch to resistance mode.
3. Connect the negative test lead to the cathode.
4. Connect the positive test lead to the anode.
5. Short circuit the gate to the anode using a jumper wire. The DMM should read almost 0 Ω. Remove the jumper wire. The low-resistance reading should remain.
6. Reverse the DMM leads so the positive lead is on the cathode and the negative lead is on the anode. The DMM should read high resistance.

7. Short circuit the gate to the anode with a jumper wire. The high-resistance reading should remain. This simulates the operation of the SCR by triggering the gate in reverse- and forward-bias.

The SCR is defective and must be replaced if it does not respond as indicated for each step. Triacs and SCRs that are installed back-to-back in sealed, modular units must be tested according to manufacturer's specifications.

Transistors

A *transistor* is a three-terminal semiconductor device that controls current flow depending on the amount of voltage applied to the base. A transistor may be used to start or stop current flow (switch) or to increase current flow (amplify). **See Figure 5-18.** Most transistors are used as current control devices. The three terminals of a transistor are the emitter, base, and collector. The lead with the arrow is the emitter. Transistors consist of three pieces of semiconductor material joined in either an NPN or PNP pattern.

Transistors are available in many shapes and sizes. Some transistors have three leads while others have two. In transistors having two leads, the transistor case serves as the third lead.

Fluke Corporation

The proper test equipment must be used when testing electronic components such as transistors, SCRs, and diodes to prevent damage to the components.

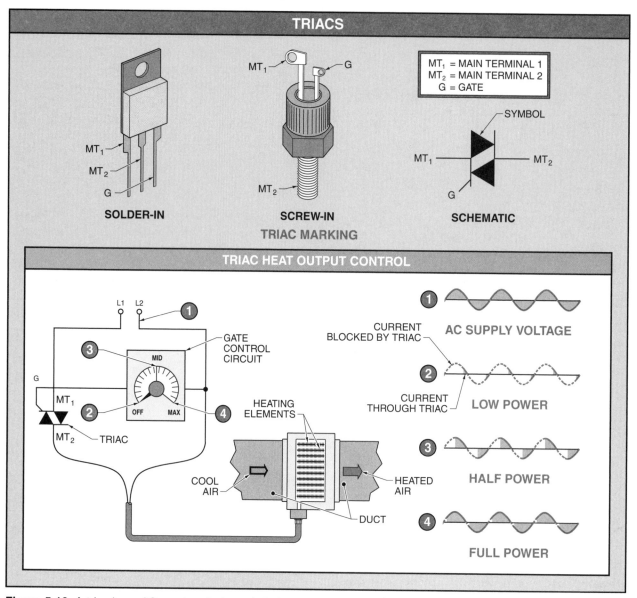

Figure 5-16. A triac is an AC semiconductor switch that is triggered into conduction in either direction and is commonly used to control the level of voltage to an electric heater.

A transistor does not allow current to flow between the emitter and collector until current is applied to the base of the transistor. A small amount of current applied to the base of the transistor allows a large amount of current to flow in the emitter-collector circuit. Current flows in both the base-emitter circuit and the emitter-collector circuit when the transistor is switched ON. The base-emitter current is very low (approximately 5%) compared to the emitter-collector current. Therefore, a small amount of current can turn ON and amplify current flow to a larger device. Today, transistors and transistor functions are being replaced by the miniature circuits contained in integrated circuits.

Integrated Circuits

An *integrated circuit (IC)* is a circuit composed of multiple semiconductor devices, providing a complete circuit function, in one small semiconductor package. ICs can incorporate the functions of thousands of transistors and other devices to create complex circuits. ICs receive inputs and send outputs through connecting pins. **See Figure 5-19.**

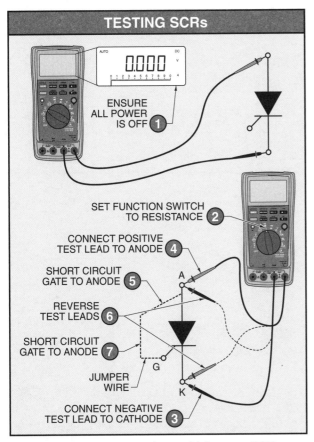

TESTING SCRs

AUTO   0.000   DC   V

ENSURE ALL POWER IS OFF ①

SET FUNCTION SWITCH TO RESISTANCE ②

CONNECT POSITIVE TEST LEAD TO ANODE ④

SHORT CIRCUIT GATE TO ANODE ⑤

REVERSE TEST LEADS ⑥

SHORT CIRCUIT GATE TO ANODE ⑦

JUMPER WIRE

CONNECT NEGATIVE TEST LEAD TO CATHODE ③

A

G

K

Figure 5-17. Low-power and some high-power SCRs may be tested using a DMM set to measure resistance.

ICs are very small, are extremely reliable, generate little heat, and are relatively inexpensive. ICs are used in low-current control circuits rather than high-current power circuits because they cannot handle large current flow. ICs cannot have the power for a motor running through them, but they can control the operation of the motor or other high-current or high-voltage loads through relays or similar devices. ICs cannot be repaired, so they are replaced when damaged.

The development of ICs has enabled the extreme miniaturization of electronic devices. Complex industrial processes are controlled by several ICs mounted on PC boards. The boards are assembled in clean, climate-controlled rooms to prevent contamination. The components are placed on the board and soldered by computer-controlled robotic equipment. Due to their complexity, PC boards cannot usually be repaired. They are usually replaced when defective, even if the failure is due to only one component, because PC boards are easily damaged further when a repair attempt is made.

Solid-State Relays

A *relay* is an electrical switch that is actuated by a separate circuit. A *solid-state relay (SSR)* is an electronic switching device that has no moving parts. **See Figure 5-20.** SSRs do not contain contacts, resist shock and vibration, create no arc when switching, can switch thousands of times a second, and are sealed, tamperproof units.

Many SSRs use optocouplers to electrically isolate the input circuit from the output circuit. An *optocoupler* is a device that converts electrical input signals to light signals and the light signals back to electrical signals on the output side. An optocoupler uses an LED in the input circuit and a photodetector (photodiode or phototransistor) in the output circuit. A *photodiode* is a diode that produces current when absorbing light. A *phototransistor* is a device that combines the effect of a photodiode and the switching capability of a transistor.

The LED emits light when it receives forward-bias input voltage. The light crosses a short space and is received by the photodetector. The light striking the photodetector causes it to conduct (close), energizing the load. SSRs are tested by measuring input and output voltage because they are sealed, non-repairable units.

SSR Leakage Current and Voltage Drop. Solid-state relays and other solid-state devices have no moving contacts. Their operation is described as if they contain contacts because they mimic electromechanical devices that have moving contacts. The opening and closing of the solid-state contacts is a result of changes in the molecular activity in the device. For this reason, when a solid-state device is open (not conducting), it is not completely open electrically. For example, a DMM set to measure voltage placed across solid-state contacts reads less than full source voltage when the device is not conducting. *Leakage current* is the small amount of current flowing through a solid-state device when it is not conducting. **See Figure 5-21.**

A voltage drop exists across the contacts of an SSR when the contacts are closed (conducting). The voltage drop is the voltage used in forcing the current through the solid-state devices. Mechanical contacts should offer nearly 0 V drop when closed. Solid-state contacts in SSRs and other electronic devices offer some voltage drop. Exercise care when taking voltage, resistance, and current measurements in circuits containing electronic devices because these readings can be confusing. Manufacturer's specifications and circuit schematics must be consulted when taking circuit measurements.

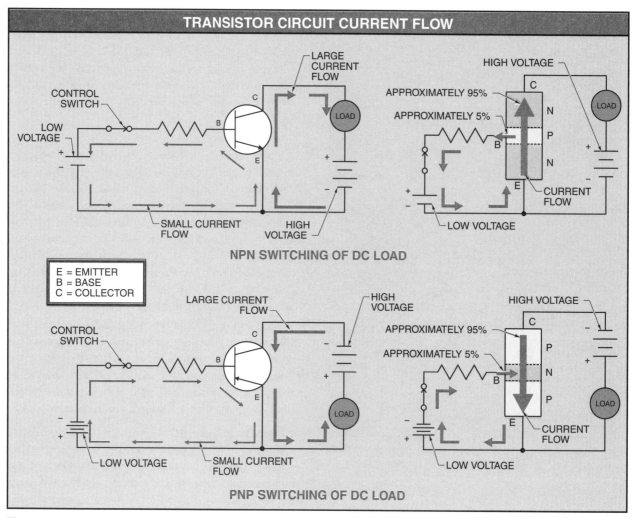

Figure 5-18. A transistor is a three-terminal device that controls current flow depending on the amount of voltage applied to the base.

Transducers

A *transducer* is a device used to convert physical parameters such as temperature, pressure, and weight into electrical signals. A *thermistor* is a transducer that changes resistance in response to a change in temperature. A thermistor may be used in a room thermostat to control the temperature of the air. **See Figure 5-22.** As the temperature of the air in the room changes, the resistance of the thermistor changes, allowing a control signal from 4 mA to 20 mA to pass through the thermistor to the heating element controller.

The thermistor circuit operates based on the exact level of DC input voltage and responds to changes in resistance according to Ohm's law. As the temperature rises, the resistance decreases, allowing more current to pass through the thermistor to the heating element

controller. The electronic circuits in the controller receive the current signal, amplify it, and send it to the heating element controller. The signal arriving at the heating element controller determines the heat output of the heating element.

A decrease in temperature measured by the thermistor causes its resistance to increase, decreasing current flow to the heating element controller. At a specific temperature, the thermistor has a specific resistance and allows a specific level of current to pass through to the controller. Thermistors can be used in almost any temperature control system or temperature sensing system.

Pressure, sound, vibration, light, and chemical levels can also be measured using transducers. Some transducers change resistance, others generate small voltages that are amplified electronically to control various loads.

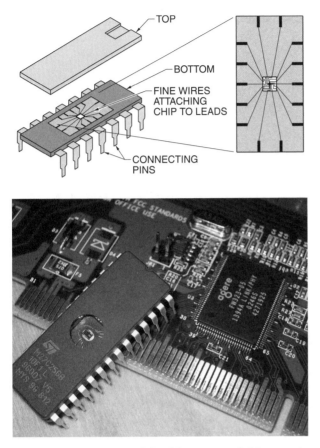

Figure 5-19. An integrated circuit is composed of thousands of semiconductor devices providing a complete circuit function in one small semiconductor package.

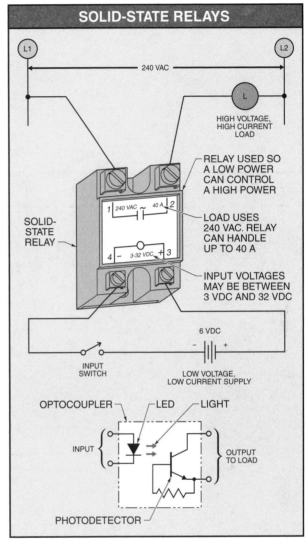

Figure 5-20. A solid-state relay is an electronic switching device that has no moving parts.

Transducers are tested by comparing their operation during changing conditions. For example, a thermistor is tested by measuring its resistance and comparing it to the manufacturer's specifications for a given temperature. The thermistor is then cooled or heated and the changing resistance is compared to the specifications. If the thermistor is not within specifications, it is replaced with an identical one or one having the same operating characteristics.

The output of a thermistor can be tested by measuring current flow in the circuit. For example, if the temperature is 68°F in a room, the system specifications may call for a current of 13 mA. The thermistor is replaced if the current is not at the correct level. The circuit is checked for incorrect input voltage, loose connections, the wrong size wire, or wiring runs that are too long if the incorrect current level continues after thermistor replacement. Transducers measuring other conditions are tested in a similar manner.

Thermocouples. A *thermocouple* is a device that produces electricity when two different metals that are joined together are heated. **See Figure 5-23.** Thermocouples consist of two dissimilar wires, such as copper and copper/nickel, that are welded at one end. This end is used to sense temperature. The other ends of the wires are connected to an electronic controller. The welded end is placed in the area where the temperature is to be measured. The application of heat causes the dissimilar wires to generate a small voltage. The small voltage produces a small current signal that is sensed by the electronic controller. The electronic controller amplifies the signal to display the temperature measurement or to operate devices that control the heat of the space in which the temperature is being measured.

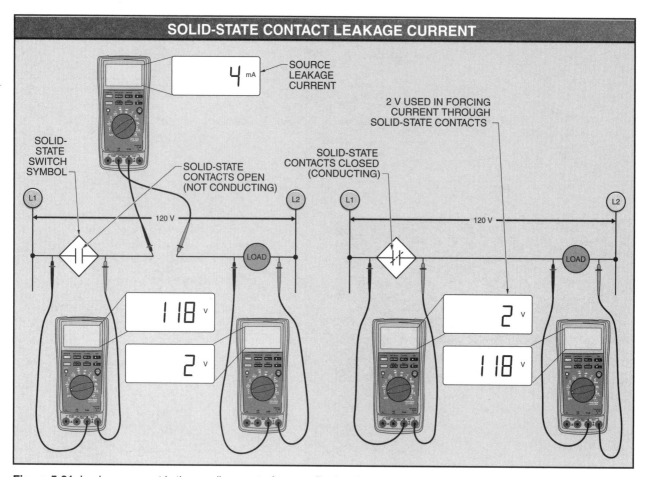

Figure 5-21. Leakage current is the small amount of current flowing through a solid-state device when it is not conducting.

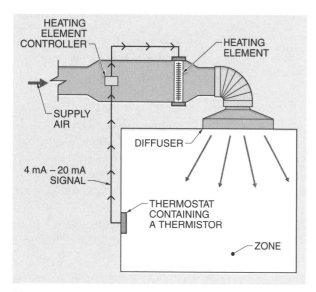

Figure 5-22. A thermistor, which contains a thermal resistor that changes resistance in response to a change in temperature, may be used in a room thermostat to control the temperature of the air.

Thermocouple wires must be matched for the temperature range to be measured. The wires must be of the correct length because wires that are too long reduce the voltage signal arriving at the controller. Thermocouples can be tested using a DMM set to measure resistance. The resistance of the thermocouple is measured and compared to the manufacturer's specifications. The thermocouple is bad if the reading is not within manufacturer's specifications, or displays overload (OL) or 0 Ω.

A DMM set to measure voltage can be used to test the operation of a thermocouple. The voltage output is measured and compared to the temperature at the heated end. The temperature and voltage measurements are compared to manufacturer's specifications. Thermocouples that are not within specifications are replaced. Spare thermocouples must be kept on hand to avoid lost time. Thermocouples should be inspected regularly for signs of wear and overheating.

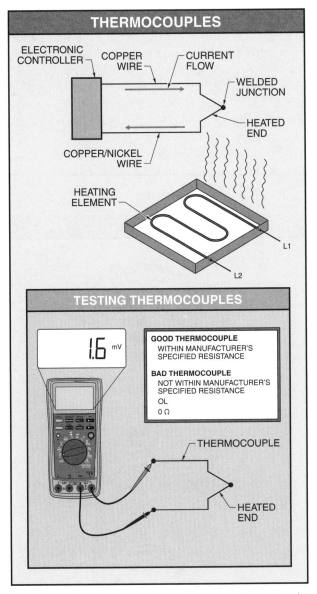

Figure 5-23. The application of heat to a thermocouple causes the dissimilar wires to generate a small voltage in the wires.

Proximity Sensors. A *proximity sensor* is a device that reacts to the nearness of a target without physical contact. **See Figure 5-24.** Mechanical limit switches depend on physical contact between the target and the switch operator to open or close the contacts. For this reason, physical contact limit switches can be easily damaged or knocked out of position. Electronic proximity sensors have replaced many mechanical limit switches because they do not require physical contact with the target for operation.

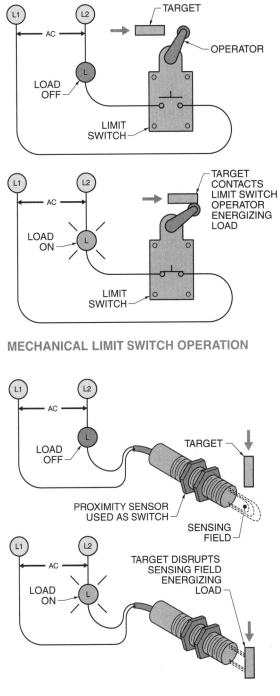

PROXIMITY SENSOR OPERATION

Figure 5-24. A proximity sensor reacts to the proximity of a target without physical contact.

TECH TIP

Always consider the sensing distance, target material, size, method of actuation, and temperature tolerance when selecting a proximity sensor for an application.

Proximity sensors operate when an object (target) enters the sensing field generated by the sensor. The sensing field is altered when the target enters the field. The altering of the sensing field causes an action in the circuit.

Photoelectric Switches. A *photoelectric switch* is a solid-state sensor that can detect the presence of an object by means of a light beam. **See Figure 5-25.** A photoelectric switch may be used to count cartons passing on a conveyor. One count is recorded by the control circuit each time the light beam generated by the photoelectric switch is interrupted.

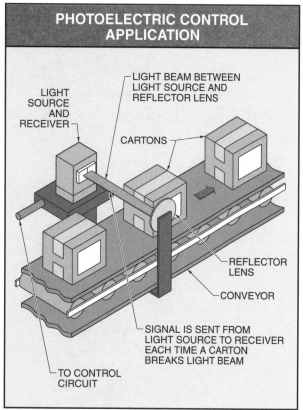

Figure 5-25. A photoelectric switch detects the presence of an object by means of a light beam.

Photoelectric switches can be used in many applications that do not allow the product to be touched. For example, photoelectric switches can be used to count hot or freshly painted parts. The most common problems with photoelectric switches are misaligned lenses and reflectors, burned-out bulbs, and dirty lenses. Keep photoelectric switch lenses clean and correctly aligned.

Timers

A *timer* is a control device that uses a preset time period as a control function. Most timers are electronic devices. Timers may have a fixed time setting (fixed duration) or an adjustable time setting, and may have one timing function or be programmable for several functions. **See Figure 5-26.** Electronic timers have replaced motor-driven and mechanically driven timers because of their accuracy, durability, and small size. The four basic timing functions include ON-delay, OFF-delay, one-shot, and recycle.

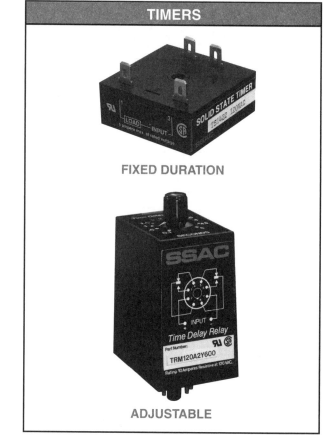

SSAC, Inc.

Figure 5-26. Timers may have a fixed duration or have an adjustable time setting.

ON-Delay. An *ON-delay timer* is a timer that switches its contacts after being energized and then a preset time period elapses. Energizing a machine that contains an ON-delay timer energizes the timer. The timer starts timing immediately, and after the time period has elapsed (timed out), the loads controlled by the timer are energized or de-energized. For example, an ON-delay timer may be used to prelubricate a large diesel engine.

When the start button is pressed, an ON-delay timer is energized but the engine does not start immediately. The start of the engine is delayed by the timer so the lubricating pump can circulate oil to lubricate the engine for easier starting. After the timer has timed out, the engine starts and the lubricating pump continues to operate.

OFF-Delay. An *OFF-delay timer* is a timer that switches its contacts after being de-energized and then a preset time period elapses. De-energizing a machine that contains an OFF-delay timer energizes the timer. After the timer has timed out, the load controlled by the timer is energized or de-energized. For example, when the stop button controlling a large diesel engine is pressed, the motor stops immediately but the lubricating pump continues to operate for a few minutes. The signal that de-energizes the engine energizes the OFF-delay timer. The OFF-delay timer keeps the lubricating pump operating long enough for the bearings to cool and for the engine to coast to a stop without damaging the bearings due to lack of lubrication. After the timer has timed out, the lubricating pump stops.

One-Shot. A *one-shot timer* is a timer that switches its contacts immediately and remain changed for the set period of time after the timer has received power. One-shot timers energize or de-energize a load once. The timer is energized, then after timing out, the load is energized or de-energized. The timer must be re-energized for the timing to repeat. A time-controlled safety shower is an example of a one-shot timer application. Turning ON the shower starts the timer and the water flow. After the timer has timed out, the water flow stops and is not restarted until the shower is turned ON, usually by adding money to the shower control.

Recycle. A *recycle timer* is a timer that switches its contacts open or closed repeatedly at a set interval once the timer has received power. A recycle timer energizes and de-energizes the load repeatedly after the timer has been energized. The cycling of the contacts continues until power is removed from the timer. Traffic control lights are often controlled by recycle timers.

Harmonic Distortion

Harmonic distortion is the addition of voltage and/or current in a power line that is a multiple of the fundamental line frequency. For example, the fundamental line frequency in the U.S. is 60 Hz. The 60 Hz fundamental line frequency may have harmonic distortion composed of 180 Hz (third harmonic) and 300 Hz (fifth harmonic) frequencies. **See Figure 5-27.** Harmonic distortion is caused by electronic equipment whose high-speed switching creates distortions in the voltage and current sine waves. These distortions can cause overheating in transformers and neutral conductors, motor burnout, tripped circuit breakers, and blown fuses. The amount of harmonic distortion is determined using a power quality meter.

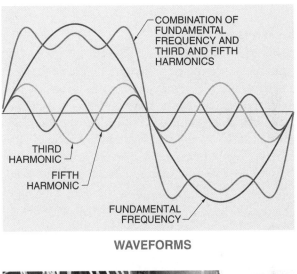

WAVEFORMS

Fluke Corporation

Figure 5-27. Harmonic distortion is caused by electronic equipment in which high-speed switching creates distortions in the voltage and current sine waves.

Harmonic distortion can be partially corrected by adding harmonic filters to circuits. The heating effect of harmonic distortion can be corrected by installing large neutral conductors, balancing loads so that even current flows in each circuit, and derating transformers. To derate a transformer is to operate it at lower than full-rated capacity. For example, a transformer rated at 1 kVA could have loads removed from it so that the transformer never carries more than 0.75 kVA. A higher rated transformer may be installed if it is not possible to remove loads. Locating and correcting harmonic distortion requires the help of electronic specialists.

Inverters

An *inverter* is a device that changes DC voltage into AC voltage of any frequency. Inverters are one of the most common industrial circuits and are used to control motor operation. Inverters contain computers that monitor speed and torque and constantly adjust the frequency and voltage level applied to a motor. Speed is controlled by adjusting the frequency of the voltage applied to the motor. The higher the frequency, the higher the speed. Frequencies below 60 Hz cause a motor to run more slowly than its rated speed. Frequencies above 60 Hz cause a motor to run faster than its rated speed.

Torque is controlled by adjusting voltage within the safe operating range of the motor. The motor cannot be operated at low or high voltage for very long before it overheats. Therefore, overload circuits are included in inverters to stop a motor if it cannot generate the needed torque for the load. Generally, the higher the voltage, the higher the torque. Voltage is controlled by devices such as SCRs.

Inverters can also create soft starts, in which the voltage and frequency of the motor are increased smoothly from startup to full speed. *Ramping* is evenly raising or lowering the voltage supply to a device. Ramping causes a motor to start or stop gradually. Voltage can also be increased in stages. *Stepping* is the pulsing of the voltage supply to a device. Stepping causes a motor to rotate in individual steps. This can be used to accelerate a motor or move it into a specific position/step. Inverters can also incorporate braking circuits that apply DC current to an AC motor to stop it quickly.

An inverter controlling a 3ϕ motor consists of a rectifier, filter, braking section, control section, and inverter. **See Figure 5-28.** The AC power input is rectified by a 3ϕ rectifier. A 3ϕ rectifier consists of three 1ϕ rectifiers

containing diodes or SCRs. The filter smooths the pulsations of the rectified DC and produces a constant DC voltage. The braking section stops the motor. The control section contains a computer that operates the inverter. The control section can be monitored and controlled by a remote computer using a keypad. The inverter converts the constant DC voltage into pulsating DC that simulates AC at any frequency to control motor speed.

Troubleshooting Inverters. Most inverters are controlled by computers that offer a wide variety of self-diagnostic capabilities. The self-diagnostic capabilities are usually displayed as fault or error codes that indicate computer malfunctions, high or low input voltage, a shorted motor, electrical interference, or loss of sensor signal from the motor. Common causes of inverter problems are excessive heat, incorrect input voltage, mechanical vibration, excessive braking, heavily loaded starts, and damage to the inverter program.

Inverters must be equipped with voltage surge and noise suppressors. Inverters cannot be repaired without following the manufacturer's recommendations. The motors controlled by inverters often require additional cooling and higher quality insulation to resist overheating. The internal fan mounted on the motor shaft does not move quickly enough to cool the motor properly when the motor runs at slower-than-normal speeds. Always consult the inverter and motor manufacturer to select the correct motor for the application.

PROGRAMMABLE LOGIC CONTROLLERS

A *programmable logic controller (PLC)* is a solid-state control device that is programmed and reprogrammed to automatically control an industrial process or machine. A PLC consists of a power supply, input and output modules (I/O modules), inputs (pushbuttons, limit switches, or other switching devices), outputs (magnetic motor starter coils, lights, solenoids, or other loads), a processor (CPU), programming terminal, monitor, and programs (data files). **See Figure 5-29.**

TECH TIP

Electronic components have many advantages including low cost, high speed, and small size. Their main disadvantage is that they cannot control high voltages and currents.

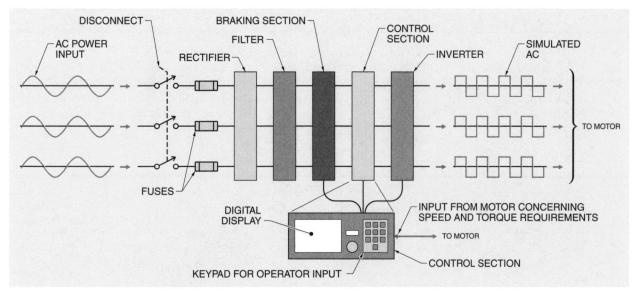

Figure 5-28. An inverter controlling a 3φ motor consists of a rectifier, filter, braking section, control section, and inverter.

PLC Operation

The power supply provides power to the CPU and I/O modules. The power supply contains rectifiers, surge suppressors, and voltage regulators because precise voltage is required for correct PLC operation. The input module connects inputs such as pushbuttons, limit switches, and electronic transducers to the CPU. Signals from the inputs are converted to low-voltage DC signals used by the CPU. Each input is connected to a specific terminal of the input module. Each input and output receives an address. An *address* is an identification number assigned to a specific input or output. Addresses are locations where the CPU scans for information from the inputs and outputs. During each scan, the computer senses the inputs that are activated or deactivated and the outputs that are energized or de-energized. The output module connects the CPU to the loads.

The CPU scans the input addresses many times a second, checking for a signal that has arrived at or been removed from an address. When a signal arrives or is removed from an input address, the CPU checks its program for the action to be done in response to the signal change. For example, when the start pushbutton connected to Input 1 (IN 1) is closed, the CPU is programmed to turn ON the magnetic motor starter coil M1 connected to Output 1 (OUT 1). The signal arriving at IN 1 is 24 VDC. The input module converts the 24 VDC into 5 VDC for use inside the CPU. The CPU sends a 5 VDC signal to OUT 1. Output 1 sends a 120 VAC signal

to M1, which starts the motor. The voltages can vary from system to system depending on the application.

The CPU is operated by programs that are created or modified by typing instructions on the programming terminal. The programming terminal can be permanently attached to the CPU or can be connected only when needed. Programming terminals may be full computer keyboards or portable keypads. Programmable logic controller programs use various computer languages, but most display the program visually using line diagram symbols similar to electrical circuit line (ladder) diagrams. **See Figure 5-30.** Some computer programs allow the addition of text to explain the operation of the program.

Rockwell Automation, Allen-Bradley Company, Inc.
PLC output modules receive low-power signals from the CPU and convert them into high-power signals to control the system loads.

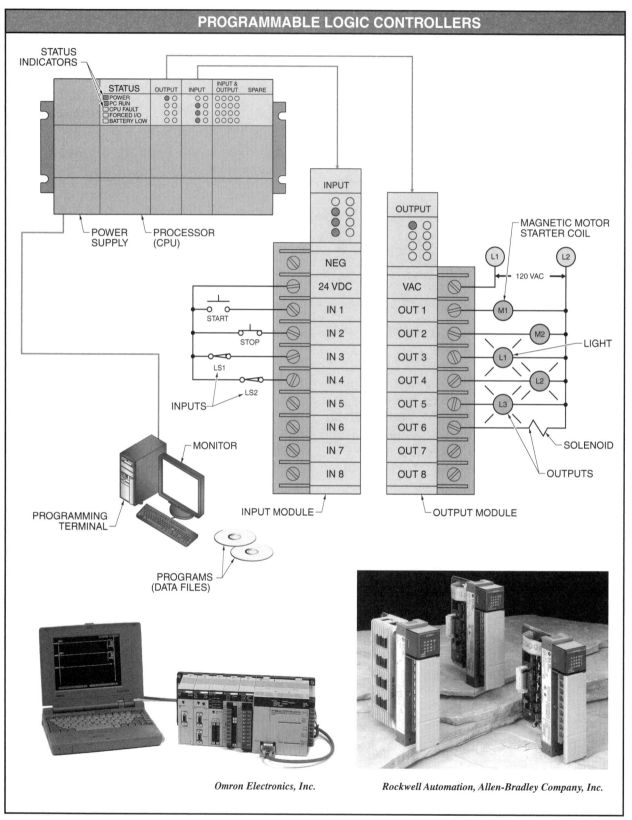

PROGRAMMABLE LOGIC CONTROLLERS

STATUS INDICATORS

STATUS
- POWER
- PC RUN
- CPU FAULT
- FORCED I/O
- BATTERY LOW

OUTPUT INPUT INPUT & OUTPUT SPARE

POWER SUPPLY

PROCESSOR (CPU)

INPUT

OUTPUT

MAGNETIC MOTOR STARTER COIL

L1 L2

120 VAC

NEG

24 VDC

IN 1
START

IN 2
STOP

IN 3

IN 4
LS1

IN 5
LS2

IN 6

IN 7

IN 8

INPUTS

VAC

OUT 1 M1

OUT 2 M2

OUT 3 L1 LIGHT

OUT 4 L2

OUT 5 L3

OUT 6 SOLENOID

OUT 7 OUTPUTS

OUT 8

INPUT MODULE

OUTPUT MODULE

MONITOR

PROGRAMMING TERMINAL

PROGRAMS (DATA FILES)

Omron Electronics, Inc.

Rockwell Automation, Allen-Bradley Company, Inc.

Figure 5-29. A PLC is a solid-state control device that is programmed and reprogrammed to automatically control an industrial process or machine.

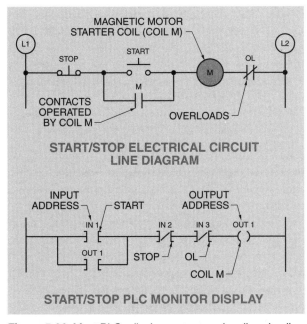

Figure 5-30. Most PLCs display programs visually using line diagram symbols similar to electrical circuit line diagrams.

After the line diagram logic program is created or modified and installed in the CPU, it is copied (backed up) on two separate sets of computer data storage. One set is stored on-site, the other is stored in another location in case the on-site data is lost or damaged. If the line diagram logic program in the CPU is lost or damaged, it can be easily reinstalled from the back-up data. *Backing up* is the process of saving computer data to more than one storage device so it is secure.

Caution: Always back up PLC programs.

A PLC receives inputs and activates loads. Inputs can also activate internal PLC loads such as relays, timers, and counters. These internal loads exist only in the program of the CPU but act exactly like the physical timers, relays, and counters that the PLC replaces. Internal loads take up no space, need no maintenance, are extremely reliable and accurate, are included in the cost of the program, and their settings can be modified without changing any wiring or replacing components. This is a major advantage of PLCs over electromechanical equipment.

PLC Applications

PLCs are commonly used in industrial electrical systems that are designed to manufacture a product. PLCs are useful in increasing production and improving overall plant efficiency. PLCs can control individual machines and link the machines together into a system. The flexibility of PLCs has found many applications in manufacturing and process control.

Industrial control has gone through many changes. In the past, industrial controls were mostly manual. Flow, temperature, level, pressure, and other control functions were monitored and controlled at each stage by production workers. Today, using PLCs, an entire process can be automatically monitored and controlled with few or no workers involved. PLCs can be used to automate processes and systems such as manufacturing and assembly processes, industrial robots, material handling machines, and fluid power systems. **See Figure 5-31.**

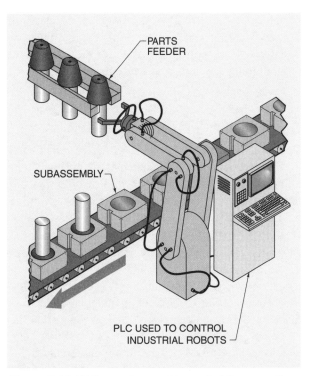

Figure 5-31. PLCs can be used to control the operations of an industrial robot.

Controls must be synchronized when machines are linked together to form an automated system. Several machines may be controlled by a PLC, with another controller synchronizing the operation. This is likely if the machines are purchased from different manufacturers. In this case, each machine may include a PLC to control all the functions on that machine only. If the machines are purchased from one manufacturer or designed in-plant, it is possible to use one large PLC to control each machine and synchronize the process.

Adjustable speed controls are available to control the speed of AC and DC motors. These controls are normally manually set for the desired speed, but many allow for automatic control of the set speed. AC drives can accept frequency and direction commands from a PLC output module.

Batch processing blends sequential, step-by-step functions with continuous closed-loop control. Process control is systems control, and systems are made up of many parts. Individual PLCs can be used to control each part and step of the process, with additional PLCs and computers supervising the total operation.

An operator interface is used for instrumentation or other monitoring functions. This may be in the form of an instrumentation and process control station, a human-machine interface (HMI), or any other type of interface. To aid in interfacing and monitoring a programmable-based system, a serial port is used for monitoring and programming a system with a computer.

Troubleshooting PLCs

Operating and troubleshooting a PLC is accomplished using PLC manuals, system schematics, and documentation. Even different models or versions of PLCs from the same manufacturer can have different programming instructions.

A *status indicator* is a light, number, or word that shows the condition of the components in the programmable logic controller. Status indicators are checked immediately when a PLC malfunctions. The troubleshooter proceeds according to the recommendations of the PLC manuals. Status indicators are usually mounted on the processor. **See Figure 5-32.** Many PLC problems can be solved with a DMM, spare input and output modules, and the PLC manuals. Common PLC problems include total shutdown, partial faults, and input and output problems.

Total Shutdown. A *total shutdown* is the malfunction of an entire machine. Check the power status indicator if a PLC has suffered a total shutdown. Check the source voltage if the power status indicator is OFF. The computer has a programming problem or is receiving a faulty input if the power status indicator is ON but the PC run status indicator is OFF and the CPU fault status indicator is ON. Access the PLC program using the programming terminal and use its diagnostic programs to determine if the problem is in the CPU or the I/O modules.

The faulty input could come from a damaged I/O module if the PC run status indicator turns ON when a module is replaced according to the manufacturer's instructions. Follow the manufacturer's instructions and/or contact the manufacturer's service department if the problem is in the programming terminal or CPU. In some cases, specially trained programmers are called in to repair a problem with the PLC program.

Partial Faults. A *partial fault* is the malfunctioning of only a section or several sections of a machine. If only some of the PLC loads are malfunctioning, the problem can be solved using the program terminal and monitor, a DMM set to measure voltage, or a combination of both. Some systems use the programming terminal and monitor to observe the operation of the circuit. Many systems allow inputs and outputs to be forced and disabled. *Forcing* is the activation of an input or output by a command entered on a PLC programming terminal keyboard rather than through an input device or the program functions. *Disabling* is the prevention of the activation of an output.

Warning: Never force or disable an output before all equipment or people are safely clear of the output(s) to be forced. Forcing is dangerous because the individual doing the forcing may not see the operation of the output. When troubleshooting a PLC, it is safest to disable all outputs once they have been checked because testing an output may cause it to operate unexpectedly.

Many PLCs illustrate circuit operation by highlighting the energized portion of the circuit on the monitor. **See Figure 5-33.** The rung leading from an input brightens or is highlighted on the monitor when the input is activated or forced. This indicates that the signal is being received by the PLC and is being passed along the rung. For example, IN 1 and IN 2 are activated and highlighted. The load (OUT 1) is not energized because IN 3 is not activated or highlighted. When IN 3 is activated, IN 3 and OUT 1 are highlighted, indicating that the load is energized. Actual system operation can be compared to the highlighting of the PLC rungs on the monitor. Compare status indicators on I/O modules and I/O voltage signals with the monitor highlighting to locate problems.

A DMM can be used to locate faulty inputs, outputs, and I/O modules. **See Figure 5-34.** With a partial fault, study the system prints to determine which I/O modules are controlling the outputs. Check that the modules are firmly connected to the CPU by gently pushing on the module. Check all wire connections on the suspect modules then check the inputs and outputs.

STATUS INDICATOR CONDITION			
Status Indicator	**Problem**	**Possible Cause**	**Corrective Action**
POWER PC RUN CPU FAULT FORCED I/O BATTERY LOW	No or low system power	Blown fuse, tripped CB, or open circuit	Test line voltage at power supply. Line voltage must be within 10% of the controller's rated voltage. Check for proper power supply jumper connections when voltage is correct. Replace the power supply module when the module has power coming into it but is not delivering the correct power.
POWER PC RUN CPU FAULT FORCED I/O BATTERY LOW	Programmable controller not in run mode	Improper mode selection for system operation	Place in run mode. Ensure that all personnel are clear before placing the system in run mode.
POWER PC RUN CPU FAULT FORCED I/O BATTERY LOW	Fault in controller	Faulty memory module, memory loss, or memory error, normally caused by a high-voltage surge, short circuit, or improper grounding	Turn power OFF and restart system. Remove power and replace the memory module when fault indicator is still ON. Load backup program on new memory module and reboot system.
POWER PC RUN CPU FAULT FORCED I/O BATTERY LOW	Fault in controller due to inadequate or no power	Loss of memory when power was OFF and battery charge was inadequate to maintain memory	Replace battery and reload program.
POWER PC RUN CPU FAULT FORCED I/O BATTERY LOW	System does not operate as programmed	Input or output device(s) in forced condition	Monitor program and determine forced input and output device(s). Disable forced input or output device(s) and test system.
POWER PC RUN CPU FAULT FORCED I/O BATTERY LOW	System does not operate	Defective input device, input module, output device, output module, or program	Monitor program and check condition of status indicators on the input and output modules. Reload program when there is a program error.

Figure 5-32. Status indicators are checked immediately when a PLC malfunctions.

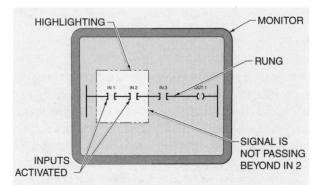

Figure 5-33. Many PLCs illustrate circuit operation by highlighting the energized portion of the circuit on the monitor.

Input and Output Problems. To troubleshoot inputs and outputs, operate the input device(s) that control the inoperative output(s). Check the status indicator on the input module. Check the input device voltage source if the status indicator does not light. Restore the correct voltage if it is not correct. Check the operation of the input device and its pathway and connections if the voltage at the input module is correct.

The input module is replaced according to the manufacturer's recommendations when the input circuit is operating correctly but the input module status indicator does not light. Check for the correct output status indicator on the output module if the input status

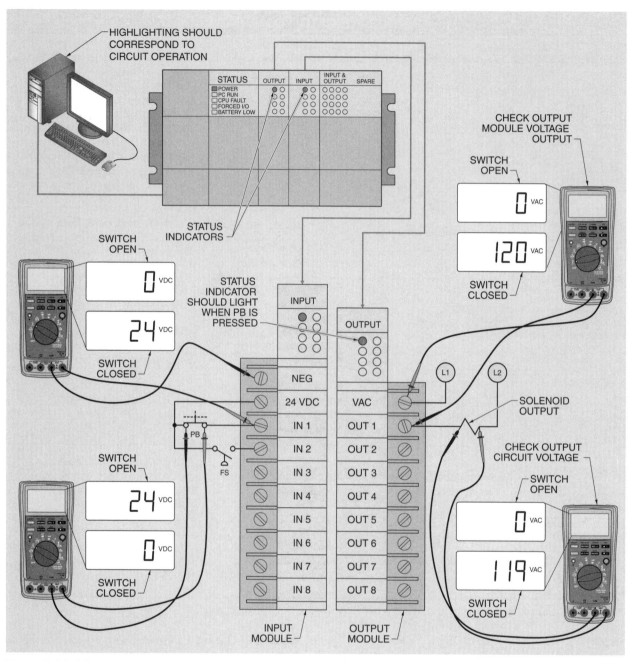

Figure 5-34. A DMM can be used to locate faulty inputs, outputs, and I/O modules.

indicator lights as desired. Check the voltage output of the output module if the correct output status indicator on the output module lights.

Faulty voltage should be traced back to the source voltage for the module. Check the output circuit, including connecting wires and connections leading to the output(s), if the output signal is correct. This is accomplished using standard electrical troubleshooting procedures for finding opens, short circuits, excessive current, and excessive resistance. Replace the module if the output status indicator lights but the voltage signal out of the module is not correct. The programming terminal is used to check the operating program or the manufacturer's service center is contacted if no problems are found in the input devices, I/O modules, or outputs. Most PLC problems occur in the input devices and with the outputs. The modules and CPU are extremely reliable, as is the PLC program.

Developing PLC Programs

Many industrial machines are purchased with the operating program already created and installed. Maintenance personnel may have to modify the program or create a new program if the machine or process is modified. **See Figure 5-35.** No rewiring is required if there is no change in inputs or outputs. This is a major advantage of PLCs, because any change in an electromechanical circuit requires rewiring.

To create a PLC program, begin with writing a description of the complete machine operation. Determine the exact function of the equipment or process and any sequencing of the operation. From this description, develop a line diagram. Develop a legend that details each input and output and identifies its function. Sketch a PLC logic diagram as a series of rungs.

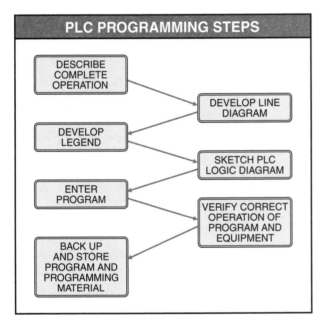

Figure 5-35. A general process is used for creating a complete PLC program.

Each rung contains one or more inputs and the output (or outputs) controlled by the inputs. The rung relates to the machine or process controls, and the programming instructions communicate the desired logic to the processor. Sketching helps to clarify the exact programming steps. The goal is to create a program that fulfills the operational needs of the machine. During sketching, input and output addresses are added to the legend. Basic logic functions are used to enter the circuit's logical operation into the processor section.

Programming PLCs

Programming a PLC follows a logical process. Inputs and outputs are entered into the controller in the same manner as if connecting them by hard wiring. The difference in programming is that, although a circuit is the same, each manufacturer has a different method of entering that circuit.

PLC software uses different types of symbols, letters, and numbers to designate each component. Components such as inputs, outputs, relays, timers, and counters each have their own symbol and addressing (assigned values and numbers). Standard symbols used in programming a circuit include normally open inputs, normally closed inputs, and standard outputs. Expanded (special) components, such as timers, counters, logic functions, and common control functions, can also be programmed into a circuit. **See Figure 5-36.**

When programming a PLC circuit, basic rules must be followed if the circuit is to be accepted by the software before downloading.

> **TECH TIP**
> *Computer operation can be adversely affected by momentary power interruptions. Install an uninterruptible power supply (UPS) to prevent costly data loss and ensure clean, continuous power for critical systems.*

Fluke Corporation

PLCs can be used to monitor and control automated packaging and materials-handling applications.

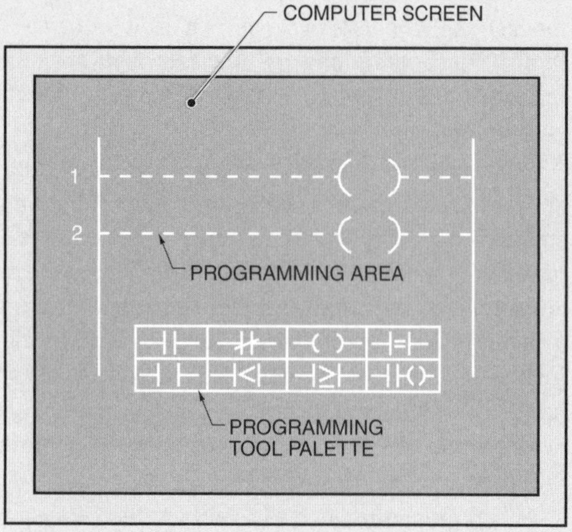

EXPANDED/SPECIAL PROGRAMMING SYMBOLS	
Component	**Symbol**
Equal-To Contact	—⊢=⊢—
Not-Equal-To Contact	—⊢≠⊢—
Greater-Than Contact	—⊢≥⊢—
Less-Than Contact	—⊢<⊢—
Addition	—(+)—
Subtraction	—(–)—
Multiplication	—(×)—
Division	—(÷)—
End	—(END)—
Set or Latch	—(SET)— —(L)—
Reset or Unlatch	—(RSET)— —(U)—
Timer	—(TMR)— or —[TMR]—
Counter	—(CNT)— or —[CNT]—

BASIC PROGRAMMING SYMBOLS		
	Component	**Symbol**
XIC	Input Device (Normally Open)	—⊢ ⊢—
XIO	Input Device (Normally Closed)	—⊬ ⊬—
	Standard Output	—( )—

Figure 5-36. Circuit functions and logic are used to create a complete PLC program.

Basic PLC circuit programming rules include the following:

Rule 1: Inputs (normally open, normally closed, and special) are placed on the left side of the circuit between the left rung and the output. Outputs are placed on the right side of the circuit.

Rule 2: Only one output can be placed on a rung. This means that outputs can be placed in parallel but never in series.

Rule 3: Inputs can be placed in series, parallel, or in series/parallel combinations.

Rule 4: Inputs can be programmed at multiple locations in the circuit. The same input can be programmed as normally open and/or normally closed at multiple locations.

Rule 5: Standard outputs cannot be programmed at multiple locations in the circuit. There is a special output called an "or-output" that allows an output to be placed in more than one location but only if the "or-out" special function is identified when programming the output.

Enter the program by typing on the programming terminal. If the system uses a monitor, compare the monitor diagram of the line diagram with the sketch made during the planning process. **See Figure 5-37.** Back up the program frequently during this process. When the program is installed, verify the correct operation of the program and equipment. Modify the PLC monitor diagram and the legend to reflect the actual program. Save all written program and programming material for future reference.

When modifying a program, work from the material created during the original programming. Write in and sketch in the changes to be made in the program. Make the program changes and operate the PLC to check that it is working as desired. Back up the modified program and store the material for future use.

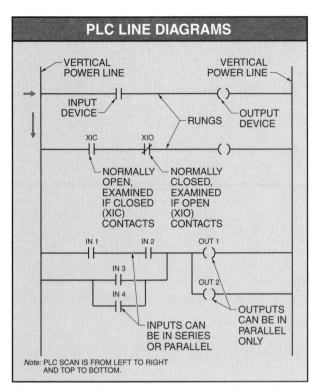

PLC LINE DIAGRAMS

Note: PLC SCAN IS FROM LEFT TO RIGHT AND TOP TO BOTTOM.

Figure 5-37. PLC diagrams follow the same basic rules as standard hard-wired line diagrams

PLC Communication Networks

PLCs may be used as stand-alone control devices or be connected through communication ports to other devices such as human-machine interfaces (HMIs), PCs, variable speed drives, and other PLCs.

In a typical network system, the PLCs become part of a large control system. Field input devices (limit switches, photoelectric switches, etc.) and output devices (solenoids, motor starters, etc.) are connected to the PLCs. The PLC is then connected to the local area network (LAN) system. Smart (intelligent) input and output devices can be directly connected to the LAN system. The LAN system is a collection of data and power lines that are used to communicate information among individual devices and to supply power to individual devices connected to the system.

Each network device monitors variables such as time, temperature, speed, weight, voltage, current, power, flow rate, level, volume, density, color, brightness, and pressure. These variables can be controlled, measured, displayed, and recorded.

Electronic Equipment Maintenance

All electronic equipment must be kept clean, dry, cool, and protected from physical damage. A *heat sink* is a device that conducts and dissipates heat away from a component. **See Figure 5-38.** Heat sinks are often attached to or are part of electronic devices. Heat sinks contain fins that increase their surface area to allow for quick heat transfer from the electronic device. Heat sinks must be kept clean to function correctly. They must make contact with the solid-state component or its case. The connecting surface should be protected by thermal grease that promotes heat transfer.

The operation of cooling fans used to cool electronic equipment must be checked regularly. Any filters used with cooling fans must be examined and changed as needed. All voltage regulators and monitors must be installed correctly, tested, and inspected according to the manufacturer's recommendations because electronic equipment is very sensitive to excessive voltage. For example, a plug-in surge suppressor may operate the equipment if its protection and monitoring circuit is defective and the equipment being protected may operate until damaged by overvoltage. Ground circuits must be maintained as direct, low-resistance pathways to ground for safety and to ensure proper voltage regulation of electronic equipment. All manuals and system prints must be stored for easy retrieval. All service bulletins that detail needed maintenance or operational problems must be stored properly.

Siemens Corporation
PLCs can be used with computer software programs to develop complex control systems for robotic operations.

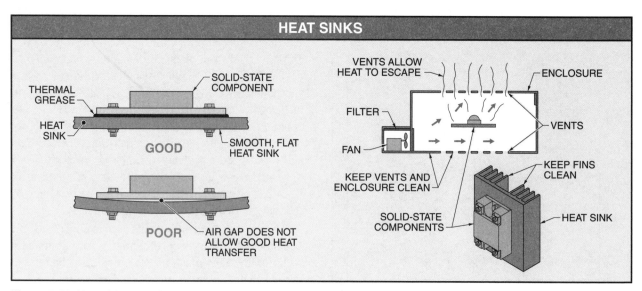

Figure 5-38. Heat sinks are often attached to or are part of electronic devices to allow for quick heat transfer from the device.

Refrigeration Systems

Refrigeration systems are found in all types of facilities. The cooling and dehumidification of room air (air conditioning) is accomplished by refrigeration systems. Food products must be kept at the correct temperature. Specialized production, medical, and chemical facilities use refrigeration systems to preserve materials and as part of the manufacturing process. The most common refrigeration systems operate on the same principles and use many of the same components. Maintaining and troubleshooting refrigeration systems is an integral part of maintenance tasks.

REFRIGERATION

Refrigeration is the process of moving heat from one area to another by use of a refrigerant in a closed system. A *refrigeration system* is a closed system that controls the pressure and temperature of a refrigerant to regulate the absorption and rejection of heat by the refrigerant. A *refrigerant* is a chemical that vaporizes (boils) at low temperatures. A refrigeration system moves heat from inside a cooled space and rejects it outside the cooled space. **See Figure 6-1.**

Temperature is a measurement of the intensity of heat. The temperature of a substance is reduced by removing heat. *Heat* is a form of energy identified by a temperature difference or change of state. All substances exist in either a solid, liquid, or gaseous state. *Change of state* is the process that occurs when enough heat is added to or removed from a substance to change it from one physical state to another. *Sensible heat* is heat energy that can be measured by a change in temperature. Sensible heat does not involve a change of state. *Latent heat* is heat energy that causes a change of state but no temperature change. Latent heat is heat added to ice that changes it to water or heat added to

water that changes it to steam. For example, a piece of ice remains at 32°F until all the ice has melted. Heat transfer in refrigeration systems includes sensible and latent heat. Adding heat increases the energy content of a substance.

MECHANICAL COMPRESSION REFRIGERATION

Mechanical compression refrigeration is a refrigeration process that produces a refrigeration effect using a compressor and a pressure control device. A mechanical compression refrigeration system consists of refrigerant inside the system, a compressor, condenser, metering device, evaporator, and accessories such as filter-dryers, sight glasses, accumulators, and liquid receivers. **See Figure 6-2.**

✚ **SAFETY TIP**

Never use oxygen for pressure-testing a system. Oxygen can explode in the presence of oil. Use nitrogen from a tank equipped with a proper pressure regulator.

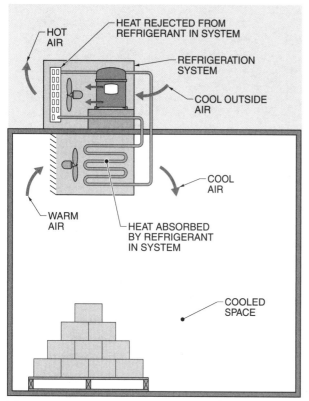

Figure 6-1. A refrigeration system controls the absorption and rejection of heat by refrigerant to move heat from inside a cooled space to outside the cooled space.

Mechanical Compression System Operation

In a mechanical compression refrigeration system, liquid refrigerant is metered into the evaporator by the metering device. An *evaporator* is a heat exchanger through which heat is transferred to low-pressure refrigerant liquid. A *metering device* is a component that controls the flow rate of refrigerant into an evaporator. The low-pressure refrigerant liquid vaporizes into a low-pressure vapor as air or water from the cooled space passes around the outside of the evaporator. Heat flows from the warm air or water to the cooler refrigerant. Heat energy in air or water from the area to be cooled fuels the evaporation of the refrigerant in the evaporator.

The refrigerant vapor is drawn out of the evaporator by the compressor and sent to the condenser. A *compressor* is a mechanical device that compresses gas. A *condenser* is a heat exchanger that removes heat from high-pressure refrigerant vapor. Inside the condenser, the refrigerant vapor condenses into a liquid, giving up heat absorbed in the evaporator. As the refrigerant condenses, latent heat in the refrigerant is carried away by air or water passing around the condenser coils. Heat flows from the hot refrigerant to the warm air or water. The refrigerant leaving the condenser is in liquid form and is sent to the metering device where the cycle begins again.

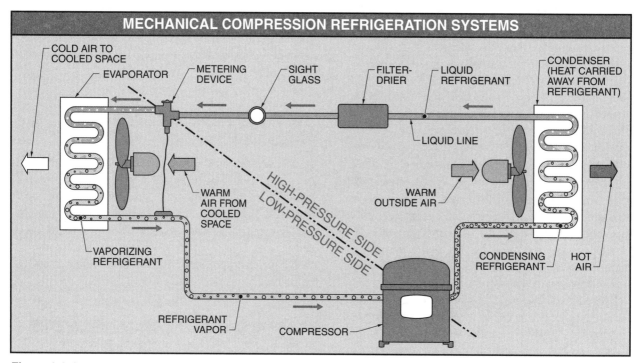

Figure 6-2. In a mechanical compression refrigeration system, a compressor is used to produce the refrigeration effect.

High- and Low-Pressure Sides

Mechanical compression refrigeration systems have a high-pressure side and a low-pressure side. A *division point* is a point in a refrigeration system where there is a significant pressure change. Division points are produced by the compressor and metering device. High and low pressures are required to control the saturation temperature of the refrigerant. *Saturation temperature* is the temperature at which a refrigerant changes state by vaporizing or condensing. Increasing the pressure raises the saturation temperature. Reducing the pressure lowers the saturation temperature. A refrigeration system requires high- and low-pressure sides to operate. Pressure is low in the evaporator so that the refrigerant vaporizes at the low temperature of the cooled space. Pressure is high in the condenser so that the refrigerant vapor can be condensed by air or water at normal ambient temperatures (approximately 70°F) passing across the condenser coils.

For example, at 26 psi, liquid HFC-134a refrigerant vaporizes at a temperature above 30°F. If air passing across an evaporator is above 30°F, the refrigerant in the evaporator vaporizes by absorbing heat from the air. The heat is transferred from the warm air into the cooler evaporator. This increases the temperature of the refrigerant and reduces the temperature of the air. The cool air is then used to cool a space.

The evaporator is kept cool by constantly removing the refrigerant vapor, which carries latent heat that was absorbed during vaporization. The refrigerant vapor is replaced by refrigerant liquid, which vaporizes, absorbing additional heat. The heat absorbed by the refrigerant in the evaporator is rejected in the condenser. The compressor raises the pressure and saturation temperature of the refrigerant vapor and sends it to the condenser. Heat flows from the hot refrigerant vapor to the warm air or water passing through the condenser. The high temperature of the refrigerant vapor enables the refrigerant to be condensed at ambient air temperatures. For example, HFC-134a refrigerant vapor pressurized to 100 psi condenses at a temperature below 124°F. If the air or water passing through the condenser coils is below 124°F, HFC-134a refrigerant vapor condenses into a high-pressure liquid.

Refrigerants

Refrigerants used in modern refrigeration systems are derived from methane (CH_4) or ethane (C_2H_6). Fluorine (F) or chlorine (Cl) atoms are used to replace the hydrogen atoms in methane or ethane to give the chemicals different properties. These refrigerants are known as halocarbon refrigerants. Halocarbon refrigerants operate at pressures easily attained in a refrigeration system, are nontoxic, nonflammable, and relatively safe. Refrigerants are identified by numbers. The numbers are assigned according to the physical and chemical composition of a refrigerant.

Most refrigerants have exact pressure and saturation temperature relationships. At any pressure, there is only one saturation temperature. A refrigerant that is vaporizing or condensing is at its saturation temperature. *Saturated liquid* is liquid at a certain pressure and temperature that vaporizes if the temperature increases. *Saturated vapor* is a vapor at a certain pressure and temperature that condenses if the temperature decreases. A saturated liquid is at its vaporization temperature and a saturated vapor is at its condensing temperature. For this reason, if the saturation temperature of a refrigerant is known, the pressures and temperatures inside the evaporator and condenser are known. Refrigerant vapor pressure charts list the temperature and pressure relationship of refrigerants and are used when testing refrigeration system performance. **See Figure 6-3.** For example, when refrigerant HCFC-22 is at 10°F, its saturation pressure is 32.8 psi.

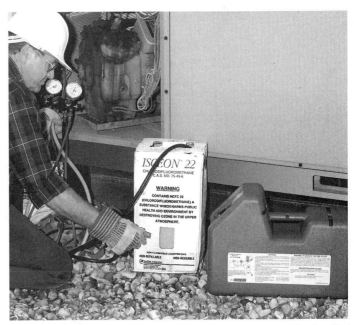

Refrigerants are stored in pressurized containers and should be handled with care.

REFRIGERANT VAPOR PRESSURE*			
Temp†	HCFC-22	HCFC-123	HFC-134a
−180	—	—	—
−170	—	—	—
−160	—	—	—
−150	29.4‡	—	—
−140	29.0‡	—	—
−130	28.5‡	—	—
−120	27.7‡	—	—
−110	26.6‡	—	—
−100	25.0‡	—	28.1‡
−90	23.0‡	—	27.1‡
−80	20.2‡	—	25.9‡
−70	16.6‡	—	24.1‡
−60	12.0‡	29.4‡	21.8‡
−50	6.1‡	29.1‡	18.7‡
−40	0.5	28.8‡	14.8‡
−30	4.9	28.3‡	9.9‡
−20	10.2	27.7‡	3.7‡
−10	16.5	26.9‡	1.9
0	24.0	25.8‡	6.5
10	32.8	24.5‡	11.9
20	43.0	22.8‡	18.4
30	54.9	20.7‡	26.1
40	68.4	18.1‡	35.1
50	83.9	15.0‡	45.5
60	101.4	11.2‡	57.5
70	121.2	6.6‡	71.2
80	143.3	1.3‡	86.8
90	168.0	2.5	104.4
100	195.4	6.1	124.3
110	225.8	10.3	146.5
120	259.3	15.1	171.3
130	296.1	20.6	198.9
140	336.5	26.8	229.4
150	380.8	33.9	263.0
160	429.2	41.8	300.1
170	482.0	50.6	340.8
180	539.4	60.4	385.6
190	602.1	71.3	434.7
200	670.3	83.4	488.7
210	—	96.6	548.3

* in psi
† in °F
‡ in inches of mercury vacuum

Figure 6-3. Refrigerant vapor pressure charts list the saturation temperature and pressure of refrigerants.

Compressors

A compressor creates pressure in a mechanical compression refrigeration system. The compressor creates the low pressure that pulls the refrigerant from the evaporator, then compresses the refrigerant and forces it through the condenser. Compressors include reciprocating, vane, centrifugal, and screw compressors. **See Figure 6-4.**

Reciprocating. A *reciprocating compressor* is a compressor that uses pistons moving back and forth to increase fluid pressure. In a reciprocating compressor, the crankshaft is turned by a motor, causing the piston to reciprocate inside a cylinder. On the suction stroke, the suction valve opens and low-pressure refrigerant vapor from the evaporator is drawn into the cylinder. At the same time, the discharge valve is pulled closed. On the compression stroke, the suction valve is pushed closed, compressing the refrigerant vapor and discharging it at high pressure through the discharge valve to the condenser. Refrigeration compressor valves are made from thin pieces of flexible metal that cover and uncover inlet and discharge ports.

Vane. A *vane compressor* is a compressor that has multiple vanes located in an offset rotor. The vanes form a seal as they are forced against the cam ring. The offset of the rotor in the cam ring produces different distances between the rotor and cam ring at different points inside the compressor. As the rotor rotates, its offset position allows the vanes to slide out and draw refrigerant from the inlet port. As the rotor continues to rotate, the volume between the vanes and the cam ring decreases, pushing the vanes into their slots in the rotor. The decreasing volume compresses the refrigerant vapor and forces it out of the outlet port.

Centrifugal. A *centrifugal compressor* is a compressor that uses centrifugal force to increase fluid pressure. In a centrifugal compressor, an impeller wheel turns inside a housing. The inlet port for refrigerant vapor is located in the side of the housing near the center of the impeller wheel. The outlet port of the compressor is located on the outer perimeter of the housing.

⊕ SAFETY TIP

Never touch or probe compressor power terminals before removing the refrigerant charge. The wiring terminals can blow out under pressure if they have been damaged. This damage may not be visible.

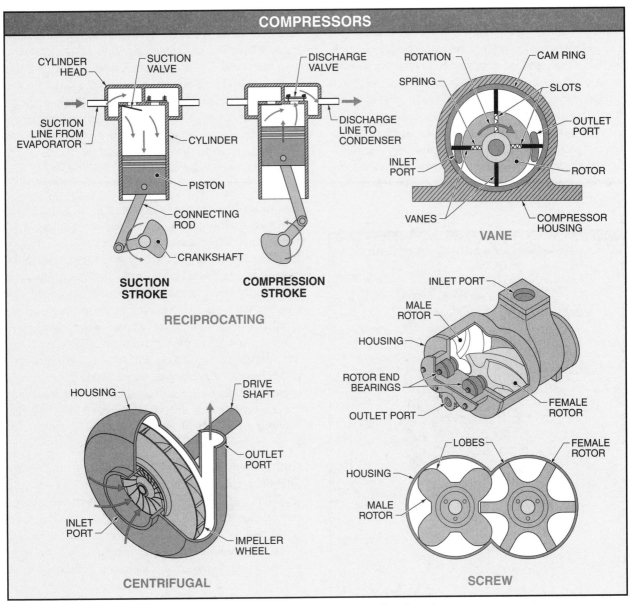

Figure 6-4. Refrigeration compressors include reciprocating, vane, centrifugal, and screw compressors.

The impeller wheel turns at a high speed and draws refrigerant into the center of the impeller wheel. The refrigerant is compressed as it flows through the vanes on the impeller wheel and is thrown off the tips of the impeller wheel by centrifugal force. The speed of the refrigerant increases as it is thrown to the perimeter of the impeller wheel. As the refrigerant leaves the outer rim of the impeller wheel, the speed is converted to pressure because the refrigerant is forced into a smaller opening.

Screw. A *screw compressor* is a compressor that contains a pair of screw-like rotors that interlock as they

rotate. The rotors are located in a tight-fitting housing and are rotated by a motor. As the rotors rotate, refrigerant is drawn into the inlet port and is forced through the housing by the interlocking lobes of the rotors. The refrigerant is compressed as the opening between the rotors becomes smaller. The compressed refrigerant is then discharged through the outlet port.

Compressor Configuration. Compressors are available in different configurations based on the applications and size of the system. Compressors are available in hermetic and semi-hermetic configurations. **See Figure 6-5.** A *hermetic compressor* is a compressor that is completely sealed inside a welded case. The compressor motor is cooled by refrigerant passing around the motor windings. Hermetic compressors are usually not repairable and are replaced when defective. Hermetic compressors must be disposed of according to local and national hazardous waste regulations because they contain refrigerant oil. A *semi-hermetic compressor* is a sealed compressor that can be serviced through removable access plates. Semi-hermetic compressors are also known as serviceable compressors because the housing can be opened and the components can be serviced on the job site.

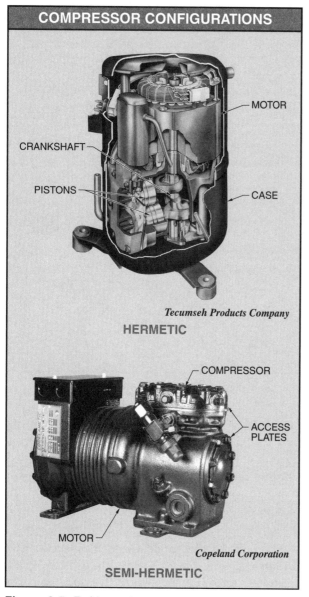

COMPRESSOR CONFIGURATIONS

MOTOR

CRANKSHAFT

PISTONS

CASE

Tecumseh Products Company

HERMETIC

COMPRESSOR

ACCESS PLATES

MOTOR

Copeland Corporation

SEMI-HERMETIC

Figure 6-5. Refrigerant compressors are available in hermetic and semi-hermetic configurations.

Refrigerant Oil. *Refrigerant oil* is a specialized liquid lubricant for refrigeration systems. Refrigerant oil lubricates at very low and high temperatures in a refrigeration system. For example, refrigerant oil lubricates the moving parts of the hot compressor and a small amount of oil vapor is carried out of the compressor to lubricate the cold metering device. Only a small amount of oil should circulate in the system. Excessive oil circulation can clog metering devices and coat the inside of the evaporator and condenser, slowing the transfer of heat and lowering the system efficiency.

Foaming. *Foaming* is the formation of a foam in a refrigerant/oil mixture due to rapid evaporation of refrigerant. Foaming occurs when a compressor is stopped long enough to cool. Refrigerant in the compressor crankcase condenses and mixes with the oil. When the compressor starts, crankcase pressure drops suddenly, causing the liquid refrigerant in the oil to vaporize rapidly. This causes the oil to foam. Foaming disrupts compressor lubrication and can force oil droplets out of the compressor. Foaming is prevented by placing a low-power heater in the compressor crankcase or wrapping a heater around the outside of smaller compressors. When the compressor is de-energized, the heater is energized, supplying just enough heat to prevent the refrigerant from condensing and mixing with the oil.

Compressor Maintenance. Compressors require little maintenance but must be kept clean and dry with sufficient airflow over them to keep them cool. Large compressors can be equipped with gauges and oil level sight glasses. An *oil level sight glass* is a window located on the compressor crankcase that indicates the level of oil in the compressor. Operating conditions such as pressure and temperature readings are recorded in a maintenance log. Some manufacturers recommend annual oil samples be taken to check for moisture and acid formation in the compressor.

Compressor burnout occurs when the compressor motor overheats, melting the motor insulation. Compressor burnout contaminates the oil and refrigerant, damaging the entire system. A burned-out compressor must be replaced and the system and refrigerant must be completely cleaned. Compressor burnout is caused by contaminants in the system, high discharge pressure, continuous running, excessive starting and stopping, or electrical problems. The cause of the burnout must be corrected to prevent the problem from recurring. Compressor burnout is a costly and time-consuming problem.

Condensers

A condenser rejects heat absorbed in the evaporator and created by compressing the refrigerant vapor in the compressor. Hot high-pressure refrigerant vapor enters the condenser coils from the compressor. The temperature of the refrigerant is greater than the temperature of the water or air flowing across the outside of the coil. Heat passes from the refrigerant to the condensing water or air because of the temperature difference between the hot refrigerant and the cooler water or air. Condensers used on refrigeration systems include air-cooled, water-cooled, and evaporative condensers.

Air-Cooled. An *air-cooled condenser* is a condenser that uses air as the condensing medium. **See Figure 6-6.** Air-cooled condensers are the most common condenser used in refrigeration systems. Air is blown across finned condenser coils and the refrigerant vapor is cooled and condenses into a high-pressure liquid, giving up its heat to the air. All the vapor should condense by the time the refrigerant passes approximately two-thirds of the way through the condenser. In the final third of the condenser, additional heat is removed from the liquid, subcooling it.

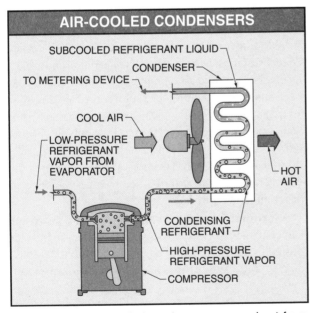

Figure 6-6. An air-cooled condenser removes heat from high-pressure refrigerant vapor by air blown across the condenser coils.

Subcooling is the cooling of a substance to a temperature that is lower than its saturated temperature at a particular pressure. For example, water at any temperature below 212°F is subcooled. Subcooling improves system efficiency by removing heat from the refrigerant before it vaporizes in the evaporator. This allows the refrigerant to absorb more heat in the evaporator. The amount of subcooling is an indication of the efficiency of the condenser. The amount of subcooling is the difference between the temperature of the refrigerant liquid leaving the condenser and the saturation temperature of the high-pressure side. For example, if an R-22 system has a high-pressure side reading of 195 psi, its saturation temperature is 100°F. If the temperature at the condenser outlet is 70°F, the subcooling is 30°F. The actual amount of subcooling that is considered normal should be determined from the manufacturer's specifications. Subcooling can vary depending on the temperature of the air.

Water-Cooled. A *water-cooled condenser* is a condenser that uses water as the condensing medium. Heat is transferred from the refrigerant to the water. Water-cooled condensers are available in shell-and-coil, shell-and-tube, and tube-in-tube configurations. **See Figure 6-7.** In a shell-and-coil condenser, water flows through a coil and refrigerant vapor circulates in the shell surrounding the coil. In a shell-and-tube condenser, water flows through tubes and the refrigerant vapor circulates in the shell surrounding the tubes. In a tube-in-tube condenser, refrigerant flows through a small pipe inside a large pipe that contains water. The refrigerant enters as a vapor and leaves as a liquid. In all water-cooled condensers, cool water enters the condenser and hot water leaves the condenser.

Evaporative. An *evaporative condenser* is a condenser that uses the evaporation of water from the outside surface of the coils to remove heat from refrigerant. **See Figure 6-8.** In an evaporative condenser, water is sprayed down over the condenser coils as air flows upward. Heat from the condenser causes some of the sprayed water to evaporate off the coils, which draws heat from the condenser coils and causes the refrigerant vapor to condense. The airflow carries away the water vapor and the heat it contains. Makeup water is added to the condenser as required.

+ SAFETY TIP

An outdoor condenser and indoor evaporator are connected with pipes to carry the refrigerant through building walls. These pipes are easily damaged, so they must be protected from any structural work nearby.

WATER-COOLED CONDENSERS

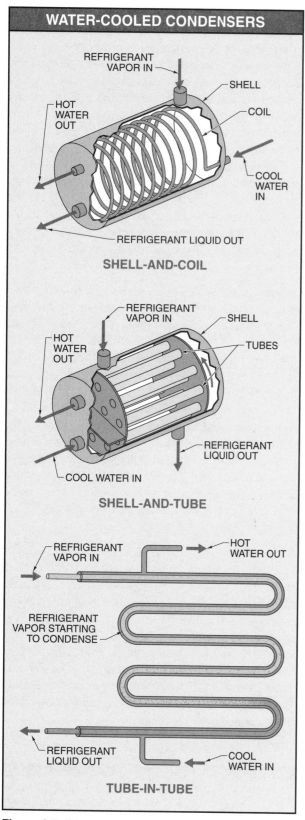

SHELL-AND-COIL

SHELL-AND-TUBE

TUBE-IN-TUBE

Figure 6-7. Water-cooled condensers transfer heat from refrigerant vapor to water.

EVAPORATIVE CONDENSERS

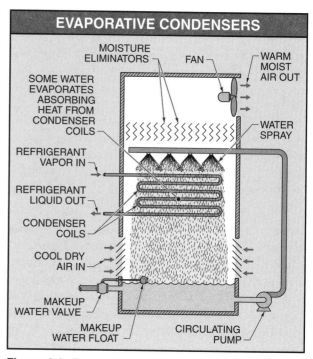

Figure 6-8. Evaporative condensers reject heat through the evaporation of water.

Condenser Maintenance. Most air-cooled condenser problems result from dirt buildup on the coils or restricted airflow that reduces heat transfer. Air-cooled condensers must be kept clean. Fan operation and airflow levels should be checked regularly. Condenser coils can be cleaned with low-pressure compressed air (less than 15 psi), brushes, and chemical sprays. Fins that become bent should be straightened using a fin comb. **See Figure 6-9.** A *fin comb* is a device used to straighten the edges of the thin metal that extend from a condenser or evaporator coil. A fin comb has different teeth corresponding to a different number of fins per inch.

Warning: Condenser tubes and fins are sharp and can be damaged easily. Always wear gloves to provide protection from the sharp edges. In addition, the sharp edges of condenser tubes and fins are often contaminated with bacteria that can infect a wound.

Water-cooled condensers should have approximately a 10°F temperature difference between the water inlet and outlet. A temperature difference of less than 10°F indicates excessive water flow or a dirty condenser. The condenser could be dirty and require cleaning if the pressure on the high-pressure side is high and water flow is correct. Water-cooled condensers require water treatment to prevent the buildup of deposits that restrict heat transfer on the tubes. No cooling water should leak into the refrigerant. Evaporative condensers must be kept clean to prevent the growth of dangerous molds and bacteria.

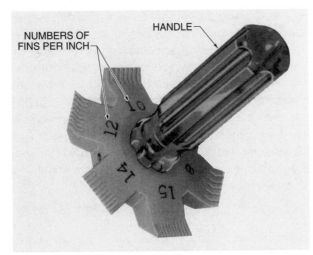

Robinair Division, SPX Corporation
Figure 6-9. A fin comb is used for condenser maintenance to straighten damaged or bent fins, which limit airflow and reduce condenser efficiency.

Metering Devices

A metering device controls the flow rate and pressure of refrigerant flowing into an evaporator. The flow rate and pressure are controlled by forcing refrigerant liquid though a small restriction. Metering devices include thermostatic expansion valves, automatic expansion valves, and capillary tubes.

Thermostatic Expansion Valves. A *thermostatic expansion valve (TXV)* is a valve that uses temperature readings at the evaporator outlet to control the rate of refrigerant flow into the evaporator. A TXV controls the flow of refrigerant into an evaporator by maintaining a constant superheat setting. *Superheat* is sensible heat added to a substance after it has turned to vapor. The amount of superheat is the difference between the temperature of the refrigerant leaving the evaporator and the saturation temperature of the low-pressure side. For example, if an R-22 system has a low-pressure side reading of 33 psi, its saturation temperature is 10°F. If the temperature of the pipe at the outlet of the evaporator is 20°F, the superheat is 10°F. The *superheat setting* is the temperature difference between the point immediately after the TXV and the outlet of the evaporator. **See Figure 6-10.** Toward the end of its passage through the evaporator, the refrigerant has completely evaporated into vapor and is superheated

before leaving the evaporator. Superheating the vapor in the evaporator ensures that all liquid refrigerant is evaporated before leaving the evaporator.

A TXV is opened and closed by pressure exerted by a remote bulb acting on the top of the valve diaphragm. The pressure on the top of the diaphragm is created by the evaporation and condensation of refrigerant contained in a remote bulb. **See Figure 6-11.** The remote bulb is attached directly to the outlet tubing of the evaporator. A capillary tube connects the remote bulb and the TXV. A *capillary tube* is a long, thin tube that resists fluid flow, which causes a pressure decrease. The temperature of the remote bulb is the same as the outlet tubing of the evaporator. A high temperature in the outlet tubing of the evaporator from increased superheat causes the pressure in the bulb to increase. Increased superheat causes the refrigerant to vaporize more vigorously and superheat the vapor earlier in the evaporator. This allows more time for superheating the vapor. The evaporator needs a greater flow of liquid refrigerant to replace what is being quickly vaporized. Each TXV is factory-set to maintain a specific superheat, usually between 8°F and 12°F.

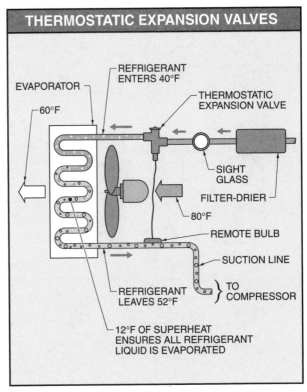

Figure 6-10. A thermostatic expansion valve uses temperature readings at the evaporator outlet to control the rate of refrigerant flow into the evaporator.

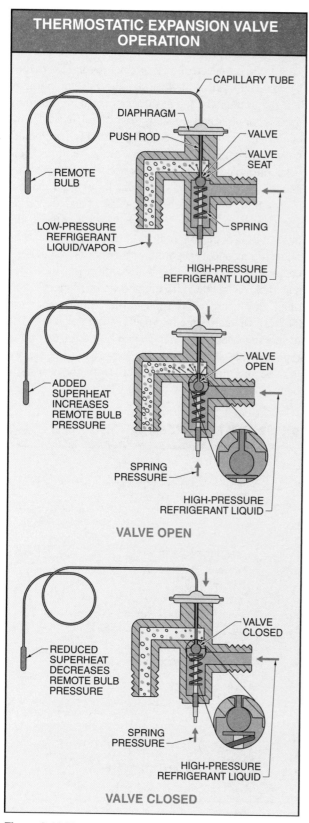

THERMOSTATIC EXPANSION VALVE OPERATION

CAPILLARY TUBE

DIAPHRAGM

PUSH ROD

VALVE

VALVE SEAT

REMOTE BULB

LOW-PRESSURE REFRIGERANT LIQUID/VAPOR

SPRING

HIGH-PRESSURE REFRIGERANT LIQUID

ADDED SUPERHEAT INCREASES REMOTE BULB PRESSURE

VALVE OPEN

SPRING PRESSURE

HIGH-PRESSURE REFRIGERANT LIQUID

VALVE OPEN

REDUCED SUPERHEAT DECREASES REMOTE BULB PRESSURE

VALVE CLOSED

SPRING PRESSURE

HIGH-PRESSURE REFRIGERANT LIQUID

VALVE CLOSED

Figure 6-11. The opening and closing of a thermostatic expansion valve is controlled by the pressure in the remote bulb.

The increased pressure in the remote bulb is transmitted to the top of the diaphragm, which moves, opening the valve and allowing more refrigerant to enter the evaporator. The additional refrigerant flow causes the temperature at the evaporator outlet to fall because the refrigerant is vaporizing further along in the evaporator. Refrigerant that vaporizes further along in the evaporator allows less time for superheating the vapor and lowers the temperature at the evaporator outlet. The lower temperature at the evaporator outlet cools the remote bulb, decreasing the pressure in the bulb. The decreased pressure in the bulb decreases the pressure on the diaphragm, closing the valve and restricting refrigerant flow. The valve opens and closes attempting to maintain a constant superheat.

The superheat fluctuates as the heat load on the evaporator changes. Heat that enters the cooled space rapidly requires more refrigerant to be evaporated to maintain the desired temperature. This occurs when a door in a cooled space is opened and closed frequently or when warm material is brought into the cooled space. The superheat fluctuates little and the valve remains at one position when the load on the system is relatively constant. The remote bulb must be firmly and directly attached to the tubing at the evaporator outlet. The capillary tube connecting the bulb and the TXV is coiled to keep it out of the way so it cannot be damaged by nearby activities.

Automatic Expansion Valves. An *automatic expansion valve* is a valve that is opened and closed by the pressure in the line ahead of the valve. Automatic expansion valves maintain a constant pressure in the evaporator to control the flow and temperature of refrigerant. **See Figure 6-12.** As the evaporator pressure falls, the valve opens to allow a greater flow of refrigerant to replace the refrigerant that is evaporated. As evaporator pressure rises, the valve closes to restrict flow of refrigerant. The opening and closing of the valve occurs as evaporator pressure is applied to a diaphragm through an internal port. Automatic expansion valves are used on refrigeration systems that have constant cooling loads. Automatic expansion valves cannot compensate for changes in the cooling load or condensing medium temperature.

Capillary Tubes. A capillary tube is a long, thin tube that resists fluid flow, which causes a pressure decrease. A capillary tube is used as a metering device on small refrigeration systems, such as refrigerators or window-mounted air conditioners. Refrigerant liquid is forced through the capillary tube by pressure from the compressor. **See Figure 6-13.** The high pressure is reduced as the liquid is forced through the small diameter and long length of the capillary tube. At the end of the tube, the pressure of the refrigerant is reduced to the desired evaporating point in the evaporator.

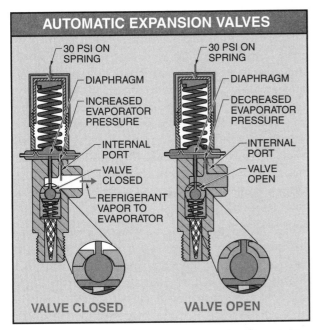

Figure 6-12. An automatic expansion valve controls the temperature of the refrigerant by controlling the pressure in the evaporator.

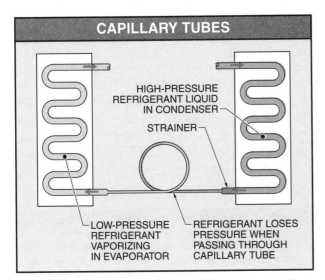

Figure 6-13. As refrigerant is forced through the capillary tube, it loses pressure until it is at the desired evaporator pressure.

Capillary tubes are equipped with a strainer to trap dirt that may block the tiny opening. Even a partial blockage can change the pressure drop through the tube, changing the pressure and temperature of the evaporator. A complete blockage causes very high pressures that damage the system. Systems using capillary tubes require an exact amount of refrigerant in the system to operate correctly because capillary tubes depend on their exact length and diameter to create the correct pressure drop. Capillary tubes should remain coiled to keep them out of the way. A capillary tube removed from a system should be replaced by a new tube of the same length and diameter.

Evaporators

An evaporator is the component in which low-pressure refrigerant liquid vaporizes into a low-pressure vapor. The most common evaporator is the air-cooled evaporator. **See Figure 6-14.** An air-cooled evaporator uses a fan to circulate air from the cooled space across the evaporator coils. Heat in the air from the cooled space flows across the cool evaporator coils, vaporizing the refrigerant and cooling the air. The low-pressure refrigerant vapor is sent to the compressor, carrying away the heat that caused the refrigerant to vaporize. The constant flow of refrigerant through the evaporator keeps it cool. Air-cooled evaporators are usually classified as dry evaporators. By the time the refrigerant flows to the end of the evaporator, all the liquid has vaporized and the refrigerant is dry (contains no liquid refrigerant). Most evaporator problems result from dirt on the coils or restricted airflow across the coils. Evaporator coils must be kept clean. Fan operation and airflow levels should be checked regularly. Coils can be cleaned with low-pressure compressed air and straightened using a fin comb.

Evaporator Defrost. *Defrosting* is the process of removing frost or ice that builds up on evaporator coils. Air carries moisture. The amount of moisture the air can carry depends on the air temperature. Warm air can carry more moisture than cool air. The amount of moisture in the air is referred to as relative humidity. *Relative humidity (RH)* is the amount of moisture in the air compared to the amount of moisture the air would hold if it were saturated. Saturated air carries as much moisture as possible before the moisture forms into water droplets. For example, air at 70°F with a relative humidity of 80% holds 80% of the maximum moisture it can hold at 70°F. Air at 100% RH holds the maximum amount of moisture it can hold at its given temperature. A reduction in air temperature causes the moisture to condense out of the air. Air that has 100% RH is at its dew point. *Dew point* is the temperature below which moisture in the air begins to condense.

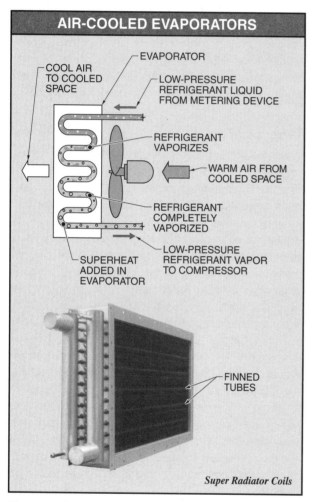

AIR-COOLED EVAPORATORS

COOL AIR TO COOLED SPACE

EVAPORATOR

LOW-PRESSURE REFRIGERANT LIQUID FROM METERING DEVICE

REFRIGERANT VAPORIZES

WARM AIR FROM COOLED SPACE

REFRIGERANT COMPLETELY VAPORIZED

LOW-PRESSURE REFRIGERANT VAPOR TO COMPRESSOR

SUPERHEAT ADDED IN EVAPORATOR

FINNED TUBES

Super Radiator Coils

Figure 6-14. An evaporator vaporizes low-pressure refrigerant liquid into a low-pressure vapor.

As air cools, the relative humidity of the air rises even though no moisture is added. This is because cool air can hold less moisture than warm air. When moist air contacts a cool surface such as metal or glass, some of the moisture condenses and forms liquid water (dew). Condensation occurs when warm, moist air passes through an air-cooled evaporator. Moisture in the air condenses onto the cold metal surface of the evaporator and freezes if the evaporator temperature is below freezing. A buildup of frost and ice can block airflow

and act as insulation that slows heat transfer. Evaporator coils must be defrosted and the melted water drained to prevent this. Evaporator defrost is accomplished by turning OFF the system for a short period of time and/or using a low-temperature heater.

In small refrigeration systems, a timer is used to turn OFF the compressor, condenser, and evaporator. Accumulated frost melts as the evaporator temperature rises and the water drains into a drain pan under the evaporator. Many refrigeration systems include a low-temperature heater under the evaporator to melt the frost more quickly during the defrost cycle and to thaw the drain pan so the melted water can drain. After the timer has timed out, the heater is de-energized and the system resumes operation. The accuracy of the time clock must be checked frequently, especially after a power outage. The length of the defrost cycle should be just long enough to melt any frost or ice buildup. Defrost cycles that are too long can damage temperature-sensitive items that warm up while the system is not operating. All drain lines must be kept open.

Hot-Gas Defrost. *Hot-gas defrost* is evaporator defrosting using hot gas from the compressor. Hot-gas defrost is used in large refrigeration systems. **See Figure 6-15.** When the timer starts the defrost cycle, a solenoid valve is opened in an additional line running between the compressor discharge and the evaporator. The hot refrigerant vapor flows from the compressor to the evaporator, bypassing the condenser and metering device. The hot refrigerant vapor from the compressor flows to the evaporator, melting the frost from the inside. The hot refrigerant vapor gives up its heat to the cold evaporator and condenses. The liquid refrigerant flows to a reevaporator, where it is evaporated into vapor by warm air blown across the reevaporator coil that is located outside the cooled space. During defrost, the defrost heater is energized to speed the defrost and melt the ice in the drain pan so the water can drain quickly. After defrosting, the system resumes normal operation. The defrost heater, reevaporator fan, and bypass solenoid valves are de-energized, allowing the system to resume cooling.

TECH TIP

Maintenance personnel are responsible for operating and maintaining refrigeration equipment so it provides the intended benefits at the lowest possible cost. Specific operating schedules should be developed to meet the cooling requirements and keep hours of operation to a minimum.

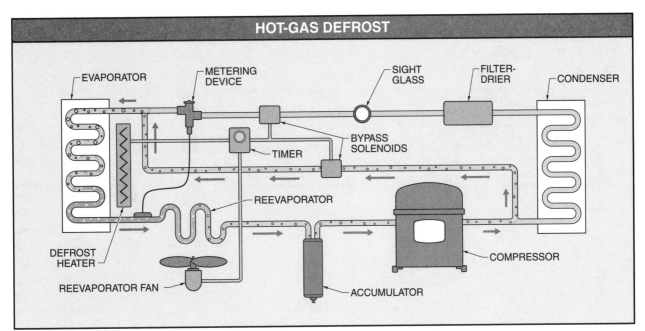

Figure 6-15. A hot-gas defrost uses hot gas from the compressor to melt frost on the evaporator.

Evaporator Pressure Regulators. An *evaporator pressure regulator* is a valve that allows two evaporators running from the same compressor to maintain different temperatures. An evaporator pressure regulating valve is used when a cold freezer and a relatively warm cooler are located side-by-side and operate from the same compressor. **See Figure 6-16.** The two different temperatures are maintained by an evaporator pressure regulating valve (hold-back valve) on the suction line of the warmer evaporator. The evaporator pressure regulating valve holds back the refrigerant and maintains a higher pressure in the warmer evaporator. This creates the warmer temperature because the saturation pressure has been raised. The compressor draws a low pressure directly from the freezer evaporator to maintain its cold temperature.

Accessories

Accessories are used for maintaining and controlling the flow of refrigerant in a refrigeration system. Accessories include filter-dryers, sight glasses, accumulators, and liquid receivers. **See Figure 6-17.**

Filter-Dryers. A *filter-dryer* is a combination filter and dryer located before the TXV that removes dirt and moisture from the refrigerant. A filter-dryer prevents contaminants from circulating through a refrigeration system.

Sight Glasses. A *sight glass* is a refrigerant line fitting located before the TXV that contains a small glass window for observing refrigerant flow. Bubbles in the sight glass indicate vaporized refrigerant caused by low pressure in the liquid line. The low pressure indicates insufficient refrigerant to completely fill the refrigerant tubing. This condition can also occur when a system has just started, so observations should be made after the system has reached normal operating conditions. Most sight glasses also incorporate a moisture indicator that changes color if moisture is present inside the system. Even a drop of water in a refrigeration system can cause serious operating problems such as oil contamination and system freeze-up.

Accumulators. An *accumulator* is a metal container that catches refrigerant liquid that escapes from the evaporator coil before the refrigerant reaches the compressor. The liquid refrigerant falls to the bottom of the accumulator and vaporizes. The vapor rises and leaves the accumulator through the outlet piping. Liquid refrigerant that enters a compressor can cause damage to the cylinders or valves. *Slugging* is a condition in which liquid refrigerant enters a compressor and causes hammering. A liquid cannot be compressed. Therefore, when the compressor tries to compress the refrigerant liquid, the valves or the piston connecting rods are damaged. Slugging can destroy a compressor.

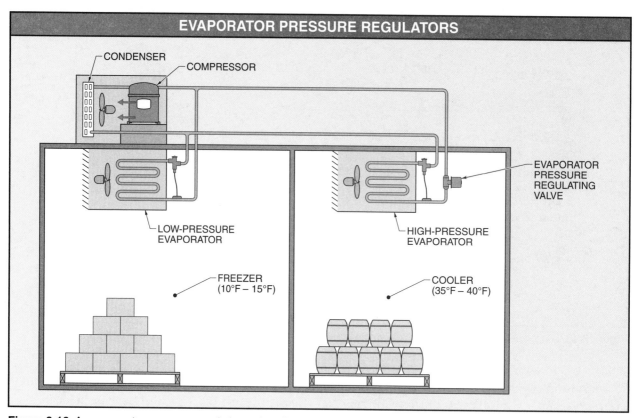

EVAPORATOR PRESSURE REGULATORS

CONDENSER

COMPRESSOR

EVAPORATOR PRESSURE REGULATING VALVE

LOW-PRESSURE EVAPORATOR

HIGH-PRESSURE EVAPORATOR

FREEZER (10°F – 15°F)

COOLER (35°F – 40°F)

Figure 6-16. An evaporator pressure regulating valve allows two evaporators running from the same compressor to maintain different temperatures.

Liquid Receivers. A *liquid receiver* is a storage tank for liquid refrigerant that is located after the condenser. A liquid receiver stores excess liquid refrigerant when the refrigeration system is operating. A liquid receiver allows for changes in the volume of the refrigerant that occur when the cooling load changes or when the temperature of the condensing medium changes. The liquid line from the condenser is connected at the top of the device. The liquid line from the receiver to the metering device is connected at the bottom. Only liquid refrigerant flows out of the tank to the metering device.

Temperature Control

Refrigeration systems are designed to maintain temperatures within a range rather than at a specific temperature. For example, a refrigeration system may be set to maintain a temperature between 35°F to 40°F instead of exactly 37°F. It is difficult and very expensive to maintain a specific temperature in a cooled space. Heat is always entering the cooled space when doors are opened, when warm material is placed in the cooled space, and

when heat enters through the walls of the cooled space. Therefore, most refrigeration systems are designed to maintain temperature within a range.

Thermostats. A *thermostat* is a temperature sensor inside a temperature-controlled space that sends signals to a control system in order to maintain a set temperature. For example, a thermostat located in the cooled space de-energizes the compressor and condenser at the lowest desired temperature and energizes the compressor and condenser at the highest desired temperature. The thermostat must be located correctly in the cooled space to ensure the proper temperature throughout the cooled space.

Pressure Switches. A *pressure switch* is a switch operated by the amount of pressure acting on a diaphragm, bellows, or electronic element. Pressure switches control refrigeration system temperature through changes in system pressure. A system not controlled by a thermostat has its temperature controlled by a pressure switch connected to the low-pressure side of the system. **See Figure 6-18.**

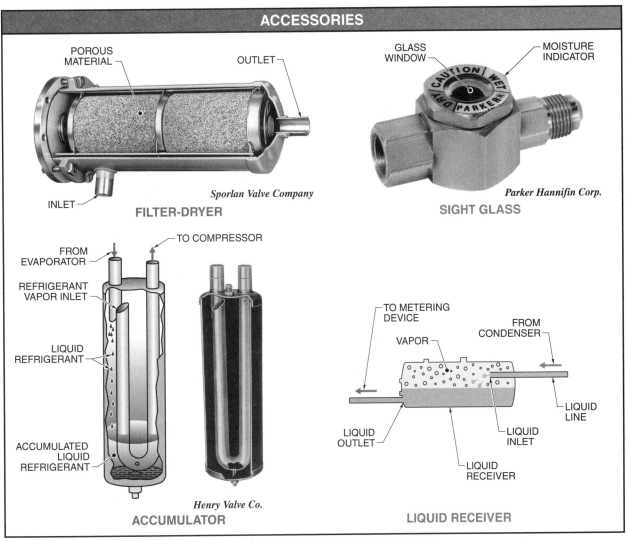

ACCESSORIES

POROUS MATERIAL

OUTLET

INLET

Sporlan Valve Company

FILTER-DRYER

GLASS WINDOW

MOISTURE INDICATOR

CAUTION WET DRY PARKER

Parker Hannifin Corp.

SIGHT GLASS

FROM EVAPORATOR

TO COMPRESSOR

REFRIGERANT VAPOR INLET

LIQUID REFRIGERANT

ACCUMULATED LIQUID REFRIGERANT

Henry Valve Co.

ACCUMULATOR

TO METERING DEVICE

VAPOR

FROM CONDENSER

LIQUID OUTLET

LIQUID INLET

LIQUID LINE

LIQUID RECEIVER

LIQUID RECEIVER

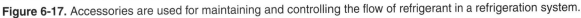

Figure 6-17. Accessories are used for maintaining and controlling the flow of refrigerant in a refrigeration system.

A pressure switch de-energizes the compressor when the suction pressure reaches the cut-out pressure in the evaporator. The cut-out pressure is the pressure that corresponds to the temperature difference below the lowest desired temperature in the cooled space. For example, if the lowest temperature desired is 30°F, the lowest temperature of the evaporator is 20°F with a temperature difference of 10°F. The evaporator is kept 8°F to 20°F cooler than the cooled space so heat flows from the cooled space into the evaporator. Heat flows only from a warmer temperature to a cooler temperature.

While the compressor is OFF, the cooled space and evaporator warm. The warming of the evaporator raises its pressure. The compressor should start (cut-in) at the pressure corresponding to the highest operating temperature of the system temperature range. After the compressor has been operating for a short time, the suction pressure should hold at the temperature difference established for the system. After continued operation, the temperature difference drops to the cut-out point as the temperature in the cooled space drops.

For example, a system using HFC-134a refrigerant with a temperature range of 35°F to 45°F and a temperature difference of 10°F cuts in the compressor when the cooled space rises to 45°F. The evaporator is also at 45°F, with a pressure of 40 psi. A few minutes after cut-in, the cooled space temperature is still approximately 45°F but the evaporator is now at 35°F. The heat flows from the cooled space air into the colder evaporator. The evaporator pressure falls as the temperature in the cooled space falls until the cooled space reaches a temperature of 35°F. At this point, the evaporator temperature reaches 25°F with a pressure of 30 psi. The compressor is stopped (cut-out) by the pressure switch, which is set to this cut-out pressure.

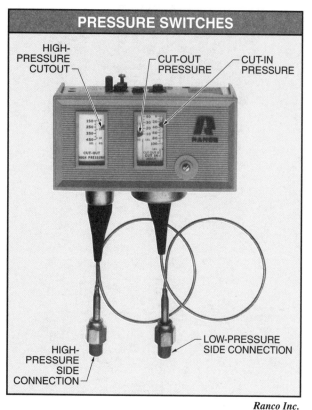

PRESSURE SWITCHES

HIGH-PRESSURE CUTOUT

CUT-OUT PRESSURE

CUT-IN PRESSURE

HIGH-PRESSURE SIDE CONNECTION

LOW-PRESSURE SIDE CONNECTION

Ranco Inc.

Figure 6-18. Pressure switches control refrigeration system temperature through changes in system pressure.

The differential is the difference between the cut-out and cut-in pressure. Short cycling results if the differential is set too close. *Short cycling* is the increase in the frequency of system operation due to improper feedback. Short cycling results from trying to maintain a temperature range that is too small. For example, short cycling may occur with a differential set at 35°F to 37°F. If not stopped, short cycling can destroy a compressor due to overheating. Too large a differential results in excessive temperature fluctuation that could damage products that are sensitive to cold or warm temperatures.

Pressure switches often incorporate a high-pressure cutout connected to the high-pressure side of a system. The high-pressure cutout de-energizes the compressor if the pressure rises enough to damage the compressor or rupture the system. Some systems also require a separate low-pressure cutout that de-energizes the compressor if evaporator pressure falls

too low, causing dangerously low temperatures in the cooled space. Extreme low pressures can also damage some compressors.

Pump-Down Control. A *pump-down control* is a control system that uses a thermostat and pressure switch to control the temperature in the cooled space. In a pump-down control, a solenoid valve is installed on the liquid line before the metering device. When the cooled space temperature is at its lowest point, the solenoid valve is closed by the thermostat, blocking refrigerant flow to the evaporator. The compressor continues to pull refrigerant from the evaporator until it is de-energized by the pressure switch reaching its cut-out pressure. After the space has warmed to the highest desired temperature, the thermostat opens the solenoid valve and refrigerant rushes into the evaporator. The pressure in the evaporator quickly rises to the cut-in pressure of the pressure switch, which energizes the compressor.

Unloading. *Unloading* is the process of varying the amount of refrigerant pumped by a compressor. Unloading can be accomplished by cylinder unloading or opening inlet control valves. Cylinder unloading varies the number of cylinders in a compressor that pumps refrigerant. All cylinders pump refrigerant if the heat load is high and maximum cooling is required. Fewer cylinders pump refrigerant if less cooling is required. When less cooling is required, the unloader holds open the inlet valves to some of the cylinders. The refrigerant is drawn into these cylinders and then exhausted back out the open inlet valves. No refrigerant is compressed from the unloaded cylinders, thereby reducing the cooling capacity of the system. When more cooling is required, the unloader closes, allowing the inlet valves to operate normally so all the cylinders compress the maximum amount of refrigerant possible to achieve the greatest amount of cooling. Rotary screw compressors use a slide valve that restricts refrigerant flow into the compressor when less cooling is required. A thermostat can be used to control the operation of unloading controls.

Ammonia Systems

An *ammonia system* is a refrigeration system that uses ammonia as the refrigerant. **See Figure 6-19.** Ammonia systems are used in applications that require the removal of large quantities of heat, such as dairies and ice making plants. Ammonia is capable of moving much more heat per pound of liquid vaporized than other refrigerants.

Unfortunately, ammonia can cause death and serious injury if inhaled and can also damage food products. Therefore, many components in ammonia systems are used to contain and control the release of ammonia.

Due to high system pressures, ammonia systems are fitted with relief valves. All relief valves are vented to a roof diffuser or water tank because a release of ammonia can be extremely dangerous. If a relief valve opens and ammonia is released, the diffuser disperses the ammonia to prevent the formation of a dangerous concentration of gas. Ammonia vented into a tank of water is absorbed into the water and not released into the atmosphere. In a fire, the ammonia can be released from the system to prevent an explosion. An emergency release valve is also connected to a diffuser or exhausts into a water tank.

Large ammonia compressors operate at high temperatures and require a water cooling system similar to that on an automobile engine. Ammonia carries large quantities of oil from the compressor. An oil separator located between the compressor and condenser changes the direction of ammonia vapor flow, causing the heavy oil

particles to fall out of the vapor flow. The oil drains back into the compressor crankcase. Due to high discharge pressures, most ammonia compressors have a valve arrangement that allows for reduced load at startup.

Due to the hazards associated with ammonia, it is usually used in a secondary refrigeration system. A *secondary refrigeration system* is a refrigeration system that uses a secondary coolant (usually antifreeze) to cool the cooled space. The antifreeze rejects its heat into the ammonia evaporator. The antifreeze leaves the evaporator to cool the cooled space. The antifreeze absorbs heat from the cooled space and returns to the evaporator, where it releases the heat to the ammonia in the evaporator. In applications in which people could be exposed to ammonia leaks, monitoring systems are installed to warn of high ammonia concentrations. The ammonia monitoring systems are also required when food products are being cooled because ammonia can contaminate food. Any plant using ammonia must take precautions to prevent the release of ammonia and have emergency evacuation and containment plans if a release should occur.

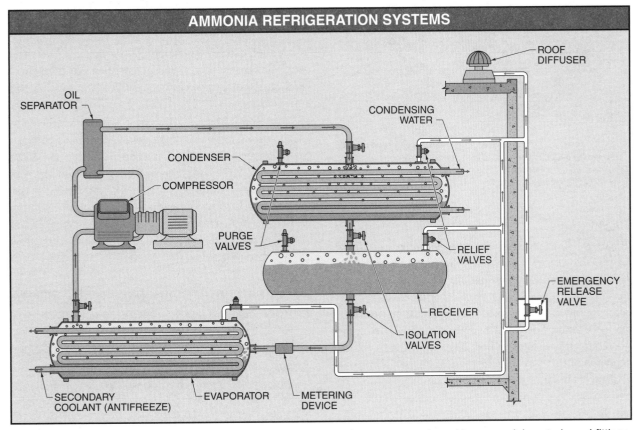

Figure 6-19. Ammonia systems operate at high temperatures and pressures and must have special controls and fittings to control the release of ammonia gas.

Purge valves are located on the condenser and receiver in an ammonia system to remove the carbon dioxide that can be produced during the normal operation of an ammonia system. A *purge valve* is a valve that removes unwanted gases from a system. Carbon dioxide is purged because it increases the pressure in the system and does no useful work in moving heat. Purge valves are often located on the receiver because the refrigerant is not moving quickly in that part of the system. Receivers are found in systems using ammonia as well as other refrigerants. Isolating valves are located throughout an ammonia system so the refrigerant can be held in part of the system while work is completed in another part of the system. Receivers are often fitted with isolation valves so the refrigerant charge can be stored in the receiver.

Lennox Industries Inc.
The heating or cooling produced by a heat pump is controlled by a reversing valve which determines the direction of refrigerant flow through the heat pump.

Heat Pumps

A *heat pump* is a mechanical compression refrigeration system that can reverse the flow of refrigerant, switching between heating and cooling modes. The heating and cooling is accomplished by operating the heat pump as a refrigeration system to cool the space, then reversing the refrigerant flow to provide heat. **See Figure 6-20.** Reversing the refrigerant flow is accomplished by using a reversing valve. A *reversing valve* is a four-way valve that reverses the flow of refrigerant in a heat pump. The reversing valve consists of a spool and a cylinder that has four refrigerant line connections. A solenoid shifts the spool to cover or uncover ports allowing refrigerant to flow in the desired direction.

In the cooling mode, the reversing valve directs refrigerant flow so the indoor coil functions as an evaporator. Heat is absorbed at the indoor coil and rejected at the outdoor coil, which functions as a condenser. In the heating mode, the reversing valve directs refrigerant flow so that the outdoor coil functions as an evaporator. Heat is absorbed at the outdoor coil and rejected at the indoor coil, which functions as a condenser. The coils are not referred to as the evaporator or condenser because both coils can have either function.

Heat Pump Classification. Heat pumps are classified as air-to-air or water-to-air. **See Figure 6-21.** In an air-to-air heat pump, air circulates across the indoor and outdoor coils. In a water-to-air heat pump, air circulates across the indoor coil and the outdoor coil is immersed in water. Heat pumps can move heat from one part of a building to another or from outside to inside a building. Heat pumps can be designed as split systems with one coil inside and one coil outside, or as packaged units with both coils in one enclosure.

Air-to-air is the most common heat pump configuration. Air-to-air heat pumps are most effective in regions that get as cold as 10°F but seldom go below 0°F. Less heat is available in cold air, so heat pumps used in cold climates require an auxiliary heat source such as electric heating elements or a gas furnace. The auxiliary heat source is energized when additional heat is required due to the low amount of heat available in the outdoor air.

The refrigerant lines in a heat pump are referred to as the vapor and liquid lines. The vapor line is larger in diameter than the liquid line. The vapor line should be cool when the heat pump is in the cooling mode and hot when the heat pump is in the heating mode. The liquid line remains warm to the touch whether the heat pump is in the heating or cooling mode. Metering devices and filters used in heat pumps must allow refrigerant flow in two directions.

TECH TIP

Heat pumps can use air, water, or the ground as the heat sink or source. Ground source heat pumps are also known as geothermal heat pumps. At a depth of about 6′, the temperature of the ground remains relatively constant year-round. These systems bury long loops of pipe that contain a heat transfer solution (usually a water-antifreeze mix). Cooling is provided by rejecting excess heat into the ground loop, where it is absorbed by the ground. Heating is provided by reversing the heat pump operation and extracting heat from the ground.

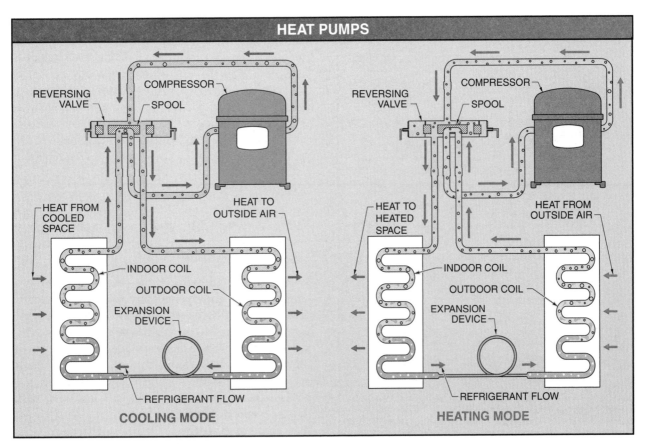

Figure 6-20. The direction of refrigerant flow in a heat pump is controlled by a reversing valve.

Heat Pump Maintenance. The outdoor coil of a heat pump must be positioned so water drains away and does not freeze on it. The support pad for the outdoor coil should drain water so ice does not form on the pad. The outdoor coil should be protected from direct gusts of wind that can lower efficiency. Any barriers erected to shelter the outdoor coil must not restrict airflow through the coil. Outdoor coils should be kept clean and free of debris. Water-to-air heat pump problems are usually related to the water source. Problems include scale formation on the coils and corrosion. The water source and water in the coils must be prevented from freezing. The indoor coil of a heat pump must be kept clean and air registers must not be covered or blocked. Filters must be changed as required to prevent reduced airflow.

> **TECH TIP**
> *The most important maintenance action of a heat pump is a regular check of the outdoor coil because a large amount of frost is formed on the coil as heat is removed from the low-temperature air passing over the coil.*

Chilled Water Systems

A *chilled water system (chiller)* is a refrigeration system that cools water that is used to cool air. **See Figure 6-22.** Chillers are used to cool large buildings such as hotels and stadiums. Chillers are classified as secondary refrigeration systems because the chilled water does the actual cooling of the building. Water is cooled to about 45°F in the chiller evaporator and pumped to the building where it cools the air.

Dunham-Bush
The physical characteristics and cooling capacity should be considered when selecting a chiller for an application.

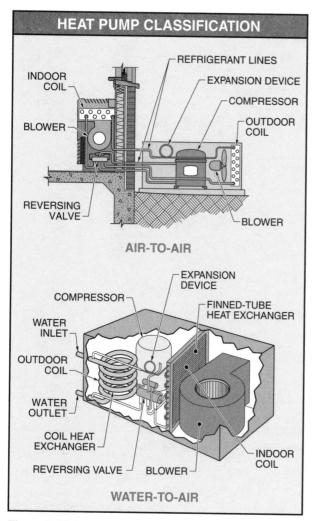

HEAT PUMP CLASSIFICATION

REFRIGERANT LINES
EXPANSION DEVICE
COMPRESSOR
OUTDOOR COIL
INDOOR COIL
BLOWER
REVERSING VALVE
BLOWER

AIR-TO-AIR

EXPANSION DEVICE
COMPRESSOR
FINNED-TUBE HEAT EXCHANGER
WATER INLET
OUTDOOR COIL
WATER OUTLET
COIL HEAT EXCHANGER
REVERSING VALVE
BLOWER
INDOOR COIL

WATER-TO-AIR

Figure 6-21. Heat pumps may use air or water as the heat source.

Hot air from the building is blown across coils containing the chilled water. Heat flows from the hot air to the cool water coils. This cools the air in the building and warms the water in the coil. The warm water (approximately 55°F) is pumped back to the chiller and its heat is used to fuel the vaporization of the refrigerant in the flooded evaporator. A *flooded evaporator* is an evaporator that is full of liquid refrigerant. The warm water from the building flows through coils in the evaporator, which are surrounded by liquid refrigerant. The level of liquid refrigerant is controlled by a metering device such as a fixed orifice placed between the condenser and the evaporator or a float valve. Fixed orifice metering devices control the rate of flow by using the pressure differential across the orifice. When the chiller is moving more heat, there is a larger pressure difference across the orifice, which

forces more refrigerant into the evaporator. In a system containing a float valve, the refrigerant flows from the condenser to a float chamber. The float chamber indicates the level of refrigerant in the evaporator. The float rides up and down on the liquid refrigerant in the float chamber and opens or closes a port that controls refrigerant flow into the evaporator.

The cooled water from the flooded evaporator is sent back to the building to absorb heat and cool the building air. The heat taken in by the evaporator is rejected in the condenser. The condenser is usually cooled by water that is cooled in a cooling tower. A *cooling tower* is a chiller component that uses evaporation and airflow to cool water. Chiller compressors may be reciprocating, vane, or centrifugal compressors.

Low-Pressure Chillers. A *low-pressure chiller* is a chiller that uses an evaporator that operates in a vacuum. A *vacuum* is pressure lower than atmospheric pressure. Low-pressure chillers use centrifugal compressors. A rotating impeller in the compressor draws in low-pressure refrigerant vapor rising from the evaporator. The vapor is accelerated by the rotating impeller and sent to a water-cooled condenser, where it condenses and flows back into the evaporator. Low-pressure centrifugal chillers use refrigerants that vaporize at relatively high temperatures. For example, refrigerant HCFC-123 vaporizes at 80°F at atmospheric pressure. To make HCFC-123 vaporize at between 45°F and 50°F, the pressure in the evaporator must be in a vacuum. The condenser in a chilled water system operates at about 9 psi, which is the reason the chiller is classified as low-pressure. Some chillers use different refrigerants and compressors that also operate at high pressures.

Because low-pressure chillers operate at below atmospheric pressure, any leak in the evaporator draws in air, contaminating the refrigerant. Centrifugal chillers are fitted with purge valves that remove air from the condenser. Air must be purged because it raises system pressure and does no useful work in moving heat. A *rupture disc* is a nonreusable device that bursts at a specific pressure. A rupture disc fitted on the evaporator or suction line bursts to vent refrigerant to the atmosphere if the system pressure rises too high due to a buildup of air in the system or overheating.

A pressure-relief valve located after the rupture disc prevents all the refrigerant from escaping. The pressure-relief valve opens until pressure falls to a safe level if the rupture disc bursts. The pressure-relief valve closes, preventing the loss of all the system refrigerant. Rupture discs are generally set to operate at 15 psi.

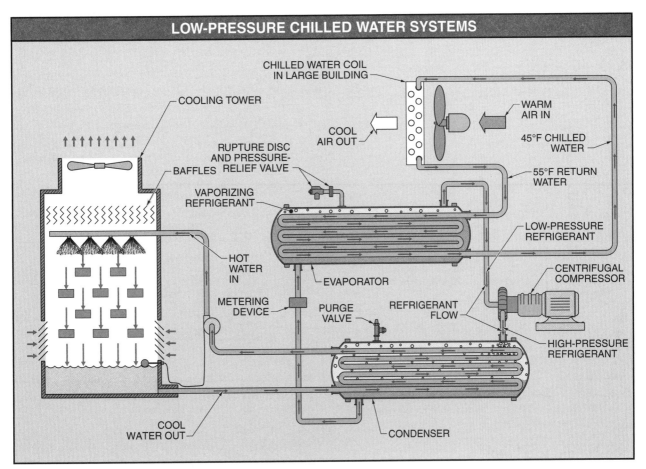

LOW-PRESSURE CHILLED WATER SYSTEMS

Figure 6-22. Chillers use chilled water to cool large building spaces.

Chiller Maintenance. Keeping records and operational logs is critical to operating chillers. Values monitored are the temperature differences between water entering and leaving the condenser and evaporator. A small temperature difference indicates low efficiency in the system due to dirty tubes or problems with the circulating pumps. The water side of the tubes must be kept clean to promote proper heat transfer. Vibration analysis and oil analysis are used to monitor the condition of the centrifugal impeller bearings. Refrigerant piping joints and couplings must be kept tight to prevent refrigerant leaks from the condenser, or of air into the evaporator.

Cooling Towers

A cooling tower uses evaporation and airflow to cool water. A cooling tower cools water from the condenser by evaporating some of the water as it cascades through the tower. **See Figure 6-23.** Cooling towers are usually located outside a facility. A fan forces air upward through the tower as warm condenser water is sprayed down and drips through rows of tiles. Some of the water evaporates as it falls, removing heat by evaporation and cooling the water left behind. The warm water vapor is carried away by the air rising through the tower. Moisture eliminators (louvers or baffles) remove water droplets before the air leaves the tower. The cool water falls to the bottom of the tower for reuse in the condenser. Makeup water is added to replace the water lost to evaporation. All cooling towers require water treatment to prevent algae growth. Cooling towers must be kept clean and fans must be inspected regularly due to the wet conditions in which they operate.

> **TECH TIP**
> *Cooling towers should be cleaned regularly. Algae growth should be cleaned from all parts of a cooling tower and water treatment put into use to prevent regrowth and to protect piping and equipment.*

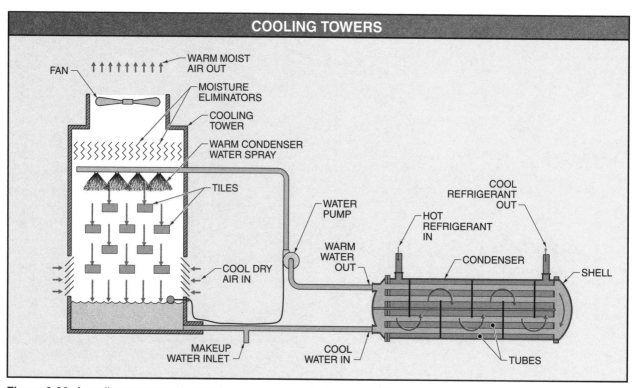

Figure 6-23. A cooling tower cools water from a condenser by the evaporation of water as it cascades through the tower.

ABSORPTION SYSTEMS

An *absorption system* is a refrigeration system that uses the absorption of refrigerant by another chemical to facilitate heat transfer. Absorption systems have a generator and absorber in place of the compressor to raise system pressure. Absorption systems are used in small refrigerators in mobile homes and in industrial settings where large quantities of surplus heat are available to fuel the generator. A solution of absorbant and refrigerant is heated in the generator. The heat source can be steam, exhaust from diesel engines, electricity, or an oil- or gas-fired furnace built into the generator. **See Figure 6-24.** The absorbant is a liquid capable of absorbing and transporting the refrigerant. For example, water absorbs and transports ammonia. Water is the absorbent and ammonia is the refrigerant. The refrigerant is vaporized as a high-pressure vapor in the generator and condenses in the condenser. The refrigerant flows through the metering device to the evaporator. The refrigerant leaves the evaporator as a low-pressure refrigerant vapor and flows into the absorber, where it mixes with the absorbant from the generator. The solution of absorbant and refrigerant is pumped back to the generator, where the cycle begins again.

TROUBLESHOOTING AND MAINTAINING REFRIGERATION SYSTEMS

Small refrigeration systems are maintained using breakdown maintenance and are not serviced until the cooled space is at an undesirable temperature. The equipment receives little or no maintenance before it breaks down.

However, cleaning the evaporator and condenser improves efficiency and can prevent many problems. Evaporator and condenser coils are cleaned with specially designed soft bristle brushes, cleaning solutions, or low-pressure compressed air. Care must be taken not to damage coil fins and tubing during cleaning. Large systems that are fitted with gauges and electronic readouts require additional maintenance. Maintenance personnel monitor temperatures and pressures in a large system and do maintenance or repairs as changing conditions indicate that problems are developing.

> ✚ *SAFETY TIP*
> *Use gloves and safety glasses when handling refrigerants and used compressor oil because burned-out hermetic compressors can contain harmful acids.*

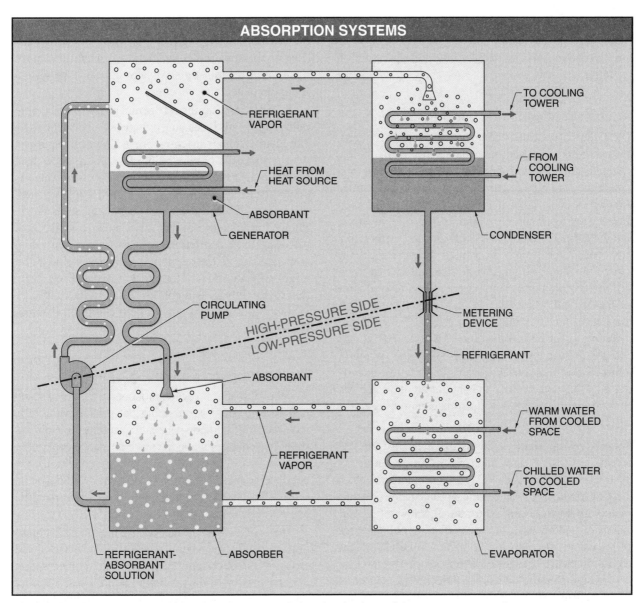

Figure 6-24. Absorption systems use a generator and absorber in place of the compressor to raise system pressure.

For example, condenser tubes may be dirty if a small temperature difference exists between cooling water entering and leaving a water-cooled condenser. The small temperature difference indicates that further testing or cleaning is required. An adequate amount of cooling air or water must be kept flowing to the evaporator, condenser, and compressor. As long as the system is operating correctly, there is little need to attach gauge manifolds to service valves to take pressure readings. Attaching gauge manifolds risks losing refrigerant or contaminating the system. Refrigeration systems are generally extremely reliable because their internal parts are enclosed and the system is sealed.

A refrigeration system problem usually involves a temperature in the cooled space that is too hot or too cold. Therefore, accurate records must be kept of the temperature range the system is supposed to maintain. When responding to a complaint, measure the temperature in the cooled space to determine if it is within temperature specifications. If the temperature is within specifications, ensure that no drafts or other conditions are making individuals uncomfortable. The request is passed to a supervisor if individuals require a change in the temperature specifications. Do not change system operation without authorization. A system designed to act as a cooler cannot operate as a freezer without extensive modifications. Simply lowering the temperature setting could destroy the compressor. Refrigeration system troubleshooting may be required on non-operating or operating systems.

Troubleshooting Non-Operating Systems

The electrical controls, compressor motor, and refrigerant charge are checked immediately if a refrigeration system is not operating. After the likely components are tested, they are repaired and/or replaced. For example, a defective thermostat is replaced. Burnt compressor motor windings require the compressor to be replaced and the entire refrigeration system cleaned. After making a repair, the system is operated to determine what caused the failure. Often, an electrical failure that stops system operation is the result of a refrigeration problem such as low refrigerant charge or system blockage. The root cause of the problem must be located. For example, if a compressor burns out because of a low refrigerant charge, the leak must be repaired before the system is returned to operation. A system that is close to failing may be short cycling and should be inspected for common problems such as low refrigerant charge or a defective pressure switch.

Short Cycling. Short cycling is the frequent ON/OFF operation of a compressor. Short cycling may be caused by the frequent opening and closing of the compressor, electrical overloads due to an electrical problem, the pressure switch differential being set too small, a low refrigerant charge caused by leaks, or dirty evaporator or condenser coils causing extremely low or high pressures. A system with a complete blockage may also short cycle because the low- or high-pressure cutouts turn the compressor ON and OFF frequently. Short cycling can cause the compressor to burn out, making the system inoperative.

A complete system blockage may cause a system to fail for long periods of time before restarting and lead to increased cooled space temperatures. The TXV is a possible location for a complete blockage due to freezing. Even a small amount of moisture in the refrigerant can freeze a TXV, especially on a very hot day. This occurs because moisture trapped in the filter-dryer is driven out of the dryer by the hot conditions. The water freezes at the cold TXV opening, blocking the opening and shutting down the compressor due to excessively high or low pressure. The TXV warms when the compressor stops and the ice blockage melts. The pressures return to normal operating conditions and the compressor restarts. The TXV is defrosted with a low-temperature

hair dryer and the moisture indicator is inspected when the system is operating if a frozen TXV is suspected. The system must be completely evacuated to remove the water if the moisture indicator indicates the system is contaminated with moisture.

A TXV that loses the refrigerant charge in its remote bulb usually fails by closing completely and becoming a complete system restriction. Some TXVs are equipped with replaceable remote bulbs. The whole TXV must be replaced if the remote bulb section of the valve is not replaceable. A plugged or kinked capillary tube metering device can also cause the system to shut down. Plugged or kinked capillary tubes must be replaced. A partial high side restriction can cause a temperature drop from one side of the restriction to the other side, but it might not cause the system to shut down. The restriction might appear as a cold or frosted spot caused by vaporizing refrigerant due to the pressure drop across the restriction.

Troubleshooting Operating Systems

The cooled space and evaporator conditions are checked first if the system is operating but the temperature is incorrect. The cooled space and evaporator conditions should be checked when the system has been operating long enough to be at normal operating conditions. Do not observe these conditions immediately after system startup or just after a large quantity of warm material has entered the cooled space.

The cooled space is checked for any unusual conditions such as warm material recently placed in the cooled space, frequent opening of doors to the cooled space, or deterioration of the insulation or door seal. On very hot days, the system may not be able to maintain the desired temperature range because the system cannot remove enough heat. In addition, high relative humidity can cause a system to be unable to maintain the desired temperature because high relative humidity is an increased heat load on the system. High relative humidity increases the heat load because when the moisture carried in the air condenses on cool evaporator coils, the latent heat used to evaporate the moisture into the air in the first place is transferred into the evaporator. The additional heat must be rejected in the condenser.

The temperature is checked at the thermostat and at several locations in the cooled space. Cool airflow from the evaporator should be reaching all locations of the cooled space. Any restrictions or blockages to airflow should be removed.

Refrigerant Charge. A system that is low on refrigerant charge will not produce enough cooling and might be short cycling on the low-pressure control or possibly due to compressor overheating. The sight glass is checked for bubbles. A system low on refrigerant charge has bubbles in the sight glass located before the TXV. If the metering device is a capillary tube, the cooling should begin at the end of the capillary tube. A frost spot or cool spot away from the end of the capillary tube indicates low refrigerant charge. The low system pressure caused by the lack of refrigerant causes the refrigerant to start vaporizing before the end of the capillary tube.

Other symptoms of low refrigerant charge can include low compressor discharge pressure, low compressor suction pressure, high superheat, and low subcooling. The most common cause of low refrigerant charge is leaks. If a leak is suspected, look for oil stains on the equipment where the refrigerant is leaking or use leak-detection dyes. Locate and repair the leak(s), then evacuate and recharge the system. *Charging* is adding refrigerant to a system.

An overcharged system will not produce enough cooling and might be short cycling on the high-pressure control. Other symptoms can include high discharge pressure, high suction pressure, low superheat, and high subcooling. Compressors are often damaged by system overcharges. Frosted or sweating suction lines are signs of overcharge. Overcharge usually occurs because too much refrigerant is added to the system.

Evaporator Problems. The evaporator is checked for ice and dirt buildup, correct fan operation, and proper superheat temperature. The superheat temperature should be measured according to manufacturer's recommendations.

The evaporator drains must be open. Blocked drains prevent the water created by the defrost cycle from draining and ice could form on the lower part of the evaporator. The defrost cycle and metering device are checked for proper operation.

A system with a dirty evaporator will not produce enough cooling and might be short cycling on the low-pressure control or possible due to overheating of the compressor. Always inspect evaporators for dirt or ice buildup and confirm that all fans are working correctly. Reduced airflow can cause an evaporator to become covered in frost or ice. Symptoms of a dirty evaporator can include low discharge pressure, low suction pressure, low superheat, and high subcooling.

Condenser Problems. The condenser is checked for cleanliness, proper fan operation, sufficient airflow across the condenser coils, excessive noise, and for whether the condenser is cooling the refrigerant to the correct temperature.

A system with a dirty condenser might not produce enough cooling and might be short cycling on the high-pressure control or possibly due to overheating of the compressor. Always inspect condensers for dirt buildup and confirm that all fans are working correctly. Symptoms of a dirty condenser can include high discharge pressure, high suction pressure, low superheat, and low subcooling.

Compressor Problems. The compressor is checked for proper operating temperature, short cycling, or continuous operation. A system with a defective compressor will not produce enough cooling and might be short cycling on the low-pressure control or possibly due to compressor overheating. The compressor will likely be very hot and drawing very low current. If the compressor valves are not sealing properly, the current will not rise significantly when the condenser fans are turned off.

Fluke Corporation
Refrigeration systems, like most other systems, often involve making electrical measurements when troubleshooting.

The suction line leading to the compressor should not be cold or wet from condensation. If the suction line is cold or wet, the problem could be a malfunctioning evaporator or TXV or a refrigerant overcharge. The sweating indicates that liquid refrigerant is leaving the evaporator and vaporizing in the suction line. Liquid refrigerant that reaches the compressor can destroy the compressor valves. Suspect refrigerant overcharge if the system has recently been charged with refrigerant.

A compressor seldom fails due to old age but can be damaged by other conditions such as low charge or dirty condensers and evaporators. Symptoms of these problems can include low discharge pressure, high suction pressure, high superheat, and high subcooling.

Taking System Pressure Readings

After conducting these checks of system operation and finding no obvious problems, it is necessary to take system pressures and then observe system operation.

Gauge Manifold. System pressure readings are taken using a gauge manifold. A *gauge manifold* is a set of valves and pressure gauges used to determine refrigerant system pressures and add or remove refrigerant. Refrigerant gauge manifolds include three- and four-hose manifolds. **See Figure 6-25.** The left-hand gauge and valve are blue and are used to take low-pressure side pressure and vacuum readings. This gauge is referred to as a compound gauge because it can register positive pressure and vacuum readings. The right-hand gauge

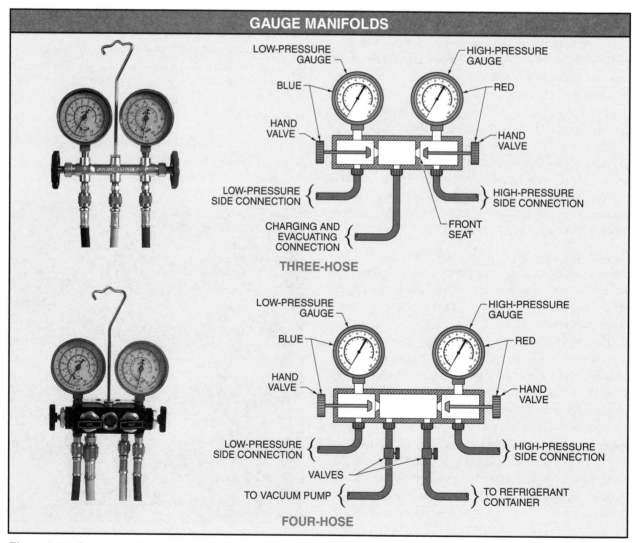

Figure 6-25. Gauge manifolds are used to take pressure readings, add or remove refrigerant, and remove air from a system before it is filled with refrigerant.

and valve are red and are used to take high-pressure side readings. The hoses attached to each side of the manifold are color-coded to match the low-pressure side and high-pressure side gauges. Only the red and blue hoses are used when taking pressure readings. The third hose, which is usually yellow, is left attached to the manifold, as is the fourth hose on a four-hose gauge manifold. These hoses are used to evacuate and charge systems. *Evacuation* is the removal of all air from a system before charging it with refrigerant.

The blue and red hoses are attached to the service port outlets of system service valves. A *service valve* is a three-way manually operated valve used to charge or remove refrigerant or monitor system pressure. Service valves are usually located on or near the compressor, though they can be located at other points in large systems. Service valves are placed in the front-seated, mid-seated, or back-seated position. **See Figure 6-26.**

The front-seated valve position has the front of the valve disk in contact with the front valve seat. The valve stem is turned completely in. This blocks the system and is used to isolate system components prior to repairing them. The mid-seated valve position has the valve disk positioned in the middle of the valve. Mid-seating the service valve is the correct position for taking system pressures, evacuating, and charging the system. The back-seated valve position has the back of the valve disk contacting the back valve seat. When back-seated, no refrigerant can escape from the service port outlet. This is the normal operating position. The correct tool for positioning a service valve is a service wrench.

A *service port* is a service valve used on small refrigeration systems that is only opened or closed by fitting a gauge manifold hose onto the port. Service ports are sealed with Schrader valves, which are the same valves used to seal automobile tires. Refrigerant can be charged into or removed from the system when the valve stem is depressed. Small systems may include a crimped and sealed stub of tubing on the compressor for servicing. A removable piercing valve is used to create a hole in the tube so pressure readings can be taken. When the need for the piercing valve is over, it is replaced with a permanently attached valve that is soldered onto the piping, or the tube is recrimped and sealed, and the section with the hole cut off.

System pressure readings are taken by connecting the blue hose to the low-pressure side service valve and the red hose to the high-pressure side service valve. The third hose remains connected to the gauge manifold. If a four-hose manifold is used, the fourth hose also remains connected to the manifold. Both gauge manifold valves remain front-seated so no refrigerant enters the third or fourth hose. **See Figure 6-27.** The service valves are opened slightly to purge air from the hoses at the manifold connection, allowing a small amount of refrigerant to escape. This removes any air from the hoses, which can lead to false readings. No air should be introduced into a refrigeration system.

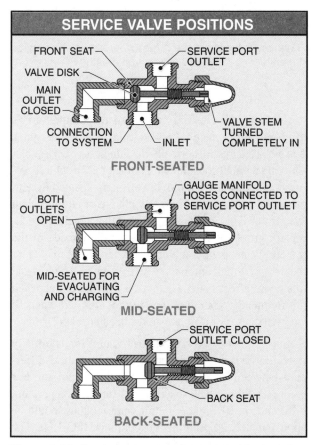

Figure 6-26. Service valves are front-seated for isolating parts of the system, mid-seated for adding or removing refrigerant or taking system pressures, and back-seated during normal operation.

⊕ SAFETY TIP

Safety precautions must be taken when working on refrigeration systems. Be aware of high-pressure refrigerant in the system while connecting gauge hoses. High-pressure refrigerant can freeze the eyes and skin. Always wear eye protection and gloves when charging a system or taking pressure readings.

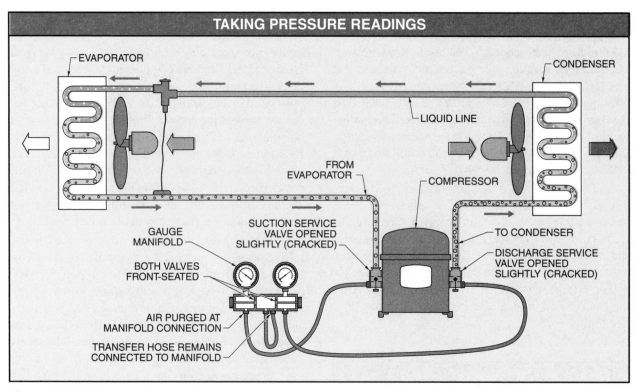

TAKING PRESSURE READINGS

Figure 6-27. Refrigeration system pressure readings are taken by connecting the blue hose to the low-pressure side service valve and the red hose to the high-pressure side service valve with both gauge manifold valves front-seated.

The system pressures can be read on the gauges after the air is purged. With the gauges attached, the operation of a pressure switch is checked along with the high- and low-pressure cutouts. The compressor should energize and de-energize at the correct pressure readings on the gauge manifold. The pressure switch is adjusted to the correct settings or it is replaced if the compressor does not turn ON and OFF at the correct pressures.

When taking pressure readings, the red hose fills with liquid refrigerant that must be returned to the system to maintain the correct charge. After pressure readings are complete, the high-pressure side system service valve is back-seated and the red and blue hand valves are cracked open. Liquid refrigerant becomes a vapor as it flows from the red hose into the blue hose and into the low-pressure side of the system. When the high- and low-pressure gauges equalize at the low pressure of the system, all the liquid refrigerant has been removed from the red hose and returned to the system. The low-pressure side system service valve is back-seated and the hoses can be removed from the system service valves.

System pressures are compared to manufacturer's troubleshooting material and normal system pressures. System pressures are normally meaningless when troubleshooting without consulting manufacturer's troubleshooting material. For example, lower-than-expected system pressures could result from low charge or an inefficient evaporator. This is why gauges are attached after checking for problems that are visible on the outside of the system. When gauges are attached, air or dirt may enter the system. Small amounts of refrigerant will be lost and this could affect system operation if hoses are attached too frequently. Therefore, gauges are attached when all external causes of problems have been checked.

In general, low-pressure side pressures for a normally operating system should be at the temperature difference between the cooled space and the evaporator. For example, if the cooled space is 35°F and the temperature difference is 10°F, then the pressure reading should be approximately 22 psi if the refrigerant is HFC-134a. The high-pressure side pressure (head pressure) of an operating air-cooled condenser should correspond to about 35°F above ambient air temperature. For example, if the air temperature is 70°F outside, the high-pressure side pressure of an HFC-134a system should be about 135 psi. This is the saturation pressure of HFC-134a at 105°F. The refrigerant in the condenser should be about 35°F hotter than the ambient air so heat flows from the hot refrigerant to the cooler air. Each system's specifications should be consulted for exact figures because high-pressure side pressures can vary depending on the efficiency rating of the condenser.

Head Pressure Controllers. A *head pressure controller* is a refrigeration system component that prevents the high-pressure side pressure from falling too low. In cold conditions, the condenser can work too efficiently and lower the refrigerant temperature and pressure so much that not enough refrigerant is forced into the evaporator. This occurs when there is not enough pressure forcing refrigerant through the metering device. The evaporator becomes starved for refrigerant and the cooled space temperature rises because there is not enough vaporizing refrigerant to absorb heat in the evaporator. This situation is prevented by using a head pressure controller that is connected to the high-pressure side tubing. **See Figure 6-28.**

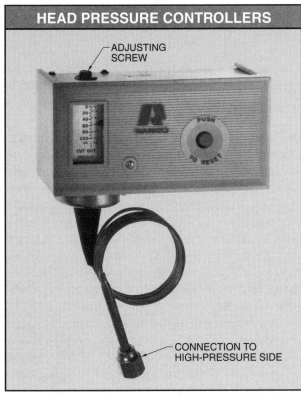

HEAD PRESSURE CONTROLLERS

ADJUSTING SCREW

CONNECTION TO HIGH-PRESSURE SIDE

Ranco Inc.

Figure 6-28. A head pressure controller prevents the condenser pressure from falling too low and starving the evaporator for refrigerant.

The electrical contacts in the head pressure controller stop and start the fan motor based on pressure changes. When the head pressure is too low, the head pressure controller de-energizes the condenser fan or closes louvers on the outside of the condenser housing, reducing airflow across the condenser. The lack of or reduced airflow causes the condenser to heat up, raising head pressure to normal levels. The head pressure controller keeps the condenser fan OFF until the pressure is within the correct operating range. For this reason, condenser fans do not always operate when the compressor is running, especially in cold weather.

REFRIGERANT REGULATIONS

The chlorine in some refrigerants contributes to the depletion of the Earth's ozone layer. The ozone layer protects the Earth from the sun's harmful ultraviolet rays. As the ozone layer is depleted, more ultraviolet rays reach the Earth's surface. Increased exposure to these rays can cause cancer in humans and diminish crop yields. In addition, some refrigerants may be carcinogenic (cancer-causing) or contribute to global warming.

National and international laws now regulate the production and handling of refrigerants in an effort to minimize their impact on the ozone layer. The Clean Air Act (CAA) of 1990 sets the requirements for refrigerant handling for the United States, which are enforced by the Environmental Protection Agency (EPA).

Section 608 of the Clean Air Act

Section 608 of the CAA promotes minimizing emissions and maximizing recycling of ozone-depleting substances. **See Figure 6-29.** Under Section 608 of the CAA, the EPA has established regulations that do the following:

- Require service practices that maximize recycling of ozone-depleting substances such as chlorofluorocarbons (CFCs) and hydrochlorofluorocarbons (HCFCs) during the servicing and disposal of air conditioning and refrigeration equipment.
- Set certification requirements for recycling and recovery equipment, technicians, and reclaimers.
- Restrict the sale of refrigerants only to certified technicians.
- Require persons or technicians servicing or disposing of air conditioning and refrigeration equipment to prove to the EPA that recycling and recovery equipment being used is in compliance with EPA rules.

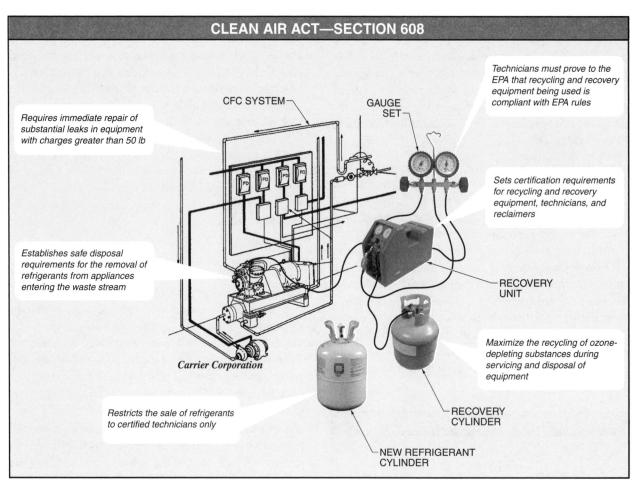

CLEAN AIR ACT—SECTION 608

Requires immediate repair of substantial leaks in equipment with charges greater than 50 lb

Establishes safe disposal requirements for the removal of refrigerants from appliances entering the waste stream

Carrier Corporation

Restricts the sale of refrigerants to certified technicians only

CFC SYSTEM

GAUGE SET

Technicians must prove to the EPA that recycling and recovery equipment being used is compliant with EPA rules

Sets certification requirements for recycling and recovery equipment, technicians, and reclaimers

RECOVERY UNIT

Maximize the recycling of ozone-depleting substances during servicing and disposal of equipment

RECOVERY CYLINDER

NEW REFRIGERANT CYLINDER

Figure 6-29. The EPA has established regulations under Section 608 of the Clean Air Act to regulate the handling of ozone-depleting substances.

- Require the immediate repair of substantial leaks in air conditioning and refrigeration equipment with a charge greater than 50 lb.
- Establish safe disposal requirements to ensure the removal of refrigerants from goods such as motor vehicle air conditioners, home refrigerators, and other equipment entering the waste stream with the refrigerant charge intact.

Service Practice Requirements

EPA regulations affect the way equipment is serviced. Refrigerant leak testing, recovery, evacuation, and charging must be performed with the most recent regulations in mind. All procedures that are in place for refrigerant recovery, evacuation, and charging have the intent of minimizing any system contamination and refrigerant loss.

Evacuation Requirements. Technicians are required to evacuate refrigeration equipment to established vacuum levels. The high vacuum levels ensure that the maximum amount of refrigerant is recovered. The required vacuum levels are based on the date the evacuation equipment was manufactured. Newer equipment must achieve deeper vacuum levels. An EPA-approved equipment testing organization must certify all recovery and recycling equipment. Technicians who are adding refrigerants to top off appliances or systems are not required to evacuate the systems.

Reclamation Requirements. Refrigerant must be recovered when removed from a system. *Refrigerant recovery* is the process of removing refrigerant from a system and capturing it in a recovery cylinder, with no cleaning of the refrigerant. Recovery is accomplished using a recovery unit. **See Figure 6-30.** Refrigerant that is lightly contaminated can be recycled by passing it through a filter in a recovery machine that is capable of recycling refrigerant.

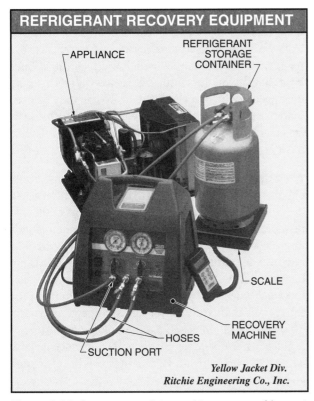

REFRIGERANT RECOVERY EQUIPMENT

Yellow Jacket Div.
Ritchie Engineering Co., Inc.

Figure 6-30. A recovery unit is used to recover refrigerant for reuse.

Refrigerant recycling is the process of removing dirt, oil, and moisture from lightly contaminated refrigerant that has been removed from a refrigeration system. The EPA has established that refrigerant recovered or recycled can be returned to the same system or other systems owned by the same individual without restriction. If the refrigerant changes ownership, the refrigerant must be reclaimed unless the refrigerant was used only in a motor vehicle air conditioner or similar appliance and will be used in the same type of appliance. *Refrigerant reclaiming* is the process of cleaning heavily contaminated refrigerant. Recovery and recycling are accomplished using special equipment. Reclaiming is accomplished at specialized facilities.

Technician Certification

Refrigerants may be charged, recovered, or recycled only by certified technicians. A *certified refrigerant technician* is a person who has special knowledge and training, and has passed one or more EPA-approved tests in the charging, recovery, and recycling of refrigerants for air conditioning and refrigeration systems. This certification ensures that technicians know the dangers of releasing refrigerants into the atmosphere and that they are familiar with procedures to prevent and limit the release of refrigerants during normal operations. The certification allows technicians to buy refrigerant, remove or add refrigerants to a system, and perform operations that may allow the release of refrigerants if not completed correctly. The EPA has four types of certification:

• servicing small appliances (Type I)
• servicing or disposing of high-pressure or very high-pressure appliances, except small appliances and motor vehicle air conditioners (Type II)
• servicing or disposing of low-pressure appliances (Type III)
• servicing all types of equipment (Universal)

Technicians are required to pass an EPA-approved test given by an EPA-approved certifying organization to become certified under the mandatory program.

Refrigerant Leaks

Section 608 of the CAA prohibits individuals from knowingly venting CFC and HCFC ozone-depleting substances into the atmosphere while maintaining, servicing, repairing, or disposing of air conditioning or refrigeration equipment (appliances). However, a tiny amount of refrigerant release is almost unavoidable during normal operations. Leaks must be limited by using proper procedures, but four types of releases are permitted.

Minimal quantities of refrigerant may be released in the course of making good faith attempts to recover, recycle, or safely dispose of refrigerant.

Refrigerant releases during the normal operation of air conditioning and refrigeration equipment are permitted, such as from mechanical purging or leaks. However, the EPA does require the repair of leaks above a specific size in large equipment. **See Figure 6-31.**

CFCs or HCFCs that are not used as refrigerants may be released. Any heat transfer fluids are considered refrigerants. For example, mixtures of nitrogen and R-22 that are used as holding charges or as leak test gases may be released, because the ozone-depleting compound is not used as a refrigerant. However, technicians may not avoid recovering refrigerant by adding nitrogen to a charged system.

Small releases of refrigerant that result from purging hoses or from connecting or disconnecting hoses to charge or service appliances are also not considered violations of the prohibition on venting.

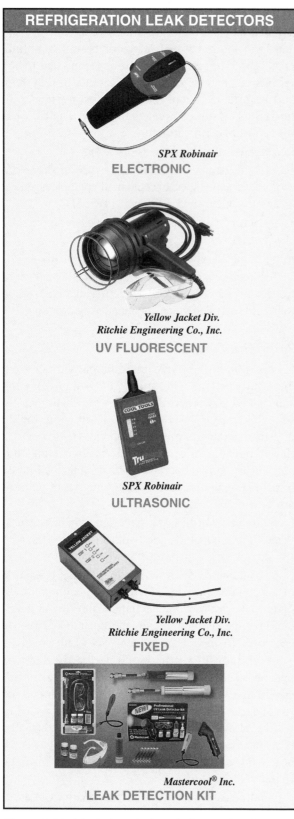

REFRIGERATION LEAK DETECTORS

SPX Robinair
ELECTRONIC

Yellow Jacket Div.
Ritchie Engineering Co., Inc.
UV FLUORESCENT

SPX Robinair
ULTRASONIC

Yellow Jacket Div.
Ritchie Engineering Co., Inc.
FIXED

Mastercool® Inc.
LEAK DETECTION KIT

Figure 6-31. A leak detector is a device used to detect refrigerant leaks in air conditioning or refrigeration systems.

For example, when purging hoses, only enough refrigerant is released to ensure that all air is removed from the hoses. One way to conserve refrigerant is to use refrigerant hoses equipped with low-loss fittings. Once hoses having low-loss fittings are filled with refrigerant, they need not be purged. These hoses can only be used with one refrigerant, however, because mixing refrigerants can lead to system failure and refrigerant mixtures must be disposed of as hazardous waste.

Leak Repair Requirements. Owners of equipment with charges greater than 50 lb are required to repair leaks in the equipment when the total leakage results in the loss of more than a certain percentage of the equipment's charge over a year. The repair requirement is the leak rate instead of the total quantity of refrigerant lost. For commercial and industrial process refrigeration sectors, leak rates that would release 35% or more of the refrigerant charge over a year must be repaired within 30 days. For all other sectors, including comfort cooling, leaks must be repaired when the appliance leaks at a rate that would release 15% or more of the refrigerant charge over a year.

The EPA mandates that owners of air conditioning and refrigeration equipment with more than 50 lb of charge must also keep records of the quantities of refrigerant added to equipment during servicing and maintenance procedures.

Owners are normally required to repair leaks within 30 days of discovery. The 30-day requirement may be waived if, within 30 days of discovery, an owner develops a one-year retrofit or retirement plan for the leaking equipment. Owners of industrial process refrigeration equipment may qualify for additional time under certain circumstances. For example, if an industrial process shutdown is required to repair a leak, owners have 120 days to repair the leak.

Safe Disposal Requirements

Under the rules of the EPA, equipment that is typically dismantled on-site before disposal, such as retail food freezers, residential central air conditioners, chillers, and industrial process refrigeration systems, must have the refrigerant recovered for disposal in accordance with the requirements of the EPA. However, equipment that typically enters the waste stream with the charge intact, such as motor vehicle air conditioners and household refrigerators, freezers, and room air conditioners, is subject to special safe disposal requirements.

Under the safe disposal requirements, the final person in the disposal chain (scrap metal recycler or landfill owner) is responsible for ensuring that refrigerant is recovered from equipment before the final disposal of the equipment occurs. However, technicians upstream can remove the refrigerant and provide documentation of the refrigerant removal to the final person if this is more cost-effective.

Planning for the Future

Observing the refrigerant recycling regulations of Section 608 is essential to conserving existing stocks of refrigerants, as well as complying with the Clean Air Act. However, owners of equipment that contains CFC refrigerants must look beyond the immediate need of maintaining existing equipment in working order. Owners are advised to plan for the replacement of existing equipment with equipment that uses alternative refrigerants. One possible refrigerant alternative is the hydrocarbon-based refrigerants.

REFRIGERANT HANDLING

Common maintenance tasks a refrigerant technician must perform in an industrial setting are system evacuation, leak testing, and system charging.

System Evacuation

Evacuation of a refrigeration system involves removing all air from inside the system. Air enters the tubing and components when a system is opened for repairs or after new construction. All the air and moisture contained in a system must be evacuated from the system before charging the system with refrigerant. This is accomplished with a vacuum pump that creates a deep low pressure that draws all the air from inside the system. **See Figure 6-32.** Before beginning evacuation, the vacuum pump oil must be changed according to manufacturer's instructions to prevent the formation of sludge in the pump. Only vacuum pump oil should be used and the oil level must be checked before operating the pump.

TECH TIP

Care must be taken not to vent or spill refrigerants on soil or water. Solids removed from or contaminated by refrigerants can be hazardous waste.

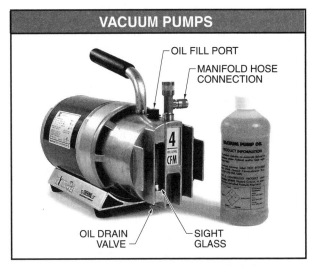

VACUUM PUMPS

OIL FILL PORT

MANIFOLD HOSE CONNECTION

OIL DRAIN VALVE

SIGHT GLASS

Figure 6-32. A vacuum pump removes all air from a refrigeration system.

The vacuum pump is connected to the system service valves using a gauge manifold and hoses. **See Figure 6-33.** As the air is removed by the vacuum pump, the pressure inside the system falls below atmospheric pressure and any water in the system vaporizes and is removed. The vaporization point of water is controlled by the pressure exerted on it. Water vaporizes (boils) at 212°F at sea level, but at higher elevations, the pressure is lower because there is less air pressing on the surface of the water. As pressure is reduced, water vaporizes at a lower temperature. In a complete vacuum, water vaporizes at a temperature above –90°F. **See Appendix.**

As the vacuum pump removes the air from the system, the low-pressure needle on the manifold gauge falls below zero and into the vacuum scale that is read in inches of mercury (in. Hg). Almost all the air is removed when the manifold low-pressure gauge needle reaches 30 in. Hg. It is relatively easy to achieve the first 29 in. of vacuum using a vacuum pump. The last inch of vacuum is much more difficult to achieve, but necessary to vaporize all moisture in the system.

A *micron gauge* is an electronic instrument that indicates the depth of vacuum in microns. A micron gauge is also used because the low-pressure gauge is not accurate enough to measure the final inch of vacuum. The final inch of vacuum is rated in microns. A *micron* is an alternate term for micrometer, or one millionth of a meter. If the vacuum inside a system is 29 in. Hg, the vaporization point of any water trapped in the system is approximately 76°F. If the vacuum can be reduced to 1000 μm, the vaporization point of water is approximately 4°F. At normal room temperatures, any water in the system would be vaporizing rapidly at this deep vacuum level. A *deep vacuum* is a vacuum between 400 μm and 700 μm. The water vapor is drawn from the system by the vacuum pump.

Micron gauges can indicate when the system is completely free of moisture. The vacuum pump is isolated and stopped when the micron gauge reading reaches between 400 μm to 700 μm and holds for about five minutes. The pressure may jump a few hundred microns then stabilize in a system that is completely dry. If there is a leak or moisture in the system, the pressure will rise slowly, then stabilize. It may take many hours of drawing a vacuum to remove the remaining moisture. Large systems may require 48 hours to achieve the desired deep vacuum. The process can be sped by using a two-stage pump or a deep vacuum pump. If a refrigeration system is equipped with Schrader valves, they should be removed so as not to slow the flow of vapor from the system. Large-diameter hoses help speed evacuation. Heat lamps can be applied to the system to speed vaporization as long as the temperature does not rise above 90°F. Shaking or tapping the compressor can release moisture trapped in the oil. The system must be tested and the leak repaired if a leak is suspected.

Leak Testing

The system probably has a leak if the vacuum does not reach the desired vacuum during evacuation and holds at about 1500 μm for over half an hour with the pump operating. It is also possible that the vacuum pump is losing its efficiency. Vacuum pumps should be tested for efficiency according to manufacturer's recommendations. Testing a system for leaks while it is under vacuum is not always effective. The maximum pressure exerted on the outside of the tubing is about 14.7 psi. The leak may not occur until a much higher pressure is reached during operation. In addition, the leak could be closed under vacuum and open when subjected to positive pressure during operation.

Leaks in an empty system are found using a standing pressure test. **See Figure 6-34.** Refrigerant vapor is added to raise the system pressure to 10 psi. Dry nitrogen gas is then added to the system to raise the pressure to about 150 psi. If the pressure does not fall after a few hours, the system does not have a leak. If the pressure falls, an electronic leak detector can be used to locate the general location of the leak.

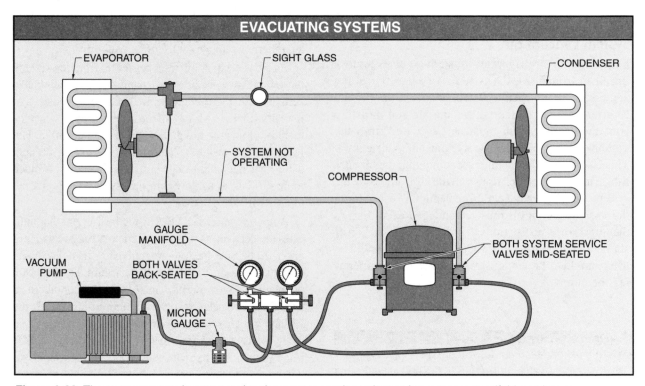

Figure 6-33. The vacuum pump is connected to the system service valves using a gauge manifold and hoses.

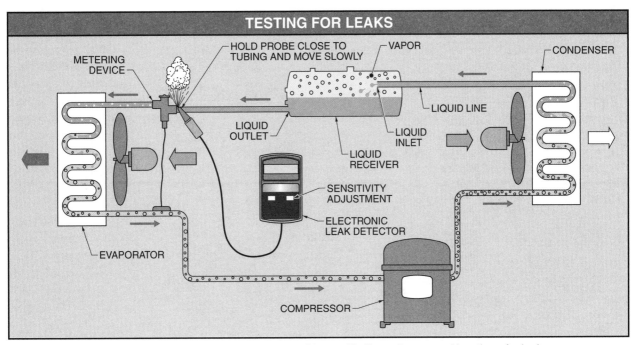

Figure 6-34. Electronic leak detectors are extremely sensitive and indicate the general location of a leak.

An *electronic leak detector* is a device that gives an audible tone or visual signal when the detector senses refrigerant. The detector is moved slowly around joints and fittings, which are the locations most likely to leak. When a leak is detected, a solution of soapy water is applied and the escaping gas causes bubbles to appear. This indicates the exact location of the leak. Electronic leak detectors and bubble testing can also be used to locate leaks in operating systems that are charged with refrigerant. In addition, leak-detecting dye can be added to the compressor oil of an operating compressor. The dye accumulates at the leak site and glows when exposed to fluorescent light. When a leak is found, it is repaired by resoldering or tightening connections.

System Charging

A system is charged with refrigerant after it has been completely evacuated. Liquid refrigerant is usually charged into the high-pressure side when a system is completely empty. Smaller systems and some refrigerants require that refrigerant be added through the low-pressure side. A special fitting is attached to the red charging hose or the refrigerant container. This fitting contains a small opening that causes the liquid refrigerant to flash into vapor so that only vapor is added to the system, even though it is being charged

from the liquid side of the refrigerant container. The system is not operated during liquid charging and the refrigerant must be weighed into the system. **See Figure 6-35.** The system is operated when the correct amount of refrigerant has been added or when the refrigerant stops flowing because pressure in the system equals pressure in the refrigerant container. If additional refrigerant is required to fill the system, it is charged into the low-pressure side as a vapor while the system is operating. An operating system that is low on refrigerant is also charged at the low-pressure side as a vapor. The low pressure of the system's low-pressure side allows the high-pressure refrigerant from the container to flow into the system. Charging is complete when the correct amount is added by weight or when the bubbles disappear from the sight glass and system pressures and temperatures are within normal operating specifications.

The refrigerant container is placed upside down to charge with liquid and right side up to charge with vapor. Four-hose gauge manifolds do not require the removal of hoses during evacuation and charging and do not require purging. This prevents refrigerant release and prevents air from entering the system because no air enters the hoses once the evacuation is complete. Low-pressure chillers and ammonia systems require special evacuating and charging procedures.

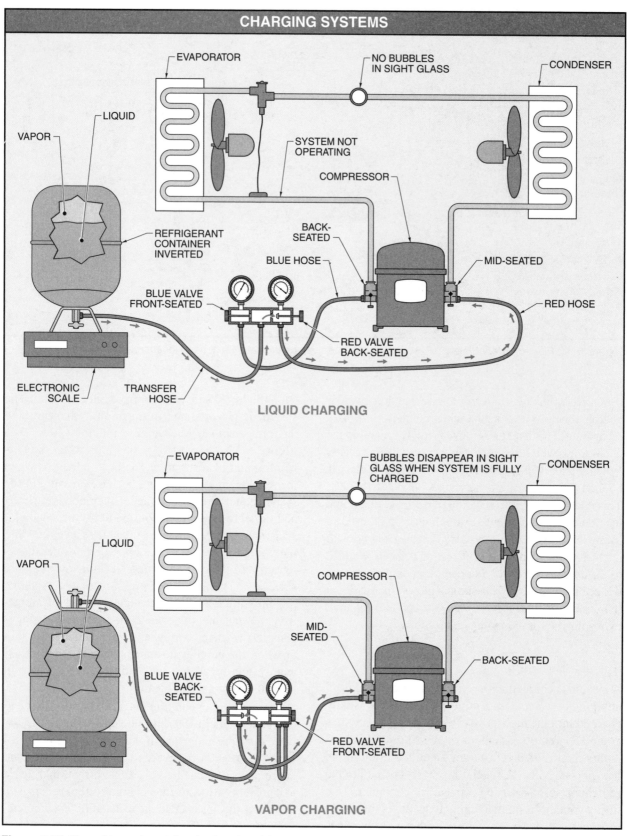

Figure 6-35. The refrigerant container is placed upside down to charge with liquid and right side up to charge with vapor.

Verifying System Charge

Modern refrigeration and air conditioning systems require an exact charge for efficient operation. It is critical that manufacturer's instructions be followed to determine if the system charge is correct. One of the most common problems with refrigeration and air conditioning systems is incorrect charge.

Subcooling or superheat calculations are used when a system is charged or when troubleshooting a system. Refrigerant is weighed into the system as it is being charged. Once the proper weight of refrigerant has been added, the superheat or subcooling is checked. Superheat is the amount of heat energy added to the refrigerant vapor by the evaporator after it is completely vaporized. Subcooling is the amount of heat energy removed from the refrigerant liquid by the condenser after it is completely condensed. The superheat or subcooling varies according to the refrigerant charge and the ambient operating conditions. Therefore, system manufacturers supply a procedure for calculating superheat or subcooling.

The subcooling calculations are used when the system metering device is a TXV. The high-pressure side temperature is measured using a contact digital thermometer placed directly on the piping. Infrared thermometers are not accurate for these types of readings unless they are carefully calibrated for taking close readings of copper piping. Then the high-pressure side pressure is measured at the service valve using a gauge manifold. Tables and graphs in the manufacturer's literature are studied to determine the required discharge temperature at a specific location on the high-pressure side. (The location varies by type of system.) If the required temperature is within ±3% of the measured temperature, the amount of refrigerant charge is correct. Refrigerant is added or removed as needed to achieve this tolerance.

The superheat process calculation is used when the metering device is not a TXV (a capillary tube, for example). Both the wet bulb temperature of the air entering the evaporator and the dry bulb temperature of the air entering the condenser are measured. The *wet bulb temperature* is a temperature reading that takes into account the amount of water vapor in the air. The *dry bulb temperature* is a temperature reading that does not take into account the amount of water vapor in the air. A *psychrometer* is a thermometer that can measure both dry and wet bulb temperatures. The dry and wet bulb temperatures are compared, and the required superheat is determined from the manufacturer's tables. The low-pressure side temperature is measured near the service valve, and the low-pressure side pressure is taken using a gauge manifold. The manufacturer's literature is used to determine the required low-pressure side temperature. If the required low-pressure side temperature is within ±5% of the measured temperature, the amount of refrigerant charge is correct. Refrigerant is added or removed as needed to achieve this tolerance.

It is important to follow the manufacturer's instructions for each system and account for details such as the length and height of piping. Using the subcooling and superheat processes accounts for system operating conditions and fine tunes the level of charge for peak efficiency.

7

Boiler Systems

Maintenance personnel commonly serve as boiler operators in a plant. Boilers are designed and manufactured for specific purposes and are unique in design, maximum allowable working pressure, boiler horsepower, fuel, and fabrication location. All steam boilers operate using a feedwater system, combustion system, steam system, and draft system. Boiler troubleshooting must quickly remedy the problem and restore steam to all loads.

BOILER APPLICATIONS

A *boiler* is a closed metal container (vessel) in which water is heated to produce steam or heated water. Boilers and their related equipment provide steam and/or hot water for heat and industrial processes. Steam is commonly used for industrial heating or process applications. For example, food processing plants use steam for cooking, pasteurization, and sterilization. Steam is also used to run turbines for electrical power generation. Paper mills use steam extensively in pulp processing and drying operations. Oil refineries use steam for distillation of petroleum products in the refining process. Hot water produced in a boiler is primarily used for building space heating applications. The boiler system supplies hot water to the heating system at a temperature between about 180°F and 220°F.

Maintenance personnel commonly serve as boiler operators and are responsible for operating the boiler plant for maximum safety and efficiency. **See Figure 7-1.**

Boiler operation requires training, and in most cases, licensing. Licensed boiler operators have passed a licensing examination. A boiler operator's license documents that the holder is qualified to safely operate and maintain steam boilers and related equipment. Approximately one-third of the states in the U.S. have licensing laws regulating the requirements for certification of boiler operators.

States that do not issue a statewide license may still have jurisdictions that establish licensing requirements within the state. For example, Pennsylvania does not have a statewide licensing requirement, but the city of Philadelphia does. Information about boiler operator licensing may be obtained from the authority having jurisdiction (AHJ).

An AHJ often has specific grades of licensing examinations for different sizes of boilers and related equipment. Some jurisdictions require experience in the field prior to taking a licensing examination. Maintenance personnel commonly hold more than one boiler operator's license.

Figure 7-1. Maintenance personnel commonly serve as boiler operators and may require licensing.

BOILER OPERATION PRINCIPLES

In a steam boiler, the closed container holds the water, transfers the heat to the water to make steam, and collects the steam that is produced. Heat is most commonly generated by the combustion of a fuel. To produce steam, heat is applied to the container. **See Figure 7-2.** Steam is produced as water is heated to 212°F and the water boils and changes from a liquid state to a gaseous state. A pressure vessel is created when a lid seals the container. A pipe in the lid can be used to direct steam to where it is needed.

Cleaver-Brooks
A boiler operator's license documents that the holder is qualified to safely operate and maintain steam boilers and related equipment.

Boilers are designed to use heat efficiently to minimize the cost of producing steam. *Thermal efficiency* is the ratio of heat absorbed to the heat available. An increase in thermal efficiency in boilers is accomplished by increasing the heating surface. The *heating surface* is the part of a boiler that has heat and gases of combustion on one side and water on the other. Increasing the heating surface increases the size of the combustion chamber, which also allows fuel to burn more efficiently as it mixes with a greater quantity of air. Greater thermal efficiency is achieved through the evolution of boiler design.

In a steam heating system, steam produced in the boiler leaves the main steam line and enters the steam header. **See Figure 7-3.** The *main steam line* is a line that connects a boiler to the steam header. The *steam header* is the part of a steam heating system that distributes the steam to branches. From the steam header, branch lines direct steam to the heating units. At the heating units, heat is transferred to the building space. As the steam releases heat to the building space and cools, it changes back to water (condensate). Condensate passes through the condensate return line to the condensate receiver tank. The condensate is used as feedwater, and is pumped by the feedwater pump back to the boiler to repeat the cycle. *Feedwater* is water that is supplied to the boiler at the proper temperature and pressure.

In a hot water heating system, hot water produced in the boiler is used to transport heat energy to building spaces. Hot water supplied commonly ranges from 180°F to 220°F. Water is heated in the boiler and circulated in the system by the circulating pump. Branch lines direct hot water to the heating units. At the heating units, heat is transferred to building spaces. The water is then circulated back to the boiler to repeat the cycle. A compression tank allows water to expand in the system without increasing the overall system pressure. Water is added to the system by the makeup water supply line.

Boiler Systems

Boilers used in industry are designed to minimize the cost of producing steam. Boiler designs vary, but all steam boilers operate using a feedwater system, fuel system, steam system, and draft system. **See Figure 7-4.** Each system requires properly operating components which must be controlled for maximum safety and efficiency.

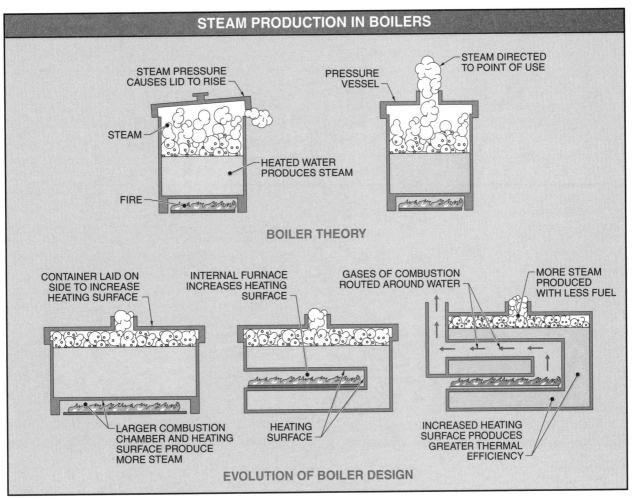

Figure 7-2. In a steam boiler, water is heated and turns from a liquid state to a gaseous state (steam).

Feedwater System. A *feedwater system* is a system that supplies water to the boiler at the proper temperature and pressure. Feedwater is treated and regulated automatically to meet the demand for steam. Valves are installed in feedwater lines to permit access for maintenance and repair. The feedwater system must be capable of supplying water to the boiler in all circumstances.

Fuel System. A *fuel system* is a system that supplies fuel in the proper amount to the boiler. The fuel is burned in order to change water to steam in the boiler. The fuel system includes all equipment used to provide fuel in order to generate the necessary heat in the boiler. The specific equipment required in the fuel system depends on the fuel used. All fuels are combustible and dangerous if safety precautions are not followed. Fuels commonly used are fossil fuels such as natural gas, fuel oil, and coal. Other methods of generating heat include electricity and nuclear fission.

Steam System. A *steam system* is a system that collects, controls, and distributes the steam produced in the boiler. Steam is directed through piping to the point of use. Throughout the steam system, steam pressure is regulated with valves and monitored using steam pressure gauges. The steam and feedwater system share some components.

Draft System. A *draft system* is a system that regulates the flow of air to and from the boiler. For fuel to burn efficiently, the proper amount of oxygen must be provided. Air must also be provided to direct the flow of air through the furnace to discharge the gases of combustion out of the furnace to the breeching. The *furnace* is the combustion chamber of a boiler. The *breeching* is the duct connecting the boiler to the chimney. The draft system provides air for combustion and discharges the gases of combustion into the atmosphere in a nonpolluted condition.

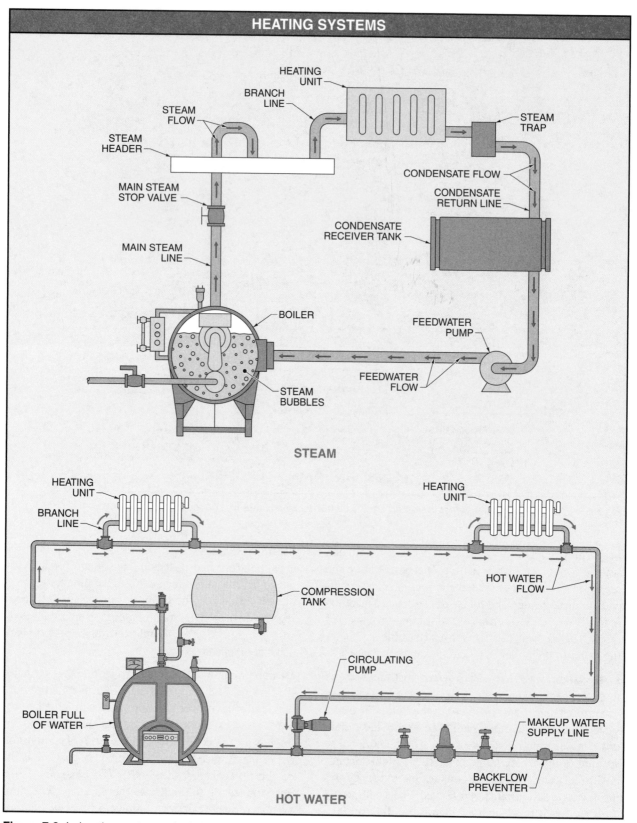

Figure 7-3. In heating systems, the boiler is used to produce steam or hot water which is used to transport heat energy to building spaces.

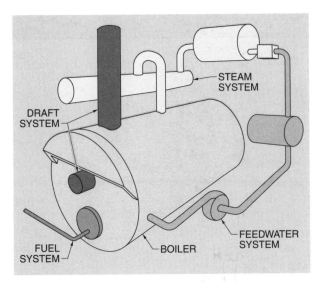

Figure 7-4. Boiler systems work together to produce steam safely and efficiently.

BOILERS

Boilers are manufactured for specific purposes and are unique in design, maximum allowable working pressure (MAWP), boiler horsepower (BHP), fuel, and fabrication location. Boilers are commonly classified by design as firetube and watertube boilers. A *firetube boiler* is a boiler in which the hot gases of combustion pass through tubes surrounded by water. A *watertube boiler* is a boiler in which water passes through tubes surrounded by the hot gases of combustion. **See Figure 7-5.**

The *maximum allowable working pressure (MAWP)* is the recommended maximum pressure at which a boiler can safely be operated. The MAWP is determined by the design and construction of the boiler in accordance with the code developed by ASME International®, formerly the American Society of Mechanical Engineers. The ASME Code regulates the materials and methods of boiler construction. The boiler MAWP is listed on the boiler nameplate. The boiler MAWP can be lowered by the AHJ based on boiler age and condition.

Boilers, fittings, and accessories are identified with an official ASME symbol stamp. **See Figure 7-6.** ASME symbol stamps use letters to identify boiler components. For example, the ASME symbol stamp for a safety valve is the letter V, and is included on the data plate of the safety valve.

Boilers are classified as low-pressure boilers or high-pressure boilers depending on the MAWP. A *low-pressure boiler* is a boiler that has an MAWP of up to 15 psi.

The MAWP for low-pressure boilers may vary in some states and should be verified with local authorities. Low-pressure boilers are used primarily for heating buildings such as schools, apartments, warehouses, factories, and residences.

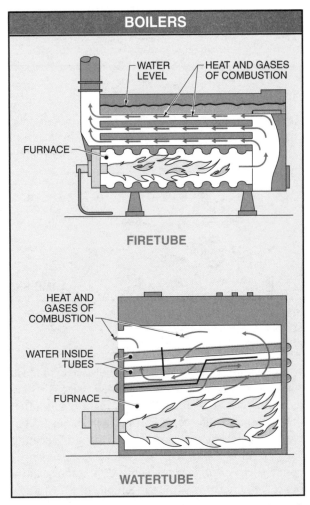

Figure 7-5. Heat and gases of combustion pass through tubes surrounded by water in a firetube boiler. Water passes through tubes surrounded by gases of combustion in a watertube boiler.

A *high-pressure boiler* is a boiler that has an MAWP above 15 psi and over 6 BHP. A *boiler horsepower (BHP)* is the power available from the evaporation of 34.5 lb of water per hour at a feedwater temperature of 212°F. An adequate supply of water must be available to evaporate or produce the required amount of steam. One BHP is equivalent to 33,472 Btu/hr. High-pressure boilers are commonly used for industrial process loads with an MAWP ranging from 75 psi to 700 psi.

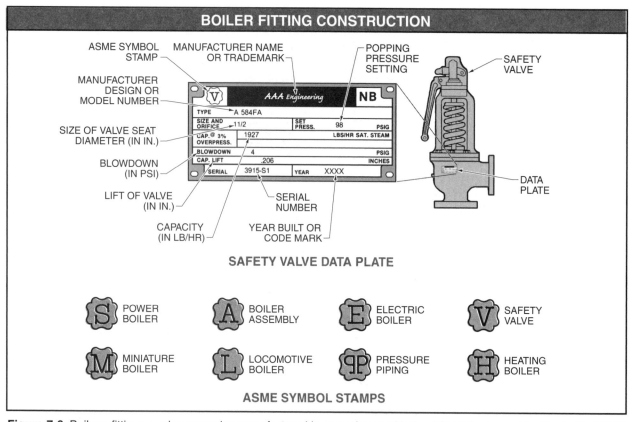

Figure 7-6. Boilers, fittings, and accessories manufactured in accordance with the ASME Code are identified with ASME symbol stamp letters.

Boilers are also classified by the fuel used to produce steam. Fuels commonly used include fossil fuels such as natural gas, fuel oil, and coal. A *fossil fuel* is a fuel that is formed in the earth over millions of years from plant or animal remains. Fuels are rated by heating value. *Heating value* is the amount of heat energy per unit of fuel. A *British thermal unit (Btu)* is the quantity of heat required to raise the temperature of 1 lb of water 1°F. **See Figure 7-7.** The heating value of natural gas is expressed in therms. A *therm* is a unit of heat energy equivalent to 100,000 Btu. The heating value of fuel oil is expressed in Btu/gal. The heating value of coal is expressed in Btu/lb.

A boiler is a heat moving device. A large amount of heat energy is needed to boil water to steam. As the water is heated to the boiling point, 1 Btu is required for each pound of water heated 1°F. *Sensible heat* is heat energy that can be measured by a change in temperature. When the water reaches the boiling point, the temperature does not change as heat is added and steam is produced. *Latent heat* is heat energy that causes a change of state and no temperature change. For example, 1 lb of water

in an open container located at sea level requires 970 Btu to change the 1 lb of water at 212°F into 1 lb of steam at 212°F. When the steam condenses back to water, the same amount of heat energy is released. In this manner, a steam boiler can transport large amounts of heat from the boiler to where it is needed.

In some applications, an electric boiler is required. An *electric boiler* is a boiler that produces heat with electrical resistance coils or electrodes. Electric boilers operate similar to boilers fired by fossil fuels. An electric boiler is less complex than boilers fired by fossil fuels because no combustion equipment is required. In electric resistance coils, electricity flows through a coiled conductor. Resistance created by the coiled conductor generates heat. Electric resistance coils are used in low-capacity boilers and are not as common as electrode electric boilers.

In electrode electric boilers, heat is generated by electric current flowing from one electrode to another through the boiler water. The conductivity of the boiler water affects the flow of electricity and the amount of heat generated.

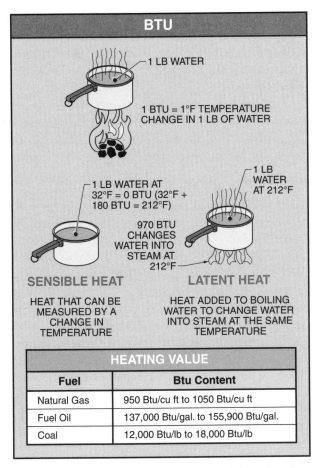

Cleaver-Brooks

Figure 7-8. Package boilers are preassembled at the factory as a self-contained unit and include all components for hookup and operation.

Warning: Electric boilers operate at up to 16,000 V. The main electrical disconnect must be locked out and tagged before maintenance work is performed on an electric boiler.

Electric boilers are typically one-fourth to one-half the size of fossil fuel boilers with similar heat output. Electric boilers are 100% emission free and are used as an alternative where fossil fuel combustion and the handling of combustion by-products is unacceptable.

Boilers are available as package boilers or field-erected boilers. A *package boiler* is a boiler that is preassembled at the factory. All components for hookup and operation are in place. A package boiler is a self-contained unit complete with feedwater pump, fuel pump, burner assembly, and combustion controls. **See Figure 7-8.** At the installation site, package boilers only require connection of feedwater lines, steam lines, fuel lines, electrical supplies, and the draft system. Package boilers may be firetube or watertube boilers.

A *field-erected boiler* is a boiler that is constructed on-site. Field-erected boilers are large-capacity boilers used when load requirements exceed the capacity of available package boilers. Field-erected boilers often reach as high as six stories. Like package boilers, field-erected boilers must be inspected by the AHJ and must comply with all applicable codes and standards.

Firetube Boilers

In a firetube boiler, the hot gases of combustion pass through tubes surrounded by water. Firetube boilers were originally designed for use on ships and were referred to as scotch marine boilers. Firetube boilers can be high- or low-pressure boilers. Firetube boilers are horizontal cylindrical vessels with the flame in the furnace and gases of combustion inside the tubes. **See Figure 7-9.** The furnace and tubes are housed within a larger vessel that contains water and steam.

Heat from the flames and gases of combustion passes through the tubes while the water surrounds the tubes. The gases of combustion make two, three, or four passes, depending on the design. The more passes, the greater amount of heat transferred to the water in the boiler. Tubes in a firetube boiler are measured by the outside diameter of the tube in contact with water.

Figure 7-7. Fossil fuels used in boilers have different heating values based on the Btu content per unit of fuel.

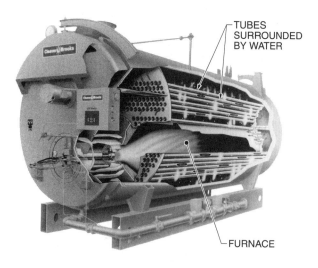

TUBES
SURROUNDED
BY WATER

FURNACE

Cleaver-Brooks

Figure 7-9. In a firetube boiler, the furnace and tubes are housed within a larger vessel that contains water and steam. The heat and gases of combustion make two, three, or four passes, depending on the boiler design.

Firetube boilers contain a large amount of water. Firetube boilers can respond to load changes with minimal variation in steam pressure because of their vessel size. As water is heated in a firetube boiler, it increases in volume and becomes lighter. The warm water rises, and cool water takes its place. Heated water causes steam bubbles to break the surface of the water and enter the steam space.

Tubes in the boiler create a large amount of heating surface in contact with the water in the boiler. The large heating surface results in more heat transfer, which increases water circulation and formation of steam bubbles. The result is high thermal efficiency for this type of boiler. An internal furnace inside the boiler shell also contributes a great amount of heating surface.

Firetube boilers are available as dryback or wetback boilers. A *dryback firetube boiler* is a boiler that has a refractory-lined chamber outside of the vessel that directs the gases of combustion from the furnace to the tube bank. *Refractory* is material that retains its strength at very high temperatures. A *wetback firetube boiler* is a boiler that has a water-cooled turnaround chamber that directs gases of combustion from the furnace to the tube bank. A *tube bank* is an assembly of tubes in a boiler.

Firetube boilers are used for hot water and low- and high-pressure steam for heating and process applications. Steam pressure in a firetube boiler is generally limited to approximately 350 psi and 800 BHP.

A watertube boiler is generally used if greater pressures and horsepower are required. In addition to the scotch marine firetube boiler, the firebox boiler and horizontal return tubular (HRT) boiler are found in industry. **See Figure 7-10.**

A *firebox boiler* is a compact firetube boiler within a round top and flat sides. This compact design provides high outputs in a smaller space than a scotch marine boiler. The firebox boiler has water legs that extend down the sides of the boiler. A *water leg* is the area around the furnace of a boiler that is filled with water to transfer heat from the furnace area. Steam produced in the steam space is routed out of the steam outlet at the top of the boiler. Firebox boilers are used for low-pressure steam or hot water heating applications.

A *horizontal return tubular (HRT) boiler* is a firetube boiler suspended over a furnace. The boiler drum is supported over the furnace by suspension slings from steel beams. The gases of combustion are directed from the furnace under the drum through the tubes secured into the front and rear tube sheets. Baffles in the furnace direct the flow of the gases of combustion for the greatest efficiency. Tube sheets are secured to the drum using diagonal stays. The stays provide support to the tube sheets in areas not supported by the tubes.

Watertube Boilers

In a watertube boiler, water passes through tubes surrounded by gases of combustion. In a watertube boiler, heat from the flames and gases of combustion is transferred to the water passing through the tubes. Baffles direct the gases of combustion for multiple passes to obtain maximum efficiency. Watertube boilers were developed to meet the demand for higher steam pressures than could be produced in firetube boilers. Firetube boilers, to produce comparable pressures, would require structural member thicknesses that are not practical. Watertube boilers can be broadly classified as industrial, commercial, or cast iron watertube boilers.

+ SAFETY TIP

The boiler industry is closely regulated by the American Society of Mechanical Engineers (ASME) and the ASME Code, which governs boiler design, inspection, and quality assurance. In addition, the insurance company insuring the facility or boiler may dictate additional fitting and/or accessory requirements for added safety.

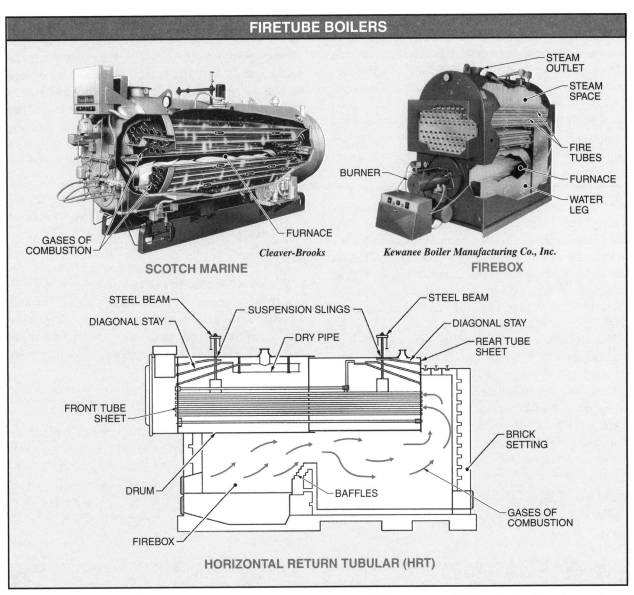

Figure 7-10. The firetube boiler design required depends on the installation space available and output requirements.

Industrial Watertube Boilers. An *industrial watertube boiler* is a watertube boiler that produces steam or hot water for industrial process applications. An industrial watertube boiler consists of multiple drums. A steam and water drum is located high in the boiler and contains both steam and water. A *mud drum* is the lowest part of the water side of a watertube boiler and collects sludge or mud. The steam and water drum and mud drum are connected by tubes. **See Figure 7-11.** Gases of combustion are directed around the tubes by baffles. Water in the tubes is heated and rises to the

steam and water drum, where steam is released. Cooler water descends to the mud drum.

Industrial watertube boilers have a fast steaming capability because there is less water to heat than in a firetube boiler. The low water content allows fast response to changing load demands. Industrial watertube boilers are used for high-temperature hot water and high-pressure steam process applications. They are rated in pounds of steam per hour (lb/hr) at operating conditions. Industrial watertube boiler ratings range from 10,000 lb/hr to 134,000 lb/hr.

Commercial Watertube Boilers. A *commercial water-tube boiler* is a watertube boiler that produces steam or hot water for commercial or medium-size applications. Commercial watertube boilers include straight-tube watertube boilers, bent-tube watertube boilers, and membrane watertube boilers. **See Figure 7-12.**

A *straight-tube watertube boiler* is a boiler that has straight tubes surrounded by gases of combustion. Steam and water enter the steam and water drum through an internal feedwater line. Straight-tube watertube boilers operate similarly to bent-tube watertube boilers.

A *bent-tube watertube boiler* is a boiler that has shaped tubes surrounded by gases of combustion. Bent-tube watertube boilers are used for steam or hot water applications. Baffles direct gases of combustion for multiple passes around the shaped tubes. Bent-tube watertube boilers are available in sizes ranging from 2 therms/hr to 9 therms/hr input, and up to 250 BHP. They are commonly used in heating applications using steam or hot water because of their resistance to thermal shock.

A *membrane watertube boiler* is a boiler that directs water through formed metal membrane tubes connected to upper and lower drums. The boiler vessel is gas tight, insulated on the outside, and covered by casing constructed of removable, formed steel. Membrane watertube boilers are used where space is limited for hot water or low- or high-pressure steam requirements.

Cast Iron Boilers. A *cast iron boiler* is a boiler that has modular cast iron sections that function similar to the watertubes in a watertube boiler. **See Figure 7-13.** The modular cast iron sections function as vessels to contain and direct the water. The water is heated in the sections and converted to steam. Additional sections can be added to increase boiler capacity as required because each section is modular. Cast iron boilers are limited to low-pressure steam or hot water for heating and process applications.

BOILER FITTINGS AND ACCESSORIES

Boiler fittings and accessories are required for each of the boiler systems to ensure safe and efficient operation of the boiler. A boiler fitting is directly attached to a boiler and is used for safety and/or efficiency. An accessory is not directly attached to the boiler but necessary for its operation. Boiler fittings and accessories include safety valves, water fittings and accessories, and steam fittings and accessories. All boiler fittings and accessories must be constructed in accordance with the ASME Code.

Safety Valves

A *safety valve* is a fitting that prevents the boiler from exceeding its MAWP. The safety valve is the most important fitting on a boiler. The safety valve opens to release pressure when pressure inside the boiler exceeds the MAWP. The safety valve functions as the last pressure regulation safeguard device on the boiler.

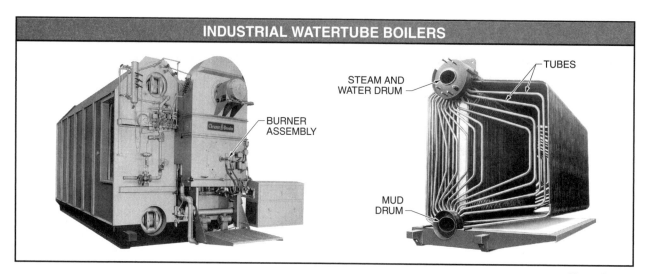

INDUSTRIAL WATERTUBE BOILERS

BURNER ASSEMBLY

STEAM AND WATER DRUM

TUBES

MUD DRUM

Cleaver-Brooks

Figure 7-11. In an industrial watertube boiler, water in the tubes is heated and rises to the steam and water drum, where steam is released. Cooler water descends to the mud drum.

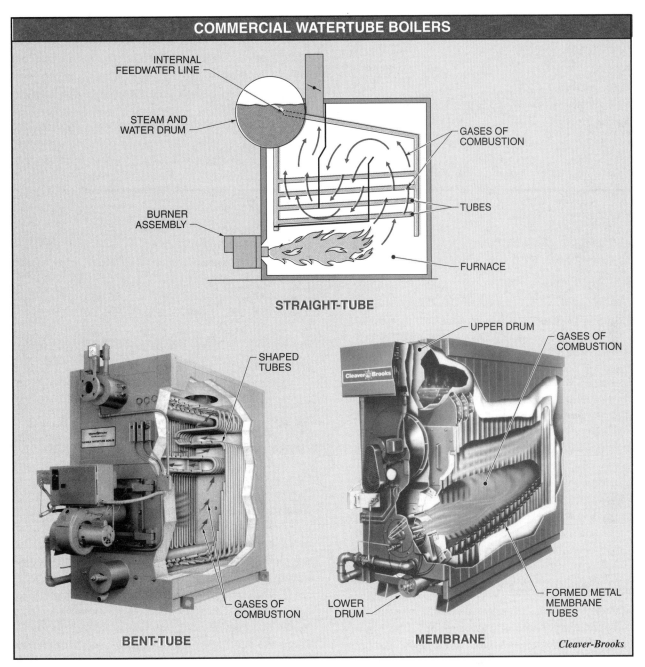

Figure 7-12. Commercial watertube boilers are used to produce steam or hot water for commercial or medium-size applications.

Safety valve capacity is measured as the amount of steam that can be discharged per hour. Once popped, the safety valve remains open until sufficient steam is released to drop the pressure to a specified level. This drop in pressure is the blowdown of the safety valve. Safety valve capacity and blowdown are listed on the safety valve data plate. Spring-loaded safety valves are the most common safety valve used on boilers. A spring exerts pressure on the valve against the valve seat to keep the valve closed. When pressure inside the boiler exceeds the set popping pressure, the pressure forces the valve to crack open. The steam passes through the partially open valve and hits the larger surface area attached to the valve seat. This creates extra force, which causes the safety valve to pop open suddenly, releasing a steady flow of steam. The steam flow holds the valve open until the pressure has fallen below the pressure setting required.

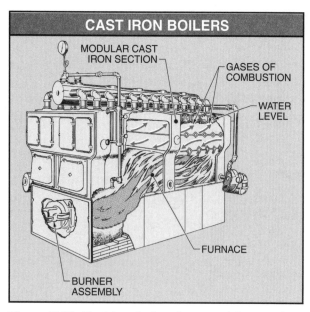

Figure 7-13. Cast iron boilers have modular cast iron sections, which function similar to the watertubes in a watertube boiler.

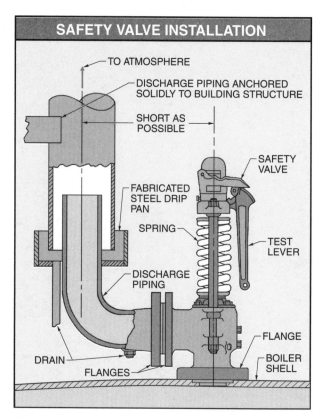

Figure 7-14. High-pressure boiler safety valves commonly have flange connections with piping to safely route discharged steam.

The ASME Code specifies the design, materials, and construction of safety valves. **See Figure 7-14.** In addition, the number of safety valves required and the frequency and procedures for testing safety valves are also specified by the ASME Code. Adjustment or repairs to safety valves must be performed by the manufacturer or an authorized representative of the manufacturer. Safety valves are installed at the highest part of the steam side of the boiler. No other valves may be installed between the boiler and the safety valve. High-pressure boiler safety valves commonly have flange connections.

Water Fittings and Accessories

Water fittings and accessories control the amount, pressure, and temperature of water supplied to and from a boiler. Water in a boiler must be maintained at the normal operating water level (NOWL). Low water conditions can damage the boiler and may cause a boiler explosion from overheated and weakened boiler metal. High water conditions can cause carryover. *Carryover* is the inclusion of small water droplets in steam lines, which can cause water hammer. *Water hammer* is a banging caused by rapid water movement in steam lines. Severe water hammer can damage equipment. Water fittings and accessories include feedwater valves, water columns, condensate return tanks, makeup water accessories, low water fuel cutoff, feedwater regulators, pumps, heaters, and bottom blowdown valves.

Feedwater Valves. A *feedwater valve* is a valve that controls the flow of feedwater from the feedwater pump to the boiler. **See Figure 7-15.** Feedwater stop valves are globe valves located on the feedwater line used to isolate the boiler from feedwater accessories. A *globe valve* is a valve that controls flow by raising or lowering a circular disc. The globe valve disc is moved manually by the operator. Turning the handle counterclockwise opens the valve. Turning the handle clockwise closes the valve.

A feedwater stop valve is positioned closest to the boiler to stop the flow of water out of the boiler for maintenance or if the check valve malfunctions. A *check valve* is a valve that allows flow in only one direction. The feedwater check valve is located next to the feedwater stop valve and prevents feedwater from flowing from the boiler back to the feedwater pump. The feedwater check valve opens and closes automatically by the swinging of a disc. The check valve disc swings open when water is fed to the boiler. The disc is forced closed if water tries to flow back from the boiler.

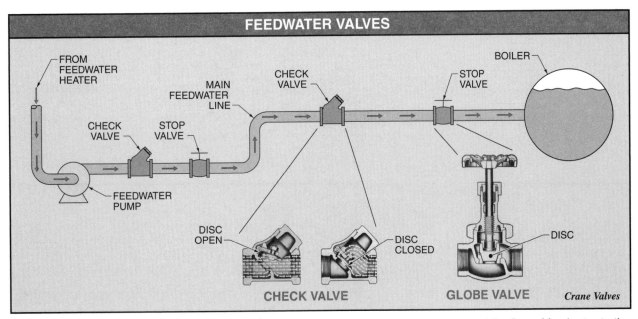

Figure 7-15. The feedwater line to the boiler includes stop valves and check valves to control the flow of feedwater to the boiler.

Water Columns. A *water column* is a boiler fitting that reduces water turbulence in the gauge glass to provide an accurate water level reading. A *gauge glass* is a boiler fitting that indicates the water level in the boiler. Water columns are located at the NOWL with the lowest part of the water column positioned at least 3″ above the heating surface. Water columns consist of the main column and three try cocks. **See Figure 7-16.** High and low water indicator lights, alarms, and/or whistles may be attached to the water column with sensors at the top and bottom try cocks.

Isolation valves located at the top and bottom of the gauge glass permit the replacement of the gauge glass. A gauge glass blowdown valve at the bottom of the gauge glass is used to remove sludge and sediment. Tubular gauge glasses are used for pressures up to 400 psi. Flat gauge glasses are used for pressures above 400 psi.

All boilers must have two methods of determining boiler water level. The gauge glass serves as the primary method of determining boiler water level. Try cocks are used as a secondary means of determining boiler water level. A *try cock* is a valve mounted on the water column that can be used to determine boiler water level. The middle try cock is located at the NOWL. Water and steam should be discharged from the middle try cock if the boiler water is at the NOWL. Water discharged from the top try cock indicates a high water condition. Steam discharged from the bottom try cock indicates a low water condition. Try cocks are effective up to approximately 250 psi. Above 250 psi, it is difficult to distinguish between water and steam discharged from the try cock.

All boilers are fitted with a vent, which is often located on the highest part of the water column or boiler. The vent is opened when a boiler is started to allow any air to escape from the boiler. It is closed when a steady flow of steam flows from the vent. The vent is opened when a boiler is shut down to allow air into the boiler to prevent the formation of a vacuum inside the cooling boiler.

Chemgrate
Floor grating systems allow drainage and access to lines in the floor connecting the boiler.

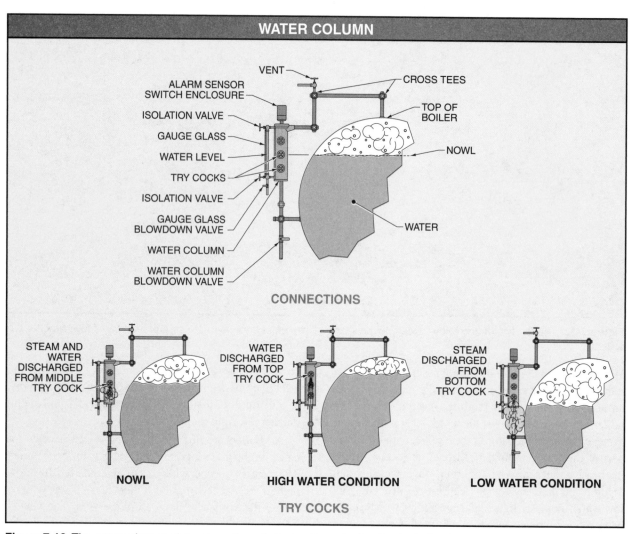

Figure 7-16. The gauge glass on the water column indicates the water level in the boiler. Try cocks are used as a secondary means of determining boiler water level.

Condensate Return Tanks. A *condensate return tank* is a boiler accessory that collects condensate returned from the steam system. Condensate that is not contaminated in the process is pumped back to the feedwater heater. A *feedwater heater* is a boiler accessory that heats feedwater before it enters the boiler. The condensate return tank and pump are installed at the lowest point in the building where steam is used. **See Figure 7-17.** A vent allows pressure in the tank to vent to the atmosphere. A condensate pump draws condensate from the condensate tank and discharges it to the feedwater heater. A pump controller regulates the flow of condensate to the feedwater heater.

Makeup Water Accessories. *Makeup water* is water that is used to replace boiler water lost from leaks.

Makeup water is fed through the condensate return tank using manual and/or automatic systems. The manual system always bypasses the automatic if the boiler has both systems. Maintenance personnel must know how to supply makeup water quickly to the boiler in the event of a low water condition. Manual systems feed makeup water with a hand-operated valve. **See Figure 7-18.** Automatic makeup water feeders control the flow of makeup water with a float-controlled valve located slightly below the NOWL. As the boiler water level falls, the float drops and the valve in the makeup water line is opened to allow water flow. Makeup water is mixed with condensate returned from the system and fed to the boiler. As the boiler water level rises, the float rises to shut the valve OFF to prevent water flow.

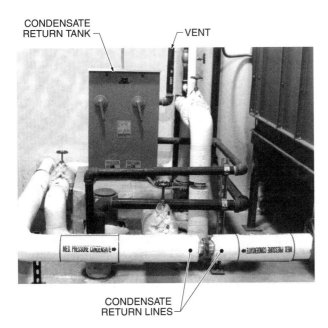

CONDENSATE RETURN TANK —

— VENT

CONDENSATE RETURN LINES —

MED. PRESSURE CONDENSATE

Figure 7-17. The condensate return tank collects condensate from condensate return lines and is located at the lowest point in the building where steam is used.

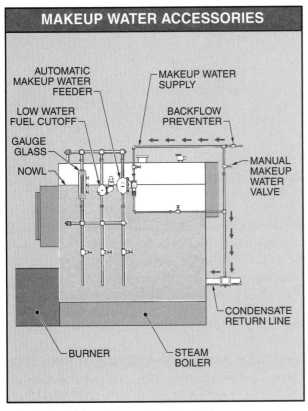

MAKEUP WATER ACCESSORIES

AUTOMATIC MAKEUP WATER FEEDER

MAKEUP WATER SUPPLY

LOW WATER FUEL CUTOFF

GAUGE GLASS

NOWL

BACKFLOW PREVENTER

MANUAL MAKEUP WATER VALVE

CONDENSATE RETURN LINE

BURNER

STEAM BOILER

Figure 7-18. Makeup water is fed to the boiler automatically or manually to replace boiler water lost from leaks and/or lack of condensate returned to the boiler.

Low Water Fuel Cutoff. A *low water fuel cutoff* is a boiler fitting that shuts OFF the burner and trips an alarm if a low water condition occurs in the boiler. **See Figure 7-19.** The low water fuel cutoff point is approximately 1″ above the lowest gauge glass water level. Low water fuel cutoffs are available with or without a water column. Low water fuel cutoffs must be tested daily, or more often depending on plant procedures and requirements.

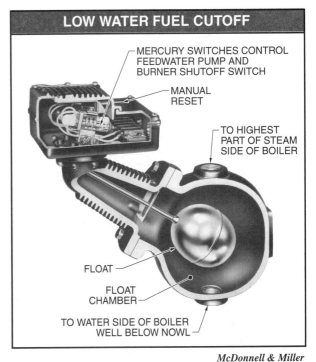

LOW WATER FUEL CUTOFF

MERCURY SWITCHES CONTROL FEEDWATER PUMP AND BURNER SHUTOFF SWITCH

MANUAL RESET

TO HIGHEST PART OF STEAM SIDE OF BOILER

FLOAT

FLOAT CHAMBER

TO WATER SIDE OF BOILER WELL BELOW NOWL

McDonnell & Miller

Figure 7-19. A low water fuel cutoff is a boiler fitting that shuts OFF the burner if a low water condition occurs in the boiler.

Low water fuel cutoffs use a float sensor or electric probe. The float sensor senses a drop in water level with a mercury switch. The electric probe senses the absence of water to activate a shutoff switch. Both sensors are wired to the burner control switch to shut OFF the burner when the water level drops in the chamber and the alarm sounds.

Feedwater Regulators. A *feedwater regulator* is a boiler accessory that maintains the NOWL in a boiler by controlling the amount of feedwater pumped to the boiler. Feedwater regulator designs vary, but all include a sensing element located at the boiler NOWL. **See Figure 7-20.** A signal is sent to the feedwater valve to increase feedwater flow when the sensing element detects a decrease in boiler water level. On some package boilers, the feedwater regulator functions as an ON/OFF switch to control the feedwater pump. Boiler water level is maintained automatically with a feedwater regulator. However, the boiler water level must still be checked periodically by maintenance personnel.

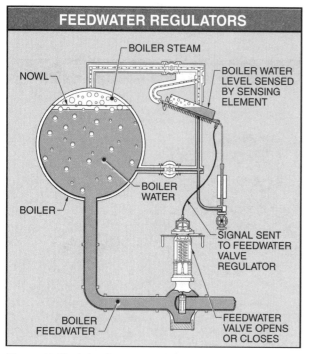

Figure 7-20. A feedwater regulator maintains the NOWL in a boiler by controlling the amount of feedwater pumped to the boiler from the condensate return tank.

Feedwater Pumps. A *feedwater pump* is a boiler accessory controlled by the feedwater regulator to supply feedwater to a boiler. Pressure produced by the feedwater pump must overcome boiler water pressure and system resistance to maintain the NOWL in the boiler. The ASME Code states that boilers having over 500 sq ft of boiler heating surface must have at least two means of supplying feedwater to the boiler. For example, a plant having one steam-driven feedwater pump must have an electric backup feedwater pump. Feedwater pumps may be centrifugal, turbine, or reciprocating.

A *centrifugal pump* is a pump that uses centrifugal force to increase the pressure of a fluid. *Centrifugal force* is the outward force produced by a rotating object. Centrifugal feedwater pumps are the most common feedwater pumps. **See Figure 7-21.** Centrifugal feedwater pumps are electric motor or steam-driven. Centrifugal force moves water to the outside edge of the rotating impeller. The housing directs water from

the impeller to the outlet. Turbine feedwater pumps operate similarly to centrifugal feedwater pumps but use diffusing rings to direct the flow of water instead of the housing in centrifugal feedwater pumps. Turbine feedwater pumps have a higher operating efficiency and can overcome greater pressure than centrifugal feedwater pumps. Reciprocating feedwater pumps use a piston to discharge water to the feedwater line.

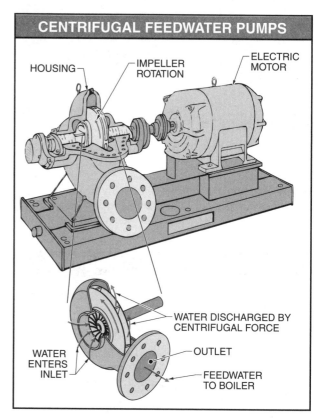

Figure 7-21. The centrifugal feedwater pump housing directs water discharged from the rotating impeller.

Feedwater Heaters. The feedwater heater heats feedwater before it enters the boiler. Feedwater heaters may be required on some boilers to remove oxygen and other gases that cause the corrosion of boiler metal. Maintaining the proper feedwater temperature also reduces the amount of scale-causing material from settling on the heating surfaces of the boiler.

⊕ **SAFETY TIP**

Before boiler start up, feedwater, transfer, and condensate pump suction and discharge port piping connections should be checked. Provisions should be made for proper expansion to prevent undue strain on the pump housing.

Feedwater heaters are either open or closed. **See Figure 7-22.** An *open feedwater heater* is a feedwater heater in which steam and water come into direct contact (mix) to raise the temperature of the water. Open feedwater heaters are located above the feedwater pump to produce a positive pressure on the suction side of the pump. A *closed feedwater heater* is a feedwater heater in which steam and water do not come into direct contact. The feedwater is directed through a large number of tubes inside an enclosed steel vessel. Steam flowing through the vessel heats the tubes. The heat is transferred to the feedwater flowing through the tubes. Closed feedwater heaters are located on the discharge side of the feedwater pump. Closed feedwater heaters achieve higher water temperatures than open feedwater heaters.

Bottom Blowdown Valves. A *bottom blowdown valve* is a valve used to release water from the bottom of a boiler in order to reduce the water level, remove sludge and sediment, reduce chemical concentrations, and/or drain the boiler. The two valves used to perform a bottom blowdown may be a quick-opening valve and a slow-opening (screw) valve, or two slow-opening valves. **See Figure 7-23.** Water is discharged through the bottom blowdown valves to the blowdown tank. A *blowdown tank* is a boiler accessory that collects water discharged through the bottom blowdown valves. The water is allowed to cool to prevent hot boiler water being sent directly to the sewer.

Steam Fittings and Accessories

Steam fittings and accessories remove air, control steam flow, and maintain the required steam pressure in a boiler and steam system. Steam fittings are also used to direct steam to various locations for heating and process. All steam fittings and accessories on a boiler are designed and manufactured for specific functions. Steam fittings and accessories include steam pressure gauges, steam valves, and steam traps and strainers.

Steam Pressure Gauges. A steam pressure gauge indicates pressure inside the boiler. A mechanical steam pressure gauge uses a Bourdon tube to sense the amount of steam pressure. **See Figure 7-24.** A *Bourdon tube* is a curved, hollow tube closed on one end that straightens when pressure is applied in the tube. The steam pressure gauge range should be 1½ to 2 times the MAWP of the boiler. For example, on a low-pressure boiler, the maximum steam pressure on the pressure gauge is 30 psi because the boiler MAWP is 15 psi or less.

✚ **SAFETY TIP**

Most valves under pressure should be opened slowly. An exception to this rule is the operation of a safety valve. The safety valve disc is designed to open quickly (pop) and remain open. During normal conditions, the safety valve disc is held closed by spring force.

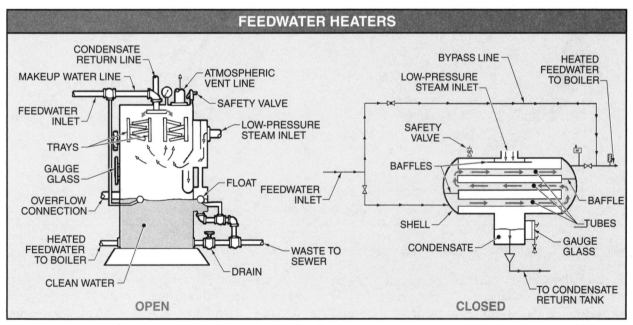

Figure 7-22. A feedwater heater uses steam for heating feedwater to remove oxygen and other gases which may cause corrosion.

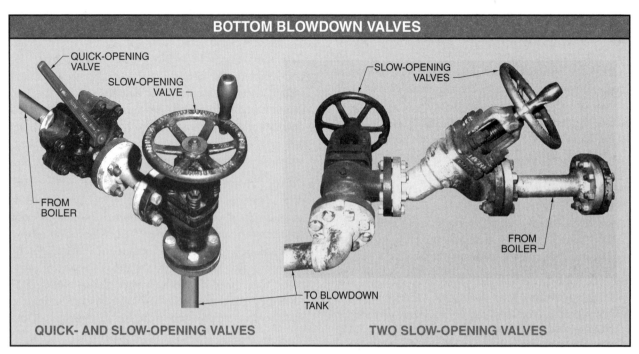

Figure 7-23. Bottom blowdown valves control the release of water from the bottom of the boiler to reduce the water level, remove sludge and sediment, reduce chemical concentrations, and/or drain the boiler.

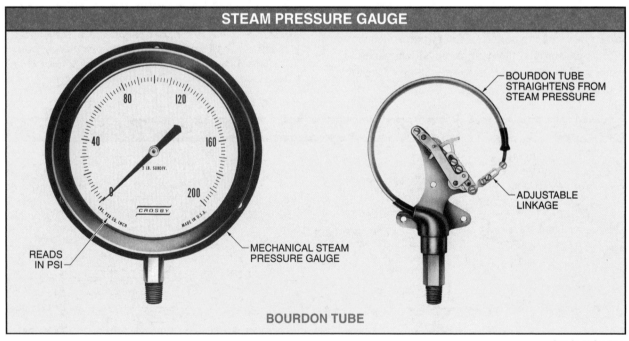

Crosby Valve Inc.

Figure 7-24. A steam pressure gauge indicates pressure inside a boiler.

Steam Valves. Steam valves commonly used include the main steam stop valve and the globe valve. A *main steam stop valve* is a valve that controls the flow of steam from the boiler. The main steam stop valve is opened to cut the boiler in on-line by allowing steam to flow from the boiler. The valve is closed to stop the flow of steam from the boiler to take it off-line. The main steam stop valve is a gate valve. A *gate valve* is a valve that controls flow by raising or lowering a wedge-like gate. A gate valve offers no restriction to flow when open and must be wide open or fully closed. An outside stem and yoke (os&y) valve is commonly used for the main steam stop valve. **See Figure 7-25.** The position of the stem indicates whether the valve is open (stem up) or closed (stem down).

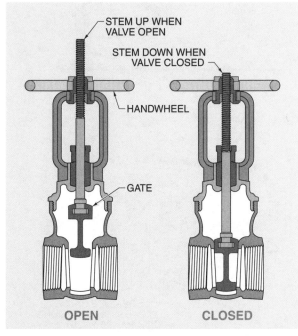

STEM UP WHEN VALVE OPEN

STEM DOWN WHEN VALVE CLOSED

HANDWHEEL

GATE

OPEN   CLOSED

Figure 7-25. The position of the stem on an outside stem and yoke (os&y) valve indicates whether the valve is open (stem up) or closed (stem down).

When open, the gate valve allows a free flow of steam. A gate valve should not be used as a regulating valve. A globe valve may be used to regulate the amount of steam flow. In a globe valve, steam flows under the valve seat through the valve. The change in direction of steam causes a decrease in steam pressure. A globe valve should never be used as a main steam stop valve.

Steam Traps and Strainers. A *steam trap* is a boiler accessory that removes air and condensate from steam lines and heat exchangers. **See Figure 7-26.** A steam trap traps steam until it has lost its heat and becomes condensate. Steam traps work automatically and increase boiler plant efficiency. They also prevent water hammer by expelling air and condensate from the steam lines without loss of steam. Steam traps are located after the main steam header and throughout the system after each device where steam is used.

A *nonreturn steam trap* is a steam trap that removes condensate from steam lines and heat exchangers and

delivers it to the condensate return tank. Nonreturn steam traps commonly used include thermostatic, float thermostatic, inverted bucket, and thermodynamic steam traps.

A *thermostatic steam trap* is a steam trap that has a bellows filled with a fluid that boils at steam temperature. As the fluid boils in the bellows, vapor expands the bellows to push the valve closed. The bellows contracts to open the valve and discharge condensate when the temperature drops below steam temperature. A variation of the thermostatic steam trap is the float thermostatic steam trap. A *float thermostatic steam trap* is a steam trap that has a float that opens and closes depending on the amount of condensate. Condensate is drawn out by return vacuum.

An *inverted bucket steam trap* is a steam trap in which steam enters the bottom and flows into an inverted bucket. When the trap is filled with water, steam inside the bucket causes it to float up, closing the discharge valve. When condensate fills the bucket, it loses buoyancy and sinks to open the discharge valve. Lower pressure in the discharge line draws condensate from the steam trap. When the bucket fills with steam again, it floats and closes the discharge valve. A small hole at the top of the bucket vents any trapped air.

A *thermodynamic steam trap* is a steam trap that has a single movable disc that raises to allow the discharge of air and cool condensate. Hot condensate creates steam at higher pressure on top of the disc and lower pressure under the disc, causing it to close and trap pressure in the chamber.

A *steam strainer* is a boiler accessory that removes scale or dirt from steam. A steam strainer is located on the inlet side of each steam trap because scale or dirt can clog their discharge orifices. Steam strainers must be cleaned regularly.

The main steam stop valve is normally an os&y valve that is opened to allow steam to flow from a boiler.

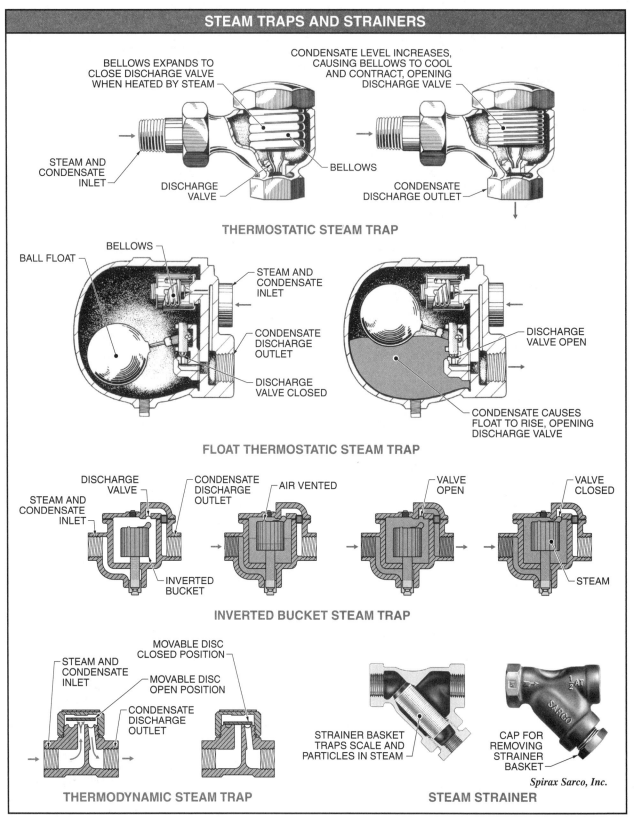

STEAM TRAPS AND STRAINERS

BELLOWS EXPANDS TO CLOSE DISCHARGE VALVE WHEN HEATED BY STEAM

CONDENSATE LEVEL INCREASES, CAUSING BELLOWS TO COOL AND CONTRACT, OPENING DISCHARGE VALVE

STEAM AND CONDENSATE INLET

DISCHARGE VALVE

BELLOWS

CONDENSATE DISCHARGE OUTLET

THERMOSTATIC STEAM TRAP

BALL FLOAT

BELLOWS

STEAM AND CONDENSATE INLET

CONDENSATE DISCHARGE OUTLET

DISCHARGE VALVE CLOSED

DISCHARGE VALVE OPEN

CONDENSATE CAUSES FLOAT TO RISE, OPENING DISCHARGE VALVE

FLOAT THERMOSTATIC STEAM TRAP

DISCHARGE VALVE

CONDENSATE DISCHARGE OUTLET

AIR VENTED

VALVE OPEN

VALVE CLOSED

STEAM AND CONDENSATE INLET

INVERTED BUCKET

STEAM

INVERTED BUCKET STEAM TRAP

MOVABLE DISC CLOSED POSITION

STEAM AND CONDENSATE INLET

MOVABLE DISC OPEN POSITION

CONDENSATE DISCHARGE OUTLET

STRAINER BASKET TRAPS SCALE AND PARTICLES IN STEAM

CAP FOR REMOVING STRAINER BASKET

Spirax Sarco, Inc.

THERMODYNAMIC STEAM TRAP

STEAM STRAINER

Figure 7-26. Steam traps remove condensate from steam lines automatically. Steam strainers are located on the inlet side of steam traps to remove scale or dirt from the steam.

COMBUSTION

Combustion of fuel provides the heat required to generate steam in a boiler. Combustion is perfect, complete, or incomplete. *Perfect combustion* is combustion of all fuel using only the theoretical minimum amount of air. Perfect combustion is achievable only in a laboratory setting and cannot be achieved in a boiler. *Complete combustion* is combustion of all fuel using an amount of air above the theoretical minimum. Complete combustion provides the greatest practical efficiency and minimizes environmental impact. *Incomplete combustion* is combustion that occurs when not all fuel is burned, resulting in soot, smoke, and unburned gases.

Air required in the combustion process is primary air, secondary air, or excess air. *Primary air* is air in the combustion process that regulates the rate of combustion. *Secondary air* is air in the combustion process that controls combustion efficiency by controlling how completely the fuel is burned. *Excess air* is air in the combustion process that is above the theoretical amount required to burn the fuel.

Combustion of fuels is optimum when the turbulent mixture of air and fuel is held in the furnace long enough to burn completely. This requires the proper mixture of air and fuel, the atomizing (breaking up) of the fuel into particles, and the proper temperature and time required in the furnace to achieve complete combustion.

Fuels

Natural gas, fuel oil, and coal are the three fuels most commonly used in boilers. Alternative fuels, including sawdust, waste oil, or municipal solid waste (MSW), are sometimes used depending on availability and environmental restrictions. Electricity or nuclear energy are also used in boilers. Each fuel requires special accessories for storage, handling, and combustion.

Natural Gas Accessories

Natural gas is combustible fossil fuel gas found in pockets trapped underground. Natural gas requires less preparation and equipment for burning than coal or fuel oil. Gas systems may be low- or high-pressure. Gas is supplied to the plant at a pressure that varies depending on the source of incoming gas. In a low-pressure gas system, incoming gas pressure is reduced to slightly above atmospheric pressure. The combustion air supplied by the blower in the front of the boiler draws the gas into the furnace. In a high-pressure gas system, gas is supplied to the burner at pressures above atmospheric pressure. The gas flows into the furnace because it is at higher pressure than the air entering the furnace.

A typical gas line contains two manual shutoff valves, couplings, a low gas pressure switch, a regulator, motorized valves, a vent valve, a high gas pressure switch, a butterfly valve, and a pilot line. **See Figure 7-27.**

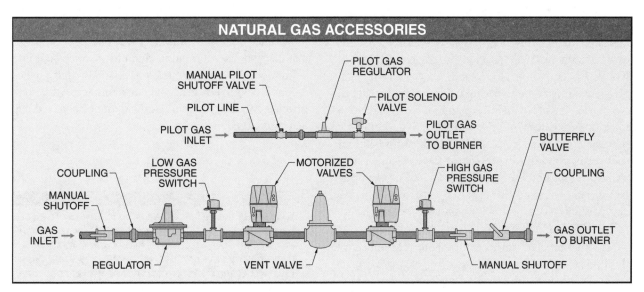

Figure 7-27. Natural gas pressure and flow are controlled by fuel system accessories.

The manual shutoff valves allow the operator to stop gas flow to the boiler. Couplings at both ends of the gas line allow for complete removal and replacement. The low gas pressure switch shuts down the burner if incoming gas pressure is too low to maintain a stable flame in the furnace. The regulator lowers the gas pressure to the level required by the furnace.

The two motorized valves provide important safety functions. Both valves open slowly during furnace ignition. This prevents a sudden inrush of explosive gas. Both valves close instantly when the burner is shut down. This prevents leakage of gas into the boiler after the burner is extinguished. The vent valve between the motorized valves exhausts the small amount of gas trapped between the motorized valves when they are closed. The vent valve is closed when the motorized valves are open.

The high gas pressure switch shuts down the burner if the gas pressure is too high for safe furnace operation. The butterfly valve regulates the amount of gas flowing into the furnace. This valve is often mechanically linked to fan and damper operation. When more gas is needed in the furnace, more air is added to ensure complete combustion. The pilot line consists of a manual pilot shutoff valve, a pilot gas regulator, and a pilot solenoid valve that opens to allow gas flow when the pilot light is to be ignited. Gas lines are similar on all gas burning equipment.

Fuel Oil Accessories

Fuel oil is liquid fossil fuel made from crude oil. *Crude oil* is a mixture of semisolids, liquids, and gases formed from the remains of organic matter that has been changed by pressure and heat over millions of years. Fuel oil is classified by grades into No. 1, No. 2, No. 3, No. 4, No. 5, and No. 6 fuel oil. Fuel oil grades most commonly used for boilers are No. 2, No. 4, No. 5, and No. 6. Fuel oil grades No. 5 and No. 6 must be preheated. In colder climates, No. 4 must be preheated or specially blended. The heating value of fuel oil is normally expressed in Btu/gal. and varies with different grades. **See Figure 7-28.** For example, No. 2 fuel oil has a heating value of 141,000 Btu/gal. compared to No. 6 fuel oil, which has a heating value of 150,000 Btu/gal.

The fuel oil accessories required are determined by the fuel oil used and the climate. For example, No. 6 fuel oil is heated with a fuel heater in the storage tank for ease in pumping and burning in cold climates.

Steam and electric fuel oil heaters are available. Fuel oil pumps discharge fuel oil from the tank to the burner at the regulated pressure. Pressure gauges and a relief valve are located on fuel oil lines. The relief valve is located between the pump and the discharge valve. The suction and discharge valves must be opened before the pump is started. Fuel oil filters located on the suction side of the fuel oil line remove foreign matter.

Fuel oil burners atomize fuel oil into a fine spray that readily mixes with oxygen in the air for maximum combustion efficiency. **See Figure 7-29.** Fuel oil burners commonly used include atomizing and rotary cup burners. An *atomizing burner* is a burner that uses steam, air, or fuel oil pressure to atomize fuel oil. A *rotary cup burner* is a burner that atomizes fuel oil using centrifugal force from the outer surface of a spinning cup. Rotary cup burners are used with No. 4, No. 5, and No. 6 fuel oil at temperatures ranging from 100°F to 180°F. Rotary cup fuel oil burners can accumulate deposits of carbon or noncombustible elements and must be serviced regularly. Combination burners use fuel oil and natural gas simultaneously or separately.

Coal Accessories

Coal is a solid black fossil fuel. Coal is classified by grade and rank. Coal grade is related to size, heating value, and ash content. Coal rank is related to hardness. Coal is fed to a boiler by stokers, pulverizers, or hand-firing.

A *stoker* is a mechanical device used to feed coal pieces into a furnace. Stokers commonly used include screw-feed, chain grate, and ram-feed stokers. A *screw-feed stoker* is a stoker that uses a long rotating screw to advance coal into a furnace. A *chain grate stoker* is a stoker that uses a rotating chain to advance coal into a furnace. A *ram-feed stoker* is a stoker that pushes feeder blocks to advance coal into a furnace. All stokers provide control of the amount and rate of coal fed and ash discharged.

> **TECH TIP**
>
> *Natural gas, fuel oil, and coal used in boilers are fossil fuels consisting primarily of the elements hydrogen (H) and carbon (C), which form hydrocarbon molecules. In the combustion process, HC molecules are broken down into H and C atoms and combine with atmospheric oxygen.*

FUEL OIL GRADES				
Characteristics	**No. 2**	**No. 4**	**No. 5**	**No. 6**
Type	light distillate	light distillate or blend	light residual	residual
Color	amber	black	black	black
°API at 60°F	32	21	17	12
Specific Gravity	0.8654	0.9279	0.9529	0.9861
Heating Value (Btu/gal.)	141,000	146,000	148,000	150,000
Heating Value (Btu/lb)	19,500	19,100	18,950	18,750

Figure 7-28. The heating value of fuel oil varies with different grades.

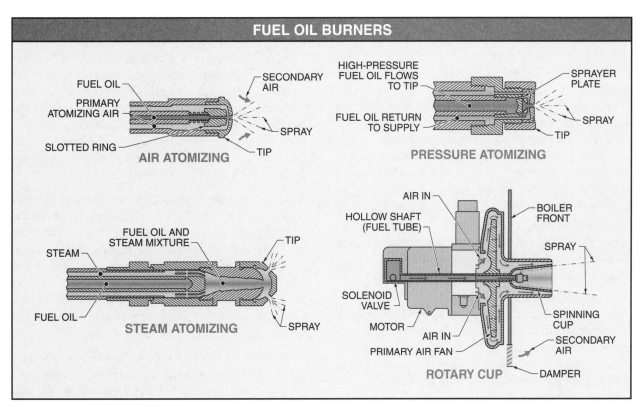

Figure 7-29. Fuel oil burners atomize fuel oil into a fine spray which readily mixes with oxygen in the air for maximum combustion efficiency.

A *pulverizer* is a mechanical device used to feed coal that has been ground to a fine powder. The powder is blown into the burner and burned while suspended in the air. Pulverized coal powder is commonly used in high-pressure boilers. *Hand-firing* is the feeding of coal pieces into a furnace by hand. Hand-firing is rarely used for feeding boilers today because of the limitations of the amount of coal fed and loss of heat when opening the firing door.

Combustion Controls

Combustion controls are boiler controls that optimize efficiency by regulating fuel supply, air supply, air-to-fuel ratio, and the removal of gases of combustion. The amount of fuel supplied to the burner is proportional to the steam pressure and the quantity of steam required. A drop in steam pressure or increase in steam demand requires an increase in the fuel supplied to the burner. Conversely, an increase in steam pressure or drop in steam demand requires a decrease in the fuel supplied. Also, any change in the fuel supply requires a corresponding change in the air supply to the burner to ensure complete combustion.

Automatic Combustion Control Systems. Automatic combustion control systems use electromechanical equipment that automatically controls the amount and frequency of firing in the burner to maintain the required steam pressure. Automatic combustion control systems reduce the risk of human error and aid overall plant efficiency. The four basic types of automatic combustion control systems are the ON/OFF control system, modulating control system, positioning control system, and metering control system.

An ON/OFF control system starts and stops the burner frequently to maintain steam pressure within a set range. A modulating control system modulates the burner between high fire and low fire to maintain a constant steam pressure. In the positioning control system, compressed air is used to control combustion rate. A positioning control system is used in plants that have a steam demand large enough to require the burners to be fired continuously. The metering control system is used in plants that require a constant steam pressure and is much more sensitive to steam pressure variation than the positioning control system. A metering system precisely adjusts the flow of fuel and air based on the flow of steam. Each control system provides the steam required based on the demand.

A programmer regulates the sequence of events for lighting and shutting down a burner. **See Figure 7-30.** The programmer operates whenever the boiler starts and stops, either through normal operation or in the event of a problem. The programmer begins a normal firing cycle when steam pressure drops below a preset level.

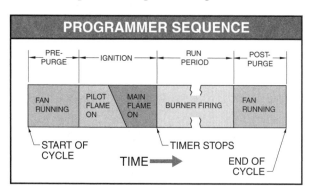

Figure 7-30. The programmer controls the sequence of events involved in a startup and shutdown of a boiler furnace.

When the programmer receives a signal to light the burner, a fan purges any unburned gases from the furnace, preventing a boiler explosion. The length of this prepurge depends on boiler size. If the fan does not start or not enough air is supplied to the furnace, the programmer stops the sequence and sounds an alarm.

If airflow is sufficient, then fuel flows to the pilot burner and is ignited from the pilot light. If the pilot flame does not ignite or is too weak, a signal from the flame scanner causes the programmer to stop the sequence and sound an alarm. When the pilot flame is established, the main fuel valve opens slowly and ignites the fuel. The pilot flame shuts off and the main flame keeps burning. If the main flame fails or is weak, the signal from the flame scanner causes the programmer to secure the main fuel and sound an alarm.

When the boiler pressure reaches the cut-off pressure, the main fuel valve closes. When the main flame stops, the fan continues to operate to purge unburned gases. After this postpurge cycle, the programmer does not restart until steam pressure falls to the start-up pressure. If the programmer shuts down the boiler because of an emergency, it must be manually reset and the boiler operation checked by the boiler operator.

Burner Control Systems. Burner control systems perform all the functions of a conventional combustion control programmer. However, they also incorporate electronic controllers to provide improved communication capabilities, system self-diagnosis, and better energy conservation. Burner control logic is saved in memory and system status is displayed on an information panel. In an automatic safety shutdown event, the panel indicates lockout status and the cause of the shutdown.

Computerized Boiler Control Systems. A computerized boiler control system integrates the functions of a programmable boiler controller and burner controller with other boiler operating and accessory controls. **See Figure 7-31.** An example of a computerized boiler control system is the Cleaver-Brooks CB-HAWK® ICS boiler management and control system. A computerized boiler control system typically includes a controller, a human-machine interface (HMI), and a flame safety control.

The controller is factory preprogrammed for most boiler applications but can also be configured for specific applications. The controller logic is password protected. Individual modules for communications, processor, power supply, and inputs and outputs (I/O) are included. A variable frequency drive controls the speed of the combustion air fan motor, which improves boiler efficiency.

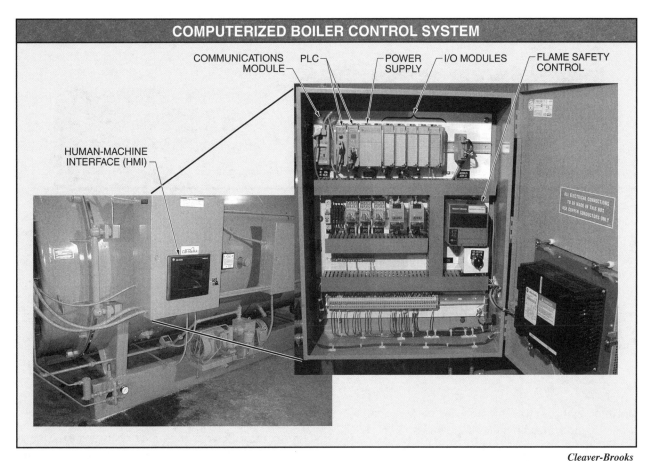

COMPUTERIZED BOILER CONTROL SYSTEM

Cleaver-Brooks

Figure 7-31. A computerized boiler control system can be interfaced with a building automation system to allow monitoring and control of the boiler within the facility and from remote locations.

The HMI displays boiler parameters, fault annunciation, and alarm history and provides easy access to boiler configuration and controls. Information such as boiler overview, burner control, diagnostics, firing rate, boiler operating parameters, alarms and limits, water level, and communications can be accessed using the display. **See Figure 7-32.**

The flame safety control controls functions such as flame detection faults, prepurge and postpurge sequences, safety shutdown, low fire start, and control of high fire. Network communication allows interfacing with various building automation systems. Through the Internet, remote monitoring of the boiler control system is possible from anywhere in the world.

Building Automation Systems. Combustion controls for boilers and related equipment can be a part of building automation systems or HVAC systems. Computerized controls in building automation systems are networked within the facility or connected to remote locations.

Building automation systems are rapidly replacing older electromechanical equipment. However, computerized control is not designed to act as the sole equipment safety device. For example, conventional controls for high- and low-temperature limits, boiler water high and low limits, and flame safeguards are still required. In addition, the installation of computerized controls does not replace scheduled testing and preventive maintenance or the need for an on-site qualified boiler operator.

Flame Scanners

A *flame scanner* is a safety device that senses if the pilot light and/or main flame are lit. This prevents furnace explosions caused by the subsequent ignition of fuel that accumulates in the burner after an extended flame failure. A *flame failure* is a boiler condition when the flame in the boiler furnace has been unintentionally lost. **See Figure 7-33.** The flame scanner verifies that

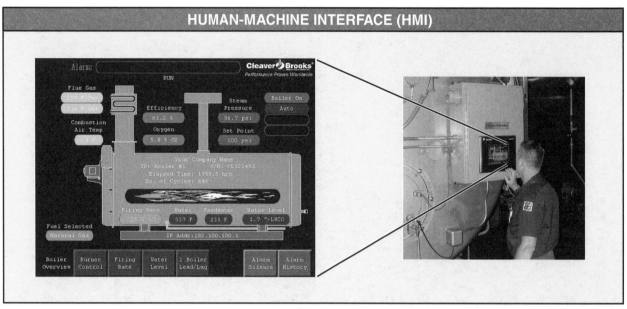

Cleaver-Brooks

Figure 7-32 The human-machine interface (HMI) graphically displays boiler operation data and control functions.

the pilot light is lit to allow the firing cycle to continue when steam demand initiates the combustion cycle. A sensor also verifies the main flame when fuel is supplied to the burner and ignited. This allows the automatic combustion controls to complete the rest of the firing cycle. The flame scanner stops fuel flow to the burner in the event of a flame failure.

Figure 7-33 A flame scanner is a safety device that shuts down the burner in the event of a flame failure.

DRAFT

Draft is the difference in pressure between two points that causes air or gases to flow. For a boiler to operate, there must be sufficient airflow to the fuel for combustion, flow of hot gases of combustion to the heating surfaces, and flow of gases of combustion up the chimney. Excessive draft causes fuel to burn too hot and decreases efficiency. Insufficient draft reduces burning efficiency, resulting in excessive carbon particles, soot, and smoke. Draft is either mechanical draft or natural draft.

> **TECH TIP**
> *An air column with a cross-sectional area of 1 sq in. extended into the atmosphere has a weight of 14.7 lb at sea level (14.7 psi) at the end of the column closest to the earth.*

Mechanical Draft

Mechanical draft is draft produced using fans. Mechanical draft may be forced, induced, or a combination of forced and induced. **See Figure 7-34.** Forced draft is mechanical draft from air pushed through the boiler with fans located in the front of the boiler furnace. Induced draft is mechanical draft from air pulled through the boiler furnace with fans located in the breeching. The breeching connects the boiler to the chimney.

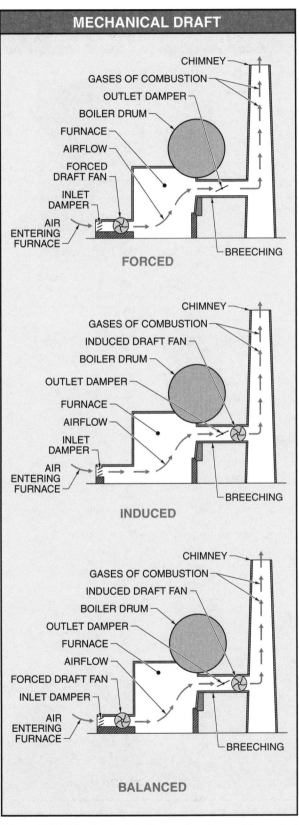

MECHANICAL DRAFT

CHIMNEY
GASES OF COMBUSTION
OUTLET DAMPER
BOILER DRUM
FURNACE
AIRFLOW
FORCED DRAFT FAN
INLET DAMPER
AIR ENTERING FURNACE
BREECHING

FORCED

CHIMNEY
GASES OF COMBUSTION
INDUCED DRAFT FAN
BOILER DRUM
OUTLET DAMPER
FURNACE
AIRFLOW
INLET DAMPER
AIR ENTERING FURNACE
BREECHING

INDUCED

CHIMNEY
GASES OF COMBUSTION
INDUCED DRAFT FAN
BOILER DRUM
OUTLET DAMPER
FURNACE
AIRFLOW
FORCED DRAFT FAN
INLET DAMPER
AIR ENTERING FURNACE
BREECHING

BALANCED

Figure 7-34. Mechanical draft uses fans to provide control of airflow used in the combustion process.

Balanced draft is mechanical draft from fans located both before and after the boiler. Mechanical draft provides greater control of the combustion process and is more widely used than natural draft. Draft accessories such as fans, blowers, and dampers produce and regulate the flow of air and gases of combustion through the boiler furnace.

Natural Draft

Natural draft is draft produced from the difference in air temperature inside and outside a chimney. Hot air rises, and the amount of natural draft produced increases with the height of a chimney. Natural draft is regulated using dampers to control the flow of air into the chimney. The amount of airflow produced with natural draft is limited, and cannot be used in plants requiring a high flow rate of air and gases of combustion. Natural draft is still used on older fuel oil and coal burning boilers.

Draft Measurement

Draft is measured in inches of a vertical column of water using a draft gauge (manometer). A *manometer* is a device that uses a liquid-filled tube to measure the difference in pressure between two locations. **See Figure 7-35.** Colored water is added to the zero marks on both legs of the U-shaped tube. A hose or piping from one leg is connected to the breeching of the draft chamber. The other leg is left open to the atmosphere. A negative pressure reading is indicated if pressure in the breeching is less than atmospheric pressure. A positive pressure reading is indicated if pressure in the breeching is greater than atmospheric pressure. For example, a $-1\frac{1}{2}''$ reading in front of an induced draft fan indicates negative pressure. A $+1\frac{1}{2}''$ reading after a forced draft fan indicates positive pressure.

BOILER OPERATION PROCEDURES

Boilers operate continuously to provide heat or steam for process. Operators responsible for the boiler work in shifts to maintain continuous operation of the plant. During a shift change, the incoming operator relieves the operator from the previous shift. Standard plant procedures are established and followed to ensure continuous safe and efficient operation of the boiler during a shift change. Although boiler operation procedures vary depending on the size and function of the plant, certain procedures are common to all plants.

Boiler Status

During a shift change, the incoming operator reports early enough to allow time to check with the operator on duty. Any unusual events or conditions that have occurred during the shift are discussed. The boiler room log is then checked for any extraordinary conditions or instructions. A brief inspection of the boiler and related equipment can identify potential problems.

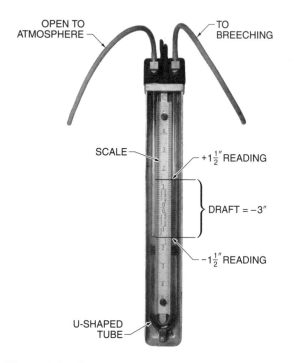

Figure 7-35. A manometer measures draft in inches or tenths of an inch of a vertical column of water.

Boiler Water Level. The water level of on-line boilers is checked first by the operator taking over the shift. The NOWL is located at approximately the center of the gauge glass. Accuracy of the water level in the gauge glass is checked by blowing down the water column and gauge glass. **See Figure 7-36.**

The blowdown valve on the water column is opened to prevent debris from being carried into the gauge glass. Water and steam should blow through the blowdown valve to a drain for 5 sec to 10 sec. The blowdown valve is then closed and water should return to the gauge glass quickly and flow without interruption. The water level shown in the gauge glass can now be trusted as an accurate reading. Debris or mud may be present in the piping if water returning to the gauge glass rises

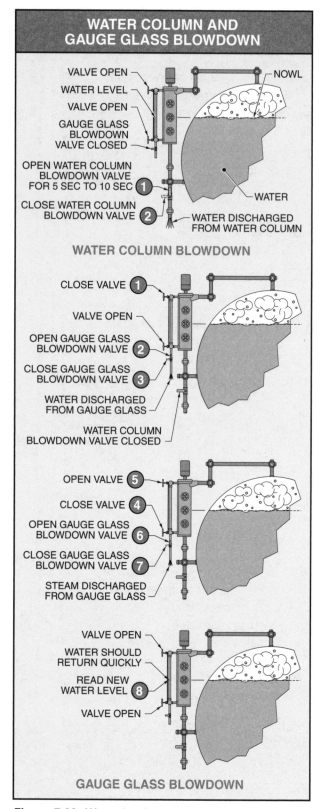

Figure 7-36. Water level accuracy in the gauge glass is checked by blowing down the water column and gauge glass.

abruptly or slowly. The water level shown in a gauge glass is not accurate and cannot be trusted if the water returns abruptly or slowly. Gauge glass blowdown procedures should be repeated until the water level provides an accurate reading.

Warning: Water must not be added to the boiler if water is not present in the gauge glass and steam comes out of the bottom try cock. Adding water may cause a boiler explosion.

After the boiler water level has been verified, the running accessories are checked. Running accessories include operating equipment not attached to the boiler such as feedwater pumps, fuel oil pumps, draft fans, and other equipment. Running accessories are checked for proper temperature, pressure, lubrication, and indications of possible malfunction. Overheating or excessive noise and vibration are indications of a potential problem. Early identification of a problem can eliminate or reduce equipment damage and repair costs.

Low Water Fuel Cutoff Testing

The low water fuel cutoff stops fuel flow to the burner if a low water condition occurs. The most accurate method of testing is to allow the boiler water level to drop to a level that trips the low water fuel cutoff. However, the low water fuel cutoff is most commonly tested for proper operation by a water column blowdown when taking over a shift. Blowdown is accomplished by opening the low water fuel cutoff blowdown valve while the burner is firing to simulate a low water level condition. **See Figure 7-37.** The float in the float chamber drops, which opens the mercury switch connected to the burner control. This stops fuel flow to the burner. The blowdown valve is closed and water returns to the float chamber to the NOWL. The float is raised and closes the mercury switch. The burner reset switch is activated to restart the burner. Blowdown of the low water fuel cutoff also discharges any accumulated sludge or sediment in the float chamber.

Boiler Start-Up Procedures

Boiler start-up procedures vary and are based on the size and type of equipment in the plant. Boiler start-up procedures are classified by the status of the boiler as a cold plant or live plant. Cold plant start-up procedures are performed on boilers that have been out of service for more than two days. Live plant start-up procedures are performed when starting a boiler with other boilers online. The boiler manufacturer's start-up procedures must be followed if plant procedures are not specified.

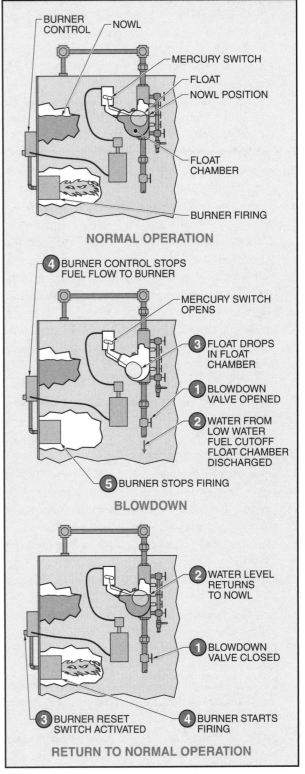

Figure 7-37. The low water fuel cutoff is tested to ensure stoppage of fuel flow to the burner if a low water condition occurs.

Cold Plant Start-Up Procedures. Cold plant start-up is considered less hazardous than live plant start-up because there is no steam pressure. However, as pressure rises from 0 psi to working pressure, packing and gaskets at flanges must be inspected for leaks. Cold plant start-up procedures for a single high-pressure boiler include:

1. Prepare the fuel for burning. For example, fuel oil may require heating. Coal must be loaded in stokers. Natural gas requires little or no preparation.
2. Check the boiler for any missing inspection covers.
3. Remove all chimney coverings and open hand-operated dampers.
4. Confirm the correct water level in the gauge glass.
5. Open the automatic nonreturn valve and the main steam stop valve in the main steam line to warm the whole plant evenly.
6. Open all drains in the main steam line and header.
7. Check the feedwater system and ensure all feedwater valves are open.
8. Inspect the feedwater pumps and regulators. Adequate water supply is required in the feedwater heater and tank.
9. Check for the proper position of all boiler valves.
10. Open the water column, gauge glass, steam pressure gauge, and boiler and steam system vent valves.
11. Purge the furnace.
12. Light the burner and maintain low fire.
13. Open the fuel valve.
14. Warm the boiler slowly to allow for uniform expansion of boiler and lines. Drains should be closed when the boiler is producing a steady flow of steam.
15. Maintain the manufacturer's recommended NOWL.
16. Close the boiler vent when the boiler is producing a steady flow of steam.
17. Test the flame scanner based on plant procedures and fuel.
18. Blow down the gauge glass, water column, and low water fuel cutoff.
19. Check the automatic combustion controls, feedwater regulator, and all boiler room accessories to ensure proper operation.
20. Switch all boiler controls to automatic when operating pressure is reached.
21. Monitor the boiler for any unusual changes in operation. Note extraordinary events in the boiler room log.

Live Plant Start-Up Procedures. Live plant start-up can be more hazardous than cold plant start-up because the boiler is started with other boilers already on-line. Live plant start-up procedures for a high-pressure plant include:

1. Follow cold plant start-up procedures to slowly warm the boiler.
2. Ensure the boiler vent and drain between the main steam stop valve and automatic nonreturn valve are open.
3. Open the equalizing valve around the main steam stop valve.
4. Open the main steam stop valve.
5. Close the boiler vent when a steady flow of steam is flowing from the vent.
6. Open the automatic nonreturn valve when steam pressure reaches between 75% to 85% of line pressure.
7. Bring boiler pressure up slowly and let the automatic nonreturn valve cut the boiler in on-line.
8. Close the drain between the automatic nonreturn valve and the main steam stop valve.
9. Check all automatic controls for proper function.
10. Test the flame scanner and low water fuel cutoff as required.

When starting up a boiler with two hand-operated main steam stop valves, the equalizing line is opened when boiler line pressure is at 85%. This warms the valve and line nearest to the boiler.

The main steam stop valve is then opened. The steam valve closest to the boiler is cracked and opened slowly when the incoming boiler pressure is slightly below line pressure. This allows steam to flow back from the header to the boiler coming in on-line. In the process, condensate is forced back to the boiler to prevent carryover. Refer to the manufacturer's specifications for the recommended startup and shutdown procedures.

Boiler Shutdown Procedures

Boiler shutdown procedures are required to safely take a boiler off-line. All loads affected by the shutdown must also be shut down or be powered by other boilers. Procedures required for boiler shutdown include:

1. Secure the fuel to the burner.
2. Reduce the draft to the furnace to allow slow cooling of the furnace.

3. Close the boiler main steam stop valve(s) and open all steam line drains.
4. Open the boiler vent when the steam pressure is just above 0 psi. This prevents a vacuum from forming in the boiler.
5. Maintain the NOWL as the boiler cools.
6. Shut down the feedwater pump and feedwater system when the boiler has cooled down sufficiently (hand can be held against bare boiler metal).

Boiler Room Log

Boiler status is documented in a boiler room log. A *boiler room log* is a record of the temperatures, pressures, and fuel consumption of the boiler and accessories in the plant over a 24-hour period. Boiler operation procedures may also be listed in the boiler room log, including procedures such as gauge glass blowdown, water column blowdown, testing the low water fuel cutoff and flame scanner, and maintenance work completed.

Boiler room logs commonly use preprinted forms with data added by the operator on duty. **See Appendix.** Extraordinary events that occur must be communicated to the supervisor using the boiler room log or other communication. Depending on the plant method, this can be done verbally or by memo, telephone message, or electronic mail (e-mail). Many plants use a computer to record boiler room log information. This information can be used for developing long-term plant operation records and/or can be used with a computerized maintenance management system (CMMS).

Accurate documentation of plant operation data is required to maintain plant records. These records provide data for assessing plant efficiency, troubleshooting activities, insurance requirements, and compliance with environmental regulations. Plant efficiency can be determined by the amount of steam generated and fuel used. Troubleshooting activities identify unusual changes in the operating characteristics of the plant. Insurance companies often require plant operation data that details boiler operation, maintenance, testing, and inspection activities. Environmental regulations may require documentation of flue gas temperature, draft, smoke, excess air, carbon dioxide, and carbon monoxide.

Some plants have burner control systems that use a computer to process information received from sensors located throughout the plant. For example, steam pressure sensors send information to the burner control. This information is recorded with the time, temperatures, and pressures present. This data may be included in the CMMS.

BOILER MAINTENANCE PROCEDURES

Boiler maintenance procedures vary with the specific plant. The operator may be assigned boiler maintenance in addition to other maintenance tasks in the plant. Boiler maintenance procedures are determined by the AHJ and plant maintenance procedures. Maintenance personnel are responsible for routine boiler preventive maintenance tasks such as repacking auxiliary pump seals, cleaning draft fan blades, and lubricating and replacing pump bearings as required. All boiler maintenance work completed must comply with established procedures for lockout, tagout, and confined space entry.

Unscheduled maintenance such as steam leaks may require special procedures. Loss of steam from leaks is costly and can cause injury. Large steam leaks are audible and can be isolated by locating the noise emitted. Small steam leaks can be isolated using a thermal imager. Detection of heat emission is identified on the captured image. **See Figure 7-38.** Routine maintenance procedures are included on a schedule. **See Appendix.**

Warning: Serious burns can occur from contact with high-pressure or superheated steam.

Cleaver-Brooks
Live plant start-up can be more dangerous than cold plant start-up because a boiler is started with other boilers already on line.

Live steam is steam that leaves the boiler directly without having its pressure reduced in process operations. Live steam may not be visible until reaching several feet from the leak source. The vicinity of any possible steam leak must be approached with caution.

Safety Valve Testing

Safety valves are tested at prescribed intervals manually or by pressure. Testing frequency should ensure proper safety valve function without wasting steam through excessive testing. Safety valve testing frequency is dictated by plant procedures and the AHJ. In general, safety valves on low-pressure boilers (up to 15 psi) should be manually tested once a month with at least 5 psi on the boiler. In addition, safety valves on low-pressure boilers should be pressure tested once a year. Safety valves on high-pressure boilers (over 15 psi) should be manually tested once a month and pressure tested once a year.

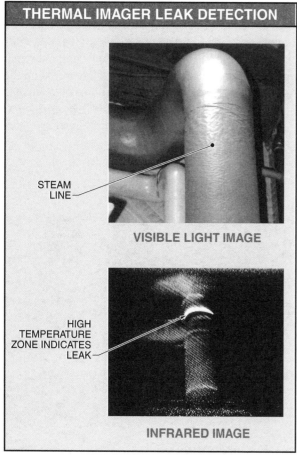

THERMAL IMAGER LEAK DETECTION

STEAM LINE

VISIBLE LIGHT IMAGE

HIGH TEMPERATURE ZONE INDICATES LEAK

INFRARED IMAGE

John Snell & Associates

Figure 7-38. Steam leaks create a high temperature zone, which can be isolated with an infrared image.

Manual testing of safety valves requires lifting the test lever to the wide open position. **See Figure 7-39.** The boiler should have approximately 75% of the safety valve popping pressure. The valve is held open for 5 sec to 10 sec. The test lever is released, allowing the safety valve to snap shut. Pressure testing of safety valves requires steam pressure to be raised to the pressure at which the safety valve should open.

Figure 7-39. Safety valves are tested to ensure proper function.

Flame Scanner Testing

A flame scanner verifies flame in the burner. The flame scanner stops the flow of fuel if fuel is not ignited during burner startup or if a flame failure occurs. Flame scanner testing frequency depends on the AHJ and plant operating procedures. **See Figure 7-40.** Defective flame scanners are replaced. The flame scanner is commonly tested for proper operation using the steps:

1. Remove the flame scanner with the burner firing and cover the scanner eye to simulate a flame failure. The main fuel valve closes and the flame failure alarm sounds. The burner control should begin a postpurge cycle to remove any residual fuel from the furnace.
2. Reinstall the flame scanner.
3. Reset the programmer and check for proper operation.

Low Gas Pressure Testing

The boiler burner must stop instantly if there is insufficient gas pressure to ensure stable flame operation. The low gas pressure switch is tested by slowly closing the manual gas valve. The burner should shut down before the manual gas valve is closed.

Low Draft Pressure Testing

A low draft failure switch constantly senses the pressure of the combustion air flowing into the furnace. If air pressure is too low, the low draft failure switch should shut down the boiler furnace immediately. Disconnecting the tubing connecting the switch and the furnace should cause the fuel flow to stop immediately.

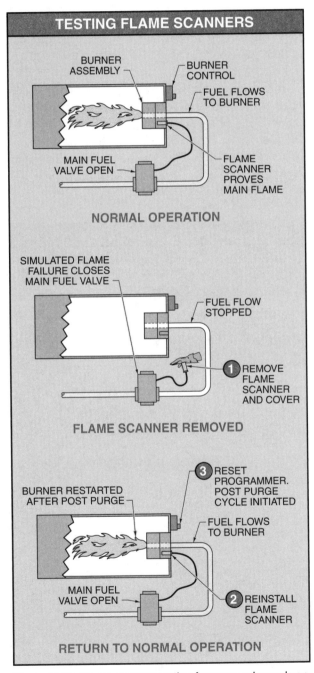

TESTING FLAME SCANNERS

BURNER ASSEMBLY
BURNER CONTROL
FUEL FLOWS TO BURNER
MAIN FUEL VALVE OPEN
FLAME SCANNER PROVES MAIN FLAME

NORMAL OPERATION

SIMULATED FLAME FAILURE CLOSES MAIN FUEL VALVE
FUEL FLOW STOPPED
1 REMOVE FLAME SCANNER AND COVER

FLAME SCANNER REMOVED

3 RESET PROGRAMMER. POST PURGE CYCLE INITIATED
BURNER RESTARTED AFTER POST PURGE
FUEL FLOWS TO BURNER
MAIN FUEL VALVE OPEN
2 REINSTALL FLAME SCANNER

RETURN TO NORMAL OPERATION

Figure 7-40. Flame scanner testing frequency depends on the AHJ and plant operating procedures.

Boiler Blowdown Procedures

Boiler blowdown is the discharge of boiler water. This is performed for various reasons, including removing sludge and sediment, controlling high water and chemical concentration, or draining the boiler for cleaning and inspection. Blowdown types are surface, continuous, or bottom. **See Figure 7-41.**

Surface blowdown is the purging of boiler water at the surface to remove floating impurities and foreign matter. The amount of water removed is controlled by the boiler operator.

Continuous blowdown is the continuous purging of boiler water at a controlled rate to maintain the proper chemical concentration in the boiler water. Continuous blowdown is performed automatically and removes boiler water through a line located below the low water level.

Bottom blowdown is the purging of boiler water from the bottom of the boiler to drain the boiler or remove settled sludge and sediment. Bottom blowdown is best performed during light loads. This allows the maximum amount of solids in the water to settle to the bottom of the boiler drum. A blowdown pipe fitted with a valve is located at the lowest part of the water side of the boiler, which allows the sludge and sediment to discharge first.

Steam pressure must be maintained during bottom blowdown, unless the boiler is being drained. Boilers should be blown down according to the schedule determined by boiler water analysis and treatment procedures. Boilers operating at over 100 psi must have two bottom blowdown valves. These valves are a slow-opening valve and a quick-opening valve, or two slow-opening valves. A slow-opening valve is fully opened or closed by five full turns of the hand wheel. A quick-opening valve is opened or closed with a lever. The quick-opening valve is located closest to the boiler. The water level should be at the NOWL. The quick-opening valve is opened first when a quick-opening and slow-opening valve are used. **See Figure 7-42.** The slow-opening valve is opened slowly to the full open position. The slow-opening valve bears the wear and tear of blowing down.

The water level in the gauge glass must be carefully monitored by the operator during blowdown. The gauge glass must show water at all times during blowdown. The operator must never walk away from open bottom blowdown valves. When blowdown is complete, the slow-opening valve is closed first and the quick-opening valve is closed last.

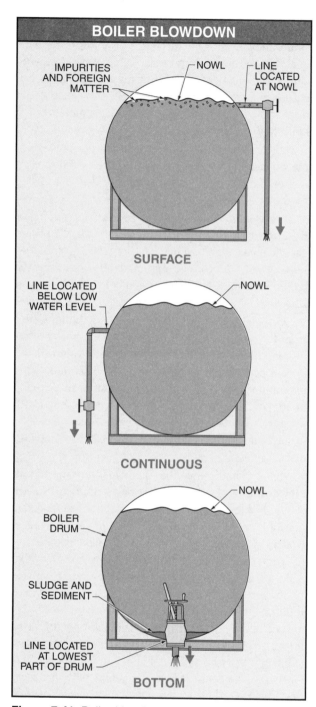

Figure 7-41. Boiler blowdown removes sludge and sediment, controls high water and chemical concentration, or drains the boiler for cleaning and inspection.

Steam Trap Testing

Steam traps should be tested at the first sign of overheating heat exchangers, a temperature increase of condensate returning to the condensate return tank, and/or pressure buildup in the condensate return tank. Dirt, rust, and scale in steam lines can lead to steam trap failure. Proper steam trap function is greatly affected by steam strainer function. Steam strainers must be routinely cleaned to prevent foreign matter from reaching the steam trap. Steam trap operation is tested by sight, temperature, and sound. Steam leaks at the trap can often be detected visually. Live steam at the discharge indicates a malfunction.

A steam trap can be tested by measuring temperatures with temperature-indicating devices such as a thermometer or a temperature-indicating crayon, or by sound analysis. The thermometer sensor is placed on the inlet and outlet side of the steam trap. The temperature readings are compared with temperature readings of a properly operating steam trap.

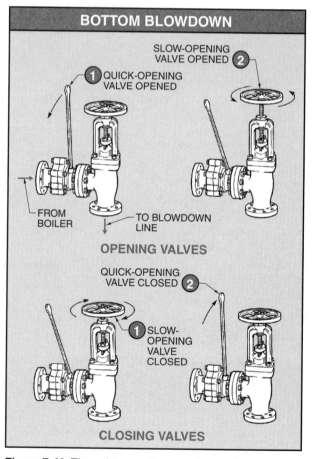

Figure 7-42. The quick-opening blowdown valve is located closest to the boiler and is opened first and closed last.

For example, a certain steam trap may produce a 10°F drop in temperature on the outlet side of the trap when operating properly. A temperature-indicating crayon with the proper melting temperature may also be used. A mark is made on the outlet side of the steam trap. The steam trap may be defective and passing steam if the mark is melted. Care must be taken when testing steam traps. Different traps have different temperature characteristics. For example, a constant discharge steam trap may have a smaller temperature differential than a steam trap that opens and closes.

Steam trap sound analysis is performed by listening for unusual steam trap sounds or by testing with an ultrasonic tester. For example, the steam trap is checked for sounds emitted under different load conditions. An ultrasonic tester is used by qualified maintenance personnel. The sound waves emitted by the suspect steam trap are compared to the sound waves emitted by a similar steam trap operating properly. **See Figure 7-43.** For example, a steam trap that fails in the open position emits the sound of steam blowing through. A steam trap that fails in the closed position emits no sound of steam and/or water flowing through.

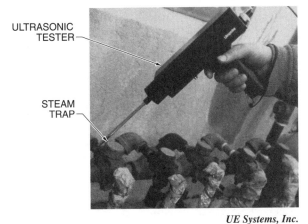

UE Systems, Inc.

Figure 7-43. An ultrasonic tester is used to test steam trap function by comparing sound waves emitted under different load conditions with sound waves emitted from a similar steam trap operating properly.

Fuel Oil Strainer Maintenance

High vacuum readings on the fuel oil return line indicate cold fuel oil and/or dirty fuel oil strainers. Duplex fuel oil strainers have two strainer baskets to allow one strainer basket to be cleaned while the other is in service. **See Figure 7-44.** Fuel oil strainers used with No. 6 fuel oil are commonly serviced every 24 hours. A strainer basket

maintenance schedule is developed based on the fuel oil, temperature, and quantity used. The strainer basket is removed and foreign matter is cleaned from the strainer basket and properly disposed. Damaged strainer baskets are replaced. Replacement strainer baskets must match the original mesh opening size.

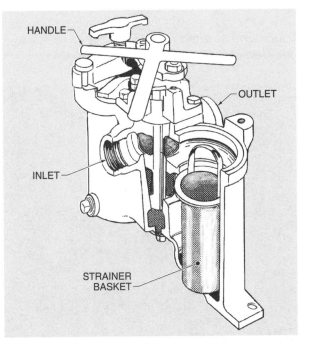

Figure 7-44. Duplex fuel oil strainers allow one strainer basket to be cleaned while the other is in service.

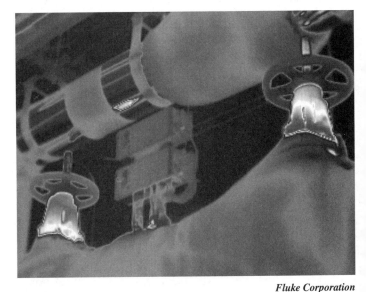

Fluke Corporation

A thermal imager can be used to check steam line and valve insulation in a boiler system.

Boiler Water Analysis

Boiler water analysis determines the condition of the water and any treatment required. **See Figure 7-45.** Boiler water is tested to determine the frequency and amount of blowdown and chemicals required. Boiler water tests and water treatment activities are recorded in a water treatment log similar to a boiler room log. Boiler water samples are tested at the plant or sent to a water treatment company for analysis. The most efficient boiler water treatment program is based on accurate daily testing by the maintenance personnel.

Figure 7-45. Boiler water analysis determines blowdown frequency and water treatment chemicals required.

Boiler Water Treatment

After analysis, water used in a boiler is treated to ensure high efficiency and to prevent damage to the boiler and accessories. All water, regardless of its source, requires treatment. Rain water accumulates chemicals and particles as it falls to the ground. Water exposed to the atmosphere accumulates solid and gaseous pollutants on its surface. Ground water has dissolved solids from minerals in the ground. Even drinking water requires removal of solids before use in a boiler. Water is treated to protect the boiler from conditions that occur from untreated water such as caustic embrittlement, scale, corrosion, carryover, and foaming.

Caustic embrittlement is the accumulation of high alkaline elements that cause boiler metal corrosion. Cracks at seams or the ends of tubes caused by metal breakdown can lead to a boiler explosion. *Scale* is the accumulation of calcium carbonate and magnesium carbonate on boiler heating surfaces. **See Figure 7-46.** Scale acts as an insulating material and reduces heat transfer efficiency, causing overheating of boiler metal. Overheated boiler metal from scale deposits causes cracks or blisters, distortion of tubes and sheets, and, ultimately, a boiler explosion. With the proper water treatment, calcium carbonate and magnesium carbonate are removed as sludge to prevent scale buildup.

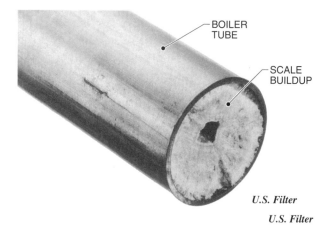

U.S. Filter

U.S. Filter

Figure 7-46. Scale buildup on boiler heating surfaces has an insulating effect, which reduces heat transfer efficiency.

Corrosion is the combining of metals with elements in the surrounding environment that leads to the deterioration of the material. Corrosion causes pitting and channeling of the metal, which results in the thinning of plates and tubes. Carryover is small water droplets carried in steam lines that can cause water hammer. Carryover causes the introduction of water into steam lines caused from high alkalinity, dissolved solids, and sludge. Severe water hammer can result in a break in steam lines.

Foaming is the formation of steam bubbles trapped below the boiler water surface. Foaming is commonly caused by a coating formed on the boiler water water from impurities and leakage of fuel oil. Leaking steam coils in the fuel oil heater allow fuel oil into the feedwater system. Boiler water contaminated with fuel oil has increased surface tension, leading to foaming. Foaming prevents proper heat transfer of heating surfaces, causing dangerous overheating. A boiler with contaminated water must be taken offline immediately, shut down, boiled out with caustic soda, and thoroughly cleaned. A water treatment company can recommend chemicals used for cleaning a contaminated boiler. Once analyzed, boiler water is treated using internal and external treatment.

Internal Boiler Water Treatment. *Internal boiler water treatment* is the treatment of boiler water after it has entered the steam and water drum. Chemicals added to the water are premixed in a tank as a treatment solution and added continuously in diluted form. With internal boiler water treatment, caustic embrittlement is prevented by analyzing and treating for proper alkalinity. Alkalinity is determined by measuring boiler water pH levels. A *pH scale* is a scale from 0 to 14 used to indicate the acidity or alkalinity of a solution. **See Figure 7-47.** A solution with a pH less than 7 is acidic. A solution with a pH greater than 7 is alkaline. A solution with a pH of 7 is neutral. Boiler water pH should be between 10 and 11.5. Boiler water below 7 is acidic and can cause thinning of boiler metal. To raise the boiler water pH (alkalinity), caustic soda or soda ash is added to the chemical treatment.

pH SCALE

NEUTRAL

ACIDIC   ALKALINE

0 1 2 3 4 5 6 7 8 9 10 11 12 13 14

pH COMMONLY REQUIRED
FOR BOILER WATER

Figure 7-47. A pH scale value indicates if a solution is acidic, alkaline, or neutral.

Scale is better prevented than treated. Scale can only be removed by scraping or acid cleaning during lay-up. Both methods remove and thin boiler metal. Caustic soda and phosphate added to water treatment chemicals change calcium carbonate and magnesium to sludge that is kept in suspension for removal by blowdown from the boiler. Corrosion is prevented by preheating the water fed to the boiler to help remove oxygen and carbon dioxide. Sodium sulfite is also used to remove residual oxygen left from heating the water. The combination of oxygen and sodium sulfite produces sodium sulfate.

+ SAFETY TIP
Face masks, goggles, rubber gloves, and protective garments must be worn when handling or mixing caustic chemicals. Do not allow caustic chemicals to contact the skin or clothing.

Carryover and foaming of boiler water are prevented by controlling the surface tension and total dissolved solids (TDS). The TDS present in boiler water is measured with a conductivity meter. A *conductivity meter* is a test instrument used to determine total dissolved solids present in boiler water. The TDS present can be controlled by continuous blowdown or bottom blowdown. Surface tension is controlled by surface blowdown. The boiler water must be tested daily in many high-pressure boilers. Once residual chemical levels are determined, chemicals are added according to specifications by the water treatment company. **See Figure 7-48.**

Figure 7-48. Chemicals are mixed in a tank and added continuously in diluted form to the boiler water once residual chemical levels are determined.

External Boiler Water Treatment. *External boiler water treatment* is the treatment of boiler water before it enters the boiler. By heating the water in a feedwater heater, oxygen and noncondensable gases can be removed and vented to the atmosphere. Heating the water also reduces thermal shock from cold water fed into the boiler. Water treated before entering a boiler has all or most of the scale-forming salts causing water hardness and gases expelled. A soap solution test indicates the hardness or softness of the water. A soap solution is added to a water sample. When shaken, a solution of soap and hard water does not produce and keep as many suds as soft water. Common treatment methods for hard water include the lime soda process water treatment or ion exchange water treatment.

Lime soda process water treatment uses hot or cold lime soda to remove chemical compounds in hard water. The hot process is performed above 212°F and the cold process is done at ambient temperature. Lime soda is added and mixed, and chemical compounds turn into sludge. The heavy sludge settles and is removed from the bottom of the treatment tank. The hot lime soda process water treatment is faster than the cold process.

The ion exchange water treatment process (zeolite process) is performed at ambient temperature in a water softener. *Zeolite* is the mineral group including silicates of aluminum, sodium, and calcium that is used to soften water. The zeolite process softens water by changing calcium and magnesium carbonates into sodium carbonates that do not settle on boiler surfaces. The unit is regenerated by passing brine (salt solution) over the zeolite in the resin bed when the sodium in the softener is depleted. **See Figure 7-49.** The ion exchange water treatment process is commonly used because it can be fully automated.

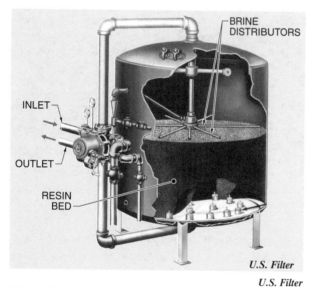

U.S. Filter

Figure 7-49. The ion exchange water treatment process uses zeolite to soften water by changing calcium and magnesium carbonates into sodium carbonates.

Boiler Fitting and Accessory Replacement

Boiler fittings and accessories are routinely replaced for malfunction, leakage problems, wear, or as part of a replacement schedule. Any replacement boiler fitting must meet the design requirements of the original boiler fitting. Removal of any boiler fitting and accessory requires a release of pressure, locking and tagging out as required, and cooling to a safe removal temperature. Boiler fitting and accessory connections vary with the specific boiler. For example, safety valves on low-pressure boilers commonly have threaded connections. Safety valves on high-pressure boilers have flange connections. The removal procedures required depends on the fitting connection. Flanged boiler fittings are unbolted and the gasket sealing surfaces are cleaned. A new gasket is installed when the original or new fitting is installed.

Gauge Glass Replacement. A gauge glass is normally attached to the water column by a screw or flanged fitting. Most are equipped with quick-closing stop valves that can be closed in the event of a glass failure. A broken gauge glass on a boiler under pressure is replaced by applying the procedure:

1. Close the lower water valve to the gauge glass.
2. Close the upper steam valve to the gauge glass.
3. Open the gauge glass blowdown valve.
4. Remove the gauge glass nuts.
5. Using gloves, remove the broken glass and washers.
6. Obtain a new gauge glass and new washers. Deduct ¼″ from the overall length for expansion from heat if the gauge glass must be cut to fit.
7. Install the gauge glass and washers.
8. Hand-tighten the gauge glass nuts and tighten a quarter turn with a wrench.
9. Open the steam valve slightly to the gauge glass to allow the glass to warm slowly.
10. Open the steam valve completely.
11. Open the water valve to the gauge glass completely.
12. Close the gauge glass blowdown valve and check for leaks.

TROUBLESHOOTING BOILER SYSTEMS

Boiler operation is critical to plant heat and process needs. Downtime must be kept to a minimum. Boiler troubleshooting must quickly identify the problem and restore steam to all loads as soon as possible. Standard troubleshooting steps are followed to isolate and remedy the problem. **See Figure 7-50.** The most common boiler problem is when the boiler does not start or restart. These conditions could be caused by a flame failure, low water condition, high water condition, and/or feedwater pump failure.

BOILER TROUBLESHOOTING – FLAME FAILURE

Problem

A plant has three boilers in battery used to provide heat for an
warm weather conditions, one boiler operates, with the other t
suddenly shuts down and sends an alarm to the maintenance

Troubleshooting Step 1 - Investigate

Operator not present - on automatic control.
No flame in boiler.
Steam pressure low.
Water level normal.
Burner control display readout - PILOT FLAME FAILURE.

Troubleshooting Step 2 - Isolate

<u>Likely Causes</u>	<u>Action/Result</u>
High or low gas pressure interlock.	Check reset on not tripped.
Dirty flame scanner.	Remove/inspec
Ignition system failure.	Reset burner co crackle heard/n
Loose igniter wire connections.	Tighten igniter change.
Faulty igniter wire.	Swap igniter wir control, start lig wire damaged

Troubleshooting Step 3 - Remedy

Replace igniter wire. Obtain new igniter wire from inventory, re
ignited. Order replacement for inventory.

Troubleshooting Step 4 - Documentation

File reports. Complete troubleshooting report and notation in

Figure 7-50. Standard troubleshooting steps are followed
when troubleshooting a boiler flame failure.

Flame Failure

Flame failure may be caused by a malfunction in the
ON/OFF pressure control, malfunction of the draft fan
switch, or fan motor failure. Flame failure can cause
furnace explosions from the buildup and ignition of fuel.
The flame scanner is tested to determine the cause of a
flame failure. The burner should shut OFF as in a flame
failure condition. The flame scanner eye is cleaned and
re-installed if the flame scanner works. The combustion
control programmer is reset for automatic control. The
combustion control programmer should properly control the burner through the firing cycle. The fuel supply
is checked if the combustion control programmer is
functioning properly.

Flame failures in gas-fired boilers are commonly
caused by insufficient pressure in gas lines. Flame failures

in fuel-oil fired boilers are commonly caused by cold
fuel oil, water in the fuel oil, air in fuel oil lines, clogged
fuel oil strainers, a clogged burner tip, and/or loss of
fuel oil pressure. Flame failures in pulverized coal-fired
boilers are caused by wet coal, loss of primary air, and
loss of coal to the pulverizer from feeder failure or
blockage.

Operators must handle a flame failure immediately.
The fuel is secured and the furnace is purged when a
flame failure occurs. A furnace explosion may occur if
the flame failure is not addressed immediately.

Low Water Condition

A low water condition occurs when the water in a boiler
is below the NOWL as indicated by the gauge glass.
Some plants have an audible alarm and light to signal
a low water condition in the boiler. A low water condition can result in damage from overheating to the boiler
drum, furnace, and tubes. The fuel must be secured immediately if no water is visible in the gauge glass.

The try cocks must be used to determine water level
if water cannot be seen in the gauge glass. The gauge
glass is not functioning properly if water comes out of
the middle try cock. Water may be safely added to the
boiler if water comes out of the bottom try cock. Water
must not be added to the boiler if water is not present in
the gauge glass and steam comes out of the bottom try
cock. Adding water could cause a boiler explosion.

The fuel to the burner of the boiler with low water
must be secured if more than one boiler is on-line and
one boiler has a low water condition. The NOWL of
all boilers on-line must be monitored. Feedwater to the
boiler in question should be secured if the water level
cannot be maintained in all boilers in the system. The
cause must be determined whenever a low water condition occurs. The boiler should be removed from service
and inspected. The feedwater supply must be checked
for proper function. The boiler drum, fittings, and accessories should be checked for leaks.

High Water Condition

A high water condition occurs when there is an excessive
amount of water in the boiler as indicated by the gauge
glass. The gauge glass should be blown down to ensure
an accurate reading. The water level is checked using try
cocks if the water level in the gauge glass is still high.
Water discharging from the top try cock confirms a high
water level in the boiler.

The immediate danger with a high water level is the possibility of carryover of water in the steam. This can result in damage to feedwater pumps, headers, and turbines. In the event of high water, a bottom blowdown is performed to bring the water level to the NOWL. The firing rate should be reduced or the fuel stopped completely. After the boiler has returned to the NOWL, the cause of the high water condition should be determined. The feedwater regulator must be checked for proper operation. Boiler troubleshooting requires the ability to act quickly in a dangerous situation and the patience to locate and repair the cause of the problem.

Feedwater Pump Failure

A feedwater pump delivers feedwater to the boiler. A low water condition can result from an inadequate supply of feedwater. Feedwater pump failure can be caused by:

• steambound feedwater pump
• worn or damaged feedwater pump impeller
• motor failure from single phasing, overheating, or other malfunction
• blockage or restriction in feedwater supply line
• worn or damaged feedwater pump bearing(s)
• leakage from feedwater pump housing

Steambound Feedwater Pump. A feedwater pump becomes steambound when pumped water gets too hot and turns to steam. The pump is designed to pump liquids, and does not work with gas or steam. To correct a steambound feedwater pump, the water fed to the pump must be cooled. Water that is drawn from a condensate return tank may be cooled by adding water from the makeup system. If the problem persists, cool water is poured directly on the pump without spilling water on the motor. Steam traps on the condensate return line should be tested for proper operation.

Furnace Explosion

A *furnace explosion* is the instantaneous combustion of flammable gases or vapors accumulated in the furnace. This accumulation can occur from leaking fuels or a flame failure. Any signs of leaking fuels must be addressed immediately. The cause of a flame failure is investigated. The boiler must be inspected for evidence of physical damage. The flame scanner is checked for proper function. Before relighting the burner, the combustion programmer must go through the purge cycle. The fuel to the burner is secured and the proper plant personnel and authorities are notified if the cause of the furnace explosion cannot be determined. Water should not be added to the boiler.

BOILER EMERGENCY

A boiler emergency can lead to serious injury and/or damage if the proper actions are not taken immediately. The operator responsible for the boiler must be thoroughly familiar with these procedures. Improper actions taken during these conditions could cause the boiler or furnace to explode. **See Figure 7-51.** Serious boiler emergencies include boiler overpressure and low water condition.

Cleaver-Brooks
Computerized boiler management control ensures that the startup of the burner follows a definite timed sequence and has the ability to work in conjunction with some energy management systems.

> **TECH TIP**
> The operator in charge of the boiler must use every opportunity to ensure boiler operation safety and efficiency. Extraordinary events are identified by comparison with normal operation. For example, when passing by the boiler, the gauge glass, flame, and burner control are routinely checked for normal condition.

Boiler Overpressure

Boiler overpressure is a condition that occurs when a boiler is operating at or above its maximum allowable working pressure (MAWP). The safety valve, when operating properly, alerts the operator to an overpressure condition. In an overpressure condition, the burner must shut down immediately. Do not open the safety valve manually as this could cause a boiler explosion. Let the boiler cool and determine why the electrically operated pressure controls failed and why the safety valve did not open at the correct pressure. If the safety valve is open but the boiler is overpressurized and still firing, the electrical pressure controls have failed and must be replaced or repaired.

Factory Mutual Engineering Corporation

Figure 7-51. Serious injury and/or damage can occur if the proper actions are not taken immediately in a boiler emergency.

Low Water Condition

The boiler fires must be stopped immediately if the boiler is still firing but there is low or no water in the gauge glass. Do not add water until the actual water level is determined. Heated feedwater can be added carefully if water flows out of the bottom try cock. Do not add water if steam flows out of the bottom try cock. If the boiler is truly low on water, it must be inspected for damage after cooling. The low water fuel cutoff and feedwater systems must be inspected and repaired.

In any boiler emergency, the fuel must be secured immediately. All boiler systems must be checked for the possible cause of the boiler emergency. The supervisor and affected plant personnel must be notified as soon as possible. The boiler operator must act quickly and effectively because boiler explosions cause tremendous damage and serious or fatal injuries.

PREPARATION FOR BOILER INSPECTION

Boiler inspection occurs as a part of plant procedures and/or requirements by the AHJ. The boiler inspector thoroughly examines the boiler for corrosion, overheating, and/or other damage. **See Figure 7-52.** Depending on the AHJ, a written report is filed and a new boiler certificate is issued after boiler inspection requirements are satisfied. An inspection date appointment is made to minimize plant downtime. The operator in charge of the boiler must be present during the boiler inspection. Before inspection, the boiler must be taken off-line and the following tasks must be completed:

- Close and tag out the main steam stop valve.

- Open the boiler vent or try cock to prevent vacuum buildup in the boiler.

- Close and tag out the feedwater line valve to the boiler. If applicable, the makeup water valve is also closed.

- Allow the boiler to cool before draining to prevent sludge and sediment from baking on hot metal surfaces.

- Open the access holes after draining, remove the cover, and completely flush and wash the water side of the vessel. Check for scale, pitting, and/or fuel oil.

- Inspect and clean the fire side. Check for signs of blisters on the heating surface.

- Remove all plugs from the water column and open the low water fuel cutoff controls.

TECH TIP

Spare fuel burner tips should be kept clean and ready to be used as necessary. Keeping a spare oil burner on hand is recommended. Combustion quality should be checked and adjustments made whenever the burner is removed and control linkage is disturbed.

Figure 7-52. Preparation for boiler inspection requires exposing boiler parts that may be affected by corrosion, overheating, and/or other damage.

BOILER LAY-UP

Boiler lay-up is the preparation of a boiler for out-of-service status for an extended period of time. *Wet lay-up* is storage of a boiler filled with warm, chemically treated water. Wet lay-up is used when the boiler may be needed on short notice. *Dry lay-up* is storage of a boiler with all water drained. Dry lay-up is used when the boiler is expected to be out of service for an extended period of time or if water in the boiler can freeze. Both wet lay-up and dry lay-up require the boiler to be thoroughly cleaned.

In wet lay-up, the boiler is closed up with new gaskets installed on handholes and access holes. The boiler is filled to the top with chemically treated water to reduce the possibility of corrosion and oxygen pitting. The water is then heated to 100°F to reduce the formation of condensation. In dry lay-up, the water side must be dry to prevent residual moisture from corrosion and pitting of metal. New gaskets are placed on the handholes. A moisture-absorbing agent (quick lime or silica gel) is placed in the boiler. The access hole is then reinstalled with a new gasket. All valves are closed to seal the boiler from moisture.

Heating, Ventilating, and Air Conditioning Systems

8

Heating, ventilating, and air conditioning systems are used in buildings to provide comfort for occupants. A feeling of comfort results when temperature, humidity, circulation, filtration, and ventilation of the air are controlled. HVAC systems may contain pneumatic or electronic control systems. HVAC systems are designed to fulfill comfort requirements with maximum efficiency.

HVAC SYSTEM OPERATION

A *heating, ventilating, and air conditioning (HVAC) system* is a system used to condition air by maintaining proper temperature, humidity, and air quality. Properly conditioned air improves the comfort and health of building occupants. Excessive airborne dirt, heat, and humidity can cause irritation and injury to people and possible damage to sensitive equipment. HVAC systems are designed to operate at optimum energy efficiency while maintaining desired environmental conditions.

An HVAC system may contain a boiler system or electric heating elements to provide heat for building spaces and a refrigeration system to provide cool air for building spaces. In addition, an HVAC system may contain air handlers and ductwork to move the air to the different building spaces, terminal air handling units to transfer heat to or from the air in the building spaces, humidifiers to increase the moisture content of the air, and filters to remove particles from the air.

HVAC systems are operated by pneumatic (pressurized air) or electronic controls that monitor air temperature, humidity, flow, volume, and pressure.

Signals are sent to control fans, electric contacts, valves, dampers, refrigeration units, steam humidifiers, and other HVAC devices.

Heating/Refrigeration Equipment

Boilers or electric heating elements produce the heat required by a building. A boiler supplies hot water or steam that is transferred to the building air by a heating coil. The cool building air passes across the heating coil that is carrying the hot water and heat moves from the hot coil to the cool air. Electric heating elements produce heat by electric current flowing through the elements. The resistance of the heating elements causes them to become hot when current flows in them. Heat from the heating elements is transferred to the cool air as it flows across the heating elements.

Cool air for building spaces is supplied by a refrigeration system. The refrigeration system may cool air or water. The building air is cooled when it passes across a cold refrigeration evaporator coil. The heat in the building air moves into the cooler evaporator. In a water-cooled system, water is cooled by a refrigeration system and is pumped to a cooling coil. Warm air blows across the cooling coil, reducing the temperature of the air. The air is then used to cool part of a building.

Air Handlers

An *air handler* is a device used to distribute conditioned air to spaces in a building. Air handlers are commonly mounted on the building roof. Air handlers include one or more fans, filters, movable dampers, and heating and cooling coils.

A *fan* is a mechanical device that creates airflow from the rotation of aerodynamic blades. Fan types include axial flow and radial flow. **See Figure 8-1.** An *axial flow fan* is a fan that produces airflow parallel to the fan shaft. A *radial flow fan* is a fan that produces airflow perpendicular to the fan shaft. A *scroll fan* is a radial flow fan contained in a scroll-shaped housing. A scroll fan has curved (backward or forward), straight, or radial-shaped blades. Different blade shapes provide different benefits. For example, backward-curved blades provide the highest efficiency and the lowest sound level for all radial fans.

Fan components include a housing, fan wheel, motor, and drive components, which depend on whether the fan is direct or indirect drive. A *direct drive fan* is a fan that has the fan wheel attached directly to the motor shaft. An *indirect drive fan* is a fan in which the fan motor is connected to the fan wheel through a belt and pulley arrangement.

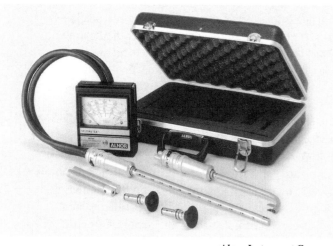

Alnor Instrument Company
A velometer can be used to measure air velocity in an HVAC system.

Fan selection and size is determined by engineers who design air handlers. Fan size is determined based on static pressure, capacity, and velocity pressure. The two most common variables are static pressure and capacity. **See Figure 8-2.** *Static pressure* is the pressure exerted by airflow in a direction perpendicular to the flow. Static pressure is measured in inches of water column (in. WC). The higher the static pressure, the less air a fan can move. *Fan capacity* is the volume of air a fan can move in a given period of time. Fan capacity is measured in cubic feet per minute (cfm). Fan capacity is affected by the physical size of the fan, fan speed, and static pressure. For example, a scroll fan with a 22″ fan wheel operating at 600 rpm and ¼″ WC moves 11,450 cfm of air. *Velocity pressure* is the pressure of airflow in the direction of flow. Velocity pressure is also measured in inches of water column. Knowledge of the fan size is necessary for proper fan replacement.

The air handler fan and dampers operate to mix outside air with return air from inside the building. **See Figure 8-3.** Mixing outside and return air ensures that enough fresh air enters the building to prevent the air from becoming stagnant, while obtaining as much natural heating and cooling from outside air as possible. *Supply air* is the mixture of outside air and return air that is conditioned for use in a building. Supply air is filtered and heated or cooled to a specific temperature and directed through supply ducts to a zone. A *zone* is a building subsection such as an auditorium, a warehouse, or a group of several rooms. Zones can be further divided into specific building spaces such as a single room.

The supply air fan mixes return air from inside the building with outside air. The supply air thermostat compares the temperature of the supply air against the setpoint temperature for the zone. *Setpoint temperature* is the temperature the HVAC system is set to maintain. For example, the setpoint temperature for a building zone may be 68°F. The system stops heating at 68°F and starts heating at 66°F, the ideal temperature being 68°F.

HVAC air handler systems can be either constant volume or variable volume. A *constant volume system* is an HVAC system that supplies a set amount of air at all times. A *variable volume system* is an HVAC system that supplies varying amounts of air. Constant and variable volume air handler systems can supply single or multiple zones and have single or double ducts.

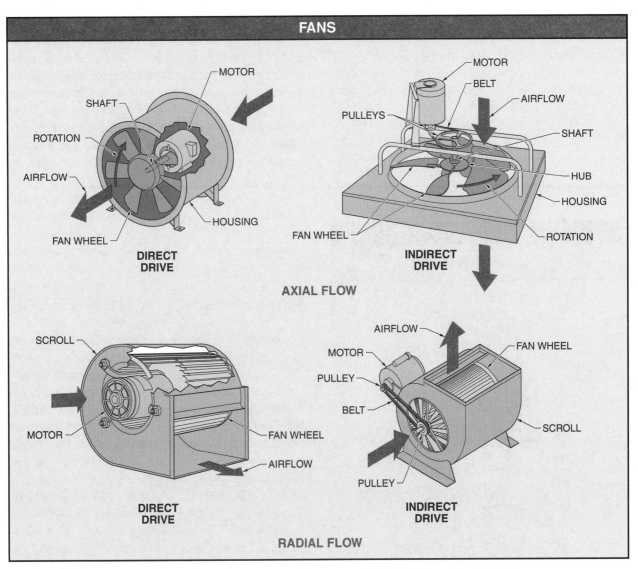

Figure 8-1. Fans create positive and negative pressures to move conditioned air through a building.

Air enters the air handler through dampers. A *damper* is a movable plate that controls airflow. Dampers are operated by electric motors or air cylinders. Most HVAC control systems use an economizer design to control the position of the outside air and return air dampers to provide the most efficient heating or cooling from the outside air. **See Figure 8-4.** Positioning is determined by comparing the setpoint with outside and inside air temperatures. For example, if the setpoint temperature is 72°F, the inside air temperature is 80°F, and the outside air temperature is 65°F, the outside air dampers are opened wider than the return air dampers to take advantage of the cooling provided by the outside air. When it is colder outside than inside and heating is required in the building, the return air damper is opened wider than the outside air damper

because it is less expensive to reheat warm indoor air than to heat the cool outside air coming into the building. When it is hot outside and the building space needs cooling, the outdoor and exhaust dampers will be at their minimum airflow position and return air dampers will be fully open. The cooling coil will be cooling supply air that is already mostly cool indoor air. The outdoor air dampers do not close completely since some fresh air is always needed to prevent the building air from becoming stale. Damper positioning is a compromise between energy efficiency and providing fresh indoor air to occupants.

The outside air damper is normally closed (NC). A *normally closed device* is a device that is closed when the signal causing it to operate is absent or at its lowest level. As signal strength increases, it pushes the damper open.

Outside air dampers are never completely closed even when the signal is at its lowest level. This prevents stale air conditions caused by lack of fresh air. Outside air and return air dampers move together because they are connected by a mechanical linkage. As the outside air damper opens, the return air damper closes. The return air damper is normally open (NO). A *normally open device* is a device that is open when the signal causing it to move is absent or at its lowest level. As the signal strength increases, the damper closes. Signals that operate the dampers are generated by the HVAC control system.

FAN CAPACITY*								
Wheel Dia†	Static Pressure‡						RPM	HP
	1/8	1/4	1/2	3/4	1	1 1/4		
22	7600	7000	5450	3500	—	—	390	2
	8700	8300	7150	5400	—	—	445	3
	10,500	10,100	9500	8350	7100	4800	530	5
	12,000	11,450	10,900	10,200	9500	8100	600	7½

* in cfm
† in in.
‡ in in. WC

Lau, a Division of Tomkins Industries

Figure 8-2. The two most common variables for determining fan size are capacity and static pressure.

After air is mixed, it is filtered and then passed through heating and/or cooling coils. A *heating coil* is a finned heat exchanger that adds heat to the air. A *cooling coil* is a finned heat exchanger that removes heat from the air. Heating coils use hot water, steam, or electric heating elements as a heat source. Cooling coils use cold water or refrigerant to remove heat from the air.

Heating and cooling coils, dampers, fans, and other devices are controlled devices. A *controlled device* is an HVAC device that is operated by signals from an HVAC control system. The signals are received from electronic or pneumatic controllers or sensors, such as thermostats. A sensor outputs a certain signal strength for a given condition, such as a pneumatic signal of 6 psi or a voltage of 5.6 VDC to represent 65°F. Electronic sensor signals are usually a current (4 mA to 20 mA) or voltage (0 VDC to 10 VDC).

In building automation systems, which focus primarily on HVAC system controls, the electronic sensors and controlled devices are connected on a common communication network. They send encoded digital signals that include the sensor or control information.

As temperature changes, so does the signal. For example, as a temperature rises to 70°F, the signal from an electronic thermostat may change from 12 mA to 14 mA. The controlled devices then change position in response to the changing signals, causing dampers to move or the flow of water, steam, electricity, or refrigerant through heating and cooling coils to increase or decrease. For example, a building's mixed air may vary from 40°F to 80°F, with a supply air setpoint of 65°F. At 40°F, the signal from the control system is at its minimum. The NO hot water valve is completely open, adding heat to the air. The NC cool water valve is completely closed. As the temperature in the building rises, so does the signal strength, slowly closing the hot water valve until it is completely closed at 65°F. The cool water valve does not begin to open until the temperature in the building rises to 72°F. If the temperature continues to rise, the increasing signal strength completely opens the cold water valve by 76°F. Air at 65°F is distributed to the building zones, where it is heated or cooled just before it enters each zone.

Terminal Units

A *terminal unit* is a device that is located close to the zone and heats or cools air flowing through it. A terminal unit usually contains a damper and a cooling and/or heating coil that conditions supply air just before it enters the zone. Air leaving an air handler is set to a general setpoint because different subsections of a zone may have slightly different heating or cooling needs. The terminal unit then raises or lowers the temperature as needed for a particular zone. Some terminal units contain fans, filters, and humidity control equipment, depending on zone needs.

Variable Air Volume Boxes. A *variable air volume (VAV) box* is a terminal unit that varies the amount of air flowing into a zone. VAV boxes may include only heating elements and dampers. **See Figure 8-5.** Therefore, the supply air from the air handler is always cool before reaching the VAV box. If a zone requires cooling, the damper is completely opened and the maximum amount of cool air flows into the space. As the space cools, the damper closes, limiting the flow of cool air. The damper does not close completely, avoiding a stale air condition. If heating is required, the damper is at its minimum open position and the electric heating element is energized. Heating in a terminal unit is often electric, which is easier to distribute to a large number of units throughout a building.

The heated or cooled air flows through a diffuser that disburses the air to prevent uncomfortable drafts. In contrast to constant air volume boxes, variable air volume boxes deliver the same volume of air to a zone.

Alnor Instrument Company

System balance and airflow distribution is verified by placing an airflow capture hood over a diffuser or grill.

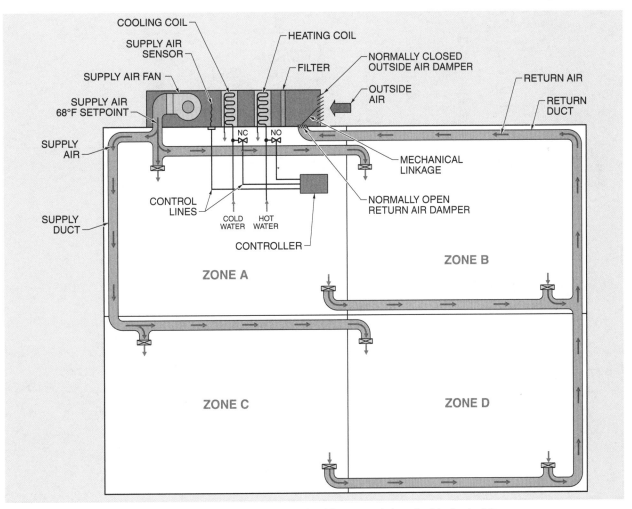

Figure 8-3. The air handler fan and dampers mix outside air with return air from inside the building.

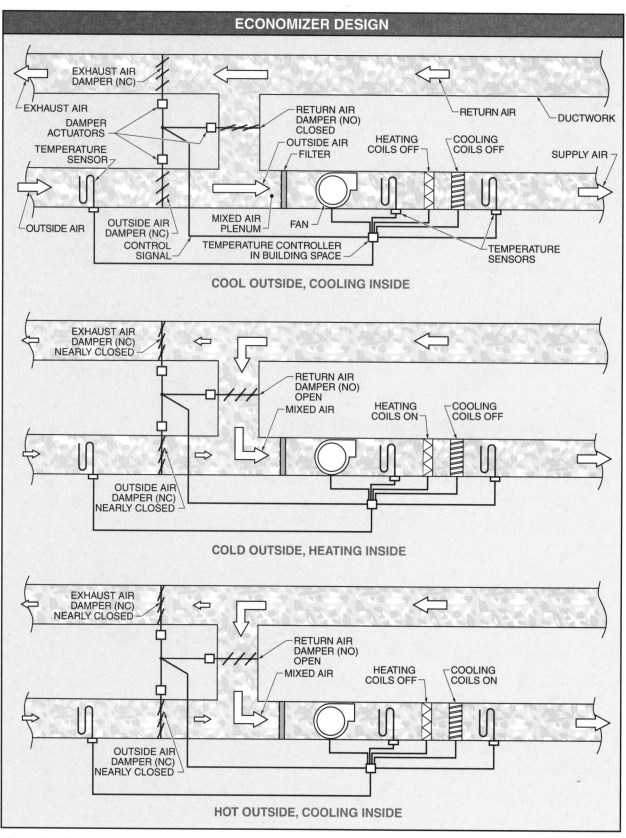

Figure 8-4. The economizer design uses damper position to balance efficient use of energy with the need for fresh air.

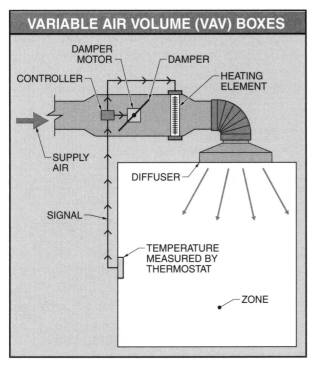

Figure 8-5. A variable air volume box varies the amount of air flowing into a zone.

Terminal Unit Control Systems. Damper position and heating element temperature are controlled by a thermostat operating in a closed-loop control system. A *closed-loop control system* is a control system in which the result of an output is fed back into a controller as an input. All components are linked by a flow of information in the form of heat or control signals. For example, air temperature is sensed by the thermostat. When heat is required, the thermostat sends a signal to a controller, which energizes the heating element. The resulting air temperature is sensed by the thermostat and signals the controller to de-energize the heating element when the temperature at the thermostat is at the setpoint. Most HVAC control systems are closed-loop systems.

An *open-loop control system* is a control system in which decisions are made based only on the current state of the system and a model of how it should work. A master/submaster control is an open-loop control system. In this system, the master thermostat is outside and the submaster thermostat is inside. The submaster thermostat commonly measures indoor temperature and controls zone temperature. The master thermostat measures outside temperature and overrides the submaster thermostat if temperature falls quickly, adding extra heat before the building cools. The changes inside have no effect on outside air temperature, the point at which the loop is open.

Building Conditions

A building is generally a closed loop or a collection of closed-loop systems. Changing conditions in one part of a building can affect other areas. For example, increasing airflow to one part of a zone to solve an odor problem can reduce the airflow to another part of the zone, causing stale air problems.

Buildings are usually designed to have a slightly higher air pressure than the outside air pressure. If air pressure inside the building is lower than the outside air pressure, doors are difficult to open and unconditioned, dirty air can leak inside. To maintain positive pressure, more air must enter the building than is being exhausted. The pressure level inside a building is usually maintained by exhaust fans that exhaust a specific amount of air. Atmospheric louvers set to open at a specific pressure are also used instead of exhaust fans to exhaust the correct amount of air. Higher inside pressure prevents unfiltered, unconditioned air from entering the building.

Air pressure can also vary inside a building. For example, a chemical storage closet should be kept under negative pressure relative to a hallway. This is accomplished by exhausting more air from the closet than enters the closet. Air flows from the hallway to the lower pressure in the closet. This prevents fumes from being drawn into the hall if a chemical leak occurs because air flows from higher to lower pressure. **See Figure 8-6.** A relative positive pressure is maintained in the hallway. Reducing airflow to the hallway could lower the pressure in the hallway below the closet pressure, causing air to flow out of the closet and creating a dangerous situation.

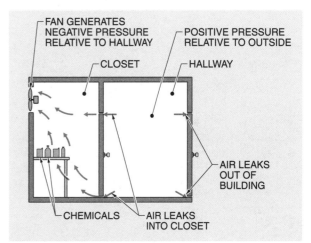

Figure 8-6. Varying air pressure inside a building controls the flow of air and fumes.

Warning: Do not change the operation of an HVAC system without determining if the changes cause problems elsewhere in the building.

The air pressure in a zone or building is controlled by pressure sensors located in the air supply ducts or throughout the building. Pressure is controlled by variable speed fan motors that change speeds to increase or decrease airflow or by dampers that open or close to allow more or less air to the supply air fan.

Relative Humidity

Relative humidity (RH) is the amount of moisture in the air compared to the amount of moisture the air would hold if it were saturated. Relative humidity is expressed as a percentage. *Saturation* is the maximum amount of moisture that the air can hold at a specific temperature. Hot air can hold more moisture than cool air.

Moisture is added or removed from air to provide comfort. *Comfort* is the condition that occurs when people cannot sense a difference between themselves and the surrounding air. An inadequate or excessive amount of humidity can cause discomfort. For example, very dry air can cause throat and lung irritation and very humid air can make people feel uncomfortably hot or cold.

Moisture can be added by steam humidifiers located in supply ducts or removed by passing the supply air through a refrigeration evaporator or cooling coil. **See Figure 8-7.** When air is cooled, some moisture condenses on the evaporator or cooling coil and collects in a pan. The water collected must be drained to eliminate puddles that promote harmful mold growth.

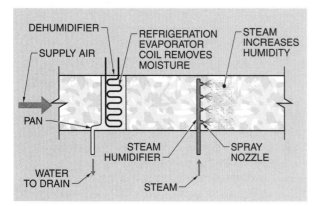

Figure 8-7. Humidity is controlled by evaporator coils or steam humidifiers.

Air Filters

An *air filter* is a porous device that removes particles from air. Filters are a critical component in an HVAC system. Filters remove particles from air before they enter the system ductwork. Filters work by trapping particles in a porous material such as fiberglass. **See Figure 8-8.**

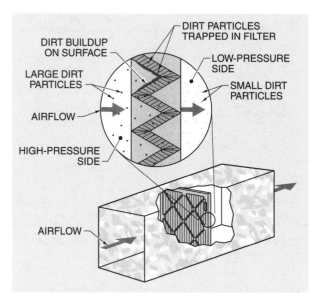

Figure 8-8. Filters consist of a porous material such as fiberglass that traps airborne particles.

The efficiency of a filter is rated by the size of particles it can trap. High-efficiency filters trap smaller particles than low-efficiency filters. Some areas, such as rooms where electronic components are assembled, require high-efficiency filters to prevent contamination of components. The use of high-efficiency filters is expanding because of the understanding that small particles can lodge in the lungs, causing potentially serious health problems.

However, high-efficiency filters restrict airflow, causing the fan motor to consume more current to force the air through the dense filter material. Changing low-efficiency filters to high-efficiency filters can increase electrical energy consumption and reduce airflow in the HVAC system. For this reason, all possible consequences must be considered before making any changes to an HVAC system.

Some HVAC systems use two filters. The first filter is an inexpensive, low-efficiency pleated filter followed by an expensive, high-efficiency bag filter. The pleated filter traps large particles and the bag filter traps small particles and holds large quantities of dirt. The low-efficiency filter is replaced more frequently than the more expensive high-efficiency filter. Selecting the efficiency, type, and quality of filter is a compromise between energy cost, purchase price, and health concerns.

Filter Replacement. Filters are replaced when damaged, wet, or clogged with dirt. A clogged filter is indicated by an increased pressure drop across the filter. **See Figure 8-9.** For example, a manufacturer may recommend that a filter be changed when the pressure drop across the filter reads 0.9″ WC on a manometer or diaphragm gauge. A *manometer* is a device that uses a liquid-filled tube to measure the difference in pressure between two locations. Manometers commonly register pressure differences in inches of water. One inch of water equals 0.0361 psi. Manometers measure pressure drop across a restriction such as a filter by allowing pressure from one side of the filter to push against pressure from the other side. The unequal pressures push on a column of water in the manometer, moving the liquid. The greater the pressure drop, the dirtier the filter, resulting in more movement of the liquid in the manometer. An increase in pressure drop is caused by air being forced through small restrictions in the dirty filter.

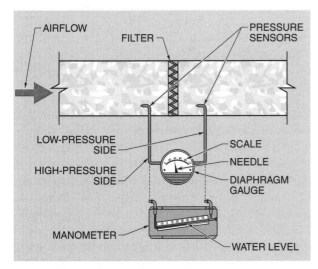

Figure 8-9. A clogged filter is indicated by an increasing pressure drop across the filter as measured on a manometer or diaphragm gauge.

A *diaphragm gauge* is a gauge that uses a flexible divider (diaphragm) to measure the pressure difference between the two sides of the diaphragm. A diaphragm gauge is commonly used to measure small changes in pressure between the two sides of a filter. The diaphragm moves as pressure changes, manipulating a needle and

registering the pressure difference on a scale. The gauge has high-pressure and low-pressure connections.

Dirt that is visible on the upstream (high-pressure) side of the filter does not necessarily mean the filter needs to be replaced. However, a filter that is clogged and needs replacement will likely be blackened on the downstream (low-pressure) side from particles trapped in the filter material. Clogged filters can restrict airflow to the conditioned space and cause supply air fans to draw greater-than-normal current. Filters are also excellent growing places for mold and bacteria, especially if they become damp or wet. Therefore, humid climates may require more frequent filter replacement. Replacement filters must fit snugly into the holder with no air gaps between the filter and frame. No air should be allowed to bypass the filter.

Warning: When handling dirty filters, try not to disturb accumulated dirt. Use proper gloves and respirators. Dispose of the filters in compliance with the authority having jurisdiction.

Electrostatic Precipitators

An *electrostatic precipitator* is a filtering system that electrostatically charges small airborne particles to remove them from the air stream. Electrostatic precipitators restrict airflow to a lesser degree than high-efficiency filters, causing very little pressure drop. Electrostatic precipitators consist of prefilters that trap large particles, ionizing plates that positively charge the remaining small particles, and negatively charged collecting plates. **See Figure 8-10.** The ionizing plates electrically charge small airborne particles with a positive charge. The particles are then attracted and attach to the negatively charged collecting plates, removing them from the air stream. Particles that build up on the collecting plates until they touch can short circuit the plates, allowing dirty air to pass. Collecting plates are cleaned on a regular basis by automatic water sprays.

Warning: Electrostatic precipitators use high DC voltages. Always lock out and tag out an electrostatic precipitator before performing any work.

SAFETY TIP
Static pressure drop measurements should be taken across filters when new because filter replacement is determined by the increase in pressure across a filter. A rule of thumb is to replace filters when the static pressure drop doubles from the reading when the filter is new.

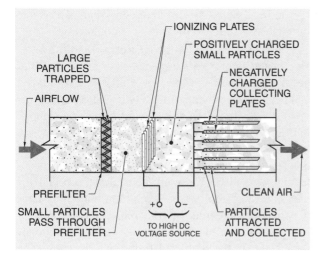

Figure 8-10. An electrostatic precipitator electrostatically charges small airborne particles to remove them from the air stream.

PNEUMATIC CONTROLS

A *pneumatic control system* is a control system that uses compressed air to send variable signals from thermostats and controllers to controlled devices. Compressed air is also used to move dampers, valves, and electrical contacts into the desired positions. **See Figure 8-11.** Pneumatic controls were used on the first automated HVAC systems. Pneumatic controls are being replaced by electronic controls, but are still found on older installations that require ongoing maintenance and repair. Pneumatic controls depend on a steady, clean, and dry supply of compressed air.

Pneumatic Control Air Supply

A pneumatic control air supply consists of two compressors and an ON/OFF pressure switch that controls the compressors as system pressure changes. Compressor size depends on the air volume required to operate the system. The compressors should be large enough to supply the required volume of air by running only one-third of the time. HVAC pneumatic control systems alternate operation between two compressors, reducing wear, keeping compressor temperatures low, and leaving one compressor running while the other can be serviced.

Low compressor temperatures reduce the amount of oil vapor pumped into the air lines because cool oil vaporizes less easily than hot oil. Additives from incorrect compressor oil can cause damage to delicate pneumatic components. Compressor oil must be kept at the correct level and changed according to manufacturer's specifications. HVAC pneumatic control compressor oil is not changed as frequently as industrial compressor oil because new oil vaporizes more easily than old oil.

Compressed air for pneumatic controls is filtered when it leaves the receiver and an automatic drain removes oil and water. Activated charcoal filters are also used to remove oil vapor. A refrigerated air dryer cools the air leaving the receiver. Cooling the air condenses moisture into an automatic drain, which must be cleaned frequently to avoid mold growth. A pressure regulator keeps the air at a steady pressure in the main lines. A pressure-relief valve is located after the pressure regulator to ensure that high air pressure does not damage the equipment downstream. The clean, dry air is sent through copper or plastic lines to thermostats, controllers, and controlled devices.

Pneumatic Thermostats. A *pneumatic thermostat* is a device that converts temperature variations into a variable pneumatic signal. A constant pressure and volume of air enters the back of a pneumatic thermostat and a variable pressure signal leaves the thermostat. The variation is created by a thin, bimetal element that flexes as temperature changes. The flexing controls the pressure sent out of the thermostat to the controllers. The controllers modify the signals and send them to controlled devices such as valves, switches, and damper actuators.

Pneumatic Operators. A *pneumatic operator* is a device that uses air pressure to position HVAC components. Air pressure fed into the operator pushes down on a bellows, diaphragm, or piston, whose motion is opposed by spring pressure. The bellows, diaphragm, or piston moves as pressure increases or decreases, causing the attached drive shaft to move accordingly. The drive shaft is connected to a linkage, which is connected to a valve or damper that opens or closes in response to the changing pressure in the operator. **See Figure 8-12.** The pressure acting on the bellows, diaphragm, or piston changes in response to variable signals from controllers and thermostats. A common problem with pneumatic operators is disconnected or plugged air lines. Plastic tubing becomes brittle and cracks or pops off fittings as it ages. Diaphragms can weaken and rupture with age. Pneumatic operators must be securely mounted and connected to their controlled devices.

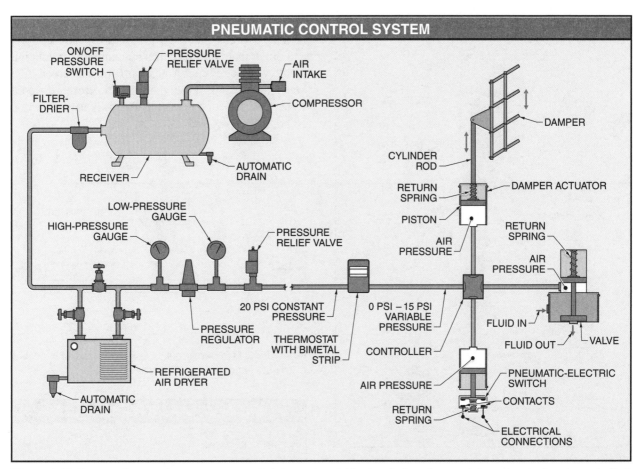

Figure 8-11. Pneumatic controls use compressed air to send variable signals from thermostats and controllers to controlled devices.

Pneumatic-Electric and Electric-Pneumatic Switches. A *pneumatic-electric switch* is a switch that uses air pressure to open or close a set of electrical contacts. Air pressure works against a spring-tensioned diaphragm. When the air pressure reaches a specific pressure, the diaphragm flexes, opening or closing contacts. An *electric-pneumatic switch* is a switch powered by a solenoid that moves a plunger, opening or closing an air line.

Electric Operators. An *electric operator* is a device that uses electrical equipment to position or switch HVAC components. Motors can position valves and dampers. Relays and contacts can de-energize heater coils, solenoids, or refrigeration compressors. Electric operators must be securely mounted and connected to controlled devices. Common problems with electric operators are loose connections or open circuits. In conversions from pneumatic to electronic control systems, the pneumatic operators are replaced with electric operators.

HVAC systems are inspected for leaks, which reduce the overall efficiency of the system.

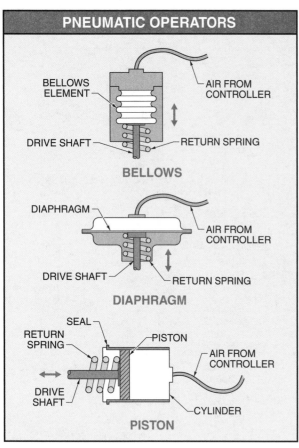

Figure 8-12. A pneumatic operator uses air pressure to position HVAC components.

Dampers and Louvers

Dampers regulate airflow and mix air to ensure even duct temperature. **See Figure 8-13.** Dampers are very reliable, but can jam mechanically or not move to the correct position. Connecting rods between dampers and damper actuators and any other linkages should be checked annually for alignment. All dampers should be operated to see that they move to the correct position when signal pressure changes. A common problem is loose fasteners that connect dampers to connecting rods. Tightness of fasteners should be checked annually. Clearly mark the fastener locations when removing or repositioning any linkage so they can be returned to the correct position. Lubricate all linkage and pivot points according to manufacturer's recommendations.

Two-Way and Three-Way Valves

Valves are used to control water or steam flow through heating and cooling coils. Valves can be either two-way or three-way valves. **See Figure 8-14.** A *two-way valve* is a valve that has two ports and controls the flow of fluid between them. Two-way valves have an inlet and outlet that controls coil temperature by regulating the flow of hot or cool water or steam. The greater the hot water flow, the hotter the coil. The greater the cool water flow, the cooler the coil. Two-way valves may be NO or NC valves.

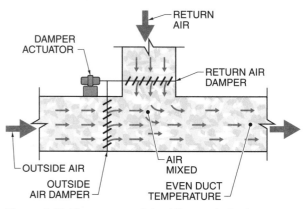
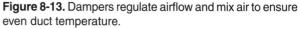

Figure 8-13. Dampers regulate airflow and mix air to ensure even duct temperature.

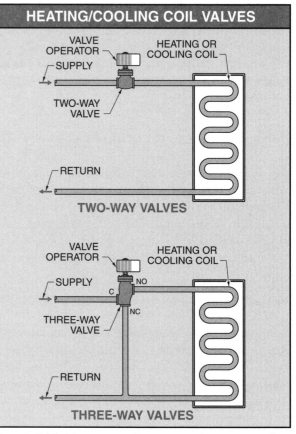

Figure 8-14. Two-way or three-way valves are used to control water flow through heating or cooling coils.

A *three-way valve* is a valve that has three ports and controls the flow of fluid from one to the other two. The three ports are labeled C, NO, and NC. Three-way valves send water through a coil or divert it around the coil to control coil temperature. Three-way valves have NO and NC valve seats mounted on one valve stem. The C port allows water to enter the valve. When the valve stem moves in one direction, the NO seat closes and the NC seat opens. When the stem moves in the opposite direction, the NO seat opens and the NC seat closes. A low signal pressure keeps the NC port closed and the NO port open, forcing all flow through the coil. As signal strength increases, the NO seat closes, allowing less supply water to the coil, and the NC seat opens, increasing bypass flow around the coil. At maximum signal strength, the NC seat is completely open and the NO seat is closed with all supply water bypassing the coil. Three-way valves can be used to control either heating or cooling coils.

Common valve problems include improper calibration, incorrect installation, leakage, and binding due to misalignment, overtightening, or poorly lubricated packing. Valves must be recalibrated after installation because their operation can be altered by water pressure and flow. Valve seats can become worn and leak. Leaking valves do not display a temperature difference from one side to the other when closed.

DIRECT DIGITAL CONTROL SYSTEMS

A *direct digital control (DDC) system* is a control system that uses low voltage and current to send variable signals from thermostats and controllers to controlled devices. Advantages of DDCs over pneumatic controls include less maintenance, better control of conditions, easier scheduling and troubleshooting, more information available about facility conditions, higher efficiency, easy computer control, and remote operation.

Direct Digital Control Operation

DDCs send and receive variable voltage and current signals that operate pneumatic or electric operators. DDC systems are often controlled by one or more desktop computers linked to sensors and controllers in building spaces, air handlers, or other controlled devices. **See Figure 8-15.** An electronic thermostat measures temperature and sends a current signal to a controller. The controller sends current signals to electric motors that position dampers and valves to open or close electric contacts and solenoids. DDC systems can also operate electric-pneumatic operators.

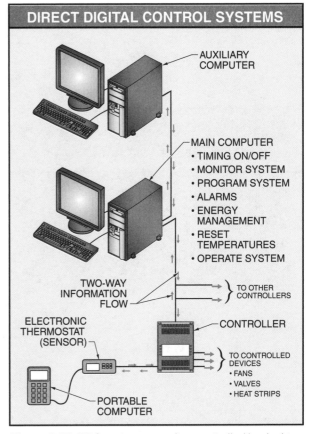

Figure 8-15. DDC systems are often controlled by desktop computers linked to sensors and controllers in building spaces, air handlers, or near controlled devices.

The system is programmed using specially designed software. System information is displayed with full-color graphics and text. Device status, air temperatures, airflow, damper and valve position, and temperature setpoint can be monitored and operated from a computer terminal. **See Figure 8-16.** The computer terminal can be connected to the system directly, or remotely using modems and telecommunications equipment. Portable computers can access and operate the system at connection points throughout the building.

Direct Digital Control Capabilities

DDC control systems contain accurate clocks that enable detailed scheduling. With customized schedules, individual rooms or whole buildings can be conditioned only when in use. Passwords and security clearance restrict access to the system. Workers not authorized to change a setting are denied computer access. Many systems generate a log of system users and the changes made.

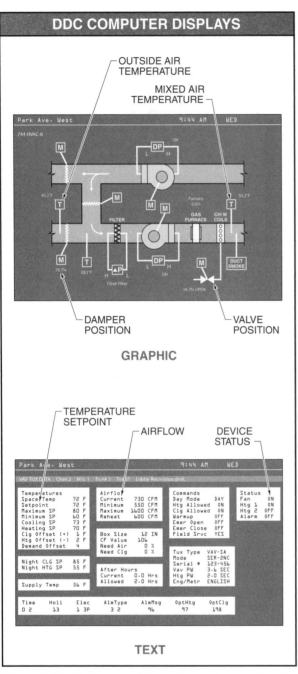

Figure 8-16. DDC system information such as device status, air temperatures, airflow, damper and valve position, and temperature setpoint can be monitored and operated from a computer terminal.

TECH TIP

The ventilation to be supplied to an occupied space is based on standards set by the American Society of Heating, Refrigerating, and Air Conditioning Engineers in the standard ANSI/ASHRAE 62.1, Ventilation for Acceptable Indoor Air Quality.

DDC systems can signal alarms for conditions such as low or high temperatures, airflow failures, sensor malfunctions, and internal problems with computer operations. The alarms are especially useful when troubleshooting. System-wide or individual room temperatures can be raised and lowered without entering the controlled space. Hours each tenant uses a building and energy use throughout the building can be recorded. DDC systems are so energy-efficient that conversion costs from pneumatic to DDCs are paid for quickly.

Direct Digital Control Energy Management

DDC systems can monitor and control energy costs in a variety of ways. For example, when a building is unoccupied during the heating season, the heating setpoint is lowered, making the building cooler. In addition, when a building is unoccupied during the cooling season, the cooling setpoint is raised, making it warmer. Setpoints are returned to normal temperatures before the facility is reoccupied. DDC systems are equipped with manual overrides when heating or cooling is required after normal hours. Some systems can track this usage and bill the tenants for extra energy consumed.

Also, equipment can be cycled ON and OFF instead of running the equipment continuously. For example, heaters can be turned ON for a few minutes at a time when the inside temperature is at the setpoint, then ON for longer periods of time if the temperature falls. When it is cool enough, the heaters operate continuously.

The most effective way to save energy is to use energy only when needed. Some DDC systems can calculate the best time to start and stop systems for daily operation by comparing outside temperatures with inside setpoints. When outside temperatures are relatively warm, the DDC system starts heating equipment near the occupied time and still reaches the setpoint in time for the building occupants. This saves money when compared to always starting the system at a fixed time each day. This strategy also allows DDC systems to stop heating equipment before the end of the occupied time and allows the stored heat to keep the building comfortable until people leave.

Load shedding is the deliberate shutting down of equipment to reduce electrical use. Load shedding can save money when electric utilities impose higher rates for greater power demand. A DDC system can monitor electrical use and, if it approaches a certain high level, turn OFF nonessential loads and turn them back ON when demand falls.

Enthalpy optimization programs compare temperature and relative humidity of outside and return air and select the air that requires the least energy (and therefore is the least expensive) to cool. *Enthalpy* is the total heat contained in a substance. Enthalpy in the air is the sum of sensible heat and latent heat. When water evaporates into the air, the water vapor contains latent heat required to make it evaporate. Air carrying water vapor condenses when cooled, giving up its latent heat. The latent heat represents a heat load on a refrigeration system, which must move and reject the heat. It takes more energy to cool hot humid air than hot dry air. For this reason, temperature is not the only consideration when cooling air.

BUILDING AUTOMATION SYSTEMS

A *building automation system* is a control system that uses digitally encoded messages shared on a common communication network to distribute information between control devices. Building automation systems are similar to DDC systems in that the controllers and sensors are electronic devices. However, the building automation controllers and infrastructure are far more sophisticated.

Each device requires only a single communication connection. The communication network is similar to those that connect computers and can transmit messages that include not only the control data, but the source, destination, time, priority, and other identifiers. This information is organized into groups known as packets, encoded in a certain protocol, and transmitted as a digital signal. **See Figure 8-17.** A *protocol* is a standardized set of rules and procedures for the exchange of digital information between two devices. This rich distribution of information allows these control systems to employ sophisticated logic to operate a facility in the most efficient way.

Building automation systems are capable of all the same functions as DDC systems, though they are even more flexible. Typically, any piece of data can be shared between any two (or more) control devices. For example, if one device reads from a sensor measuring outside air temperature, it can share that temperature with any other device, in any building system, that may have use for that information. System data and operations can be easily monitored from any point in the network and the system can be programmed for complex alarming, scheduling, and trending functions.

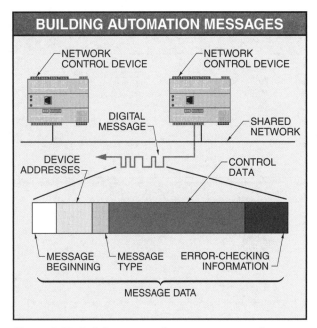

Figure 8-17. Building automation systems transmit system information within digitally encoded network messages, which contain several parts.

Each device needs only one network connection, which is part of a shared network infrastructure, so there is far less wiring required when compared to DDC systems. **See Figure 8-18.** This means that it is feasible to incorporate other systems into the network as well. Though HVAC systems are the most common type of system to automate, other implementations include lighting, electrical system, security, access, and elevator controls. For example, a building automation system can be designed to turn on certain lights and change a heating setpoint when a particular person enters a building via an access control station.

Each building automation control device includes a microprocessor, memory, and a control program. The control program identifies the data and procedures for sending messages, the data expected to be received, and the decisions it should make based on that data received. The data and control program stored on a particular device can be accessed from anywhere else on the network, such as a central maintenance computer. Programmers use a main computer to set up and modify the control programs in order to optimize the control behaviors throughout the entire network. Once established, the computer can be disconnected from the network, since all of the control logic is saved into the individual control devices.

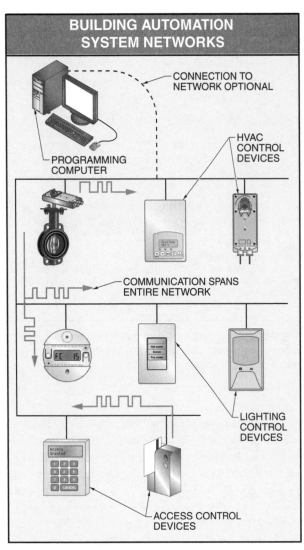

**BUILDING AUTOMATION
SYSTEM NETWORKS**

CONNECTION TO
NETWORK OPTIONAL

PROGRAMMING
COMPUTER

HVAC
CONTROL
DEVICES

COMMUNICATION SPANS
ENTIRE NETWORK

LIGHTING
CONTROL
DEVICES

ACCESS CONTROL
DEVICES

Figure 8-18. A building automation system can use a single communication network, which carries the messages to and from all control devices. Control devices only act on the messages that contain their identity as the destination.

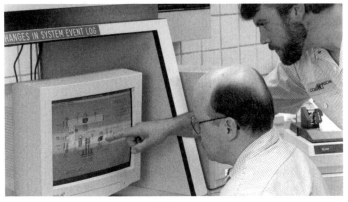

A computer terminal can be used to program and monitor building automation system operation.

TROUBLESHOOTING AND MAINTAINING HVAC SYSTEMS

The "too hot/cold" call is the most common HVAC system complaint encountered. Always respond quickly to any complaint and record it into the HVAC operations log. Note the time, who requested help, the symptoms of the problem, and the actions taken. It is critical that responses to complaints be prompt even if problems cannot be resolved quickly. A major consideration in HVAC work is good customer service.

Measure the temperature a few inches from the thermostat and record whether the system is within specifications. At the person's location, measure the temperature and check the airflow. Ensure enough heating or cooling air is reaching the individual. Look for structures such as room dividers that block or divert airflow. **See Figure 8-19.** Check for additional sources of heating or cooling that might affect the person but not the thermostat, such as a cool draft, radiant heat sources, or cold walls. Supply extra cooling or heating for the individual only after determining how this would affect others in the space.

If the temperature is incorrect at the thermostat, check the setting and calibration of the thermostat. Replace the thermostat if it cannot be calibrated. Check the airflow from the diffusers. Ensure that air is reaching the thermostat and all areas of the space. Check the relative humidity. High or low relative humidity can make individuals uncomfortable even when the temperature is correct.

Sometimes there is nothing wrong with the HVAC system except that the cooling or heating load is too great to maintain the desired setpoint. Remove heating and cooling loads whenever possible. If the system is working correctly, explain to the individuals that the system is working within specifications and make the individuals as comfortable as possible. Document all work done and report all findings to the building manager.

Pneumatic Control System Troubleshooting

Tracing problems in a pneumatic system requires a great deal of manual work. System sketches and troubleshooting manuals are used to trace air signals from the thermostat input to the corresponding operator. Pneumatic airlines are vulnerable to damage, blockage, and disconnection. Look for cracked or loose airlines and calibrate all devices in the system. **See Figure 8-20.**

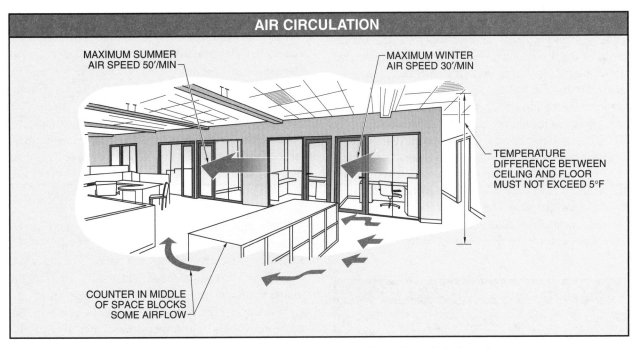

AIR CIRCULATION

MAXIMUM SUMMER
AIR SPEED 50'/MIN

MAXIMUM WINTER
AIR SPEED 30'/MIN

TEMPERATURE
DIFFERENCE BETWEEN
CEILING AND FLOOR
MUST NOT EXCEED 5°F

COUNTER IN MIDDLE
OF SPACE BLOCKS
SOME AIRFLOW

Figure 8-19. Room layout and furniture can adversely affect the flow of conditioned air through a space.

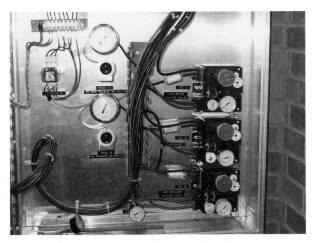

Figure 8-20. Pneumatic system controllers must be checked for loose airlines, proper calibration, and appropriate settings.

Check mechanical linkages and controlled devices for freedom of movement and calibration. For example, the amount of water flow through a valve is incorrect if the valve cannot open or close correctly. Check that the proper values of steam, cool water, hot water, or electricity are available at the controlled device. For example, the valve may be opening correctly but the water temperature may not be hot enough to heat the air to the correct temperature.

Mechanical linkages must be checked for breaks, corrosion, or obstructions that restrict their freedom of movement.

Direct Digital Control System Troubleshooting

DDC system problems are commonly located and repaired from a computer terminal. Access information about the area, including current temperature, setpoints, control status, control mode, alarms, and damper/valve position. **See Figure 8-21.** For example, the computer may show that the temperature is 83°F but the setpoint is 71°F. If the system allows, try to force the damper open by overriding room controls from the computer. If this does not work, or the system does not allow forcing, go to the area and inspect the VAV box and its mechanical controls and operation. The problem could be a damaged operator, a defective controller, or something as simple as a loose setscrew on the damper connecting shaft.

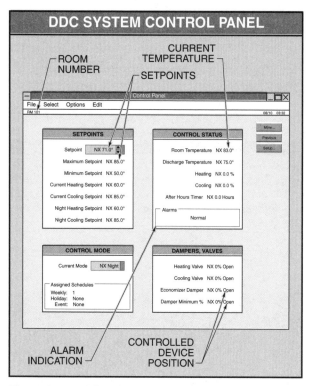

Figure 8-21. DDC system problems are commonly located and repaired from a computer screen display.

Some DDC systems self-diagnose problems and offer suggestions. For example, an alarm is displayed if there is a defective sensor or controller. Some controllers include indicator lights to aid in troubleshooting, such as indicating that the power supply is at the correct level. DDC components are commonly sealed units whose electronic circuits are not repairable in the field. **See Figure 8-22.** If a sensor or controller fails, it is replaced.

Electronic sensors change resistance or generate voltage in response to conditions in the controlled space. To check a sensor, measure the resistance across the sensor or the current output from the sensor. Use only DMMs when taking resistance, voltage, and current readings because extreme accuracy is needed. Compare these readings to system specifications or a good sensor. For example, a certain thermostat measuring 70°F should have 50 Ω of resistance. If it does not, the thermostat is defective and should be replaced or recalibrated. Test the pressure, flow, and humidity sensors in a similar fashion.

If the sensor is functioning properly, but no signal is reaching the controller, check the wiring between them. Look for loose connections and sources of unwanted resistance. The conductors should not exceed a specific resistance for their length. High resistance lowers the signal reaching the controller, affecting operation. Remove any causes of excess resistance, usually by tightening or cleaning connections. Replace the conductors if necessary, ensuring that they are the correct size and meet all system specifications.

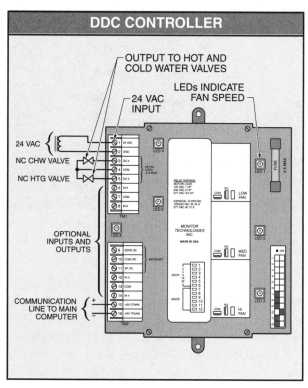

Figure 8-22. DDC controllers should be checked for tight and proper connections. LEDs may indicate self-diagnosed problems to aid in troubleshooting.

Other than a yearly operational check, DDC systems should require very little maintenance. Operate the complete system and check to see that each component is functioning properly and the function is verified by the computer. Compare building space temperatures to those on the computer readout. Operate all valves and dampers from fully open to fully closed to ensure proper operation. Always follow the maintenance and testing procedures suggested by the equipment manufacturer and system designers for DDC systems.

Never attempt to work on a DDC system without completely understanding the operation of the system and electrical troubleshooting. Tinkering or swapping conductors can damage a system. Some problems, such as corrupted system software, require outside help to troubleshoot and repair.

Building Automation System Troubleshooting

Like DDC systems, building automation systems require very little maintenance. Also, many problems that do occur can be attributed to mechanical problems unrelated to the automation system. For example, broken linkages, loose connections, leaky ducts, or blocked airflow are just as common in automated HVAC systems as those with any other type of controls. These problems are easy to troubleshoot by on-site maintenance personnel.

A major advantage of building automation systems is that they can provide a wealth of information about the sensor measurements and control actions occurring throughout the building. Most systems are designed to include at least one monitoring station that receives data from the control devices and displays it for the maintenance personnel. For ease of use, the displays usually include color graphics illustrating the general arrangements and relationships between devices and the building. **See Figure 8-23.** Graphs prominently show alarms and data changes over time and in context.

The primary disadvantage of building automation systems is that the design and programming is more complex than with other control systems. If the program is developed and verified adequately, there should be little need to make changes. However if the system operation must be changed or devices are replaced, program modifications may be necessary, which may require contractors with specialized knowledge of these systems.

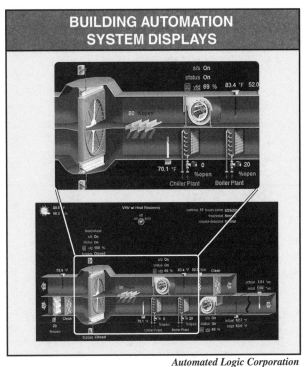

Automated Logic Corporation

Figure 8-23. Building automation systems typically support high-quality graphics for displaying current system and control device status.

ENERGY AUDITING

An *energy audit* is a comprehensive review of a facility's energy use and a report on ways to reduce the energy use through changes to buildings, equipment, and procedures. The audit can focus on the building operation, process, or both. The audit collects detailed information about energy use and prioritizes strategies to reduce the consumption, in order of cost effectiveness. The goal is to reduce costs through lower energy use without negatively impacting the facility operations, processes, occupant comfort, or safety.

Energy audits are not exclusive to HVAC systems. However, since HVAC systems are typically the most energy-consuming system in a building or facility, they are usually the primary target. The energy audit can also survey multiple types of energy consumption, including electricity, steam, hot water, chilled water, natural gas, diesel, and other fuels. Maintenance personnel are important to all phases of an energy audit. These phases include gathering system information, measuring energy use, developing reduction strategies, choosing the most cost-effective plan, implementing changes, and verifying results. **See Figure 8-24.**

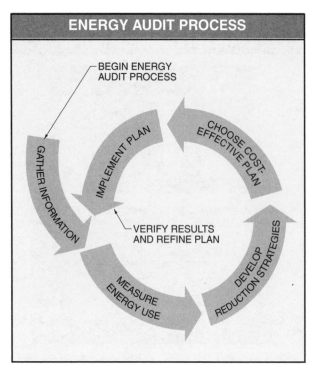

Figure 8-24. The results of an energy-use reduction plan should be verified by conducting another round of measurements. Depending on the outcome, this may lead to repeating the audit process to further refine the energy savings.

Energy Auditing Factors

An energy audit involves an investigation of a number of factors. These factors are the building envelope, equipment operations, system controls, maintenance procedures, infrastructure conditions, and occupant behavior.

Building Envelope. The building envelope includes walls, ceilings, floors, doors, windows, and skylights that separate the interior with the outside environment. For each of these components, the area and resistance to heat flow (R-value) is measured or estimated. Energy auditors look for air leaks in the building envelope, which are common around poorly sealed doors or windows, but can also occur at locations such as transitions between different building materials. This investigation provides data on the building's insulation. When compared against local climate data, auditors determine the building's thermal performance.

Equipment Operations. Investigating equipment operations involves researching the specifications on each major piece of equipment and individually measuring their energy consumption. The audit assesses the efficiency, physical condition, and operating profile of the equipment, including duty cycle, load changes, and controls. Suggested energy-use reduction measures may include changing the operation profile (such as greater load for a shorter duty cycle) or replacing older equipment with newer, more efficient models.

System Controls. Automation and integration of controls is generally associated with a high level of energy efficiency. Manual controls leave energy use up to individuals, who may let equipment run longer than necessary. Automated controls, however, make consistent and objective energy-use decisions. An energy audit should differentiate between the types of controls used for various systems and analyze the automation programming for appropriate system integration, algorithms, and manual overrides.

Maintenance Procedures. Maintenance programs have a significant effect on the integrity of a building envelope and the operating conditions of the equipment. The cleanliness of the environment influences the health of the equipment. Prompt and complete repairs reduce waste and downtime. Evaluation of the emergency, preventive, and predictive maintenance procedures determines whether a greater investment in maintenance activities can increase operating efficiency.

Infrastructure Conditions. The condition of the facility and its system infrastructures should be part of an energy audit. Auditors look for inefficiencies in the interior layout, process flows, and utility infrastructure. For example, a hot exhaust placed near a cooling unit causes the cooling unit to work harder than otherwise necessary. If the facility interior has been reconfigured in the past, hot water supply pipes may be routed in an unnecessarily complex way. These supply pipes may be very long or include too many bends, allowing the water to cool significantly and lose pressure before arriving at the point of use. As a result more energy is required to supply sufficient hot water to the process.

Occupant Behavior. The actions of building occupants can have a significant effect on energy use. For example, occupants switch lighting ON and OFF, prefer certain indoor climates, and open and close doors and windows. A comprehensive energy audit typically includes information on patterns of use based on interviews of many occupants. This information is used to suggest changes to occupancy schedules, automated controls, and instruction on energy use reduction.

Energy Audit Types

Energy auditing includes a broad spectrum of studies, ranging from a quick walk-through to a comprehensive analysis of all energy consumption. There are three common types of audit programs. The differences between them are the level of breadth and detail involved in the investigation, which affects the accuracy of the resulting analysis and the possible solutions presented.

As an audit becomes more thorough, more opportunities for savings are typically uncovered, though they are often of increasingly smaller benefit. **See Figure 8-25.** Therefore, it may be necessary to evaluate the potential efficiency gains against the growing cost of the audit at multiple times during an investigation in order to decide whether to go further. Each facility has a different situation, so the most cost-effective solution will vary.

Preliminary Energy Audit. A *preliminary energy audit* is an overview of a facility's major energy-consuming processes to identify only significant inefficiencies. This type of audit is also called a simple audit, screening audit, or walk-through audit. This is the simplest and quickest type of audit, but often provides limited benefit, as it can only uncover the most obvious energy waste or inefficiency problems.

A preliminary energy audit involves a walk-through of the facility to become familiar with the building operation, interviews with facility personnel, and a review of utility bills and other operating data. The result is a simple prioritized list of corrective measures, estimates of their costs, and potential savings.

Typically only major problems, such as leaks or inappropriate setpoints, are uncovered during this type of audit. However, depending on the situation, this may still be the most cost-effective option. Based on the results, it is then determined whether a more detailed audit is appropriate.

General Energy Audit. A *general energy audit* is a facility energy study that expands on the preliminary energy audit by collecting more information and performing a more detailed analysis. This type of audit requires a record of utility bills over a 12- to 36-month period, which reveals short- and long-term consumption patterns and rate changes. Data from any other sources of energy use over time is also gathered.

This type of audit identifies and prioritizes appropriate energy conservation and efficiency measures for the facility. Each proposed project includes a detailed financial analysis with implementation cost estimates, site-specific operating cost savings, and the customer's investment criteria. The level of detail in a general energy audit is often needed to justify significant implementation costs.

Investment-Grade Energy Audit. In large corporate facilities, the types of changes recommended by energy audits are often major investments that require executive scrutiny and significant supporting financial analysis. Budget decision-making is based on an expected return on investment (ROI) that applies to all energy and non-energy related expenditures. For the completeness and accuracy demanded in these situations, a facility may conduct an investment-grade energy audit.

An *investment-grade energy audit* is a comprehensive facility-energy study that expands on the general energy audit by developing a dynamic model of the facility's energy-use characteristics. The model is calibrated against actual utility data to ensure that it provides a realistic baseline. Then the model is used to simulate the results of proposed energy conservation and efficiency measures. This provides the most accurate estimates of potential energy savings, which are then included in the financial analysis reports. The extensive data gathering and model testing of an investment-grade energy audit earns a high level of confidence in the conclusions of the final reports.

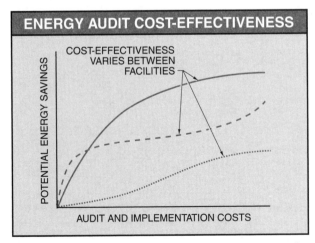

Figure 8-25. More involved and expensive energy audits typically yield greater opportunities for energy savings, but the cost-effectiveness varies depending on particular energy-use profiles of the facility.

INDOOR AIR QUALITY

Indoor air quality (IAQ) is a description of the type and quantity of contaminants in indoor air. Good IAQ means the air is relatively free of harmful contaminants. Poor IAQ means that air is contaminated beyond safe levels. Poor IAQ can affect human wellness and lead to serious health problems for building occupants. Poorly designed, installed, or maintained HVAC systems and equipment can contribute to IAQ problems.

The Environmental Protection Agency (EPA) ranks IAQ as one of the top five environmental threats to human health. In addition, people working in poor quality air are often less productive. Noticed or unnoticed, indoor air pollutants can be irritating and even life threatening. A low level of contamination may not harm some people, but it may cause serious health problems for others. IAQ problems are often difficult to identify and eliminate.

Contaminants

Indoor air can become much more polluted than outside air. Common indoor air pollutants include particulates, microorganisms, chemicals, and gases. IAQ problems may include one or more of these pollutants in different degrees. Many different factors affect the blend and concentration of pollutants. **See Figure 8-26.** IAQ problems vary between different buildings, but also from day to day and season to season.

Particulates. A *particulate* is a very small particle of solid or liquid matter. Each particle may be composed of one material, such as asbestos, or a cluster of many materials, such as dust. Particulates can land on surfaces and become airborne again when disturbed. Particulates can be easily inhaled and cause significant health problems, depending on their size. Larger particulates are generally filtered in the nose and throat and do not cause problems, but particulate matter smaller than about 10 μm can settle in the lungs and cause breathing problems. Particulates smaller than 2.5 μm penetrate into the gas-exchange regions of the lungs, and very small particles (<100 nm) may pass through the lungs to affect other organs. Particulate contaminants include dust, smoke, asbestos, and fiberglass. Manufacturing and processing systems can produce significant quantities of particulate contamination from the processed materials.

Microorganisms. Microorganism contamination includes mold, bacteria, and viruses. When mold spores are present in the air, they can cause allergic reactions, asthma episodes, irritations of the eye, nose and throat, infections, sinus congestion, and other types of respiratory health problems. Bacteria and viruses can cause illnesses, such as legionellosis (Legionnaire's disease).

COMMON PATHWAYS FOR CONTAMINANTS	
Common Pathway	**Effects**
Outdoor air intake	Polluted outdoor air or exhaust air enters building through air intake
Windows/doors Cracks and crevices	Negatively pressurized building draws air and outside pollutants into building through any available opening
Substructures and slab penetrations	Radon and other soil gases and moisture-laden air or microbial-contaminated air travel through crawlspaces and other substructures into the building
Stairwell Elevator shaft Vertical electrical or plumbing shafts	Stack effect causes airflow by drawing air toward shafts on lower floors and away from shafts on higher floors, affecting flow of contaminants
Wall openings and penetrations	Contaminants enter and exit building cavities and move from space to space
Duct or plenum	Contaminants commonly carried by HVAC system throughout occupied spaces
Flue or exhaust leakage	Leaks from sanitary exhausts or combustion flues cause serious health problems
Room spaces	Air and contaminants move within room or through doors and corridors to adjoining spaces

Source: Environmental Protection Agency (EPA)

Figure 8-26. Contaminants originate from both indoor and outdoor sources and can enter occupied areas through various mechanisms.

Warm temperatures and moisture promote microbial growth. In HVAC systems, water collected from dehumidification equipment must be drained immediately to avoid excessive moisture or standing water. All water-damaged furnishings and carpets should be dried or removed immediately. Leaks should be fixed and relative humidity kept between 40% and 60%, depending on the temperature and activities in the building.

Chemicals. The chemicals found in indoor air come from many sources, including cleansers, carpet, furniture, photocopiers, fax machines, fuels, and products such as perfumes and deodorants. Mitigation strategies include locating equipment such as photocopiers and laser printers in rooms with increased outside air flow and properly vented exhausts. The least chemically harmful supplies should be purchased for these devices. Only small amounts of chemicals should be stored in rooms under negative pressure with exhaust vented outside.

Gases. Common harmful gases include radon and carbon dioxide. Radon is an invisible, radioactive gas that slowly leaks into basements and ground floors from contaminated earth. Due to its prevalence in certain rock formations, radon hazards vary regionally. However, radon is a serious health hazard for indoor air, and a known cause of lung cancer.

High levels of carbon dioxide are also an IAQ problem and may cause drowsiness, headaches, or impaired function. Humans are the main source of indoor carbon dioxide, which accumulates in classrooms, conference rooms, and other areas of high occupancy. Carbon dioxide levels are an indicator of the adequacy of outdoor air ventilation.

Ventilation

Sources of indoor air pollutants should be minimized or removed, though they cannot be completely eliminated. Diluting and removing contaminated indoor air requires ventilation.

Ventilation is the process of introducing fresh outdoor air into an indoor space. **See Figure 8-27.** If an insufficient amount of outside air is brought into a building, contaminants generated in the building are trapped inside and pollution levels rise. Therefore, outside air and supply air dampers must not close tightly, as this limits the flow of fresh air. Both outside air and supply air dampers have a minimum open position that supplies enough fresh air to meet the occupants' ventilation needs. Inspections of HVAC systems should ensure that dampers do not close beyond their minimum opening or are otherwise blocked.

Within the indoor spaces, air is typically introduced through ceiling diffusers and exhausted through ceiling grills. Diffusers must not be blocked by furniture or other obstructions, that limit fresh airflow. Also, diffusers and grills must not be installed too close to each other, which can cause short cycling.

Short cycling is the increase in the frequency of system operation due to improper feedback. With short cycling, the supply air exits through the return air grill without reaching the occupied zone, which is 3″ to 72″ above the floor. **See Figure 8-28.** The thermostat continues to sense the stale room air and tries to correct it through repeatedly cycling the system. To avoid short cycling, ensure that air circulates through the occupied zone by properly locating diffusers and grills and avoiding any airflow obstructions.

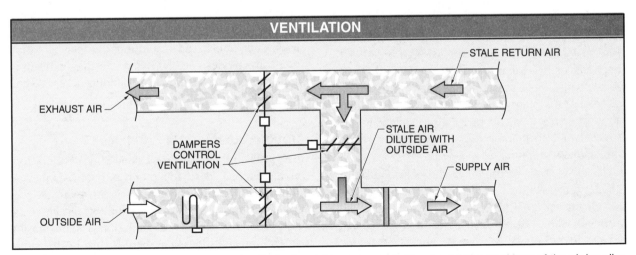

Figure 8-27. Ventilation dilutes stale return air with fresh outside air by controlling the relative positions of the air handler dampers.

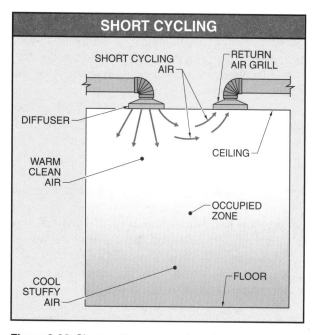

SHORT CYCLING

Figure 8-28. Short cycling occurs when air entering a space moves directly to the return air grill, resulting in poor air distribution and poor air quality.

Ventilation is only effective if the outside air is free of contaminants. If the wind blows from a particular direction, it may bring in additional pollutants. Some solutions to this problem are installing baffles to deflect air, increasing smokestack height, or installing high-quality filtration devices. Parking garages and loading bays are common sources of contaminated outside air. **See Figure 8-29.**

Pressure Regulation

Buildings are usually operated at a slightly positive pressure so contaminants do not leak in. This is especially important on lower floors due to stack effect. *Stack effect* is the rising of hot air up the center of a building through elevator shafts, stairwells, and service columns. Cooler air is drawn into lower floors, bringing in pollutants from street level. Keeping lower floors at a positive pressure limits inflow of unfiltered, cold air.

Relative negative pressures can also be used for certain areas within a facility, such as storage areas for chemicals. These areas require increased outside airflow to dilute contaminants and all return air must be vented outside. By keeping the area under negative pressure, the chemical fumes do not leak into nearby occupied areas.

HVAC System Operation and Maintenance

The operation and maintenance of an HVAC system is the primary factor in IAQ. When operating an HVAC system, start the system early enough to ensure proper ventilation by the time occupants arrive. Leave the system running long enough to ensure the building is flushed clean of contaminants. Ensure proper airflow to all parts of the conditioned space. Allow building tenants to have some control over temperature and airflow if needed after normal hours.

Maintain temperature and humidity within the range of system setpoints. Individuals who are uncomfortable because they are hot or cold are more likely to suffer from, and complain about, IAQ problems. Also, high temperatures and high relative humidity (above 60%) can promote microbial growth and increase the release of chemicals from furniture and carpets. Low relative humidity (below 40%) can cause respiratory problems and irritate mucous membranes. Maintain proper settings and operation of all humidification and dehumidification equipment.

Cleaning, maintenance, remodeling, and building repairs affect IAQ. Keep the entire building neat and clean. Avoid clutter, such as piles of papers and supplies, which are excellent breeding places for mold and other microbial contaminants. Remove or refurbish old or deteriorated furnishings. Schedule work when the building is not occupied or at low occupancy and operate the HVAC system during all building cleaning and maintenance activities. Seal construction areas and keep them under negative pressure, exhausting return air outside.

IAQ problems are avoided by operating a comprehensive preventive maintenance program that includes documentation of all system specifications, building uses, complaints, and resolutions. Proper operation of a correctly designed HVAC system should result in comfortable, safe conditions for building occupants and prevent most IAQ problems.

Troubleshooting IAQ Problems

IAQ problems can be complex and often develop from more than one cause. Symptoms of poor IAQ include sore eyes and throat, coughing, chest pain, fatigue, headache, muscle aches, and irritability. Since these are also general symptoms of many illnesses, diagnosing an IAQ-related illness can be difficult. Investigation of complaints must be done methodically to determine possible air quality problems.

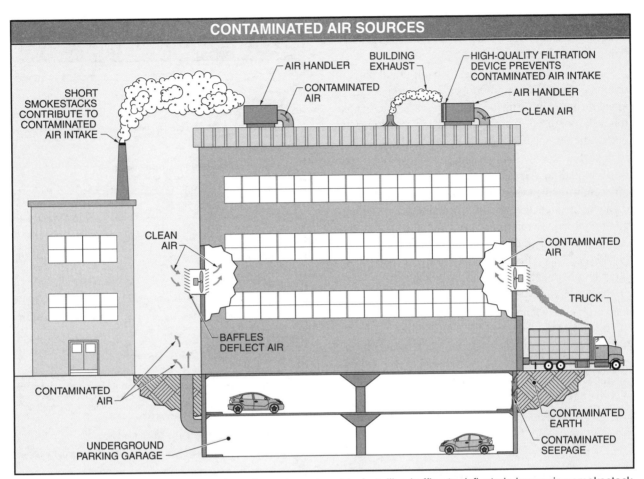

Figure 8-29. Intake of contaminated outdoor air can be reduced by installing baffles to deflect air, increasing smokestack height, or installing high-quality filtration devices.

When an IAQ complaint is received, maintenance personnel should respond quickly by talking to the person and making detailed notes on the symptoms and any obvious problems. Standard forms are available to help collect all the important information. It may be useful to ask the person making the complaint to keep a diary of their symptoms and air-quality observations. This is because air-quality problems may develop or change over time.

Maintenance personnel should inspect the area and any nearby areas for obvious problems like mold growth, reduced airflow, or chemical spills. The air-handling equipment servicing the area should be inspected. **See Figure 8-30.** Temperature, relative humidity, and airflow readings should be taken in the problem area. The airflow level should be confirmed by taking measurements with a flow hood or an anemometer and comparing them to specifications. If airflow to the area is incorrect, the system may need to be rebalanced. However, increasing

Figure 8-30. Removable doors permit maintenance on HVAC equipment that is otherwise difficult to access.

airflow to the problem area may create problems elsewhere. Balancing airflow requires specialized measuring equipment and training.

Airflow patterns can be seen by using small puffs of nontoxic smoke from smoke tubes. The smoke is carried along by the air, making the flow patterns visible. Smoke tubes can be used near supply and return air grills to determine the speed and pattern of the airflow.

If airflow and distribution appear normal, then carbon dioxide readings are taken to determine if enough fresh air is entering the space. High carbon dioxide levels indicate inadequate ventilation. Normal airflow with high carbon dioxide readings indicates that the air handling equipment is only recirculating air and not adding fresh outside air. Depending on the IAQ policy in the building, other air chemical readings might be taken.

Outside experts are often employed when serious IAQ problems occur or the problem requires specialized equipment or procedures. Maintenance personnel must methodically record all results of IAQ investigations. **See Figure 8-31.** Even if the on-site personnel cannot diagnose or remedy the problem, having a complete record of all work done in response to an IAQ complaint assists outside experts in solving the problem.

TECH TIP

Sustaining acceptable IAQ involves more than just the HVAC system. The building site, building occupants, local climate, contamination sources, process equipment, and construction methods all contribute to IAQ.

IAQ TROUBLESHOOTING—MOLD GROWTH

Problem

Many occupants in a building became seriously ill with chest c and allergic reactions. The building had an effective preventive program.

Troubleshooting Step 1 - Investigate

There were no changes in activity in the rooms of the zone
Airflow was normal.
CO_2 levels were normal.
Called in IAQ investigators to do specialized measurement

Troubleshooting Step 2 - Isolate

Likely Causes	Action/Result
Dirty filters causing mold growth.	Filters inspecte
Mold growth in building.	No mold locate
Toxic chemicals in building air.	None located d
Mold growth inside HVAC ducts.	Mold found gro

Troubleshooting Step 3 - Remedy

As a temporary solution, the ductwork was treated with mold c and the insulation liners were resealed. The replacement of th lined ducts with externally lined ducts was planned. The opera over a year and could be accomplished by moving tenants to building as their areas were renovated.

Troubleshooting Step 4 - Documentation

Filed all IAQ reports. Filed all recommendations. Filed plans a

Figure 8-31. Following standard troubleshooting steps and recording all results helps diagnose and solve IAQ problems.

Mechanical Systems

Mechanical systems support loads, transmit energy, and secure equipment in industrial facilities. Mechanical systems require lubrication to reduce friction, dissipate heat, and prevent wear. Mechanical drive systems include belt, chain, gear, cam, and variable-speed drives. Mechanical systems fail due to excessive wear, overheating, and overloading. Effective preventive, and predictive maintenance are essential to reduce and prevent failures in mechanical systems.

LUBRICANTS

A *lubricant* is a substance that separates moving (bearing) surfaces to reduce the friction and/or wear between them. The separation provided by a lubricant eliminates surface-to-surface contact. Bearing surfaces consist of asperities. *Asperities* are microscopic peaks and valleys left over from the machining process. During the break-in period, the microscopic peaks from each bearing surface slide across each other, wearing away the peaks of the asperities. **See Figure 9-1.** After the break-in period, the peaks of the asperities are removed and the valleys act as reservoirs for the lubricant. A microscopic film of lubricant separates the two bearing surfaces. Lubricant film thickness is determined by the clearance between two bearing surfaces and the lubricant viscosity. *Viscosity* is the measure of the resistance of a fluid's molecules to move past each other. Lubricants may be liquid (oil), semisolid (grease), or solid (graphite). Viscosity indicates the expected thickness of the lubricant film separating bearing surfaces.

Lubricant Selection

Lubricants are used to reduce friction, dissipate heat, absorb shock, and prevent wear, corrosion, and adhesion. Friction and wear occur simultaneously. The greater the friction, the greater the wear. Lubricants reduce sliding and rolling friction. *Sliding friction* is friction that occurs when one surface moves across another or both surfaces move in opposite directions. *Rolling friction* is friction that occurs when a rolling device (roller or ball) moves on a stationary surface.

Lubricants reduce the operating temperature of machine parts by transferring heat from the metal surfaces to the lubricant. During startup and operation, a lubricant absorbs shock caused by normal machine operation. The shock is absorbed as lubricant is trapped and compressed between bearing surfaces or mating gears. The lubricant becomes pressurized during equipment operation and cushions the mating components from damage. The proper lubricant viscosity is critical for protection against shock. The lubricant must be thin enough to slide in between the mating surfaces, but thick enough to keep from being squeezed out from between mating surfaces and absorb shock when force is applied.

Lubricants also minimize corrosion by forming a protective layer on bearing surfaces to prevent oxidation when the machine is not operating. *Oxidation* is the combination of metal and oxygen into metal oxides. Rust is an example of oxidation. The slick nature of a lubricant prevents components from adhering even though they slide across one another.

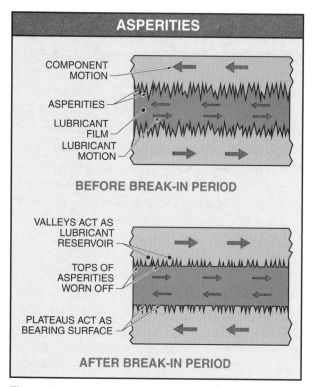

Figure 9-1. Asperities are microscopic peaks and valleys left over from the machining process.

Lubricating Oil

Lubricating oil is a liquid lubricant having a mineral, synthetic, vegetable, or animal origin. Commercial lubricating oil is classified into Groups A through F. Each group is designed for a specific application. **See Figure 9-2.** Common commercial lubricating oils do not consist of 100% oil stock. They include different percentages of additives, depending on the oil. An *additive* is a substance added to a lubricant to enhance a desired characteristic. Common oil additives include wear preventatives, oxidation inhibitors, defoaming agents, pour point depressants, rust inhibitors, anti-ash compounds, and viscosity index improvers.

The primary lubricating oil characteristic concerning maintenance personnel is oil viscosity. Viscosity is the measure of the resistance of a fluid's molecules to move past each other. Oil viscosity is affected by temperature. For example, resistance to flow and/or oil thickness increases as temperature decreases. Oil is given a Society of Automotive Engineers (SAE) viscosity rating based on its ability to flow through a specific orifice at a specified temperature, atmospheric pressure, and time period. A high viscosity rating results from a small volume of oil flowing through the orifice. This is caused by the oil's high resistance to flow. A low viscosity rating results from a large volume of oil flowing through the orifice. This is caused by the oil's low resistance to flow. The higher the viscosity rating number, the thicker the oil. For example, a 40 weight oil is much thicker than a 10 weight oil. The viscosity rating number assigned to an oil does not change, but oil viscosity may change with temperature and pressure. During startup, oil is cool and does not flow easily. As the machine and oil warm, the oil flows more easily. Therefore, most machine wear occurs during startup when the cool oil provides less lubrication.

Oil film thickness decreases with an increase in oil temperature and may be completely depleted in high operating temperatures. Oil specified for an application should provide flow at low temperatures but still protect the machine at high temperatures. Oil recommendations from the manufacturer use standards provided by the SAE, ASTM International®, and the American Petroleum Institute® (API®). These organizations provide standards in viscosity and additive packages for the majority of lubricants manufactured worldwide. Follow the manufacturer's recommendations when selecting an oil. In general, operating temperature, frequency of stopping and starting or other shock-producing actions, and any unusual conditions must be considered when selecting a lubricant.

Lubricating Grease

Lubricating grease is a semisolid lubricant consisting of a mixture of oil and thickening agents. Common lubricating grease is not suitable for operating temperatures above 325°F or speeds over 12,000 rpm. High-temperature greases that can maintain lubrication properties up to 500°F are available. Grease has a higher viscosity than lubricating oil and is used to provide a thicker lubricant film or sealing action. Grease also provides a barrier to outside contaminants.

OIL GROUPS AND APPLICATIONS	
Group A: Automotive	**Group B: Gear Trains and Transmission**
• SAE 10 • SAE 20 • SAE 30 • SAE 40 • SAE 50	General-purpose oils
Group C: Machine Tools	**Group D: Marine Propulsions and Stationary Power Turbines**
• SAE 75 • SAE 80 • SAE 90 • SAE 140 • SAE 250	Turbine oils: • Light weight • Medium weight • Heavy weight
Group E: Turbojet Engines	**Group F: Reciprocating Engines**
Aviation oils	Aviation oils

Figure 9-2. Commercial lubricating oil is classified into Groups A through F.

Thickening agents in grease include soap, synthetic oil, and other organic and inorganic ingredients. **See Figure 9-3.** Thickening agents may have lubricating qualities, but their primary function is storage of lubricating oil. During machine operation, friction and heat cause oil to bleed from the grease at a controlled rate. The *grease dropping point* is the temperature at which the thickening agent of a grease turns to a liquid. Grease is classified by thickener.

GREASE THICKENERS	
Thickener	**Maximum Usable Temperature***
Calcium, water-stabilized	175
Aluminum	175
Calcium, anhydrous	250
Sodium	250
Lithium	325
Complex soap	400
Nonsoap polyurea	400
Nonsoap organoclay	400

* in °F

Figure 9-3. Lubricating grease is a semisolid lubricant consisting of a mixture of oil and thickening agents.

Dry Lubricants

Dry lubricant is a lubricant in non-liquid form. Dry lubricants include silicone, graphite, molybdenum disulfide, and solid fluorocarbons. **See Figure 9-4.** Dry lubricants are used in extreme operating conditions. Extreme operating conditions include corrosive environments, extreme temperatures, excessive loads and weights, high speeds, vacuums, and applications that receive infrequent lubrication. Some dry lubricants provide excellent load-bearing capacities while others perform better at low temperatures. The selection of dry lubricant is determined by the application.

Lubricant Contamination

Lubricant contamination is the main cause of mechanical system failure. Dirt and other abrasive materials that contaminate a lubricant cause wear between moving components and bearing surfaces as the equipment operates. Lubricants must remain free of contaminants because of the microscopic thickness of the lubricating film. A maintenance program should be used to provide a routine schedule for lubricant changes and/or filtering. Some mechanical systems use oil purifiers to clean and recycle lubricating oil while the equipment is operating. Oil purifiers can reduce maintenance, lubrication, and disposal costs.

DRY LUBRICANT SELECTION FOR APPLICATION CONDITIONS								
Lubricant	Dirty, Dusty	Heavy Loads	Vacuum	Corrosive	Wet	Close Tolerance	Temperature Range*	Speed Range†
Graphite	Good	Good	Poor	Good	Good	Good	−40 to 750	<500 to >2000
Silicone	Poor	Poor	Poor	Good	Good	Good	−40 to 400	<500 to >2000
Fluorocarbons (Teflon)	Good	Poor	Good	Excellent	Excellent	Good	0 to 600	<500 to >2000
Molybdenum Disulfide (Moly)	Excellent	Excellent	Excellent	Good	Good	Excellent	−100 to 600	<500 to >2000

* in °F
† in rpm

Figure 9-4. Dry lubricants are used in extreme operating conditions.

Lubricants should be kept as clean as possible. Store lubricants in a clean, dust-free cabinet. Use only clean containers with lids when carrying lubricants from storage to the point of use. Use a clean, lint-free cloth when checking lubricant levels. Never mix funnels or oils. When checking lubricant level or adding lubricant to a system, always clean the area around the dipstick and filler cap before removal.

BEARINGS

A *bearing* is a machine component used to reduce friction and maintain clearance between stationary and moving parts. Bearings may be subjected to radial loads, axial (thrust) loads, or a combination of radial and axial loads. **See Figure 9-5.** A *radial load* is a load applied perpendicular to the rotating shaft. An *axial load* is a load applied parallel to the rotating shaft.

Bearings are classified as friction or antifriction bearings. A *friction bearing* is a bearing consisting of a stationary bearing surface, such as machined metal or pressed-in bushings, that provides a low-friction support surface for rotating or sliding surfaces. Friction bearings commonly use lubricating oil to separate the moving component from the stationary bearing surface. Friction bearing surfaces normally consist of a material that is softer than the supported component. This allows foreign matter to embed in the bearing material, preventing it from spreading throughout the lubrication system. Friction bearings can conform to slightly irregular mating surfaces. Friction bearings may be integrally machined, one-piece sleeve, or split-sleeve

for easy installation and removal. Split-sleeve bearings are commonly used by engine manufacturers.

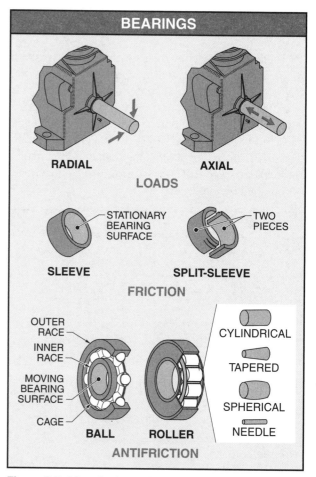

Figure 9-5. A bearing is used to reduce friction and maintain clearance between stationary and moving components.

Friction bearings are normally manufactured from nonferrous metals such as bronze, aluminum, and babbitt. A *nonferrous metal* is a metal that does not contain iron. Babbitt is commonly used as a coating on the load-bearing surface of steel split-sleeve bearings.

An *antifriction bearing* is a bearing that contains moving elements, which provide a low-friction support surface for rotating or sliding surfaces. Antifriction bearings are normally manufactured with hardened rolling elements and races. The rolling elements may be balls or rollers. The rollers may be cylindrical, tapered, spherical, or needle. A *race* is the bearing surface of an antifriction bearing that supports the rolling elements. A *cage* is an antifriction bearing component used to maintain the position and alignment of the rolling elements. Antifriction bearings reduce lubrication requirements and decrease starting and operating friction. Reduced friction results in less power required to rotate mechanical components and an increase in overall efficiency of the machine. Antifriction bearings are available separately or pre-installed in a housing. Common housings include pillow blocks, flanges, adjustable flanges, rubber cartridges, take-up units, and hanger units.

A *linear bearing* is a bearing designed to provide low-friction movement of a mechanical device that moves in a straight line. A linear bearing contains several holes drilled in the shape of a loop. **See Figure 9-6.** One side of each loop is exposed on the inside diameter of the linear bearing. Hardened steel balls fill each loop. Half of the steel balls are exposed at the inside diameter of the bearing. The other half are located inside the bearing and are not exposed.

As a shaft slides through the bearing or the bearing moves across the shaft, the exposed balls on the inside diameter of the bearing are pushed toward one end of the loop. As each ball reaches the end of the loop, it is pushed below the surface of the inside diameter, losing contact with the shaft. The balls below the surface move in the opposite direction of the shaft and return to their starting point where the cycle begins again.

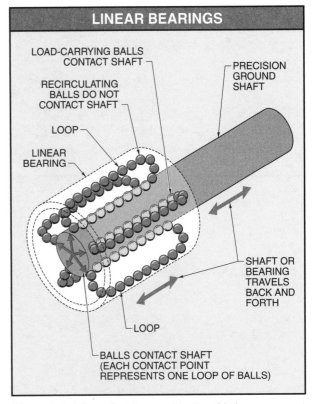

Figure 9-6. A linear bearing provides low-friction movement of a mechanical device that moves in a straight line.

Bearing Lubrication Methods

Friction and antifriction bearings normally require periodic lubrication to prevent premature failure. Sealed bearings do not require lubrication. A *sealed bearing* is an antifriction bearing that is completely sealed so the lubricant stays inside the bearing and dirt is kept out.

Many bearings are fitted with a shield that helps to contain the grease inside the bearing. Shielded bearings require regular lubrication and dirt can enter the bearing by going around the shield. Devices used for lubricating bearings include grease fittings, pressure cups, oil cups, and oil wicks. **See Figure 9-7.**

A *grease fitting* is a hollow tubular fitting used to direct grease to bearing components. The head of the grease fitting is designed to open when a grease gun is attached and close when it is removed. A *grease gun* is a lubrication tool that is attached to a grease fitting to provide grease to a bearing. Grease fittings are available with different head designs and configurations to provide easy access during lubrication. Grease fittings are attached to a bearing housing by a pipe thread. The pipe thread provides a sealed connection between the bearing housing and the grease fitting.

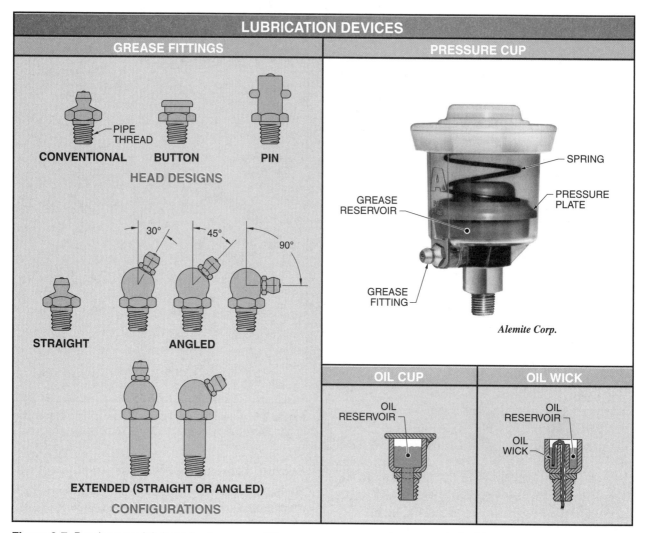

Figure 9-7. Bearings are lubricated using grease fittings, pressure cups, oil cups, and oil wicks.

A *pressure cup* is a pressurized grease reservoir that provides constant lubrication to a bearing. A pressure cup consists of a grease reservoir with an internal spring and pressure plate. As grease is pumped into the reservoir, the spring is compressed and the pressure plate is pushed toward the top of the cup. During bearing movement, grease is forced into the bearing components by pressure from the compressed spring.

An *oil cup* is an oil reservoir located on a bearing housing to provide lubrication to a bearing. Oil cups rely on gravity to feed the lubricant through a direct passage to the bearing. Lubrication continues as long as there is a supply of oil in the reservoir. Some oil cups have an oil wick. An *oil wick* is an absorbent material which serves as a conduit for oil from the oil cup to the bearing surface. The flow of oil from the cup is regulated by the maximum flow allowed through

the oil wick material. This provides a controlled flow of lubricant to bearing components over an extended period of time. To prevent contaminants from entering the lubrication system, always use clean tools and ensure grease fittings or oil cups are clean before adding the lubricant.

OIL SEALS

An *oil seal* is a device used to contain oil inside a housing. Oil seals are used to prevent oil from leaking out through the hole in a housing where the shaft protrudes. **See Figure 9-8.** Oil seals extend the life of a mechanical device by retaining lubricant and protecting it from foreign material contamination and moisture. Oil seal components include the outer case, inner case, lip, and spring.

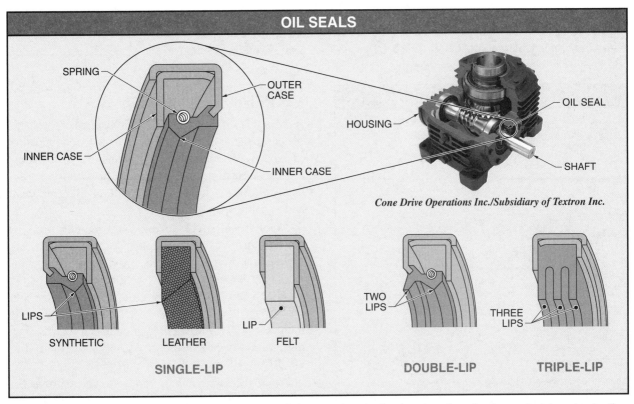

Figure 9-8. Oil seals extend bearing life by retaining lubricant and protecting bearings from foreign material contamination and moisture.

The outer case of an oil seal is made of metal and provides structural support for the oil seal. The diameter of the outer case is slightly larger than the bore into which the oil seal fits. This size difference provides a tight fit in the housing. The inner case is used with larger diameter seals to provide extra rigidity of the oil seal, but provides no sealing capabilities. An *oil seal lip* is an oil seal component that contacts the moving part of the equipment to prevent material from passing by the oil seal. The lip is the main component of an oil seal. The lip provides the actual seal to the shaft. Some seals use a spring to provide extra contact pressure to the shaft.

Oil seal lips are made from synthetic materials, leather, or felt. Most general-purpose seals have a synthetic lip molded directly to the metal case on the inside or outside diameter. Synthetic general-purpose seals may have one, two, or three sealing lips, depending on the application. The flexible synthetic lip may be spring-loaded or springless. The lip is designed to work like a squeegee, wiping oil from the shaft and preventing it from escaping. The lip also prevents contaminants from entering the bearing. The spring-less single-lip seal is considered a grease seal because

it retains thick lubricants but leaks if used with thin lubricants and high pressures. Oil seals located in dirty environments, such as in agricultural or construction equipment, require double or triple lips to provide extra protection for the bearing.

Oil seals are commonly installed next to a bearing on the outside of a housing. **See Figure 9-9.** The oil seal outside diameter is usually larger than the bearing outside diameter, allowing the seal to seat against a recess when it is pressed into the housing. The recess provides a press fit for the seal and also keeps the seal from touching the bearing.

Oil Seal Installation

Correct installation of an oil seal ensures proper performance. The lip of an oil seal must be lubricated before installing the seal. The same oil used in the equipment is used to lubricate the lip to prevent mixing of oils. Lubricating the lip allows it to slide during installation, preventing seal damage. When lubricating an oil seal lip, ensure no lubricant contacts the outside diameter of the seal or the housing bore. The seal outside diameter and housing bore must remain clean and dry, providing a leak-proof fit. The oil seal outside diameter should also be coated with a bore sealant before installation. This helps ensure a leak-proof fit between the seal and housing.

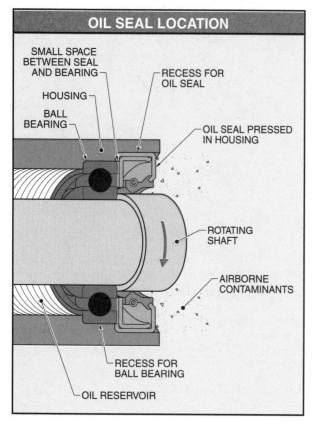

OIL SEAL LOCATION

SMALL SPACE BETWEEN SEAL AND BEARING

HOUSING

BALL BEARING

RECESS FOR OIL SEAL

OIL SEAL PRESSED IN HOUSING

ROTATING SHAFT

AIRBORNE CONTAMINANTS

RECESS FOR BALL BEARING

OIL RESERVOIR

Figure 9-9. Oil seals are commonly installed next to a bearing on the outside of a housing.

Never hammer an oil seal to install it. Hammering an oil seal normally results in a leaky seal. Special tools should be used to install oil seals. Oil seal installation tools apply force evenly and at the correct location on the seal to ensure proper installation. **See Figure 9-10.** A variety of installation tools are available for specific applications, such as proper positioning, installation when no shaft is present, and installation over a shaft. A piece of tubing (steel or heavy-wall plastic) may also be used as an oil seal installation tool. Ensure the end of the tubing is smooth, square, and the correct diameter when used for oil seal installation.

An oil seal lip must also be protected when being installed over a shaft. Sharp edges on a shaft damage the lip and must be removed. A cone-shaped installation tool, a sleeve with smooth rounded corners, or piece of shim stock rolled in the shape of a cone can be used when installing a seal over the end of a shaft. Keyways

on shafts can also damage oil seals. A sleeve is commonly used to cover the keyed area protecting the seal as it is slid on the shaft.

MECHANICAL DRIVE SYSTEMS

A *mechanical drive system* is a combination of mechanical components that transfer power from one location to another. Mechanical drive systems may also change the size, direction, and speed of the applied force. Mechanical drive systems include belt, chain, gear, cam, and variable-speed drives.

Mechanical drive systems are synchronous or non-synchronous drive systems. A *synchronous drive system* is a drive system that provides a positive engagement between the drive and driven sides of the system. There is no slippage between the drive and driven parts of the machine. Synchronous drive systems include chain, timing belt, gear, and shaft drives. Any motion that occurs at the drive side is transferred to the driven side.

A *non-synchronous drive system* is a drive system that does not provide positive engagement between the drive and driven sides of the system. Non-synchronous drive systems include flat belts, V-belts, and round belts. Non-synchronous drive systems slip if resistance to movement on the driven side exceeds the friction of the belt.

Belt Drives

A *belt drive* is a mechanical drive system that uses a belt and pulleys to transfer power between two shafts. Belt drives are one of the most common mechanical drive systems used in industry. Belt drives are relatively inexpensive, quiet, easy to maintain, and provide a wide range of speed and power.

A *belt* is a flexible loop that connects two or more pulleys for transferring rotational power. Belts are made from rubber, leather, canvas, or synthetic material. Synthetic or rubber belts consist of a synthetic or rubber core containing tensile members surrounded by a reinforced outer envelope. A *tensile member* is cording material that runs the entire length of the belt, increasing the tensile strength of the belt. The outer envelope is usually reinforced and is capable of resisting wear and abrasion. Common materials used for the core of a belt include natural rubber, neoprene, or polyurethane. The material used for the belt is normally determined by the application and its surrounding environment.

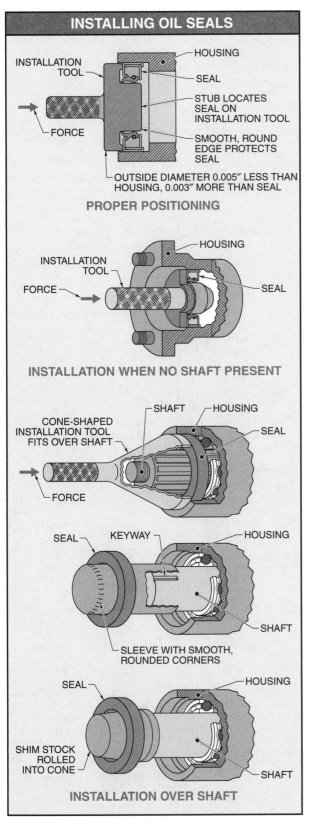

Figure 9-10. Oil seal installation tools apply force evenly and at the correct location on the seal, ensuring proper installation.

Characteristics of belt drives include belt tension, creep, and slip. Belt tension includes tight-side, slack-side, and centrifugal. **See Figure 9-11.** *Tight-side tension* is the tension on a belt when it is approaching the drive pulley. *Slack-side tension* is the tension on a belt when it is approaching the driven pulley. The total tension on the belt is equal to the tight-side tension minus the slack-side tension. *Centrifugal tension* is the tension needed to offset the centrifugal force on the belt as it engages the pulley. Centrifugal force pushes the belt away from the pulley, reducing contact with the pulley. Contact with the pulley is reduced as the speed of the belt increases due to increased centrifugal force.

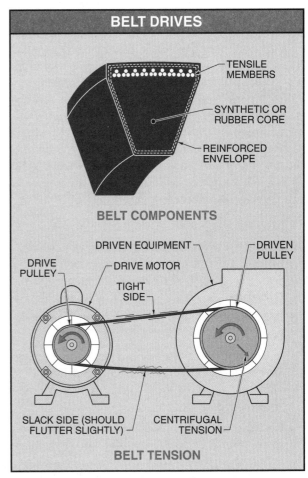

Figure 9-11. A belt drive uses a belt to transfer power between the drive pulley and the driven pulley.

✚ SAFETY TIP

Always keep drives properly guarded. Every belt drive must be guarded when in operation. The guard must be designed and installed according to OSHA standards.

Belt creep is the natural movement of the belt on the face of the pulley when it is subjected to changes in tension. For example, belt creep occurs at startup or when the load is increased or decreased due to the change in belt tension at these times.

Slip is the movement of a belt on the face of the pulley when belt tension is too loose. Slip occurs when the tension on a belt is insufficient and the belt slides on the pulley. Slip can be eliminated by increasing belt tension. Slip reduces pulley speed and causes premature wear on the belt and pulley.

All belts must operate at the proper tension. The belt slips if the tension is too low, causing rapid wear of the belt and pulley. The belt overheats if the tension is too high, resulting in rapid wear and increased load on the motor and bearings. The proper tension is just tight enough to prevent slippage at full load and is determined by the belt type, pulley size and speed, and drive ratio. The proper belt tension should result in approximately $\frac{1}{64}''$ of belt deflection for every inch of span between pulley centers. Belt manufacturers also specify belt tension requirements, which can be checked with a tensiometer. A *tensiometer* is a device that measures the amount of deflection of a belt. **See Figure 9-12.** Proper belt tension is determined using a tensiometer by applying the procedure:

1. Measure span length (t).
2. Slide the lower O-ring to the position equal to $\frac{1}{64}''$ per inch of span length on the deflection distance scale. *Note:* Read scale at the bottom edge of the O-ring.
3. Position a straightedge even with the top of the belt.
4. Apply force with the tensiometer at the center of the span with enough force to deflect the belt until the bottom edge of the lower O-ring is even with the straightedge.
5. Locate amount of deflection force on the deflection force scale. The upper O-ring slides down the scale as the tensiometer is compressed and remains there when force is released. *Note:* Read scale at the top edge of the O-ring.
6. Compare the deflection force with the range in the recommended belt deflection force table. Tighten the belt if force is less than the minimum recommended deflection force. Loosen the belt if force is greater than the maximum recommended deflection force.

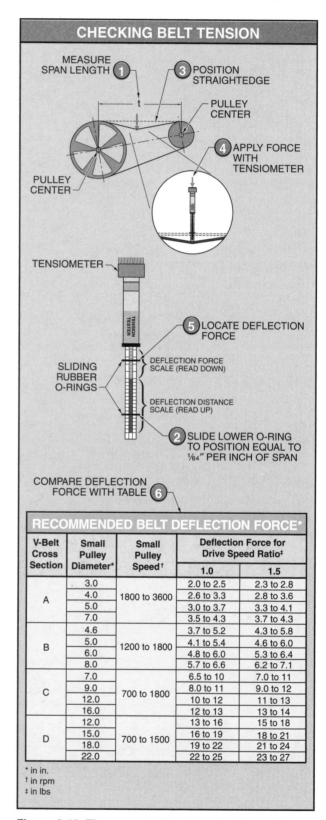

CHECKING BELT TENSION

RECOMMENDED BELT DEFLECTION FORCE*

V-Belt Cross Section	Small Pulley Diameter*	Small Pulley Speed[†]	Deflection Force for Drive Speed Ratio[‡]	
			1.0	1.5
A	3.0	1800 to 3600	2.0 to 2.5	2.3 to 2.8
	4.0		2.6 to 3.3	2.8 to 3.6
	5.0		3.0 to 3.7	3.3 to 4.1
	7.0		3.5 to 4.3	3.7 to 4.3
B	4.6	1200 to 1800	3.7 to 5.2	4.3 to 5.8
	5.0		4.1 to 5.4	4.6 to 6.0
	6.0		4.8 to 6.0	5.3 to 6.4
	8.0		5.7 to 6.6	6.2 to 7.1
C	7.0	700 to 1800	6.5 to 10	7.0 to 11
	9.0		8.0 to 11	9.0 to 12
	12.0		10 to 12	11 to 13
	16.0		12 to 13	13 to 14
D	12.0	700 to 1500	13 to 16	15 to 18
	15.0		16 to 19	18 to 21
	18.0		19 to 22	21 to 24
	22.0		22 to 25	23 to 27

* in in.
[†] in rpm
[‡] in lbs

Figure 9-12. The proper belt tension should result in approximately $\frac{1}{64}''$ of belt deflection for every inch of span between pulley centers.

Belt operation also provides clues to proper belt tension. For example, excessive fluttering of a belt between pulleys indicates that belt tension may be too low. Maintenance personnel often judge the tension of a belt by feel. The tight side of the belt should feel springy, not excessively tight or excessively loose.

Belts commonly used in industry include flat, V-, and timing (synchronous) belts. **See Figure 9-13.**

Protecting the belts from the surrounding environment is a high preventive maintenance priority. A protective cover should be designed to keep objects and foreign substances such as grease, oil, and dirt from contacting the belt or pulleys. Belts should operate in a cool, dry environment. High humidity and temperatures may cause premature belt failure.

Keeping belts clean is also a critical preventive maintenance concern. Foreign material on a belt causes glazing. *Glazing* is a slick polished surface caused by dirt and other debris being rubbed on the surface of a belt. A glazed belt has reduced friction with the pulleys, resulting in belt slippage and a loss of power transmission. Glazed belts should be replaced and the pulleys inspected for possible damage.

Fenner Drives

Link-type V-belts perform well in extremely hostile environments and can be adjusted to the length required for the application.

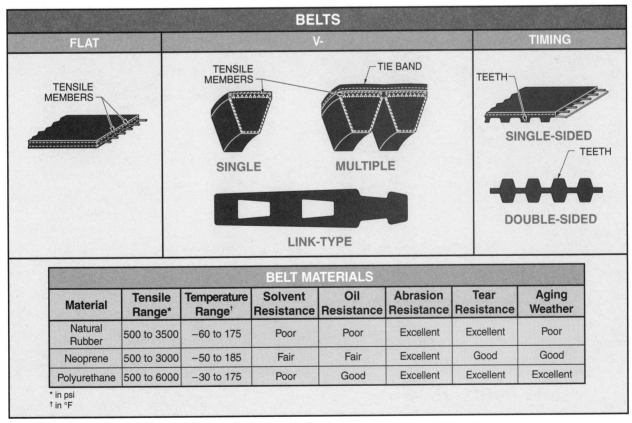

BELT MATERIALS							
Material	Tensile Range*	Temperature Range†	Solvent Resistance	Oil Resistance	Abrasion Resistance	Tear Resistance	Aging Weather
Natural Rubber	500 to 3500	−60 to 175	Poor	Poor	Excellent	Excellent	Poor
Neoprene	500 to 3000	−50 to 185	Fair	Fair	Excellent	Good	Good
Polyurethane	500 to 6000	−30 to 175	Poor	Good	Excellent	Excellent	Excellent

* in psi
† in °F

Figure 9-13. Belts used in industry include flat, V-, and timing belts.

Flat Belts. A *flat belt* is a belt that has a rectangular cross-section. Flat belts are commonly used in applications that require high speeds, low noise, and small pulley diameters. Some flat belts can operate at pulley speeds of 140,000 rpm. Flat belts are used to drive a wide range of industrial machinery and vary in width and thickness depending on the power requirements.

Flat belt material includes leather, rubber, canvas, and mylar. Leather belts are made from animal hide cut from the center part of the hide. Rubber belts are made by vulcanizing rubber around tensile members. Rubber belts are used in environments where leather deteriorates quickly. Canvas belts are used in environments where rubber and leather fail prematurely. Mylar belts are considerably thinner than other flat belts and are used for miniature drive systems, such as tape recording equipment. Mylar belts may be as thin as 0.0005″, allowing the use of small diameter pulleys (0.050″ diameter). Preventive maintenance of flat belts includes protecting the belts and keeping them tight and clean. Flat belt tension is tested using a tensiometer.

A *conveyor belt* is an extra wide flat belt made from rubber or synthetic material used for a conveyor system. A *conveyor system* is a system used to transport material from one location to another. Conveyor systems consist of a conveyor belt, pulleys, and rollers. Conveyors may be direct drive or roller drive. **See Figure 9-14.**

A *direct drive conveyor* is a conveyor that has the material in direct contact with the belt. A *roller drive conveyor* is a conveyor that has the material sitting on rollers that are rotated by the belt. A *roller* is a cylindrical device used to guide and support the conveyor belt. The roller does not drive the belt. The roller drive method is used for heavy loads.

Conveyors have a head end and a tail end. The *head end* is the end where the material exits the conveyor. The *tail end* is the end where the material enters the conveyor. A drive unit is connected to either the tail-end pulley or the head-end pulley. A *conveyor pulley* is a long cylinder that supports one end of a conveyor belt. One pulley is attached to a motor that drives the conveyor belt.

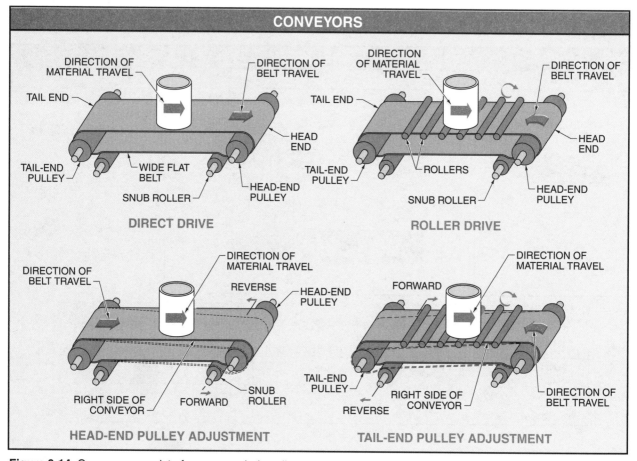

Figure 9-14. Conveyors consist of a conveyor belt, pulleys, and rollers, and may be direct drive or roller drive.

A *snub roller* is a roller used to guide (track) the conveyor belt on the conveyor pulleys. Proper belt tracking depends on the alignment of the pulleys and rollers. During operation, the conveyor belt should track in the center of the pulleys and rollers. When aligning a belt, all adjustments should be slight with time allowed for the belt to react to the adjustment. Belt adjustments may be made at the head-end snub roller or the tail-end snub roller. If a belt drifts to the right side of the head-end pulley during forward material travel, adjust the right side of the head-end snub roller in the forward direction of material travel and/or the left side of the head-end snub roller in the reverse direction of material travel. Belt adjustments may also be made at the tail-end roller. If the belt drifts to the right on the tail-end roller during forward material travel, adjust the right side of the tail-end snub roller in the reverse direction of material travel and/or the left side of the tail-end snub roller in the forward direction of material travel.

On some smaller conveyors, snub rollers are not used and belt adjustments are made by adjusting the pulleys. Belt adjustments may be done either manually or automatically. Manual adjustments can be time-consuming and are a common maintenance activity. Automatic tracking is accomplished by using a mechanical or electronic sensor along with a piston that adjusts the snub roller to maintain proper tracking. The sensor analyzes the location of the conveyor belt and actuates a piston to increase or decrease tension whenever necessary. Conveyor belt tension must be set correctly to prevent slippage, belt stretching, and excessive loading of the conveyor bearings.

V-Belts. A *V-belt* is a drive belt made from rubber or synthetic material that has a cross-section in the shape of a V. V-belts are the most frequently used belt for belt drive systems. Many flat belt systems have been replaced by V-belt systems. Compared to flat belts, V-belts require less tension, tolerate some misalignment, run with very little noise, and require less maintenance. The lower operating tension required for V-belts reduces bearing loads. V-belts have tensile members that run the length of the belt. The tensile members are encased in a rubber or synthetic material. The tensile members provide tensile strength for the belt.

V-belts transmit power to and from a pulley by contact between the sides of the belt and the pulley. V-belts should never run in contact with the bottom of the pulley. V-belts are classified as fractional horsepower (FHP), standard, and narrow and are available in a variety of lengths. Proper V-belt selection is determined by pulley diameter, groove angle, motor frame number, pulley center distance, and belt cross-section.

A *fractional horsepower (FHP) belt* is a V-belt designed for light-duty applications. FHP V-belts are often referred to as light-duty V-belts. FHP V-belts are used on mechanical drive systems that range from 3 HP to 17 HP. FHP V-belts are made from lightweight materials, allowing them to bend over small-diameter pulleys. Some small FHP V-belts are notched part way through the cross-section, allowing more flexibility in the belt.

FHP belt size is determined by the length of the belt (10″ to 100″) and the distance across the top of the belt. Different numbers and letters are used to designate V-belt sizes and shapes. Numbers are commonly used to identify the distance across the top of a belt. Each unit represents $\frac{1}{8}$″. For example, a number 3 belt measures $\frac{3}{8}$″ across the top. An FHP V-belt may be labeled 5L-25. The 5 indicates the belt is $\frac{5}{8}$″ across the top, the letter L indicates light-duty, and 25 indicates that the outside length of the belt measures 25″.

A *standard belt* is a V-belt constructed from multiple cords, providing added strength. Standard V-belts are available in five different cross-sectional sizes, indicated by the letters A through E. Standard belts range from 25″ to 660″ in length. For example, a standard V-belt may be labeled C105. The C indicates a cross-sectional measurement of $\frac{7}{8}$″ and the 105 is a nominal size designation that indicates a length of 107.9″.

A *narrow belt* is a V-belt having a smaller cross-section and a higher profile than a standard belt. The shape of a narrow belt provides more surface contact with the pulleys allowing the belt to transmit large amounts of power with a small cross-section. Narrow V-belts are often referred to as heavy-duty belts. Narrow V-belts require high operating tension and are sized by the same system as FHP belts. Cross-sectional sizes include 3, 5, and 8. For example, a narrow V-belt may be labeled 3VX300. The 3 indicates a cross-section of $\frac{3}{8}$″, the VX indicates a narrow V-belt, and the 300 indicates an outside length of 30″.

Some industrial applications use a multiple-belt drive system to transfer large amounts of horsepower. The set of V-belts must be matched and have a consistent length and cross-section. If the belts are different sizes, one belt carries the majority of the load, resulting in premature failure of the belt and an overall loss of power transfer. All belts should be replaced when any one belt in a multiple-belt drive is replaced. Newer belts are not the same size as used belts because they have not been stretched during use.

Multiple V-belts are also available with a common top (tie band) to provide a single V-belt with multiple strands. The tie band contains reinforcement materials. A *V-belt strand* is one V-shaped cross-section. Multiple V-belts may have two to five strands. Multiple V-belts may have large or small cross-sections and are used to transfer large amounts of horsepower.

A *link-type V-belt* is a V-belt consisting of individual connected links. Link-type V-belts were initially designed for use as temporary belts. Their advantages include inventory reduction, easy installation, reduced vibration, and excellent resistance to the elements. These advantages have allowed them to replace traditional V-belts in some applications. The length of the belt is determined by the number of links that are connected. This eliminates the need for large belt inventories because various belt lengths can be achieved simply by adding or removing links. The design provides easy and quick installation. The disassembly of drive components is not necessary because the belt twists together. The belts fit in existing pulleys and require no special tools for installation.

Link-type V-belts reduce the vibration generated during operation. Conventional V-belts use continuous tension cords that transmit vibrations from the drive system to the driven equipment. The method of construction of link-type V-belts reduces the transmission of vibrations, resulting in smoother torque at the driven shaft. Link-type V-belts are made from a polyester/polyurethane composite. This material outperforms traditional V-belt material when exposed to water-based lubricants, cutting oils, and extreme temperatures.

Timing Belts. A *timing belt* is a flat belt containing gear teeth. Timing belts are used for synchronous drive systems. Timing belts provide a drive system that has no slippage or creep, does not stretch, needs no lubricant, requires low belt tension, and has very little backlash. Timing belts can provide 10% more efficiency than V-belts by combining the advantages of a flat belt drive system and a positive synchronous drive system. Timing belt selection is determined by horsepower and pulley size. Timing belt teeth are evenly spaced, providing timing and positive engagement with the mating pulley.

Belt Removal/Installation

V-belts and timing belts are removed when worn or according to a preventive maintenance schedule. **See Figure 9-15.**

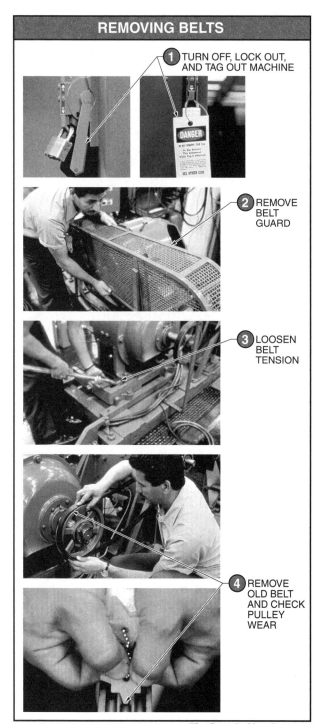

The Gates Rubber Company

Figure 9-15. V-belts or timing belts are removed when worn or according to a preventive maintenance schedule.

> **TECH TIP**
> *Abnormal wear on the bottom surface of a belt is caused by a belt that is too small for the pulleys or by worn or dirty pulleys.*

V-belts and timing belts are removed by applying the following procedure:

1. Turn OFF, lock out, and tag out machine.

2. Remove the belt guard.

3. Loosen belt tension. Motor bases normally have slotted holes to allow the motor to be moved closer or farther away from the driven pulley. To loosen the belt, loosen the motor base bolts and pry the motor toward the driven pulley using a pry bar. If the drive system has an idler instead of an adjustable motor base, loosen the idler bolts and swing the bracket out of the way.

4. Remove the old belt from the pulley. Do not pry the belt or roll it off the pulley. This can cause equipment damage or personnel injury. After the belt is removed, check pulley wear.

V-belt replacement, whether for preventive maintenance or equipment breakdown, begins with proper identification and sizing of the belt being replaced. Premature belt failure is prevented by selecting the proper belt and belt size, and by following the proper installation procedure. **See Figure 9-16.** V-belts and timing belts are installed by applying the procedure:

1. Select proper belt.

2. Move pulleys to the required center-to-center distance and slide the new belt over the drive and driven pulleys. Never pry the belt over the pulleys. Prying a belt damages internal tension members and weakens the belt.

3. Move the motor back using a pry bar until the belt has sufficient tension.

4. Check belt tension. Adjust motor position if necessary.

5. Check pulley alignment using a straightedge or string.

6. Replace guard and remove lockout/tagout. *Note:* Check the belt tension within 24 hr to 48 hr.

Chain Drives

A *chain drive* is a synchronous mechanical drive system that uses a chain to transfer power from one sprocket to another. A *chain* is a series of interconnected links that form a loop. A *sprocket* is a wheel with evenly spaced teeth located around the perimeter that engage a chain. The chain meshes with the teeth on the sprocket, creating a positive drive system. Chain types includes roller, silent, and ladder chain.

TECH TIP

Belts used in belt drive systems should be replaced if there are signs of cracking, fraying, unusual wear, or loss of teeth in a timing belt.

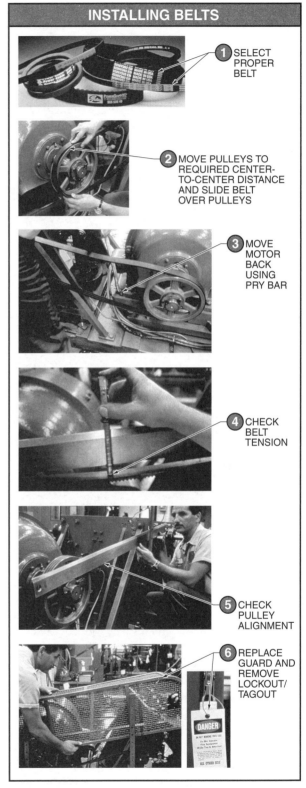

INSTALLING BELTS

1 SELECT PROPER BELT

2 MOVE PULLEYS TO REQUIRED CENTER-TO-CENTER DISTANCE AND SLIDE BELT OVER PULLEYS

3 MOVE MOTOR BACK USING PRY BAR

4 CHECK BELT TENSION

5 CHECK PULLEY ALIGNMENT

6 REPLACE GUARD AND REMOVE LOCKOUT/TAGOUT

The Gates Rubber Company

Figure 9-16. Maintenance personnel can prevent premature belt failure by selecting the proper belt and by following the proper installation procedure.

Roller Chain. *Roller chain* is a synchronous chain that contains roller, pin, and connecting (master) links. Offset links may also be used on roller chain. **See Figure 9-17.** A *roller link* is a chain link that consists of two bushings placed inside two rollers that are pressed into two side bars. A *side bar* is a steel plate with two precision holes used to connect two pins or two bushings. When the roller link is pressed together, the rollers are free to spin around the bushings, providing a pivot point.

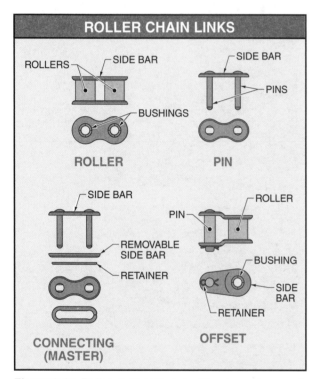

Figure 9-17. Roller chain includes roller, pin, and connecting (master) links.

A *pin link* is a chain link that consists of two steel pins pressed into two side bars with matching holes. Pin links are used to join roller links. The width of a pin link is greater than the width of a roller link. This enables the pin link side bars to fit on the outside of the roller link side bars. Assembly of the roller chain requires sliding one pin of a pin link through one roller of a roller link and the other pin through one roller of another roller link. The link is made by pressing on the second side bar of the pin link. Pin and roller links are alternately joined until the desired chain length is achieved.

A *connecting (master) link* is a modified pin link that has a removable side bar and retainer. The retainer snaps on the pin and outside the side bar to hold the side bar in place. Connecting links are used to connect

two ends of a chain, making a complete loop. An *offset link* is a chain link that is used to shorten or lengthen a chain and to connect two ends of a chain. The length of an offset link is half the length of a pin, roller, or connecting link. An offset link should only be used for short length adjustments because it is weaker than other chain links. Roller chain must be lubricated because it wears quickly. Roller chain is available in multiple sizes and strands. Roller chain includes transmission (common), conveyor, and drive chain.

A sprocket has evenly spaced teeth located around its perimeter. The teeth of a sprocket provide positive engagement with a chain transferring force from one sprocket to another. Sprockets are classified as A, B, and C sprockets. **See Figure 9-18.** An A sprocket has no hub, so it is usually mounted on a hub or flange with mechanical fasteners. A B sprocket has an integrated hub on one side of the sprocket. A C sprocket has an integrated hub on both sides of the sprocket. B and C sprockets are commonly attached to a shaft using a key and two setscrews placed 90° apart. The key fits in a keyway located in the shaft and sprocket. One setscrew is located so it contacts the key when the sprocket is in place. The size of the sprocket is determined by the required power and speed of the drive system.

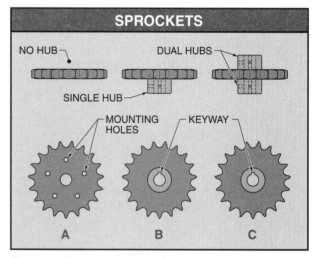

Figure 9-18. A sprocket provides positive engagement with a chain transferring force from one sprocket to another.

Roller chain is sized by the pitch, roller width, and roller diameter. **See Figure 9-19.** *Pitch* is the distance between corresponding points on an adjacent pair of evenly spaced projections. A coding system located on the side of the chain is used to identify chain size. The first digit indicates the pitch of the chain in eighths of an inch. For example, a number 50 chain indicates that the pitch is $^5/_8''$. The second digit represents the type of chain. A 0 indicates a standard roller chain, a 5 indicates a rollerless chain, and a 1 indicates that the chain is a light-duty chain. Chain and sprockets must match.

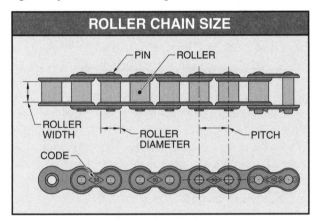

Figure 9-19. Roller chain is sized by the pitch, roller width, and roller diameter.

Silent Chain. *Silent chain* is a synchronous chain that consists of a series of links joined together with bushings and pins. The bushing and pin arrangement provides a pivot point for the silent chain. The links are flat on one side and have teeth on the other. Tooth width is determined by the number of plates. Silent chain sprockets mesh with the chain teeth, providing a positive drive system. **See Figure 9-20.**

A silent chain sprocket has a groove cut in the center of the teeth. The groove prevents the silent chain from moving laterally on the sprocket during operation by controlling the position of the guide link. A *guide link* is the center link of a silent chain that has no teeth. The guide link fits into the groove on the sprocket and keeps the chain properly aligned in the sprocket. Silent chain is more expensive than roller chain but has higher operating speeds, is more efficient, runs quieter and smoother, and provides longer life.

Ladder Chain. *Ladder chain* is a light-duty synchronous chain that is commonly used for actuating control functions of equipment, such as speed control. Ladder chain consists of a series of links that engage to teeth on a sprocket, providing a positive drive system. **See Figure 9-21.** An advantage of a ladder chain is that it does not require lubricant because it runs at very low speeds and usually operates intermittently. Ladder chain links are made from precision bent wire with a loop on each end to provide a method for connecting other links.

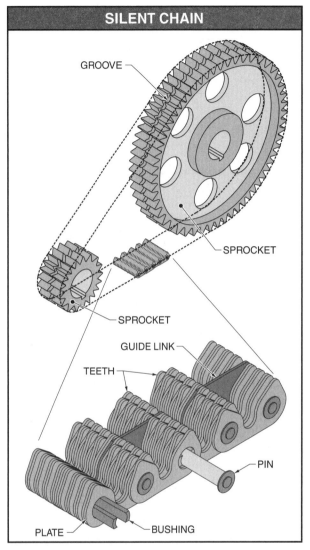

Figure 9-20. Silent chain consists of a series of links that are flat on the top and have teeth on the bottom.

The Gates Rubber Company
Change all belts on a multiple-belt drive even when only one belt requires replacement. Otherwise, the new belt carries more of the load, which shortens belt life.

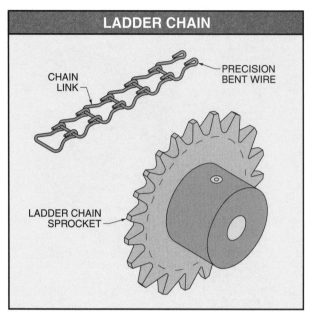

LADDER CHAIN

CHAIN LINK

PRECISION BENT WIRE

LADDER CHAIN SPROCKET

Figure 9-21. Ladder chain consists of a series of links that engage to teeth on a sprocket providing a positive drive system.

Chain Removal/Installation. Chain removal and reinstallation is required when chain becomes worn and can no longer be tensioned correctly. The procedure used for removing and installing a roller chain is similar to that used for belts. **See Figure 9-22.** Roller chain is removed and reinstalled by applying the procedure:

1. Lock out and tag out the equipment.
2. Remove chain guard.
3. Reduce tension on the chain by loosening the motor mounts and moving the motor or releasing tension on the idler sprocket or rub block.
4. Remove connecting link and chain.
5. Inspect the sprockets for wear. Badly worn sprockets should be replaced. New chain installed on worn sprockets wears quickly.
6. Wrap the new chain around each sprocket, ensuring that no links or teeth have been skipped.
7. Insert the connecting link into the two ends of the chain while holding the two ends close together. Install the side bar and retainer.
8. Tension the chain by replacing the idler sprocket or rub block and adjusting the slack or repositioning the motor. Ensure that the sprockets are closely aligned.
9. Replace the chain guard and remove the lockout/tagout.
10. Operate the chain drive to ensure proper operation.

Gear Drives

A *gear drive* is a synchronous mechanical drive system that uses the meshing of two or more gears to transfer motion from one shaft to another. A *gear* is a toothed wheel that meshes with other toothed wheels to transfer rotational power.

Gear drive systems provide positive contact between gears. The meshing of the teeth prevents significant slippage (lost motion). Gear driving and driven shafts must be located close to each other. Long distances between shafts are not economical for gear drive systems. A *pinion gear* is the smaller of two meshing gears. Any two gears that mesh rotate in opposite directions. **See Figure 9-23.** For example, a gear rotating counterclockwise causes the second gear to rotate clockwise. An added third gear rotates in the same direction as the first gear.

The driven gear is forced to move by the drive gear. The driven gear has a load attached to it while the drive gear is often attached to a motor. *Gear ratio* is the ratio of the number of teeth on the drive gear to the number of teeth on the driven gear. Since the gear diameter determines the number of teeth, this ratio works with diameters as well. The gear ratio determines how fast the driven gear rotates in relation to the drive gear. Gears with equal diameters have a 1:1 ratio and rotate at the same rate. A gear ratio of 1:1.6 means that the drive gear makes 1.6 revolutions for each revolution of the driven gear. Therefore, this drive gear also rotates more quickly than the driven gear.

To find gear ratio, the number of teeth of each gear must be known. Gear ratio is found by applying the formula:

$$M = \frac{N_D}{N_d}$$

where

M = gear ratio

N_D = number of drive gear teeth

N_d = number of driven gear teeth

For example, what is the gear ratio if the drive gear has 19 teeth and the driven gear has 38 teeth?

$$M = \frac{N_D}{N_d}$$

$$M = \frac{19}{38}$$

$$M = \textbf{1:2}$$

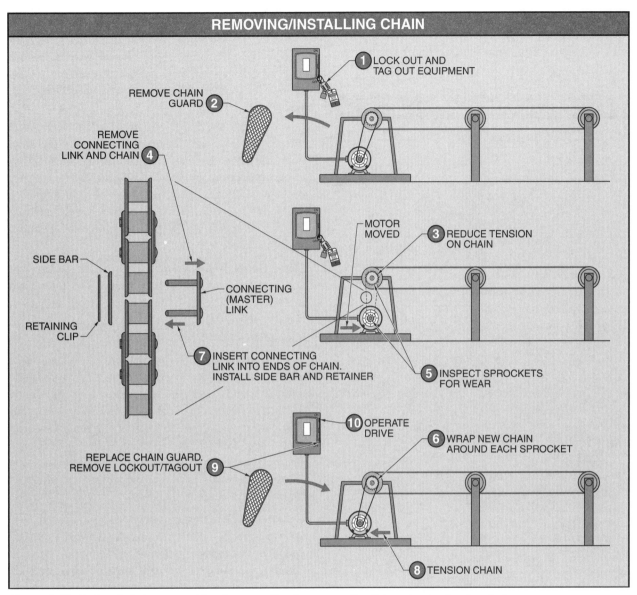

REMOVING/INSTALLING CHAIN

1 LOCK OUT AND TAG OUT EQUIPMENT

REMOVE CHAIN GUARD 2

REMOVE CONNECTING LINK AND CHAIN 4

SIDE BAR

RETAINING CLIP

CONNECTING (MASTER) LINK

7 INSERT CONNECTING LINK INTO ENDS OF CHAIN. INSTALL SIDE BAR AND RETAINER

MOTOR MOVED

3 REDUCE TENSION ON CHAIN

5 INSPECT SPROCKETS FOR WEAR

10 OPERATE DRIVE

6 WRAP NEW CHAIN AROUND EACH SPROCKET

REPLACE CHAIN GUARD. REMOVE LOCKOUT/TAGOUT 9

8 TENSION CHAIN

Figure 9-22. Roller chain is removed by disconnecting the connecting link and removing the chain.

Lubrication of a gear drive system is necessary for the drive system to function properly. The lubricant must provide a protective barrier between the mating gears to eliminate any metal-to-metal contact.

Gears are designed to have slight backlash between meshing teeth for maximum life and efficiency. *Backlash* is the amount of movement (play) between meshing gear teeth. Backlash is required to prevent full contact on both flanks of the teeth. The space created enables the flow of lubricant between the teeth flanks. Inadequate backlash can cause resistance, resulting in overheating or jamming of the meshing gears. Excessive backlash can cause problems if the direction of rotation is reversed frequently. Each time the direction of rotation is reversed, the gear teeth are subjected to excessive impact.

Gears commonly used in industry include spur, helical, herringbone, bevel, miter, and worm gears. **See Figure 9-24.**

Spur Gears. A *spur gear* is a gear with straight teeth cut parallel to the shaft axis. Spur gears are the most common gears used in industry and are used where gears are mounted on parallel shafts. Spur gears may be external, internal, or rack gears. External gears are gears with teeth cut on their outside perimeter. Internal gears are gears in which one gear meshes on the inside circumference of a larger gear. This permits a large ratio of speed reduction in a small space. A *rack gear* is a spur gear with teeth spaced along a straight line. A pinion is used with a rack to convert rotary motion into linear motion.

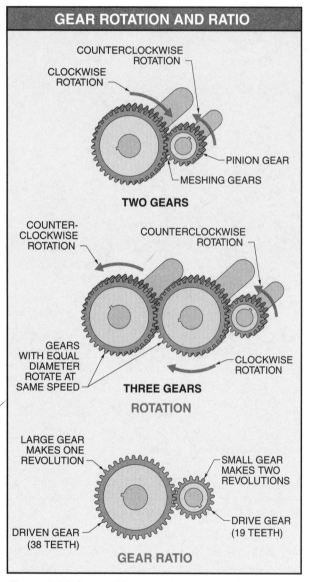

GEAR ROTATION AND RATIO

COUNTERCLOCKWISE ROTATION

CLOCKWISE ROTATION

PINION GEAR

MESHING GEARS

TWO GEARS

COUNTER-CLOCKWISE ROTATION

COUNTERCLOCKWISE ROTATION

GEARS WITH EQUAL DIAMETER ROTATE AT SAME SPEED

CLOCKWISE ROTATION

THREE GEARS

ROTATION

LARGE GEAR MAKES ONE REVOLUTION

SMALL GEAR MAKES TWO REVOLUTIONS

DRIVE GEAR (19 TEETH)

DRIVEN GEAR (38 TEETH)

GEAR RATIO

Figure 9-23. A gear drive uses the meshing of two or more gears to transfer motion from one shaft to another.

Helical Gears. A *helical gear* is a gear with teeth that are cut following a line that spirals around the shaft axis. Helical gears are manufactured in pairs so their helix axes match. Helical gears are commonly used in gearboxes because they provide smoother operation than spur gears. Helical gear teeth mesh with each other in a sliding motion. This results in more tooth contact, but causes thrust in the direction of the axes.

Herringbone Gears. A *herringbone gear* is a gear with two rows of helical teeth. Herringbone gears have parallel shafts and teeth cut at opposite angles. End thrust is avoided by the teeth being at opposite angles. Herringbone gears provide quiet and efficient operation.

Bevel Gears. A *bevel gear* is a gear with straight tapered teeth used in applications where shaft axes intersect. Bevel gears are conical in shape rather than cylindrical. Bevel gears are manufactured in pairs to ensure matching tapers. Bevel gears are primarily used in gearboxes requiring shafts at right angles to each other. A *spiral bevel gear* is a bevel gear that has curved teeth, which provide smoother operation at high speeds.

Miter Gears. A *miter gear* is a bevel gear used at right angles to transmit power at a 1:1 ratio. The miter gears in a pair must have the same number of teeth. However, more than two miter gears may be used in sets, such as in automotive differentials.

Worms and Worm Gears. Pairs of gears consisting of a worm and a worm gear are used for large speed reductions and smooth, quiet service. A *worm* is a screw thread that rotates the worm gear. A *worm gear* is a spur gear with specially cut teeth that are driven by a worm. Worm gear teeth are cut at an angle and in a concave shape to mate securely with the worm. A single-threaded worm is designed to advance the worm gear one tooth for every revolution of the worm.

Gearboxes. Gears used in gear drive systems may be exposed or encased in a gearbox. A *gearbox* is a sealed container that has an input shaft and an output shaft and houses at least one set of mating gears. **See Figure 9-25.** Gearboxes are used to change shaft rotation, speed, and/or direction of rotation, and to protect gears from the surrounding environment. Two common gearboxes are in-line and right angle.

An in-line gearbox is primarily used to reduce shaft speed. In-line gearboxes may also be used to change shaft rotation. Gears used for in-line gearboxes include spur and helical gears. Speed is controlled by the gear ratio. In-line gearboxes also provide single axis offset that can be useful for joining shafts that are on different axes.

A right angle gearbox is used to change the direction (by 90°) of the output shaft in relation to the input shaft, in addition to changing shaft speed. Right angle gearboxes may also be used to change the direction of rotation of the output shaft. Gears used in right angle gearboxes include bevel, miter, and worm gears. Worm gears provide a large speed reduction.

The lower portion of a gearbox is usually used as an oil reservoir for lubricating the internal gears. Proper gearbox maintenance includes changing the gearbox oil according to manufacturer specifications, using the proper gear oil, and visually inspecting the dirty oil for contaminants. The oil level in a gearbox must be kept at the correct level and changed at periodic intervals. Oil level is checked by looking through the sight glass or by removing a plug located on the gearbox. The oil should just begin dripping from the opening revealed by the plug. Oil that is removed may be analyzed to monitor the conditions inside the gearbox. Chemical breakdown, contamination levels, and particle type and amount indicate the level of gear wear inside the gearbox.

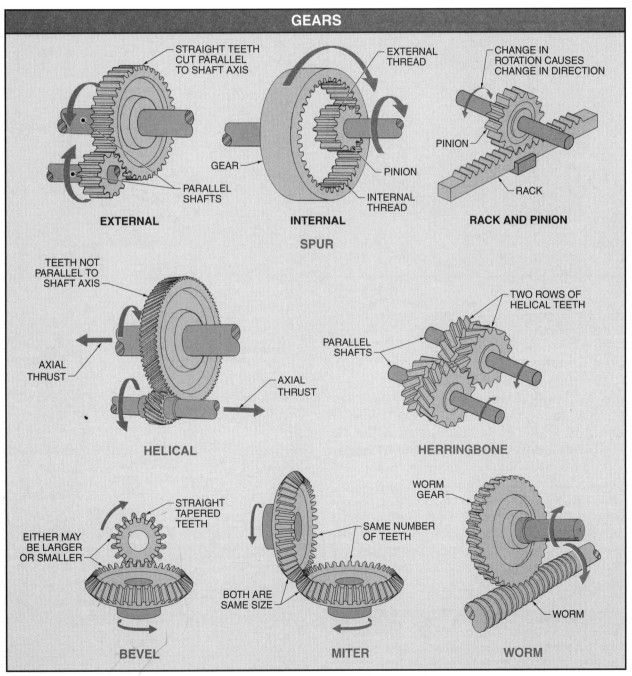

Figure 9-24. Gears commonly used in industry include spur, helical, herringbone, bevel, miter, and worm gears.

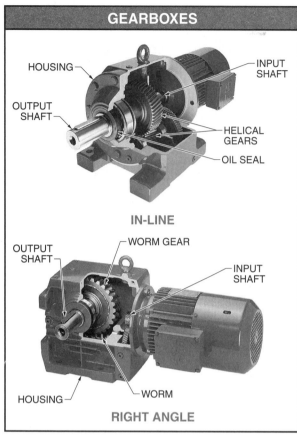

GEARBOXES

SEW-Eurodrive, Inc.

Figure 9-25. A gearbox is a sealed container that has an input shaft and an output shaft and houses at least one set of mating gears.

Cam Drives

A *cam* is a machine part that transmits motion using an irregular external or internal surface. **See Figure 9-26.** Cams are used to change rotary motion into linear motion. Cams are useful in creating motions that are difficult to reproduce using other mechanisms. Cams are used with cam followers to transfer load, change direction, or change the speed of a machine. A *cam follower* is a machine part that contacts the cam and moves in a designated path. The shape of the cam determines the direction and motion of the cam follower. Cams may be used in combination with other cams to produce a combination of motions. Common cam designs include plate, face, barrel, and yoke cams.

> **TECH TIP**
>
> *All machines vibrate from minor manufacturing defects that produce a vibration signature. Excessive machine vibration is usually caused by imbalance from worn or damaged machine parts.*

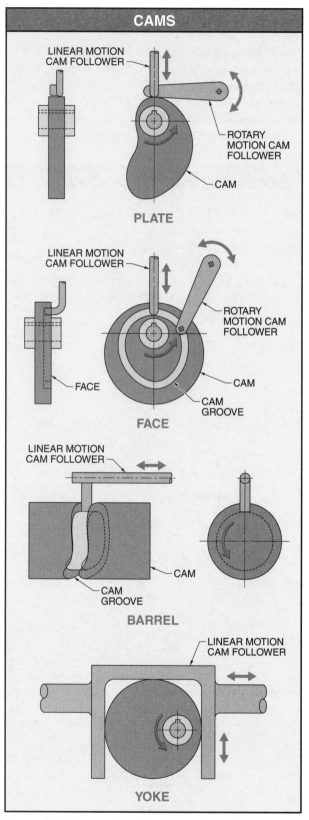

CAMS

Figure 9-26. A cam is used to transmit motion using an irregular external or internal surface.

Variable-Speed Drives

A *variable-speed drive* is a mechanical drive system that provides variable output speed without changing the speed of the drive motor. Varying the output speed is accomplished mechanically, electronically, or hydraulically. Mechanical variable-speed drives include open belt and enclosed belt drives. All mechanical variable-speed drive systems use variable diameter pulleys to achieve different output speeds. **See Figure 9-27.**

Open Belt. An *open belt variable-speed drive* is a mechanical drive system that uses two opposing cone pulleys linked together with a flat or V-belt to obtain variable speeds. Adjusting output speed is accomplished by hand when the drive system is not operating. Belt tension is released and the belt is moved to another position on the cone pulley. Changing the position of the belt changes the ratio between the drive and driven pulleys, resulting in a different output speed.

Enclosed Belt. An *enclosed belt variable-speed drive* is a mechanical drive system that uses one or two split-half adjustable pulleys and an adjusting mechanism to change output speed. Output speed is changed while the drive system is operating. The pulleys are split in half, allowing the halves to move closer or further apart, changing the overall width of the pulley. The change in pulley width forces the drive belt to ride higher or lower on the face of the pulley. This changes the operating diameter of the pulley, which changes the ratio between the drive and driven pulley and therefore the output speed. Powerful springs provide the tension to keep the pulley halves in position.

COUPLINGS

A *coupling* is a device used to connect a mechanical drive to a prime mover. Couplings are the most common and least expensive method of connecting two shafts. Mechanical couplings are classified as rigid or flexible. Couplings require accurate alignment of the mating shafts.

Rigid Couplings

A *rigid coupling* is a device that joins two shafts that are precisely aligned within a common frame. Rigid couplings are made of metal and are secured with bolts or setscrews. One advantage of rigid couplings is the ability to transmit more power than flexible couplings because of their simple design and rigidity. Rigid couplings include flange and sleeve couplings. **See Figure 9-28.** A *flange coupling* is a coupling consisting of two hubbed flanges fastened together with mechanical fasteners (bolts). Each flange has a center hole and keyway sized for mating shafts. A *sleeve coupling* is a coupling consisting of a tube with an internal keyway and center hole that fits mating shafts. Flange and sleeve couplings use setscrews and a keyway to secure the coupling to the shafts.

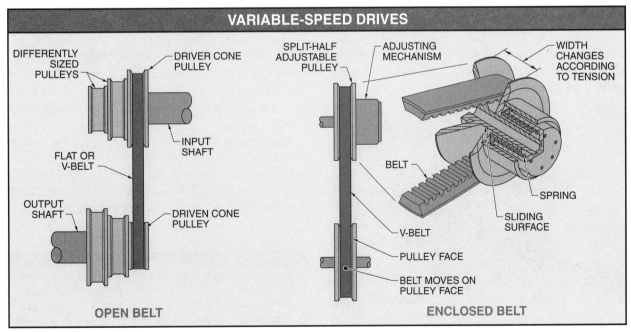

Figure 9-27. A variable-speed drive provides variable output speeds without changing the speed of the drive motor.

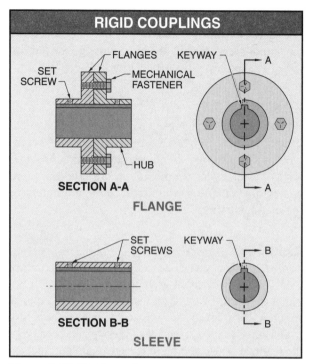

RIGID COUPLINGS

SET SCREW
FLANGES
MECHANICAL FASTENER
KEYWAY
A
A

SECTION A-A

FLANGE

HUB

SET SCREWS
KEYWAY
B
B

SECTION B-B

SLEEVE

Figure 9-28. Rigid couplings transmit rotary motion in applications where two shafts are positioned in precise lateral and angular alignment.

Flexible Couplings

A *flexible coupling* is a coupling designed to join two shafts that have a small amount of misalignment. Flexible couplings reduce the conduction of heat, sound, and electricity through the drive system. Flexible couplings also provide torsional dampening in a drive system. *Torsional dampening* is a dampening process that smoothes torque fluctuations in a drive system. Flexible couplings are classified as mechanical-flexing or material-flexing.

Computational Systems Incorporated

A laser alignment system can be used to accurately align machine shafts to ensure extended machine and coupling life.

Mechanical-Flexing Couplings. A *mechanical-flexing coupling* is a flexible coupling with components that move or slide in relation to each other to accommodate shaft misalignment. Clearance between mechanical-flexing coupling components is provided to permit movement. Common mechanical-flexing couplings include block-and-jaw, gear, grid, chain, and universal joint. **See Figure 9-29.**

A *block-and-jaw coupling* is a mechanical-flexing coupling that contains two or more jaws and a floating block (spider). Block-and-jaw couplings are used for moderate speeds at light to medium duty. The floating block is trapped between the jaws by grooves on the jaws and mating keys on the floating block. The number of jaws may be two, three, five, or seven. Shaft misalignment is accommodated by the movement of the floating block between the two jaws.

A *gear coupling* is a mechanical-flexing coupling that includes two identical hubs with external gear teeth and a sleeve(s) with mating internal gear teeth. Gear couplings are used for variable speeds and high torques. Shaft misalignment is accommodated by the clearance between the external and internal gear teeth. Some light-duty gear couplings have nonmetallic sleeves made from nylon or urethane that eliminate the need for lubrication. All other gear couplings require lubrication to ensure long life.

A *grid coupling* is a mechanical-flexing coupling that includes two identical hubs with axially cut slots along their perimeter and a wire grid. Grid couplings are used for transmitting high torque at moderate speeds. A cover protects the coupling from contaminants and retains lubricant inside the coupling. The wire grid joins the two hubs by the interlacing of the grid through the axially cut slots. Misalignment between the connected shafts is accommodated by the sliding movement of the grid in the slots. Wire grids may be one piece or are available in multiple sections. Some grids are tapered for ease in installation and removal of the grid. Excessive amounts of misalignment between the shafts may cause premature wear of the grid. Misalignment should be kept within the manufacturer's specifications.

TECH TIP

Avoid sudden shock loads during start up and operation of coupled equipment. A coupling assembly should operate quietly and smoothly. If a coupling assembly vibrates or makes beating sounds, shut the machine down immediately and recheck alignment.

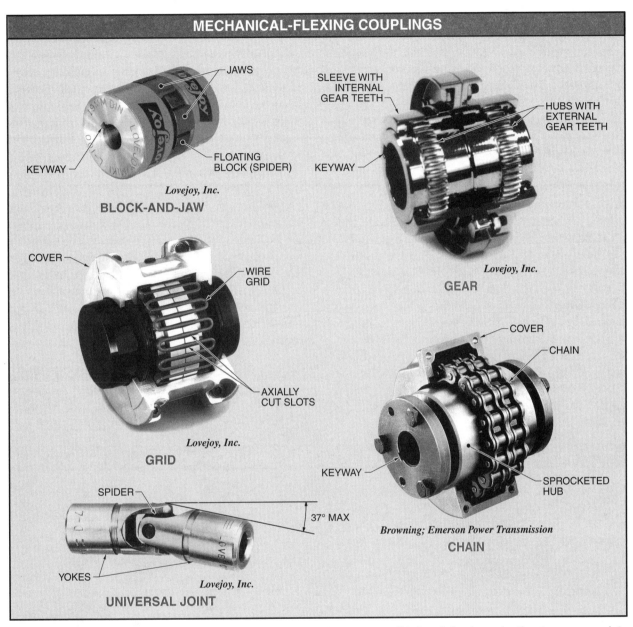

Figure 9-29. Mechanical-flexing couplings have components that move or slide in relation to each other to accommodate shaft misalignment.

A *chain coupling* is a mechanical-flexing coupling that includes two identical sprocketed hubs connected with a single-roller, double-roller, or silent chain. Chain couplings are used for transmitting high torques at low speeds. Clearances between the sprockets and chain and within the chain itself provide room for misalignment between the shafts. Special shaped rollers that provide more flexibility and reduced wear are also available. Light-duty chain couplings use nonmetallic chains and are self-lubricating. Covers for chain couplings

to protect the coupling from contaminants and retain lubricant are available. Covers on most couplings rotate with the coupling.

A *universal joint* is a mechanical-flexing coupling that includes two yokes connected by a spider. Universal joints are used for transmitting high torque at variable speeds. A *spider* is a cross-shaped member having four ends providing pivot points. Universal joints are used individually or in pairs. Mechanical drive systems using universal joints are commonly referred to as shaft drives.

Material-Flexing Couplings. A *material-flexing coupling* is a flexible coupling that uses a flexible material to accommodate shaft misalignment. Common materials used for material-flexing couplings include metal, rubber, plastic, or composite material. Material-flexing couplings include flexible disk-ring, spring, diaphragm, and elastomeric couplings. **See Figure 9-30.** Each material-flexing coupling relies on flexing material to accommodate misalignment between two shafts.

A *flexible disk-ring coupling* is a material-flexing coupling that includes two flanged hubs and flexible disks. Flexible disk-ring couplings are used for transmitting high torques at variable speeds. The flexible disks are made from metal, providing rigidity in a torsional plane with no backlash. Flexible disk-ring couplings may be single- or double-flexing. Single-flexing couplings include two flanged hubs and one laminated disk. The disk is bolted to the hubs in an alternating fashion. Single-flexing couplings accommodate angular misalignment and reduce end float. Double-flexing couplings include two flanged hubs connected to a floating center member by two flexible disks. The bolt pattern is the same as single-flexing units. The two flexible disks accommodate angular and parallel misalignment.

A *spring coupling* is a material-flexing coupling that uses the flexing of a spring to accommodate shaft misalignment. Spring couplings are used for transmitting light to medium torque at variable speeds. The spring in the coupling may be long or short and is fastened at each end to a hub that fits on a shaft. A long spring material-flexing coupling is referred to as a flexible shaft. Flexible shafts accommodate angular and parallel misalignment.

A *diaphragm coupling* is a material-flexing coupling that includes two flanges separated by a floating member. Diaphragm couplings are used for transmitting high horsepower and high torque. Diaphragm couplings are available in many sizes and styles to accommodate angular and parallel misalignment and restrict end float. The floating component may consist of one or multiple diaphragms depending on the design and the power transmission requirements of the application. Power is transferred through the diaphragm from the outside in or from the inside out.

> **◇ TECH TIP**
> *Shortly after initial operation and periodically thereafter, inspect couplings for alignment, wear, and bolt torque, and flexing elements for signs of fatigue.*

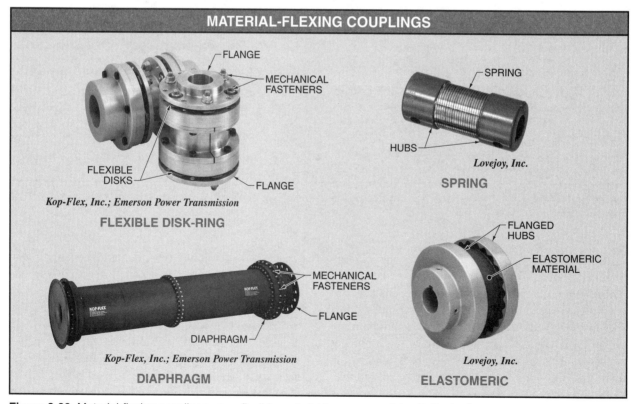

MATERIAL-FLEXING COUPLINGS

FLANGE
MECHANICAL FASTENERS
FLEXIBLE DISKS
FLANGE
Kop-Flex, Inc.; Emerson Power Transmission
FLEXIBLE DISK-RING

SPRING
HUBS
Lovejoy, Inc.
SPRING

MECHANICAL FASTENERS
FLANGE
DIAPHRAGM
Kop-Flex, Inc.; Emerson Power Transmission
DIAPHRAGM

FLANGED HUBS
ELASTOMERIC MATERIAL
Lovejoy, Inc.
ELASTOMERIC

Figure 9-30. Material-flexing couplings use a flexible material to accommodate shaft misalignment.

An *elastomeric coupling* is a material-flexing coupling that includes two flanged hubs connected and separated by elastomeric material. Elastomeric couplings are used for transmitting medium torque at moderate speeds. Materials include rubber, plastic (polyurethane or neoprene), composite, or leather. The ability of the elastomeric material to flex and distort allows the coupling to compensate for shaft misalignment. The elastomeric material may be bonded, clamped, fit, or fastened to the flanged hubs. Compression or shear forces are placed on the elastomeric material during operation. Elastomeric couplings are normally maintenance-free.

Coupling Removal/Installation. Couplings are removed and reinstalled when the couplings are damaged or when either of the coupled devices must be removed. Coupling removal and installation procedures vary depending on the coupling. Always follow proper procedures when removing and installing couplings. **See Figure 9-31.** Couplings are removed and installed by applying the procedure:

1. Lock out and tag out equipment.
2. Remove guard to provide access to coupling.
3. Loosen any equipment mounts and remove old coupling. Inspect old coupling for signs of wear caused by improper installation.
4. Reposition motor and check equipment shafts or angular and parallel misalignment. Realign if necessary.
5. Install new coupling and check for axial positioning after reviewing coupling manufacturer installation instructions. Torque coupling fasteners to the proper specifications.
6. Replace guard.
7. Remove lockout/tagout.
8. Start equipment and monitor coupling for unusual vibration and/or sounds.

PULLERS

A *puller* is a device used to remove gears, pulleys, sprockets, bearings, and couplings from a shaft or housing. Pullers may be mechanically or hydraulically operated and consist of jaws, a cross head, and a threaded shaft or hydraulic ram. A *cross head* is a metal frame that accommodates a threaded shaft and supports and positions the jaws of the puller. Cross heads may be either two-jaw, three-jaw, or combination two/three-jaw. **See Figure 9-32.** Pullers are sized by pressure (1 t to 100 t), spread, and reach.

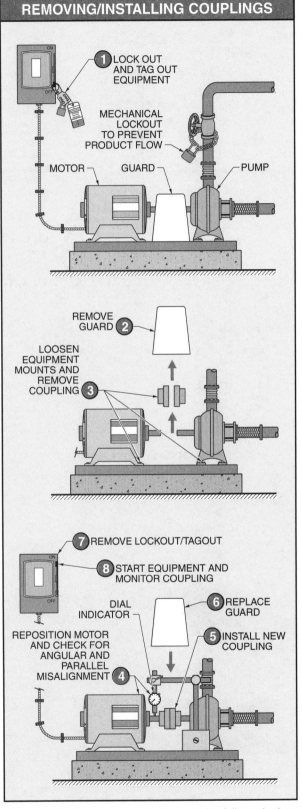

Figure 9-31. Proper procedures must be followed when removing and installing a coupling.

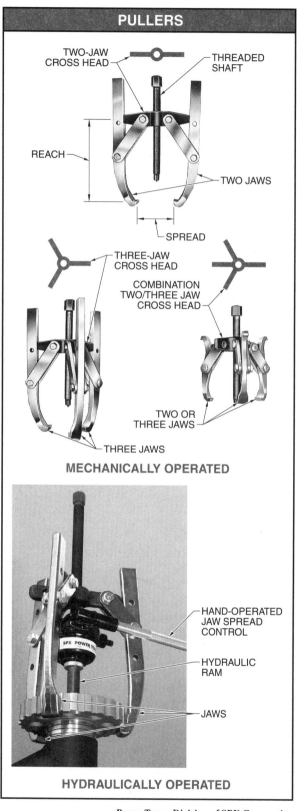

PULLERS

TWO-JAW CROSS HEAD

THREADED SHAFT

REACH

TWO JAWS

SPREAD

THREE-JAW CROSS HEAD

COMBINATION TWO/THREE JAW CROSS HEAD

TWO OR THREE JAWS

THREE JAWS

MECHANICALLY OPERATED

HAND-OPERATED JAW SPREAD CONTROL

HYDRAULIC RAM

JAWS

HYDRAULICALLY OPERATED

Power Team, Division of SPX Corporation

Figure 9-32. A puller is used to remove gears, pulleys, sprockets, bearings, and couplings from a shaft or housing.

Safety must be the highest priority when using a puller. The proper puller is selected for a specific job based on information about the job and general puller use rules.

The area of press fit determines the puller size used. All objects pressed on a shaft or in a housing have a specific contact area (press fit area) with the shaft or housing. The larger the press fit area, the more pressure required to remove the object. Shaft diameter also determines the size of puller used. For hand-operated screw pullers, the screw diameter must be at least half as large as the shaft. For example, if an object must be pulled from a 2″ diameter shaft, the screw diameter of the puller must be at least 1″ in diameter. For hydraulic pullers, the amount of tonnage exerted on the shaft should be 7 to 10 times the diameter (in in.) of the shaft. For example, if a shaft has a 2″ diameter, the tonnage of the hydraulic puller should be 14 t to 20 t. When using a puller, wear safety glasses, use the right size puller, apply force gradually, align the puller jaws and legs, and mount the puller so that the grip is tight. Do not couple legs of a puller. Keep the reach to a minimum and cover the work area with a protective blanket to protect personnel from possible flying metal parts.

SHAFT ALIGNMENT

When installing mechanical drive systems, the shafts to be connected by the drive system must be aligned to an acceptable tolerance. For example, a belt or chain drive system requires a lesser tolerance than a gear drive system, and a flexible coupling requires a lesser tolerance than a rigid coupling. Misalignment may be parallel, angular, or a combination of parallel and angular. **See Figure 9-33.** Parallel and angular misalignment should be checked using a dial indicator, laser, or specialized equipment. Misalignment must be checked when installing or performing routine maintenance on any mechanical drive system. The mechanical drive system experiences premature failure if angular and parallel misalignment occurs.

Equipment joined with a drive system should first be checked for soft foot. *Soft foot* is a condition that occurs when one or more machine feet do not make complete contact with the base plate. This occurs when a machine foot, mounting surface, or both are not flat. Soft foot places stress on components such as bearings and housings, resulting in vibration and early component failure. Soft foot may also cause false readings when trying to perform alignment procedures. Soft foot must be eliminated before performing shaft alignment. Eliminating soft foot is done by placing a shim under the rocking corner of the machine foot. A *shim* is a thin precision piece of material, usually metal, that provides an accurate spacing between two surfaces. Common shim thicknesses are available in 0.001″ increments and range from 0.001″ to 0.125″. Shims may be cut from a piece of shim stock to the desired thickness or purchased precut and organized according to thickness.

PULLEY AND SPROCKET ALIGNMENT

Proper pulley and sprocket alignment is required for correct operation of belt and chain drives. Misaligned pulleys and sprockets place lateral and twisting forces on belts and chains. These forces cause rapid wear on belts, chains, pulleys, and sprockets, resulting in premature failure. Pulleys and sprockets are aligned using the straightedge or string alignment method. Belt and chain drive systems compensate for misalignment in one axis. The main concern is to ensure that the faces of the pulleys or sprockets are parallel to each other.

The straightedge alignment method uses a straightedge placed across the faces of the pulleys or sprockets to check alignment. The straightedge must touch at four points. The pulleys or sprockets require alignment if any visible gaps exist between the straightedge and pulleys or sprockets. **See Figure 9-34.** A feeler gauge is used to determine if gaps are present when it is difficult to see an actual gap. A *feeler gauge* is a steel strip of a specific thickness. The feeler gauge is inserted between the pulley or sprocket and the straightedge. Pulleys or sprockets should be checked for alignment anytime a belt or chain is tensioned. A string may also be used to check alignment. In the string alignment method, a string is tied to one shaft behind the pulley or sprocket and wrapped around the face of both pulleys or sprockets. The loose end of the string is tied to the other shaft or held tight. Alignment is required if visible gaps exist at any of the four points. Properly aligned pulleys and sprockets prevent premature failure of belt and chain drive systems. Misalignment may be angular and/or parallel, causing side thrust and twisting of the belt or chain.

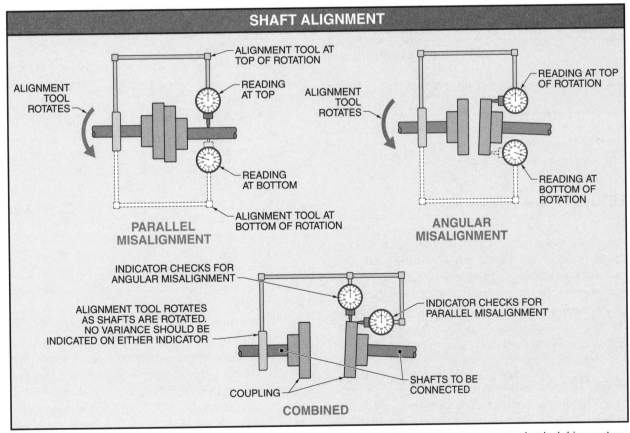

SHAFT ALIGNMENT

Figure 9-33. Misalignment must be checked when installing or performing routine maintenance on a mechanical drive system.

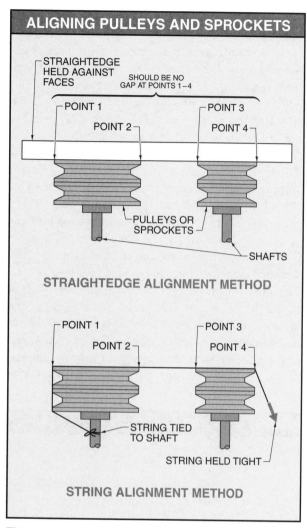

ALIGNING PULLEYS AND SPROCKETS

STRAIGHTEDGE HELD AGAINST FACES

SHOULD BE NO GAP AT POINTS 1–4

POINT 1
POINT 2
POINT 3
POINT 4

PULLEYS OR SPROCKETS

SHAFTS

STRAIGHTEDGE ALIGNMENT METHOD

POINT 1
POINT 2
POINT 3
POINT 4

STRING TIED TO SHAFT

STRING HELD TIGHT

STRING ALIGNMENT METHOD

Figure 9-34. Pulleys and sprockets are aligned using the straightedge or string alignment method.

PREDICTIVE MAINTENANCE

Mechanical system predictive maintenance uses scheduled procedures to evaluate the operating condition of equipment. Data acquired from predictive maintenance is analyzed and used to determine maintenance procedures and frequency, or when a specific piece of equipment may fail. Predictive maintenance includes manufacturer statistics, vibration analysis, thermography, and oil analysis.

Manufacturer Statistics

Manufacturer statistics include data from equipment manufacturers on the life span of their products. Manufacturer statistics on their designed product life span may be used as a predictive maintenance tool. Products may include bearings, engines, chains, or belts. Life span is commonly provided in operating hours under certain operating conditions. Equipment operating time can be documented and components replaced before they reach their expected life span when life span hours for a specific product are known.

Vibration Analysis

Vibration analysis is monitoring of individual component vibration characteristics to analyze the component condition. Any operating mechanical device displays a vibration pattern (signature) specific to the machine. Vibration analysis is generally performed by comparison against a baseline signature taken when the machine was new, or against comparable machines. For example, a new bearing produces a low level of vibration that appears as random noise. **See Figure 9-35.** As a bearing begins to fail, the vibration produced by the bearing changes due to varying friction placed on the rolling elements of the bearing. The varying friction results in pulses of vibration that repeat periodically. Equipment vibration may exist over a wide frequency spectrum. The frequency value helps isolate the problem. For example, low-frequency vibrations are caused by imbalance, misalignment, a bent shaft, etc. Medium-frequency vibrations are caused by wear and faults in gears. High-frequency vibrations are caused by faulty bearings.

Machine vibration is primarily caused by imbalance and/or misalignment. *Imbalance* is lack of balance. Imbalance is commonly caused by defective or improperly installed machine components. For example, machine castings are subject to defects in the manufacturing process that can alter their weight characteristics. This affects the part balance. Misalignment may occur between components within a machine such as meshing gears, or outside a machine such as driven belts, pulleys, shafts, and couplings. Other causes of machine vibration include component damage, loose machine components, and component contact.

Vibration measurement indicates the amount of oscillation from a position of rest. Vibration analysis is the systematic process used to determine the cause of vibration. Each vibration condition produces unique vibration characteristics that include frequency, displacement, velocity, acceleration, phase, and spike energy. **See Figure 9-36.** These characteristics are analyzed on site or at a contracted company by trained specialists.

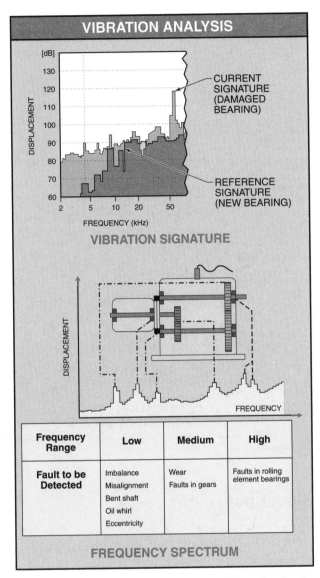

VIBRATION ANALYSIS

VIBRATION SIGNATURE

FREQUENCY SPECTRUM

Frequency Range	Low	Medium	High
Fault to be Detected	Imbalance Misalignment Bent shaft Oil whirl Eccentricity	Wear Faults in gears	Faults in rolling element bearings

Figure 9-35. Vibration analysis uses machine vibration to identify potential mechanical problems.

Thermography

Thermography is a predictive maintenance procedure that uses heat energy emitted from operating equipment to analyze the status of moving components. Moving components such as bearings and shafts overcome friction to rotate, generating heat. Under normal operating conditions, rotating components generate a specific amount of heat. If friction around moving components increases, the heat emitted also increases. Rotating components are designed to operate in a specific temperature range. Operating temperatures that exceed the specific temperature range shorten the life of the moving components. Thermography commonly uses specialized equipment, but a simpler version can be done with a simple contact thermometer.

Oil Analysis

Oil analysis is a predictive maintenance technique that detects and analyzes the presence of acids, dirt, fuel, and wear particles in lubricating oil. Oil includes lubricating oil, hydraulic fluids, transmission fluids, and grease. The oil is tested for cleanliness, moisture content, depletion of the additives added to improve the oil's performance, and particles of metal worn away from the machine.

Oil analysis involves removing an oil sample from a machine and sending it to a laboratory for analysis. A sample must be obtained that reflects the actual condition of the oil. Oil samples should be taken at the same location and on a regular schedule, such as after a certain number of operating hours. The sample location is important because it affects the composition of the potential contaminants. A straight section of pipe will not provide a good sample because the oil is traveling at high speed and any particles in the oil will bypass the sampling port. An analysis of this type of sample would not accurately show the particle contamination in the oil. Oil samples should be taken at locations where the oil is turbulent, such as pipe bends, so the sample includes all possible contaminants. Sampling should be done during normal operation, temperatures, and loads. It should not be done right after startup, or on an extremely hot day. Oil sampling ports should be located upstream of filters and downstream of gears and bearings that could contribute metal particles to the oil.

Oil must not be contaminated during the sampling process. Sampling begins by flushing five to ten times the volume to be sampled into a clean oil sampling bottle. This flushing oil is returned to the machine. Flushing removes any accumulated contaminants from the sampling port and any fittings.

A vacuum sampling pump is often used to easily draw oil from the machine without leaving a residue in the pump to contaminate future samples. A vacuum sampling pump is attached to the sampling bottle and the sample is drawn from the machine. The bottle is not filled completely so that it can be shaken at the laboratory to mix the oil with any contaminants that may have settled to the bottom of the bottle. When the bottle is filled to the correct level, the vacuum pump is removed and the bottle lid is replaced. The sampling bottle is sealed inside two new plastic bags and sent immediately to the lab for analysis. Delays in sending the sample decrease the accuracy of the analysis.

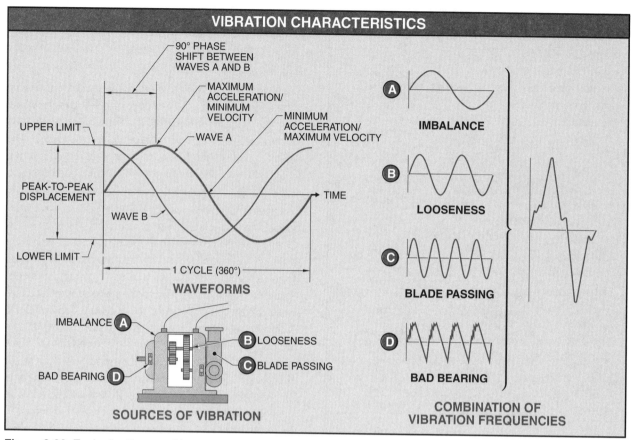

Figure 9-36. Each vibration condition produces unique vibration characteristics.

A spectrometer is used for oil analysis. A *spectrometer* is a device that vaporizes materials and records the resulting light. The light is separated into a spectrum and converted into electrical signals, which are processed and displayed by a computer. Data management software for oil analysis is available from different manufacturers. The software enables maintenance personnel to be directly connected (via a network) to an analysis laboratory where oil samples are tested and results are returned, providing quick, comprehensive analysis and trending. The analysis includes oil viscosity, particle count, wear particle concentration analysis, and wear particle analysis. Analysis information allows the user to determine if wear is occurring, which component(s) are affected, the cause of the damage, and the extent of the damage.

TECH TIP

Excessive belt drive noise is often incorrectly diagnosed as a belt problem. To determine if the belt is the problem, spray the belt with soapy water while operating. If the noise stops, the cause is isolated to the belt. If the noise continues, the problem is caused by other drive components.

PREVENTIVE MAINTENANCE

Preventive maintenance of a mechanical system prevents failures from occurring by using scheduled routine checks and service procedures. Always ensure that power is turned OFF, locked out, and tagged out before performing any preventive maintenance procedures. Careful observation requires smelling, listening, touching, and looking at the mechanical system. A preventive maintenance program for mechanical drive systems ensures long life of the mechanical drive system. Preventive maintenance for mechanical systems should include oil level, oil contamination, misalignment, tension, unusual wear, and cleanliness checks.

Belt and Chain Drives

Once power is turned OFF, locked out, and tagged out, preventive maintenance of a belt or chain drive includes inspecting the belt or chain and pulley or sprocket. In addition, the drive system is checked for misalignment, correct placement of guards, proper operation of other drive components, and proper belt or chain tension. Inspecting the belt and chain begins by observing signs of unusual wear on the belt or chain. Mark a spot on the belt or chain and gradually work around it, looking for potential problems. For belts, look for cracks, frayed spots, cuts, or unusual wear patterns. For chains, look for unusual wear on the inside or outside of each link, discoloration, proper slack, and movement of the rolling elements. Oil reservoir level should be checked at prescribed intervals and visual inspection or oil analysis should be performed on a routine basis. Dirty oil or lubricant should be replaced immediately so damage does not occur to bearing or mating components. The lubricant must meet manufacturer's recommendations. Failure to use the proper lubricant can result in failure and may void the manufacturer's warranty.

Misalignment causes premature wear of drive mechanisms such as belts, chains, pulleys, sprockets, and couplings. Any unusual wear on a belt or chain results in abnormal wear on the pulley or sprocket.

Belts and chains with improper tension may also cause the drive system to fail prematurely. Belt and chain tension should be set to manufacturer's specifications. Too little tension causes a belt to slip, resulting in a loss of power transmission, belt glazing, and premature wear of the pulley. A chain that has too little tension can jump a tooth, damaging the chain and/or sprocket, and lose timing. Excessive tension places an undesirable load on drive system bearings, which may also result in premature failure of the drive system.

Chains and belts should be protected from foreign matter such as dirt. Oil and grease on a belt can result in belt slippage. Dirt on a chain or belt acts as an abrasive material wearing away the belt, chain, and mating components. Proper guards should always be in place and secured to prevent foreign objects from entering the drive system components.

TROUBLESHOOTING MECHANICAL SYSTEMS

Troubleshooting mechanical systems should be done in a systematic manner. Troubleshooting can be done on equipment that is operational but malfunctioning or on equipment that has completely failed.

Mechanical systems fail due to excessive wear, overheating, and overloading. All mechanical systems use moving components that eventually fail due to normal use. Failed or malfunctioning equipment should be carefully inspected to help isolate the cause of the problem. Inspection or investigation requires the use of all senses. Unusual sounds, heat, vibration, motion, and smell are all indications that a problem exists. Tracing these symptoms to their source usually reveals the cause of the problem. For equipment failures, the cause of the problem can usually be found in the failed equipment. During inspection of failed equipment, extreme care must be taken so that evidence is not destroyed. Troubleshooting mechanical problems often requires monitoring machine operation over a long period of time.

Many repairs are made by replacing a failed component. For example, a failed bearing or drive coupling is replaced to bring the equipment back into service. This may bring the equipment back into service but does not address the cause of the problem. When replacing any failed component, the reason for the component failure should be determined and corrected. Ensure the component failed due to normal wear and not from an abnormal condition.

Mechanical systems are subjected to many stresses during operation that can damage the structural material of the equipment. If the stress is great enough, it can deform or fracture the material. *Deformation* is a change in the shape of a material caused by stress. Other conditions that can have serious effects on the operation of a mechanical system include shocks, impacts, extreme temperature changes, and vibration. Heavy shocks or impacts on a mechanical system should be eliminated or reduced. Temperature changes cause material to expand and contract, changing its tolerance. The changes in tolerance may be great enough to cause mechanical failure. Vibration can be strong enough or at the correct frequency to cause stress cracks or individual components to loosen.

Bearing Failure Analysis

Causes of bearing failure include dirt, improper lubrication, operating conditions, improper installation, and surface reaction. Bearing failure normally occurs over a period of time. During failure, the bearing usually generates a large amount of heat and noise. The heat may be severe enough to cause physical damage to other components. **See Figure 9-37.**

Dirt. Dirt causes approximately 45% of all bearing failures. Dirt can enter a bearing during operation, lubrication, and/or assembly. Dirt is very abrasive to bearing components and causes premature failure of the bearing. Dirt is also a poor heat conductor, which reduces the ability of a lubricant to conduct heat away from the bearing. Bearings located inside a gearbox may have dirt and metal chips introduced during manufacturing or rebuilding. During operation, the bearing may be subjected to dirt from the surrounding environment. Poor bearing seals allow dirt and other contaminants to enter the bearing components. Bearings located inside a crankcase may also be contaminated during operation due to leaky gaskets. Replace leaking gaskets or seals immediately.

Dirt large enough to damage a bearing can be invisible to the naked eye. Using a lint-free rag and keeping visible dirt away from lubricants contributes greatly to prolonging bearing life.

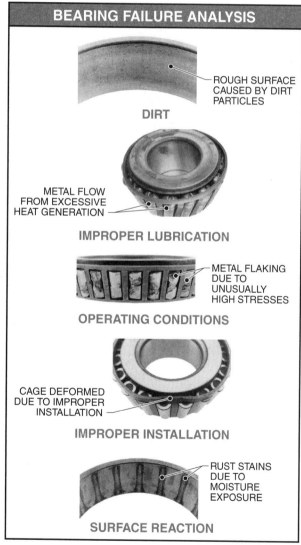

BEARING FAILURE ANALYSIS

ROUGH SURFACE CAUSED BY DIRT PARTICLES

DIRT

METAL FLOW FROM EXCESSIVE HEAT GENERATION

IMPROPER LUBRICATION

METAL FLAKING DUE TO UNUSUALLY HIGH STRESSES

OPERATING CONDITIONS

CAGE DEFORMED DUE TO IMPROPER INSTALLATION

IMPROPER INSTALLATION

RUST STAINS DUE TO MOISTURE EXPOSURE

SURFACE REACTION

The Timken Company

Figure 9-37. Causes of bearing failure include dirt, improper lubrication, operating conditions, improper installation, and surface reaction.

Dirt can contaminate a bearing during lubrication because oily or greasy areas such as oil cups and grease fittings attract and gather dirt and dust. This dirt must be removed from the fitting before the bearing is lubricated, otherwise the dirt mixes with the lubricant and enters the bearing. Lubrication tools and fill caps located on gearboxes must also be cleaned before any lubricant is added or checked. For example, the tip of a grease gun collects dirt when not in use and should be cleaned before it is placed on a grease fitting.

Dirt may also enter a bearing during assembly. Care should be taken to not contaminate the bearing, seals, or gearbox during assembly or rebuild operations. Bearing journals must also be free of dirt before the bearing is put into position. Dirt lodged behind a bearing can cause misalignment, an isolated pressure point, and incorrect location. During assembly, hands, bearings, seals, and gearboxes must be clean and work must be performed in a clean environment. Do not touch bearing surfaces with bare hands because hands can contain enough oil and dirt to damage the bearing.

Improper Lubrication. Improper lubrication is another cause of bearing failure. Improper lubrication accounts for approximately 25% of all bearing failures. Improper lubrication includes starvation, underlubrication, overlubrication, lubricant contamination, and mixing lubricants. A bearing that experiences starvation experiences wear immediately. If the bearing is underlubricated, the metal surfaces touch, causing rapid failure.

Bearings must be prelubed to ensure lubricant is present during initial startup. New bearings are generally $\frac{1}{3}$ to $\frac{1}{2}$ filled with grease. This is their normal operating level. A common mistake when greasing a bearing is to add too much grease. This causes overheating and damages the bearing. The overheating is caused when the rollers must force their way through the excess grease that is packed into an overlubricated bearing.

Contamination of the lubricant can also reduce the effectiveness of the lubricant. Common lubricant contaminants include water and dirt. Lubricant contaminated with dirt subjects moving components to a constant flow of abrasives. Water causes bearing components to rust, increasing friction and causing eventual bearing failure. Sources of water may be condensation and environments with high humidity. Periodic oil changes are necessary to remove water from the lubricant. Oil that is contaminated with water has a milky appearance. Never mix different lubricants as this can cause severe contamination.

Operating Conditions and Improper Installation. Operating conditions and improper installation are the other main causes of bearing failure. Bearings that are subjected to rough operating conditions such as impact loading, severe loading, and severe speeds normally experience premature failure. Failure due to operating conditions can be avoided by selecting the proper bearing for the application requirements.

Improper installation also results in premature bearing failure. Bearings are usually mounted by pressing or by heating or cooling the bearing. Pressing a bearing into place requires the use of mechanical or hydraulic force to push the bearing into place. **See Figure 9-38.** The bearing size must be verified before trying to press a bearing into place. Burrs and rough surfaces on the shaft over which the bearing slides must also be removed. A burr that a bearing is pressed over results in an isolated pressure area that deforms the bearing. Pressing force must be applied to the proper race or the rolling elements receive unusual pressure and damage occurs to the bearing components. For example, when pressing a bearing on a shaft, the pressing force must be applied to the inner race. A bearing pressed in a housing must have the pressing force placed on the outer race. A bearing pressed on a shaft and in a housing must have the pressing force placed on both races.

Bearings can be mounted by heating (expanding) or cooling (shrinking) the bearing, depending on the application. Heating a bearing increases the size of the bearing, allowing it to slip over a shaft. Bearings should never be heated above 300°F or damage may occur. Bearings are normally heated to 200°F. Methods of heating include oil bath and induction heating. Oil bath is the best heating method because it provides even and controlled heating of the bearing. Induction heating is extremely quick but also risks the chance of overheating the bearing. Never heat a bearing with a torch. The heat is too localized, causing distortion of the bearing. Cooling the bearing is done to reduce the size of the bearing, allowing it to fit inside a housing. Common cooling methods are dry ice and liquid nitrogen.

Surface Reactions. Surface reactions damage bearing surfaces with chemical reactions, oxide buildup, or electric current between the lubricant and the metal of the bearing. Surface reaction affects bearing surfaces and can result in bearing failure by gradually eroding the bearing surfaces. Chemical corrosion is caused by reactions between lubricants, materials, and moisture. These reactions create fluids and vapors that are extremely acidic and erode the surface of the bearing.

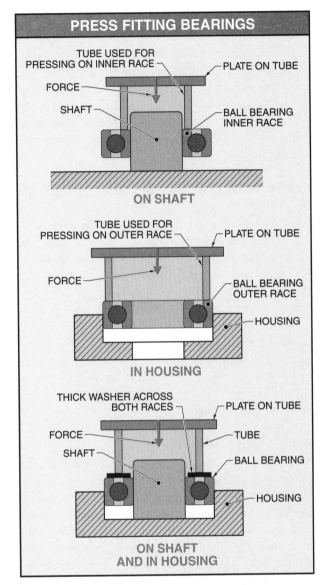

Figure 9-38. Pressing force must be applied to the proper bearing race or the rolling elements of the bearing receive unusual pressure and damage occurs to the bearing components.

Oxide buildup is caused by the presence of moisture on the bearing surfaces. The moisture creates surface rust, which flakes off, reducing bearing dimensions. The reduced dimensions of the bearing components provide slop, which allows the mechanical system to hammer itself until it fails. Moisture can enter through leaky seals and gaskets, rapid changes in temperatures, and disuse.

Electric motors, generators, welding equipment, and electrical components can cause electric current to pass through bearing material. Electric current damage produces pitting, patterns of pitting, or fluted surfaces.

Oil Seal Failure Analysis

Oil seals are designed to close spaces between moving and stationary components in mechanical systems. Oil seals seal in lubricant and seal out foreign matter, providing protection from abrasive materials, moisture, lack of lubrication, and other harmful conditions. Oil seal leakage occurs due to a failed sealing member, failed metal components, and poor shaft and/or bore conditions.

The actual sealing lip of an oil seal can fail if it is brittle or cracked, has excessive wear, has very little contact with the shaft, is nicked or scratched, or has been inverted or reversed. The metal components of an oil seal can also fail and cause leakage. Common leak symptoms include a distorted steel case, damaged sealing spring, seal cocked in housing, and loose internal components. Poor shaft and bore conditions can also cause seals to leak. Poor shaft conditions cause leaks at the inside diameter of the seal and poor bore conditions cause leaks at the outside diameter of the seal. Spiral machine grooves (outward bound), scratches, or grooves on the shaft or in the bore can cause seal leakage. Paint, shellac, or gasket cement on the shaft or in the bore can also cause seal leakage.

Belt Failure Analysis

Careful analysis of belt failure or wear can indicate the cause of the problem. Common V-belt problems include premature belt failure, severe or abnormal wear, belt turning over or coming off, noise, unusual vibration, extreme stretching, and hot bearings. Belt drive problems are caused by improper drive maintenance, environmental factors, improper belt or pulley installation, poor drive design, improper belt storage or handling, and defective drive components.

Multiple belt problems include tie band separation, worn or frayed tie band, belt coming off, and one or more ribs running out of its pulley. Timing belt problems include unusual noise, tension loss, excessive belt edge wear, tensile break, cracking, premature tooth wear, tooth shear, unusual sprocket wear, belt tracking, excessive temperature, shafts out of sync, and vibration.

Chain Failure Analysis

Common problems that result in chain failure include incorrect or insufficient lubrication, misalignment, improper tension, overloads, fluctuating loads, excessive speeds, and faulty sprockets.

Improper or insufficient lubrication can cause a chain to run hot, corrode, or wear teeth on the sprockets. Misalignment causes unusual wear on the inside of the chain side plates and on the side of the sprocket teeth. Tension may be light or excessive. Insufficient tension may cause chain whip or the chain to climb the teeth of the sprocket. Excessive tension generates an unusual amount of heat. Fluctuating loads also cause chains to fail by applying intermittent stresses on the chain. A fluctuating load jerks the chain, causing it to stretch. When the load is removed, the chain whips due to the slack in the chain. The repeating cycle of jerking and whipping eventually causes the chain to fail.

Gear Failure Analysis

Proper analysis of a failed gear can help determine the cause of failure. Gear tooth failure can be classified as surface deterioration or tooth breakage. Surface deterioration includes wear, plastic flow, and surface fatigue. Tooth breakage includes fatigue, overload, and heavy wear.

Surface Deterioration. *Surface deterioration* is the loss of a material due to one surface contacting another surface. This wear can occur at a normal or an abnormal rate. Wear includes abrasive, adhesive, and corrosive wear. *Abrasive wear* is wear that occurs when a hard surface rubs against a softer surface. The harder surface gouges grooves into the softer surface, gradually reducing the size of the softer material. *Adhesive wear* is wear that occurs when high spots on objects come in contact with each other, weld together, and tear away metal at the points of adhesion. Adhesive wear usually occurs due to the lack of lubricant between the meshing teeth. Adhesive wear is commonly referred to as scoring, scuffing, or galling. *Corrosive wear* is wear that occurs when acid or moisture deteriorates a surface. Corrosive wear usually appears as pitting or etching on the entire gear.

Plastic flow is the movement of material below the surface of an object under mechanical stress. It is caused by heavy loads or high contact stresses. Plastic flow includes ridging, rolling and peening, and rippling. *Ridging* is plastic flow that occurs due to excessive loads in localized areas. Ridging usually results in gear tooth failure unless the material has excellent work hardening capabilities. *Rolling and peening* is plastic flow that occurs due to excessive loads or impact loading. *Rippling* is plastic flow that occurs from heavy loads, vibration, or improper lubrication.

Surface fatigue is failure that occurs when repeated pressure exceeds the limits of a material. Small pieces of metal are removed, leaving little cavities on the tooth surface. Surface fatigue includes pitting and spalling. *Pitting* is localized corrosion that has the appearance of cavities (pits). Pitting can be reduced or eliminated by surface hardening the gear or by grinding and polishing the gear face, providing a smooth surface with more contact area. *Spalling* is the flaking off of a metal surface.

Tooth Breakage. *Tooth breakage* is the removal of a gear tooth or part of a tooth. Tooth breakage is caused by fatigue, overload, and heavy wear. Tooth breakage usually ceases the operation of the mechanical drive system. *Fatigue* is material failure caused by repeated bending stresses that exceed the material's limits. Fatigue breakage occurs over an extended period of time that can be observed once the tooth has been severed. *Overload* is breakage caused by one large sudden shock. Shear pins are sometimes used to protect mechanical drive systems from overloads. Instead of permanent damage being done to the gear, the shear pin breaks. *Heavy wear* is a severe wearing away of an object's cross-section. Extreme abrasive wear, corrosive wear, and/or other types of wear can be great enough to erode away the cross-section (width) of the gear tooth, weakening it and causing it to break. Heavy wear can be reduced by replacing the lubricant on a frequent basis and supplying the gearbox with the proper lubricant.

Fluid Power Systems

Fluid power systems are used to control and provide power for production processes. The fluid used in a fluid power system is a liquid or gas that can move and change shape without separating when under pressure. Fluid power systems may be small for moving delicate electronic parts, or large for moving heavy assemblies on automatic welders.

HYDRAULIC PRINCIPLES

Fluid power systems use a liquid (pressurized oil) in hydraulic systems or a gas (compressed air) in pneumatic systems for the transmission and control of energy. A *hydraulic system* is a fluid power system that transmits energy in an enclosed space using a liquid (hydraulic oil) under pressure. Liquids are noncompressible fluids that can readily flow and assume the shape of an enclosed space.

Hydraulic systems produce force, ranging from a few pounds to several thousand pounds, based on pressure and area. *Force* is anything that changes or tends to change the state of rest or motion of a body. *Pressure* is force per unit area. Pressure is the internal force present in a hydraulic system. *Area* is the number of unit squares equal to the surface of an object. Force produced by a hydraulic system is determined by the pressure applied and the area of the surface receiving the pressure. **See Figure 10-1.** When pressure and area are known, force is found by applying the formula:

$$F = P \times A$$
where
F = force (in lb)
P = pressure (in psi)
A = area (in sq in.)

For example, what is the force on an actuator piston with a fluid pressure of 5000 psi and an area of 20 sq in.?

$$F = P \times A$$
$$F = 5000 \times 20$$
$$F = \textbf{100,000 lb}$$

Force increases as pressure increases. The amount of force exerted on the top of an actuator piston is determined by the pressure on the piston. When force and area are known, pressure is found by applying the formula:

$$P = \frac{F}{A}$$
where
P = pressure (in psi)
F = force (in lb)
A = area (in sq in.)

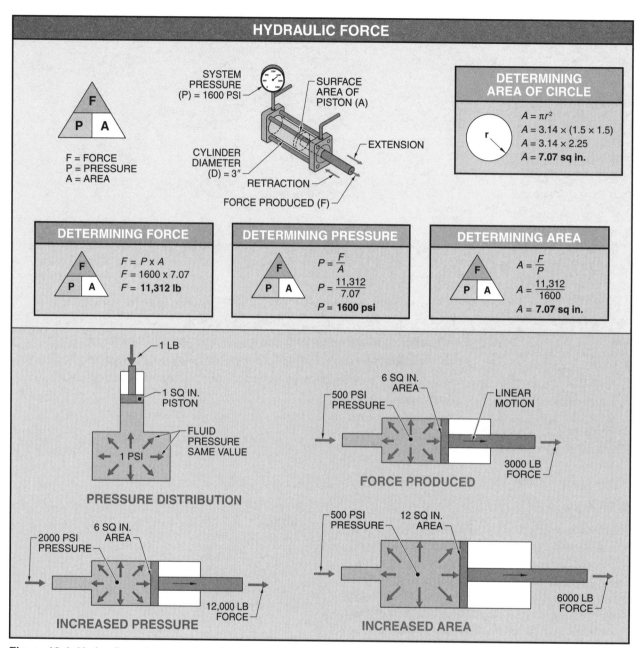

Figure 10-1. Hydraulic systems produce force ranging from a few pounds to several thousand pounds based on pressure and area.

For example, what is the pressure of a fluid if a 50,000 lb force is produced by a piston having an area of 50 sq in.?

$$P = \frac{F}{A}$$

$$P = \frac{50,000}{50}$$

$$P = \textbf{1000 psi}$$

Force also increases if the area the pressure is acting against increases. When force and pressure are known, area is found by applying the formula:

$$A = \frac{F}{P}$$

where
A = area (in sq in.)
F = force (in lb)
P = pressure (in psi)

For example, what is the area of a piston if a 300,000 lb force is produced by a cylinder having a pressure of 6000 psi?

$$A = \frac{F}{P}$$

$$A = \frac{300,000}{6000}$$

$A =$ **50 sq in.**

Pressurized hydraulic oil acting on one surface of a piston produces linear motion and force. *Linear motion* is the movement of an object in a straight line. Force is transmitted through the piston rod in a straight line to act on the load. The piston moves the load if the load offers less resistance than the force generated by the piston. The piston does not move the load if the load offers more resistance than the force generated by the piston.

One application of hydraulic force is the operation of a hydraulic jack. In a hydraulic jack, a small amount of pressure is used to create a large force. **See Figure 10-2.** A hydraulic jack can move more weight than can be moved by the jack operator alone. To raise the load a few inches, the operator must move the handle up and down many times. This is due to the law of conservation of energy which states that energy can be neither created nor destroyed, but can be changed from one form to another. In a hydraulic jack, distance and speed (repeated pumping of the handle) provide the energy to move a heavy load a short distance at a relatively slow speed.

In a hydraulic jack, lifting the pump handle raises the pump cylinder piston, opening check valve 1 and closing check valve 2. A *check valve* is a valve that allows flow in only one direction. Check valve 1 allows oil to flow from the reservoir to the pump cylinder. When the pump handle is pushed down, check valve 1 closes and check valve 2 opens, allowing hydraulic oil flow to the load cylinder. Pressure is applied on the hydraulic oil by the piston in the pump cylinder. The hydraulic oil cannot be compressed, so the same pressure is exerted throughout the system. For example, the pump cylinder piston exerts 100 psi on the 1 sq in. pump piston. The oil in the system becomes pressurized to 100 psi. The oil at 100 psi acts on the load piston, which has an area of 5 sq in., to produce 500 lb of force. The piston raises the load if the load is less than 500 lb. To lower the jack, the release valve is opened. Oil is discharged from the load cylinder from the weight of the load and the piston and flows back to the reservoir.

HYDRAULIC SYSTEMS

A hydraulic system transmits energy in an enclosed space using hydraulic oil under pressure. Hydraulic systems used on equipment are organized into circuits similar to electrical circuits. All hydraulic systems consist of a source, pathways, and load(s). **See Figure 10-3.**

Hydraulic systems are detailed in schematic diagrams. Fluid power symbols are used to identify components in hydraulic system schematic diagrams. **See Appendix.**

The source in a hydraulic system is a pump, usually driven by an electric motor or an internal combustion engine. The pathways in a hydraulic system are the hoses or pipes that direct flow from the pump through the system. A load in a hydraulic system is an actuator. An *actuator* is a component that produces motion from some other form of energy. Actuators used in hydraulic systems include cylinders and motors. Hydraulic systems are used on stationary equipment such as presses or mobile equipment such as forklifts.

Hydraulic systems have a variety of uses in industry. For example, the same principles used to move a bulldozer blade can also be controlled to perform quick, delicate tasks such as placing electronic parts on a PC board. Hydraulic systems are used in hydraulic pullers for gear and bearing removal and to position the rudders of ships.

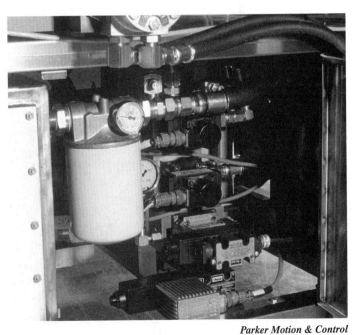

Parker Motion & Control
Hydraulic systems are often used to provide power for applications in wet environments that electrical systems may not tolerate.

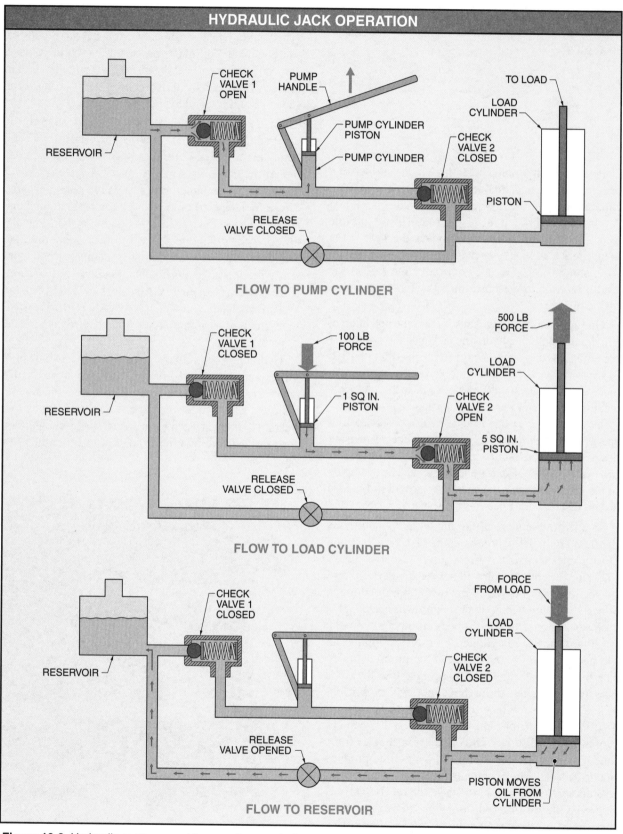

Figure 10-2. Hydraulic systems enable a small amount of pressure to create a large force.

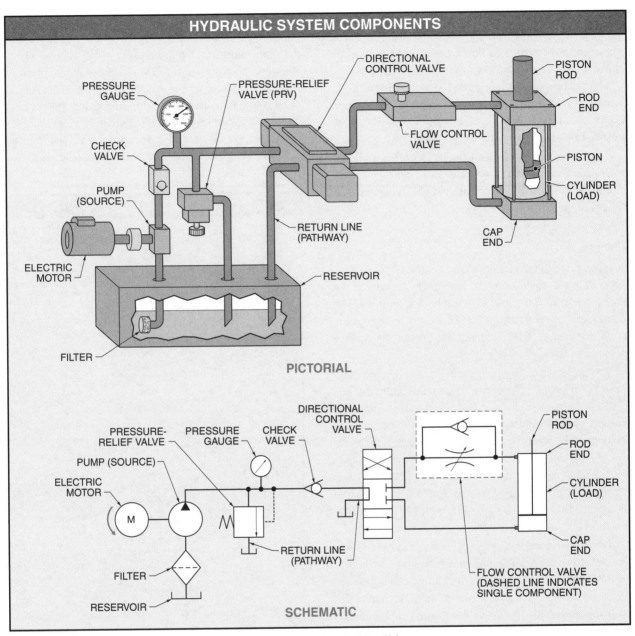

Figure 10-3. Hydraulic systems consist of a source, pathway(s), and load(s).

In a typical hydraulic system, hydraulic fluid (oil) is drawn from the reservoir, through a filter, to the pump. A *reservoir* is a container for storing fluid under little or no pressure. Valves in the pathway control the flow and pressure of the hydraulic oil. A check valve allows the flow of fluid in one direction to prevent hydraulic oil from flowing back into the pump. A *pressure-relief valve (PRV)* is a valve that sets a maximum operating pressure level for a fluid-power circuit to protect the circuit from overpressure. The PRV opens and directs hydraulic oil into the reservoir if system pressure exceeds the set pressure. Pressure gauges are installed to monitor pressure at various locations in a system.

A *directional control valve* is a valve that directs or prevents flow through selected passages. Directional control valves may be used to control hydraulic oil flow to the cap end or rod end of a cylinder. The piston rod extends when hydraulic oil flows to the cap end. The piston rod retracts when hydraulic oil flows to the rod end. Directional control valves can be controlled electrically or manually. A *flow control valve* is a valve that regulates the volume of hydraulic oil flowing to components in a system. Flow control valves may be used to control the flow rate of fluid into a cylinder to control the speed of piston movement. The flow control valve contains a check valve to prevent flow in the reverse direction.

Hydraulic systems are closed systems. In a closed system, hydraulic oil is pumped, used in the system, filtered, and reused in a continuous cycle without exposure to elements outside the system.

System components and operation are similar in hydraulic and pneumatic systems. Both include various combinations of pumps, piping, valves, cylinders, and other components.

Pumps

A *pump* is a mechanical device that causes fluid to flow. Hydraulic oil flows from the high pressure area at the pump outlet into the low pressure hydraulic lines. The resistance to flow in the pathway and load on the system creates pressure in the system. The greater the load or resistance, the higher the pressure.

Hydraulic pumps are commonly positive displacement pumps. A *positive displacement pump* is a pump that has a seal between its inlet and outlet and moves a specific volume of fluid with each revolution. Displacement is the volume of hydraulic oil moved during each revolution, stroke, or cycle of a pump. A *fixed displacement pump* is a pump that moves a fixed amount of fluid during each revolution of the pump. A *variable displacement pump* is a pump that moves a variable amount of fluid during each revolution of the pump.

Pumps are rated by flow in gallons per minute (gpm) and pressure rating. The pressure rating of a hydraulic pump is the amount of pressure the pump can withstand before it is damaged. Excessive pressure may cause the pump housing to rupture or the pump seals to burst. For maximum safety, the hydraulic pump selected for a system should be rated to withstand a pressure higher than the highest anticipated pressure in the system. This prevents damage and possible failure if an overpressure condition occurs. Hydraulic pumps include gear, vane, and reciprocating pumps.

Gear Pumps. A *gear pump* is a pump that consists of two meshing gears enclosed in a close-fitting housing. Gear pumps may be external, internal, or gerotor pumps. **See Figure 10-4.** An *external gear pump* is a pump that consists of a pair of meshing gears that seal with the pump housing. A motor drives one gear (drive gear), and the driven gear rotates freely inside the pump housing.

As the gears rotate, hydraulic oil is directed from the pump inlet into the space between the gear teeth and the pump housing. The open spaces between the gear teeth and housing are filled with oil. As the teeth mesh on the outlet side of the pump, hydraulic oil is forced out through the outlet (high-pressure area). The process continues as the gear teeth unmesh at the inlet and create a low-pressure area. Hydraulic oil is forced into the pump inlet by the higher pressure in the reservoir.

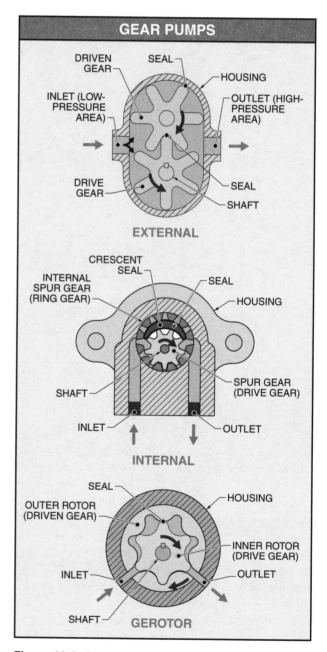

Figure 10-4. A gear pump is a pump that consists of two meshing gears enclosed in a close-fitting housing.

An *internal gear pump* is a pump that consists of a small spur gear (drive gear) mounted inside a large internal spur gear (ring gear). The gears rotate in the same direction. A crescent seal separates the low- and high-pressure areas of the pump. Hydraulic oil is trapped as the gears rotate and is discharged through the pump outlet.

A variation of the internal gear pump is the gerotor pump. A *gerotor pump* is a pump that consists of a rotating multiple-lobed inner rotor that meshes with lobes of a rotating outer rotor. Gerotor pump operation is similar to internal gear pump operation. The inner rotor has one less lobe than the outer rotor. Both rotors rotate in the same direction. One lobe is completely meshed at all times. Oil enters the low-pressure area formed at the pump inlet. As the rotors mesh, a high-pressure area is created, and oil is discharged from the pump outlet.

Vane Pumps. A *vane pump* is a pump that contains vanes in an offset rotor. **See Figure 10-5.** Like gear pumps, vane pumps use a rotating component (rotor) to produce the flow of hydraulic oil. The tips of the vanes ride on a machined surface inside the pump housing. The pump housing contains the oil between the vanes. Oil enters at the pump inlet and fills the spaces between the vanes, which become sealed chambers. As the rotor rotates, the vanes are forced in the slots in the rotor by the shape of the pump housing to create smaller spaces. This produces a high-pressure area and oil flows out of the pump outlet. Throughout the process, the vanes are held against the pump housing by spring pressure or hydraulic oil pressure.

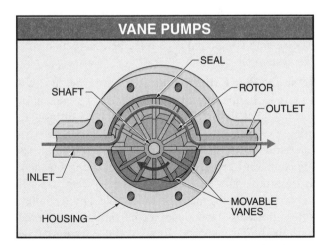

Figure 10-5. Vane pumps consist of a rotor with slots housing movable vanes.

Vane pumps may be balanced or unbalanced. A *balanced pump* is a pump that has two inlets and outlets at opposite sides of the pump. This creates a balanced load on the pump bearings and seals. Unbalanced pumps have only one inlet and outlet.

Reciprocating Pumps. A *reciprocating pump* is a pump in which fluid flow is produced from pistons moving back and forth. Reciprocating pumps are axial or radial. **See Figure 10-6.** An *axial reciprocating pump* is a pump that consists of pistons in a rotating cylinder block parallel to the drive shaft. A *radial reciprocating pump* is a pump that consists of pistons located perpendicular to the pump shaft.

In an axial reciprocating pump, the pistons move in and out of the cylinder block as the cylinder block rotates due to the angle of the swash plate. A *swash plate* is an angled plate in contact with the piston heads that moves the pistons in the cylinders of a pump. Oil enters the cylinder as the piston moves away from the inlet. Oil is discharged from the cylinder as the piston moves toward the outlet. Pump displacement is controlled by changing the angle of the swash plate. When the swash plate is in the vertical position, there is no piston movement, and no hydraulic oil is pumped even though the cylinder block rotates. The direction of flow in the pump is reversed by reversing the angle of the swash plate.

Radial reciprocating pumps contain pistons located perpendicular to the pump shaft. Radial reciprocating pumps are classified as cam or rotating reciprocating pumps. In a cam pump, a rotating internal cam moves the pistons in cylinders. The cam is shaped to push the pistons out during one half of the cam rotation and allow the pistons to retract during the other half. In a rotating pump, pistons are housed in a rotating cylinder block. The rotating cylinder block is offset inside the pump housing and rotates around a fixed shaft. Oil enters the pump inlet as the pistons extend and is discharged from the pump outlet as the pistons retract.

Component wear at the sealing points between the inlet and outlet side of the pump can allow internal leakage. Internal leakage generates heat in the hydraulic oil, causing thinning (low viscosity). Low viscosity results in a further breakdown of lubrication capabilities at sealing points and increases the wear and resulting leakage. In some pumps, internal components are manufactured as a cartridge assembly that is replaced as a unit when pump output decreases from wear or damage. Replacement cartridge components may differ from the original pump components and change the rated output of the pump.

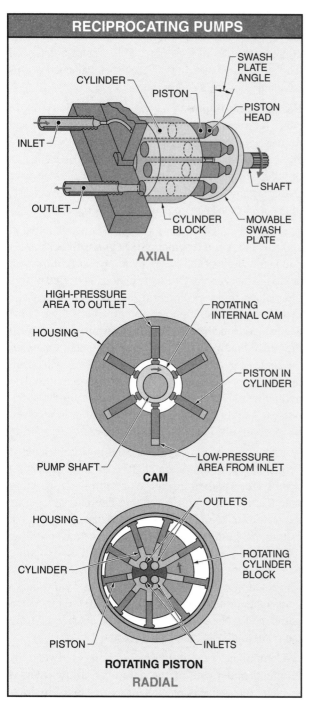

RECIPROCATING PUMPS

AXIAL

CAM

ROTATING PISTON
RADIAL

Figure 10-6. Reciprocating pumps consist of reciprocating pistons in cylinders.

Hoses and Piping

Hydraulic systems use hoses and/or pipe to transport the hydraulic oil from the pump to the actuators. A *hose* is a flexible tube for carrying fluids under pressure. Hydraulic hose is less susceptible to damage from vibration or movement than rigid pipe. Flexibility allows the dampening of pressure surges and the ability to route around obstacles in a system. Hydraulic hose is constructed of synthetic material and consists of the tube, reinforcement, and cover. **See Figure 10-7.**

The tube contains and directs the flow of oil. The reinforcement has wire or textile plies to provide resistance to system pressures. The cover protects the reinforcement layers and tube from abrasion and external conditions. Some hose cover is designed to be skived. *Skiving* is the removal of the cover from a hydraulic hose down to the reinforcement in order to attach a fitting. Thick-cover hose usually requires skiving. Thin-cover hose usually does not require skiving. Hydraulic hoses are available as braided, spiraled, and helical.

Hose selection is based on ambient temperature and system pressure, flow, and temperature requirements. In addition, the hose must be compatible with the oil used in the system. Special hose properties are listed by the manufacturer.

A *pipe* is a hollow cylinder of metal or other material of substantial wall thickness. Pipe is a rigid or semi-rigid material and is more durable than hose. Pipe is used to transport oil in parts of a system that do not require flexibility. Copper, aluminum, or plastic pipe is used only in low-pressure applications. Steel pipe is used in high-pressure applications. Manufacturer's specifications list recommended products and applications.

Pipe must be secured with brackets and clamps. Before use, the inside of the pipe must be completely cleaned. Dirt and/or foreign matter can clog small passages in system components. Threaded connectors used with pipe require sealing compound or have a built-in sealing ring to prevent leaks. A small amount of sealing compound is applied to three-quarters of male threads. Excessive sealing compound use can force the compound inside the pipe, resulting in contamination. The sealing compound used must be compatible with the system oil.

Hose and Pipe Fittings. Hose and pipe fittings are used to join hydraulic lines. Hose fittings are permanent or reusable. **See Figure 10-8.** Permanent hose fittings are crimped into the hose reinforcement to withstand system operating pressures and vibration. Replacement hoses are commonly ordered with the correct fitting attached. Hose fittings may be joined to the hose in the field if the proper crimping machine is available. For maximum safety, hoses used in industrial applications are ordered with the fittings attached. Reusable hose fittings allow the joining and removal of hose using threads, clamps, and/or special nipple designs.

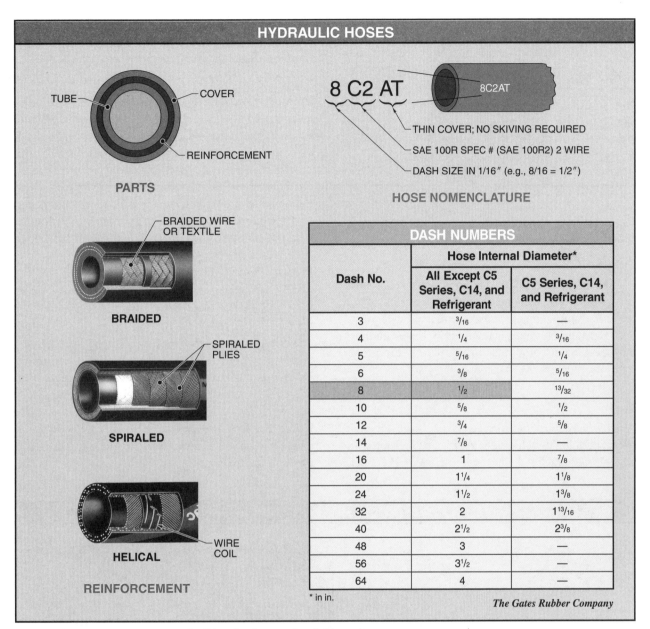

Figure 10-7. Hydraulic hose is a flexible tube for carrying fluids under pressure.

Pipe fittings vary in design depending on the pipe used and application requirements. Fittings can be joined with threads or hose couplings.

Hose Couplings. A *hose coupling* is a fitting that allows a line to be connected or disconnected by hand. A quick-disconnect coupling seals quickly to prevent oil leakage when connected or disconnected. **See Figure 10-9.** Oil flow through the coupling is controlled with poppet valves. A *poppet valve* is a valve in which the valve seating element (poppet) pops open to allow flow in one direction. The poppet valve quickly closes when

the direction of flow reverses. When disconnected for an extended period of time, the hose ends should be fitted with plugs or caps to prevent contaminants from entering the coupling. Hose coupling selection is based on flow, pressure, and temperature requirements.

🔧 **TECH TIP**

The pressure drop in a hydraulic hose is based on the fluid flow rate, length and diameter of the hose, type of hydraulic oil and viscosity, friction in the hose, and couplings/fittings in the line that change the direction of the fluid.

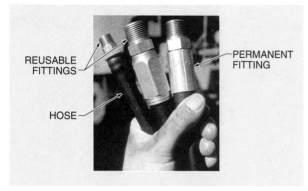

The Gates Rubber Company

Figure 10-8. Hose and pipe fittings are used to join hydraulic lines and may be permanent or reusable.

Like hose fittings, hose couplings may be permanent or reusable. Permanent hose couplings are crimped into the hose reinforcement for maximum resistance to system operating pressures and vibration. Hose couplings may be joined to the hose in the field if the proper crimping machine is available.

Valves

A *valve* is a device that controls the pressure, direction, or rate of fluid flow. Hydraulic valves are precision components manufactured with close tolerances. Premature wear can occur if valves are exposed to dirt, water, or acids, which can form in hydraulic oil. Common hydraulic valves include pressure control, directional control, and flow control valves.

Pressure Control Valves. A *pressure control valve* is a valve used to regulate pressure in a hydraulic system. Resistance produced by system components and the load determines system pressure. A pressure control valve prevents system overpressure, reduces pressure in various branches of the system, and unloads pumps when hydraulic oil flow is not needed. *Unloading* is allowing a pump to run against little or no pressure. In hydraulic systems, unloading sends oil directly from the pump to the reservoir, which is usually under low pressure. Pressure control valves include pressure-relief valves, pressure-reducing valves, pressure-sequence valves, and unloading valves. **See Figure 10-10.**

A pressure-relief valve (PRV) sets a maximum operating pressure level for a circuit to protect the circuit from overpressure. A direct-acting pressure-relief valve uses a ball under spring pressure or a poppet. The valve opens and discharges hydraulic oil to the reservoir when hydraulic oil pressure overcomes the valve spring pressure. Pressure-relief valves are commonly adjustable within a given pressure range and are used on many industrial systems.

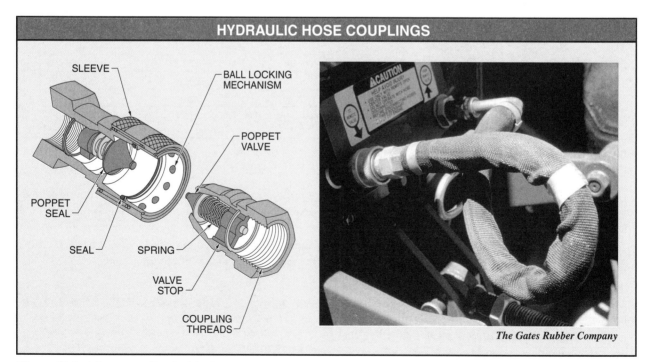

HYDRAULIC HOSE COUPLINGS

The Gates Rubber Company

Figure 10-9. A hose coupling (quick-disconnect) allows a line to be connected or disconnected without tools.

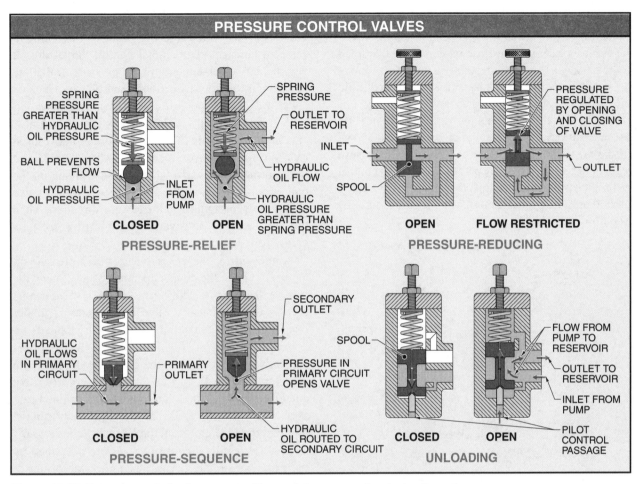

Figure 10-10. Pressure control valves are used to regulate pressure in a hydraulic system.

A *pressure-reducing valve* is a valve that limits the maximum pressure at its outlet, regardless of the inlet pressure. Pressure-reducing valves operate by pressure being sensed at their outlet. The spool is moved off its normal position, reducing or blocking working pressure when higher-than-system pressure is reached. Pressure-reducing valves are used to maintain different pressures in different branches of a system. A pressure-reducing valve is normally open and is forced closed by pressure applied from the downstream side of the valve.

A *pressure-sequence valve* is a pressure-operated valve that diverts flow to a secondary actuator while holding pressure on the primary actuator at a predetermined minimum value after the primary actuator completes its travel. The primary circuit must operate and reach the specified pressure before the secondary circuit operates. For example, a hydraulic press positions and then stamps a part. The primary circuit controls movement of the part into position. The secondary circuit controls the ram on the press. Flow to

the secondary circuit is allowed by a pressure-sequence valve when pressure in the primary circuit reaches a specific level.

An *unloading valve* is a pressure control valve that directs hydraulic oil from the pump to the reservoir after system pressure has been reached. An unloading valve is pilot operated. A *pilot-operated valve* is a valve controlled by a pressure signal in the pilot control passage connecting parts of the pressure-reducing valve. A *pilot control passage* is a small pathway used to transmit a valve pressure signal.

The pilot control passage is connected to the bottom of the spool inside the unloading valve. The pilot control passage is connected to a point beyond the unloading valve and senses system pressure. When pressure in the pilot control passage reaches a specific pressure, the spool moves and flow is diverted to the reservoir. When the pump is unloaded, it works against the lower pressure of the reservoir, reducing wear on the pump and generating less heat in the hydraulic oil.

Pilot control passages consist of small paths connected to various parts of the valve. Pilot control passages must be kept clean. System operation is disrupted if the passages become clogged with dirt. Pilot control of a valve operates because pressurized hydraulic oil exerts pressure equally and in all directions. For example, if pressure in one part of a system is 1000 psi, pressure in a pilot passage connected to that part of the system is also 1000 psi. This allows pressure signals to be sent from any part of a system to control different valves. Pilot control passages are indicated by dashed lines on hydraulic system schematic diagrams. **See Figure 10-11.**

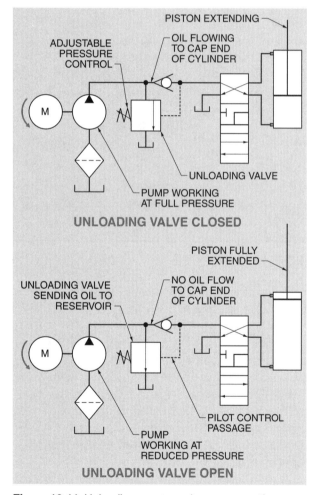

Figure 10-11. Unloading a pump reduces wear on the pump and generates less heat in the hydraulic oil.

TECH TIP

Hydraulic system oil filling tubes and funnels must be cleaned with lint-free rags. Lint from common shop rags can enter a hydraulic system and disrupt lubricating capabilities and block pilot control passages.

Directional Control Valves. A directional control valve is a valve that directs or prevents flow through selected passages. Directional control valves do not significantly change hydraulic oil pressure or the rate of hydraulic oil flow. Directional control valves include check valves and spool directional control valves.

A check valve allows flow in only one direction. A check valve consists of a ball or poppet held closed by a spring. The valve opens and hydraulic oil flows through the check valve when pressure acting on the ball or poppet exceeds the spring pressure. The valve closes, preventing flow in the reverse direction, when the flow attempts to reverse or when pressure falls below spring pressure. **See Figure 10-12.**

A pilot-operated check valve uses pilot pressure on a piston to force the poppet open and allow flow in the reverse direction. A pilot-operated check valve can be used as a counterbalance valve that prevents a cylinder from moving when hydraulic oil flow has stopped.

A *spool directional control valve* is a directional control valve that controls the flow of hydraulic oil to and from actuators. A spool directional control valve consists of a spool fitted into precision machined passages in a block of metal. The passages lead from the pump to the load and from the load to the reservoir.

Spool directional control valves are described by their number of ways. A *way* is a port into or out of a valve. Directional control valves are classified as two-way, three-way, and four-way valves. **See Figure 10-13.** The moving spool covers or uncovers the ways to direct hydraulic oil to the correct location in the system. The motion of the spool can be controlled manually, mechanically, pneumatically, hydraulically, or electrically.

Four-way directional control valves are most commonly available as open center or closed center valves. **See Figure 10-14.** Open center directional control valves allow hydraulic oil to flow from the pump to the reservoir when the spool is in the neutral position. The *neutral position* is the position of a directional control valve when no controlling devices in the system are energized and there is no force causing the spool to move from the neutral position. Closed center directional control valves block hydraulic oil flow from the pump to the reservoir when the spool is in the neutral position. A closed center directional control valve requires a pressure-relief valve to unload the pump or the pump must be shut OFF to prevent system overpressure.

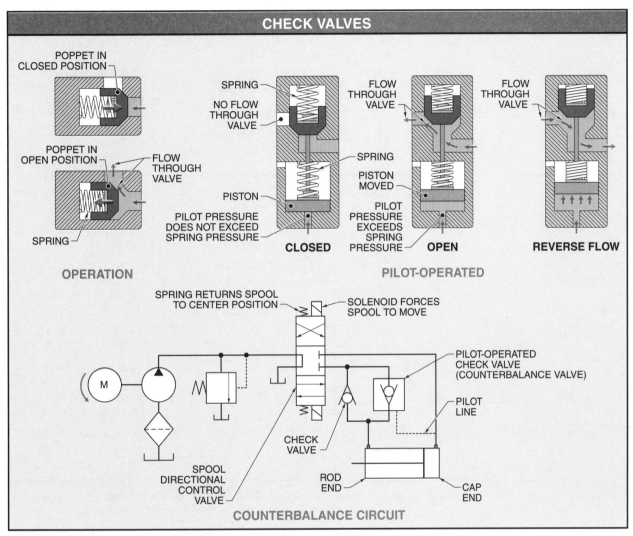

Figure 10-12. A check valve is a directional control valve that allows flow in only one direction.

Directional control valves may be stacked in groups or located in a valve body that controls hydraulic oil flow for other valve operations. Hydraulic oil flow paths in these valves can be complex. Specific operation determination requires analysis of circuit schematics and symbols marked on the valve body.

Flow Control Valves. A flow control valve regulates the volume of hydraulic oil flowing to components in a system. Flow through a flow control valve is controlled by varying (metering) the size of the opening in the valve. The amount of flow controls the operation of system components. For example, the volume of hydraulic oil flow controls the speed of the piston in a cylinder. The larger the opening in the flow control valve, the greater the flow, and the faster the piston rod moves.

Diverting flow to another circuit in the system can also control the volume of oil flow. Flow rate is commonly measured in gallons per minute (gpm). Flow control valves can be designed to compensate for changes in hydraulic oil flow due to changing temperatures and pressures in a system.

A flow control valve may be used to control flow into or out of an actuator such as a cylinder. **See Figure 10-15.** In a meter-in application, the flow control valve controls the rate at which oil enters the cylinder. A check valve built into the flow control valve body prevents unrestricted oil flow into the cylinder. In a meter-out application, the flow control valve controls the rate at which oil flows out of the cylinder. The check valve allows free oil flow when the direction of flow is reversed in both applications.

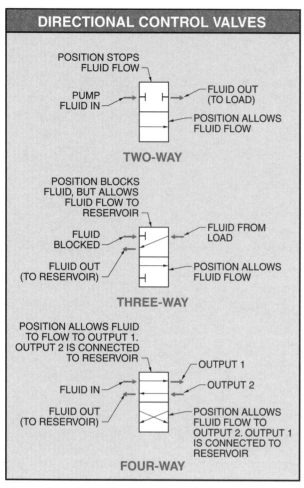

Figure 10-13. Spool directional control valves are described by their number of ways.

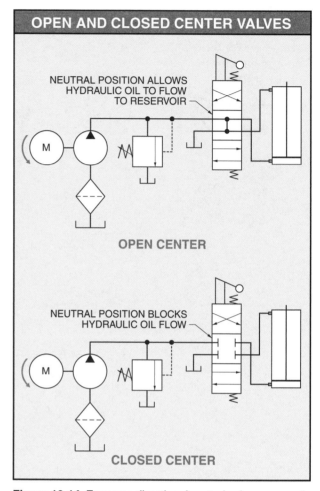

Figure 10-14. Four-way directional control valves are available as open center and closed center valves.

Cylinders

A *cylinder* is a component that converts fluid pressure into linear mechanical force. The speed of piston movement in a cylinder is directly proportional to the rate of oil flow. Hydraulic cylinders are single-acting or double-acting. **See Figure 10-16.** Single-acting cylinders are used to produce force in one direction. Single-acting cylinders have a piston in the cylinder that is extended by applied hydraulic oil pressure and retracted by the weight of the load, gravity, or a spring. Double-acting cylinders are used to produce force in two directions. Double-acting cylinders have a piston that is extended and retracted by applied hydraulic oil pressure.

✚ *SAFETY TIP*

Always wear safety glasses when skiving hose. Wear hearing protection and avoid loose-fitting clothing when power skiving.

Double-acting cylinders may be balanced or unbalanced. A balanced cylinder exerts equal force when extending and retracting because the cylinder rod extends through the piston, creating an equal surface area on both sides of the piston. Unbalanced cylinders generate a greater force extending because the piston rod is attached to only one side of the piston. This creates a greater surface area on the cap end than the rod end. All pistons have seals to keep hydraulic oil inside the cylinder and internal seals to prevent leakage of hydraulic oil around the piston. Wiper seals prevent accumulated dirt on the piston rod from entering the cylinder housing.

A *cushion cylinder* is a cylinder that slows piston movement to provide a gradual stop. **See Figure 10-17.** In a cushion cylinder, a tapered projection closes the outlet gradually and slows hydraulic oil flow out of the cylinder by forcing the hydraulic oil through a small passage located next to the main hydraulic oil passage.

The reduced flow slows the cylinder to a gentle stop. Flow through the small passage is adjusted by a needle valve that controls the flow of hydraulic oil. The larger the opening allowed by the needle valve, the quicker the piston stops.

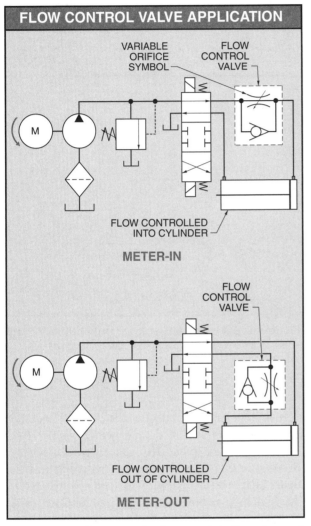

Figure 10-15. Flow control valves regulate the volume of hydraulic oil flowing to the components in a system.

A *regenerative cylinder* is a cylinder that has a piston rod sized to have a cross-sectional area equal to half of the piston area. **See Figure 10-18.** Hydraulic oil discharging from the rod end is added to hydraulic oil entering the cap end. This doubles the hydraulic oil flow to the cap end, which has twice the volume of the rod end. Therefore, the piston extends and retracts at the same speed. Twice the oil flow fills twice the volume in the cap end as the rod end because the volume of hydraulic oil flow controls speed.

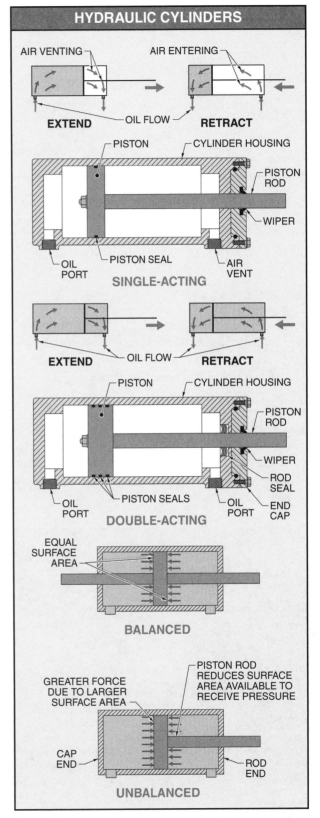

Figure 10-16. Hydraulic cylinders convert fluid pressure into linear mechanical force.

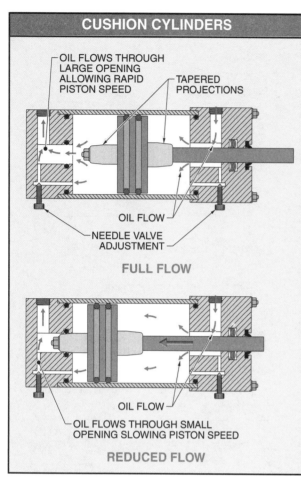

Figure 10-17. Cushion cylinders slow piston movement to provide a gradual stop.

Power Team, Division of SPX Corporation
Hydraulic systems are used in maintenance applications that require a great amount of force and accurate speed control.

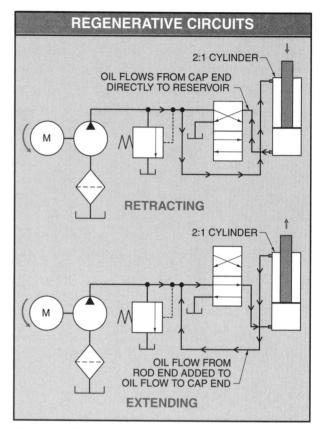

Figure 10-18. A regenerative circuit has the hydraulic oil discharged from the rod end added to the hydraulic oil entering the cap end.

Hydraulic Motors

A *hydraulic motor* is a device that converts hydraulic energy into rotation. A hydraulic motor is similar in design to a hydraulic pump, though it operates in reverse. **See Figure 10-19.** Oil is forced into the hydraulic motor under high pressure and is discharged at low pressure. The hydraulic oil flow is converted into rotational force. Hydraulic motor speed is controlled by the rate of oil flow. The greater the flow, the faster the motor rotates. Torque produced by a motor is controlled by oil pressure. The greater the pressure, the greater the torque produced. Hydraulic motors are rated for the maximum torque and/or horsepower produced.

Hydraulic motors are used in applications where electric motors are impractical. For example, hydraulic motors are used on large hoists where an excessively large electric motor would be necessary to produce the required horsepower. Hydraulic motors are less affected by adverse conditions on ships or other wet locations. In addition, hydraulic motors are often used in atmospheres where electric arcing may cause an explosion.

HYDRAULIC MOTORS

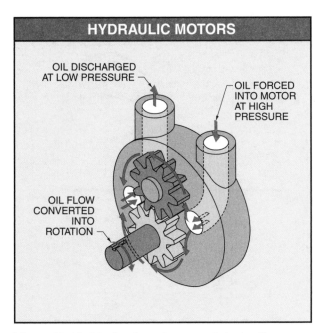

Figure 10-19. A hydraulic motor converts hydraulic energy into rotary mechanical energy.

Accumulators

An *accumulator* is a container in which fluid is stored under pressure. Most accumulators are pneumatically charged with dry nitrogen gas. The gas is located in the top of the accumulator and is separated from the hydraulic oil by a piston, diaphragm, or bladder. Other common accumulators include weight-loaded and spring-loaded accumulators.

A common function of an accumulator is to maintain the proper motor or piston speed by smoothing fluctuations in system pressure. Hydraulic oil enters the accumulator and compresses the nitrogen gas. **See Figure 10-20.** The pressure in the accumulator is the same as in the rest of the system when the system pressure is constant.

However, if the pressure in the system drops, the hydraulic oil in the accumulator is then at a higher pressure than the hydraulic oil in the system. This allows the hydraulic oil in the accumulator to flow into the system, increasing oil flow and maintaining the speed of the motor or piston. When the flow from the accumulator is not required, the accumulator is recharged by hydraulic oil flow from the pump until the pressure in the accumulator and the system are equal. Accumulators also absorb pressure shocks caused by the opening and closing of valves.

SPEED CONTROL CIRCUITS

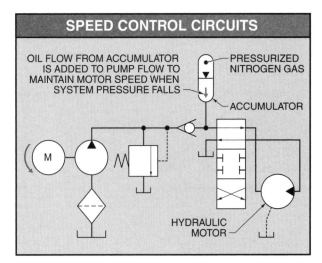

Figure 10-20. An accumulator stores pressurized hydraulic oil for use in a system.

Caution: Always discharge hydraulic oil from all accumulators before working on a hydraulic system. Stored hydraulic oil in an accumulator can cause a system to operate without warning.

Reservoirs

A reservoir stores extra hydraulic fluid. **See Figure 10-21.** In the reservoir, heat is transferred from the hydraulic oil through the reservoir walls to the atmosphere. Any air present in the hydraulic oil is separated from the oil, and contaminants settle to the bottom. The most common reservoir uses a filler cap with an air vent. The air vent has a filter that allows airflow in and out of the reservoir but prevents dirt from entering the system. Sealed reservoirs do not require air vents because air does not enter or exit the system. Sealed reservoirs are often pressurized with dry nitrogen gas and can improve pump operation.

A baffle plate inside the reservoir routes hydraulic oil in a path for maximum cooling and prevents hot hydraulic oil from flowing directly from the return line to the pump inlet (suction piping). An oil level gauge indicates hydraulic oil level in the reservoir.

Hydraulic Filters

A *hydraulic filter* is a component that traps dirt and contaminants present in hydraulic oil. Dirt and contaminants are the most common cause of hydraulic system problems, so a hydraulic filter is the most important component in a hydraulic system and must be properly maintained.

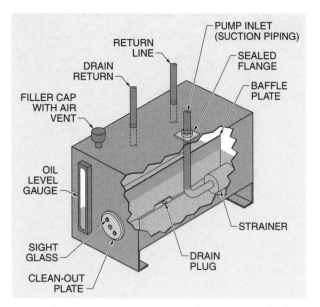

Figure 10-21. A reservoir stores hydraulic oil required for system operation.

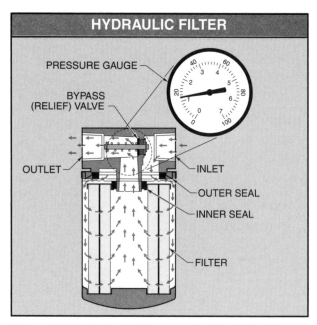

Figure 10-22. Hydraulic filters may be fitted with a bypass (relief) valve that opens to allow hydraulic oil flow if the filter becomes clogged.

Hydraulic filters are rated in microns. A *micron* is an alternate term for micrometer, or one millionth of a meter. The unit is designated with "μm" or sometimes abbreviated with just "μ". The number of microns indicates the actual size of the openings in the filter material. The smaller the micron number, the finer the filtering capability. Hydraulic filters trap most contaminants 10 μm and larger. The human eye can see a 40 μm particle without magnification. Contaminants the size of a grain of table salt are approximately 70 μm. Contaminants the size of milled flour are approximately 25 μm.

During use, the size of the filter material openings decreases as dirt becomes trapped in the filter material. Hydraulic oil flow is restricted, starving the system of hydraulic oil if the filter becomes completely clogged. Many hydraulic filters are fitted with a bypass (relief) valve that opens to allow hydraulic oil flow if the filter becomes clogged. **See Figure 10-22.** A pressure gauge located on the filter indicates when the filter should be changed based on the flow rate through the filter.

Clean oil is crucial for proper system function. Hydraulic oil provides a thin film that acts as a seal to separate metal surfaces. This seal allows the metal surfaces to slide past each other easily, which is the lubricating function of the oil. Foreign matter can contaminate and break the oil film, allowing metal-to-metal contact. This increases the wearing away of the metal. Therefore, even tiny and unseen contaminants can disrupt the sealing and lubricating properties of hydraulic oil.

Contaminants may be hard like sand or pieces of metal or soft like rubber, fibers from rags, or microorganisms such as bacteria. Water in hydraulic oil causes a chemical breakdown of the oil, which forms acids. The acids corrode metal surfaces, resulting in premature wear and bearing fatigue, which may lead to bearing failure. An increased frequency of equipment service may be caused by hydraulic oil contamination. The source(s) of contamination must be removed and the system cleaned according to manufacturer's recommendations.

Oil cleanliness is rated by ISO Standard 4406. This standard uses a three-number code to determine the cleanliness of oil. **See Figure 10-23.** The oil is examined under a microscope and the contaminant particles are counted. The first number indicates the number of particles per milliliter of oil that are larger than 2 μm. The second number indicates the number of particles per milliliter that are larger than 5 μm. The third number indicates the number of particles per milliliter that are larger than 15 μm.

Various hydraulic components require a specific fluid cleanliness. For example, a gear pump should have oil with a rating of 19/17/14, meaning that it should have no more than 5000 particles 2 μm or larger, 1300 particles 5 μm or larger, and 160 particles 15 μm or larger for each milliliter of hydraulic oil.

FLUID CLEANLINESS STANDARDS			
ISO Code	Particles/Milliliter		
	>2 μ	>5 μ	>15 μ
23/21/18	80,000	20,000	2500
22/20/18	40,000	10,000	2500
22/20/17	40,000	10,000	1300
22/20/16	40,000	10,000	640
21/19/16	20,000	5000	640
20/18/15	10,000	2500	320
19/17/14	5000	1300	160
18/16/13	2500	640	80
17/15/12	1300	320	40
16/14/12	640	160	40
16/14/11	640	160	20
15/13/10	320	80	10
14/12/9	160	40	5
13/11/8	80	20	2.5
12/10/8	40	10	2.5
12/10/7	40	10	1.3
12/10/6	40	10	.64

Figure 10-23. Oil cleanliness is rated by ISO Standard 4406.

In addition to microscopic inspection, a chemical analysis of hydraulic oil is performed to identify acid contamination or chemical breakdown. These tests are usually performed by companies specializing in oil analysis, though some plants have their own portable test kits to test critical hydraulic systems frequently. **See Figure 10-24.** The report from the testing company may list recommendations such as a complete flushing of the system, more efficient filters, and/or changes in maintenance procedures such as the use of lint-free rags for cleaning.

Hydraulic Oil Selection

Hydraulic oil is selected carefully to match system operating conditions such as ambient temperature, operating temperatures, and system pressures. For example, ambient temperature has an effect on equipment operation. Hydraulic systems may operate erratically in cold temperatures. This is caused by an increase in hydraulic oil viscosity. A heater is commonly installed in a system to minimize cold temperature problems.

Hydraulic oil contains additives to improve its performance. Anti-wear additives help separate metal surfaces to reduce friction. Detergent additives help carry away dirt. Anti-foaming additives prevent the formation of bubbles as oil moves through a system. Oxidation inhibitors help prevent the breakdown of metal surfaces and the formation of sludge and acids.

Most hydraulic oils are petroleum-based. Other hydraulic oils are synthetic, water-based, or vegetable oil-based. Industry is using a greater number of hydraulic oils that are less damaging to the environment. These oils are usually synthetic or vegetable-based oils. Hydraulic oil should remain chemically stable while in storage and service. Different hydraulic oils should not be combined. Combining different hydraulic oils in a system can result in the formation of sludge and/or corrosive acids. The oil must match system requirements for proper performance.

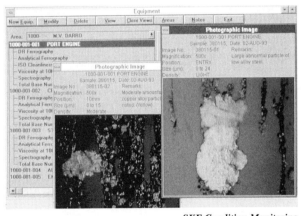

SKF Condition Monitoring

Figure 10-24. A chemical analysis of hydraulic oil is performed to identify the presence of acids and/or chemical breakdown.

Hydraulic Oil Storage

Hydraulic oil containers should be stored indoors in clean areas. All containers should be clearly labeled to avoid accidental mixing of different oils. Drums should be stored on their sides with their bungs horizontal so that they are sealed with oil. Drums that are stored upright allow water and air containing dirt and moisture to enter the drum as it expands and contracts with temperature changes. Bungs should be kept tightly sealed. Drums stored outside should be properly covered to eliminate problems from moisture and dirt.

HYDRAULIC SYSTEM MAINTENANCE

Hydraulic systems should be inspected routinely. The frequency of inspection varies with manufacturer's recommendations. Mobile hydraulic equipment should be inspected every 400 hr to 600 hr or every three months if manufacturer's recommendations are not available. Stationary hydraulic equipment should be inspected every three months if manufacturer's recommendations are not available. Inspection intervals are dependent on conditions such as operating temperatures, pressures, environment (such as shock, vibration, or abuse), and accessibility.

Hydraulic Oil Maintenance

Most hydraulic system maintenance problems are caused by contamination or breakdown of the hydraulic oil. The most important aspect of hydraulic system maintenance is to prevent contamination of the hydraulic oil. Air, water, and dirt can combine in hydraulic oil to create sludge and corrosive acids. Sludge and small particles can clog small openings such as pilot passages, causing severe operating problems. Acids corrode metal components, plastic seals, and hoses. Contaminated hydraulic oil has diminished lubricating capabilities, causing rapid wearing of metal parts.

The most important maintenance activity is to change hydraulic system filters according to the manufacturer's schedule or when indicated by the filter pressure gauge. The correct filter must be installed for the system. Dirt must not enter the system when handling the hydraulic oil. The hydraulic oil filling tube and funnels must be cleaned with lint-free rags before and after use. The reservoir cap should fit snugly. The air vent filter must keep dirt out but allow air to flow freely in and out of the reservoir.

The correct hydraulic oil must be used for the application. All hydraulic oils contain additives such as foaming and rust inhibitors. Some hydraulic oil additives can cause deterioration of certain hose and valve materials. Improper hydraulic oil viscosity can cause serious operating problems. Hydraulic oil that is too thick can cause sluggish operation. Hydraulic oil that is too thin can cause internal leakage and poor lubrication.

Hydraulic oil must maintain the required viscosity as the system warms up. Hydraulic oil that loses viscosity can cause internal leaks and reduced lubrication capabilities. Hydraulic oil must be drained and replaced according to manufacturer's recommendations. All used hydraulic oil must be disposed according to applicable environmental regulations.

Overheating can damage hydraulic oil and system components. Operating temperatures must be kept within the range suggested by the manufacturer. The reservoir and all system components should be clean so heat can dissipate easily. Kinked lines route hydraulic oil through bends and small openings, which can cause heat buildup. The pressure relief settings must be set at the recommended level. Excessive pressures generate additional heat. Avoid overspeeding or overloading a system to prevent heating hydraulic oil. Hydraulic oil coolers must be kept clean and operating efficiently.

Hydraulic Hose Maintenance

Hoses should be regularly inspected for signs of wear, cracking, or leaks. Hoses with deep cracks or splits should be replaced immediately. Suspected small leaks are located safely by passing a clean piece of paper or cardboard near the suspected leak source. Oil stains indicate a leak in the line.

Warning: Small hydraulic oil leaks are extremely dangerous as high-pressure hydraulic oil can pierce the skin and enter the body.

Hydraulic hoses can be damaged if improperly sized or installed. Elbows and adapters should be used to relieve strain on hoses. **See Figure 10-25.** Hoses must not be allowed to rub against anything that could remove the outer layer. Clamps are often required to support long hose runs or to keep the hose away from moving parts. Keep hose bends as large as possible to avoid hose collapse and flow restriction. Hoses exposed to excessive heat experience deterioration of the outer layer (cover) and add heat to the hydraulic oil. The inner layer (tube) can be damaged from overheating of the oil, rough treatment, or incompatible oil. Breakdown of the inner layer can cause an internal blockage, resulting in noisy pump operation or erratic actuator operation. Hoses should not be stretched too tightly, twisted, or kinked. Allow enough slack in the hose line in straight installations to provide for length changes that occur when pressure is applied.

Care is required when removing packaging material from a hose. A hose nicked by the sharp edge of a blade is weakened and may fail. Hoses should be stored in cool, dry areas away from direct sunlight. Hoses should never be hung on nails or hooks. Hoses should be stored in smooth, even coils to avoid kinking. Never use hoses designed for one application in another. Never kink or fold back a hose to stop flow. Ensure that there are enough shutoff valves to stop fluid flow where required.

Hose spring guards should be used to avoid flexing. When reinforced wire is deformed, it should be replaced because the reinforcing wire cannot be bent back into its original shape. Hoses stored in temperatures below freezing should be warmed before being unwound and placed in service.

Hydraulic System Service

Hydraulic valves, pumps, and actuators are precision devices and must be kept clean inside and out. Always keep hands, tools, and the work area clean when working on hydraulic components. Use sheets of lint-free paper to cover the work area, and lint-free rags to clean the components. Prepare the repair area to contain any hydraulic oil discharges during the repair. Follow the manufacturer's recommendations for the service procedure.

Clean the outside of the components before dismantling them. Use clean plastic plugs to cover openings. Clean components with an approved cleaning solvent as soon as they are removed. Dry parts with low-pressure, clean, dry compressed air. Coat the dried parts with hydraulic oil and store them in clean plastic bags. Use only recommended cleaning solvents on plastic parts.

Cylinders should be stored with piston rods retracted to prevent dirt from settling on the exposed bare metal and entering the system when the piston is moved. Piston rods that must be stored exposed should be coated with grease that can be easily removed before startup.

Clean and flush the system using only recommended fluids if required by the manufacturer. Using the incorrect flushing fluid can leave residual chemicals inside the system that can break down hydraulic oil. When flushing, operate the system to circulate the flushing fluid only if recommended by the manufacturer. Drain the flushing fluid, then refill the system with the correct hydraulic oil. The system is then operated carefully to remove any air trapped in the system.

Keep air out of the system by maintaining the correct hydraulic oil level in the reservoir. Cycle the system several times to purge air after refilling or working on the system. Before operating, always check the hydraulic oil level and condition of the oil. Start and operate a system carefully in cold weather. Cold hydraulic oil can cause problems due to high viscosity, which results in poor lubrication. The cold oil may move so slowly in cold applications that it does not lubricate properly.

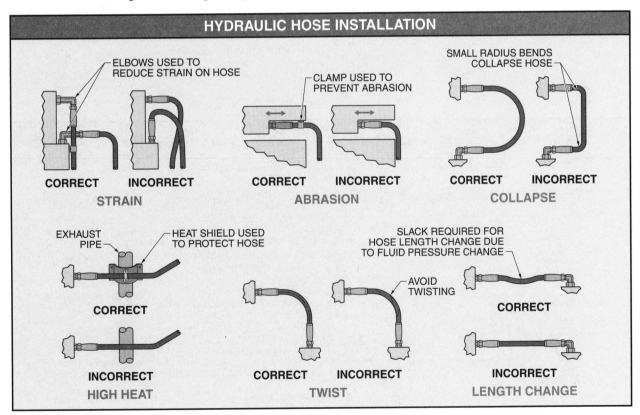

Figure 10-25. Hoses should be properly installed for maximum operating life.

Always use the correct replacement parts. Replace springs, seals, and O-rings any time a component is serviced. Ensure that the O-ring material is resistant to chemical additives in the hydraulic oil and expected system pressures. Clean the shaft or working area before installing new O-rings. Remove any nicks or sharp edges from O-ring grooves or along the shaft. Lubricate the O-ring with the same hydraulic oil used in the system. Carefully install the O-ring. Avoid contact with sharp edges and overstretching. Verify that the O-ring is the correct size for the application.

Ensure that all packing and seals are installed in the correct position. Seals normally expand to make a tighter seal when system pressure is applied. Seals should be installed using the same precautions as O-rings. Check for hardness, brittleness, or spongy consistency of the seal material that indicates damage from incorrect hydraulic oil or excessive temperatures.

Avoid damage to metal surfaces from excessive force on parts. When cleaning dirty metal surfaces, polish the metal with crocus cloth or a synthetic cleaning pad that does not contain surface grit. Grit can remove surface metal and cause internal leakage. Use approved cleaning products and materials supplied or recommended by the manufacturer.

After service, always check for leaks. External leaks usually leave stains and accumulate dirt. Internal leaks caused by worn or damaged seals or metal components are harder to detect. Internal leaks cause slippage of pistons and poor operation of valves and motors. Internal leaks generate heat at the site of the leak.

TROUBLESHOOTING HYDRAULIC SYSTEMS

Hydraulic system problems are usually related to a malfunction in force and/or speed. **See Figure 10-26.** The problem is likely related to lack of pressure in the system if the system is not producing enough force. Check system pressure settings and the operation of all pressure control valves. Check pump and flow control valve operation when equipment speed is not correct.

Internal and external leaks affect pressure and flow in a hydraulic system. External leaks are usually easy to locate if the outside of the equipment is kept clean. Internal leaks are harder to locate and are commonly caused by wear of close coupling parts or deterioration of seals or O-rings. Leaks reduce the capacity of the equipment to perform work. Serious leaks may cause systems to stall when heavily loaded. Internal leaks can be isolated using hydraulic testers.

> **TECH TIP**
> Piston rod seals and reservoir air breathers are the most common locations where dirt and foreign material can enter a hydraulic system.

TROUBLESHOOTING HYDRAULIC SYSTEMS – UNRESPONSIVE EQUIPMENT	
Possible Cause	**Probable Solution**
Air in hydraulic system	• Check fluid level. • Purge (bleed) air out of system. • Check connections on suction line (pump intake).
Pressure drop	• Review hose sizes and lengths to minimize pressure drop. Smaller hose diameters and longer lengths increase pressure drop in a line. • Replace block style couplings and adapters with bent tube style to improve laminar (streamline) flow and reduce pressure drop.
Hose tube collapse or swell	• Check fluid compatibility with tube material. • Vacuum may have exceeded hose vacuum rating. Select hose that meets requirements.
Nonfunctioning hydraulic components (pumps, valves, etc.)	• Check each hydraulic component for full function, i.e., seal may have rolled in a cylinder, causing it to bind and limit stroke.
Fluid flow blockage	• Check flow in each line and component for blockage. Eliminate source of contamination. • Check and replace filter if necessary.

Figure 10-26. Troubleshooting hydraulic systems requires identifying possible causes and probable solutions.

Flow control problems are identified using a flowmeter connected into the system. In addition, actuator operation speed can be timed and compared to system specifications. Tachometers are used to measure motor speed. Excessive heating of a system and hydraulic oil can also cause changes in pressure and flow in a system. System temperatures should not exceed the recommendations of the equipment and hydraulic oil manufacturer.

Always consult manufacturer's troubleshooting charts when working on a hydraulic system. Complex hydraulic systems are broken down into specific parts and component functions. Hydraulic testers are used to determine the proper function of specific parts of a system and to test component functions. A *hydraulic tester* is a device that measures the pressure, flow, and temperature of hydraulic oil in a system. **See Figure 10-27.** Flow and pressure are measured at various points in the circuit to test the operation of pumps, relief valves, and directional control valves.

Hydraulic systems critical to plant production must operate without interruption. Downtime of critical systems must be kept to a minimum. Hydraulic system troubleshooting must isolate and remedy the problem as quickly as possible. Standard troubleshooting steps are followed to isolate and remedy hydraulic system problems. **See Figure 10-28.**

Cavitation

Cavitation is the process in which microscopic gas bubbles expand in a vacuum and suddenly implode when entering a pressurized area. An *implosion* is an inward bursting. Cavitation is an undesirable condition caused by vapor cavities in hydraulic oil. Cavitation is caused by incomplete filling of the pump with hydraulic oil so that an excessive vacuum is created at the pump inlet. If the pump inlet pressure is too low, the hydraulic oil boils and creates vapor pockets. Vapor pockets reaching high-pressure areas inside the pump implode violently. **See Figure 10-29.** The implosions remove pieces of metal from the pump, and the vibrations cause damage to pump bearings and seals.

Cavitation can also be caused by restrictions at the pump inlet from partial blockages, clogged filters, or inlet piping that is too small or too long. Cold hydraulic oil or hydraulic oil with the wrong viscosity can also cause cavitation. Cavitation creates a loud sound as if the pump was pumping rocks along with the oil.

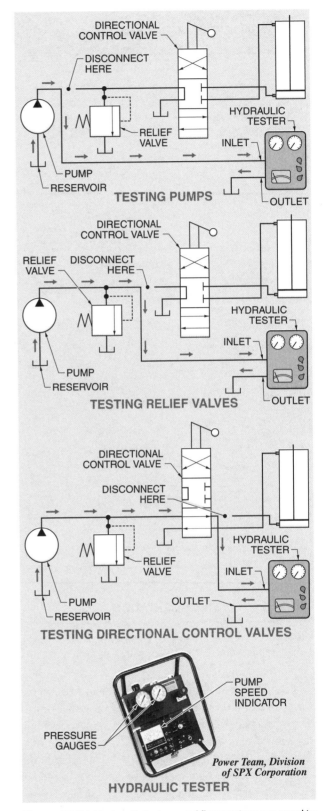

Figure 10-27. Hydraulic testers and flow meters are used to determine the proper function of specific parts of a system and to test specific components.

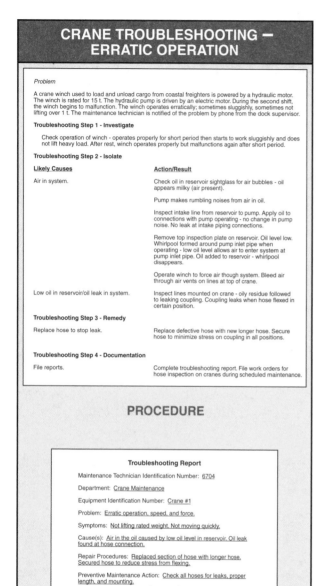

Figure 10-28. Standard troubleshooting steps are followed to isolate and remedy hydraulic system problems.

Cavitation is different than the problem of air in the oil, even though both problems produce a similar sound. Air in a system causes sluggish or erratic operation. Hydraulic oil contaminated with air appears milky. Cavitation usually leaves the hydraulic oil clear in the reservoir because the vapor bubbles have imploded and disappear in the pump. Applying hydraulic oil on the area of the leak isolates air leaks. The hydraulic oil is drawn into the pump, and the noise of the air going through the pump stops or is reduced.

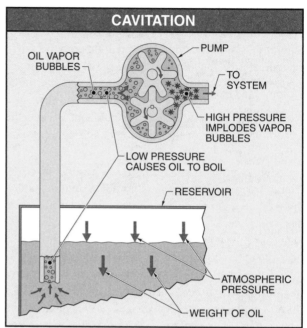

Figure 10-29. Cavitation is the process in which microscopic gas bubbles expand in a vacuum and suddenly implode when entering a pressurized area.

HYDRAULIC SYSTEM SAFETY

Hydraulic system accidents can be reduced by following basic safety rules. Hydraulic system safety rules include:

- Use all appropriate PPE.
- Remove and lock out all potential energy in a system, including mechanical, hydraulic, and electrical energy.
- Lower all elevated components. Secure elevated components that cannot be lowered.
- Bleed off all hydraulic oil pressure from the system, including the hydraulic oil in the accumulators.
- Use caution when disassembling components that may contain springs.
- Charge accumulators only with nitrogen. Never use compressed air, which can be explosive around oil. Never overcharge an accumulator.
- Do not perform work on hydraulic elevators. Although most hydraulic systems are maintained by maintenance personnel, elevators are maintained and tested by certified personnel from companies specializing in elevator installation and repair.

+ SAFETY TIP

Never attempt to locate a leak in a hydraulic system by running a hand over the suspected area. Escaping fluid can penetrate the skin and cause serious injury.

PNEUMATIC PRINCIPLES

A *pneumatic system* is a fluid power system that transmits energy using a gas (typically compressed air). The physical characteristics of gases are affected by pressure, volume, and temperature. Pressure increases as gas molecules are forced closer together through compression. *Volume* is the three-dimensional size of an object. Liquids and solids have definite volumes that do not vary significantly when compressed. However, gases have indefinite volumes that vary significantly when compressed.

The temperature of a gas increases when compressed. This occurs because the molecules are forced closer together and rub against each other. This rubbing generates friction, which causes heat. When the compressed air in a pneumatic system becomes hot, the heat must be dissipated for efficient system operation. When heated, compressed gases try to increase in volume. Therefore, the heat of compressing the air causes the air molecules to try to push further apart, causing the hot air to occupy more space in a pneumatic system. More air can be stored and used in a system if it is cooled after being compressed.

Compressed gases flow quickly from a pressurized container to a lower pressure area, which may be inside or outside the system. This can cause an explosion if a pressurized container, including a hose carrying pressurized air, bursts. However, if the control devices within a system control the release of pressurized air, the airflow from high pressure to low pressure can be used to do work.

Gas laws are the predictable relationships between the volume, pressure, and temperature of a gas. Gas laws are used to determine the change in volume, pressure, or temperature of a gas when the other two characteristics are known. For example, if temperature increases and volume remains the same, then the effect on pressure can be determined.

PNEUMATIC SYSTEMS

Pneumatic systems are similar to hydraulic systems so they are commonly classified together as fluid power systems. For example, a press may be operated hydraulically or pneumatically. A hydraulic system can handle heavy loads and generate a greater force than a pneumatic system. Instead of pressurized oil, a pneumatic system uses compressed air in an enclosed space to exert pressure. Pneumatic cylinders are generally used for low- to medium-force applications that do not require accurate speed control. Hydraulic cylinders are used for high-force or low-force applications that require accurate speed control. Accurate speed control is possible because hydraulic oil cannot be compressed.

Pneumatic system components are less costly than hydraulic system components. Pneumatic tools are used in production settings to increase assembly productivity. Common pneumatic tools include pneumatic wrenches, grinders, nail guns, and impact wrenches. Pneumatic tools are also used for medical and dental applications.

Pneumatic System Operation

A pneumatic system is not a closed system like a hydraulic system in which the hydraulic oil is reused. In a pneumatic system, air is compressed, used in the system, and exhausted to the atmosphere. Pneumatic systems usually operate at lower pressures than hydraulic systems, though many machine functions can be accomplished using either hydraulic or pneumatic energy. The keys to proper operation of a pneumatic system are the supply of clean, dry, compressed air and the prevention of air leaks.

Pneumatic System Components. Compressed air in a pneumatic system is supplied by the air compressor. An *air compressor* is a component that takes air from the atmosphere and compresses it to increase its pressure. **See Figure 10-30.** Hoses or pipes direct compressed air to actuators. Actuators in pneumatic systems include cylinders, motors, and oscillators. Pneumatic systems commonly contain various valves in the pathway to control the direction of flow and the amount and pressure of the air.

W.W. Grainger, Inc.
Compressors supply clean, dry air at the correct pressure for industrial pneumatic systems.

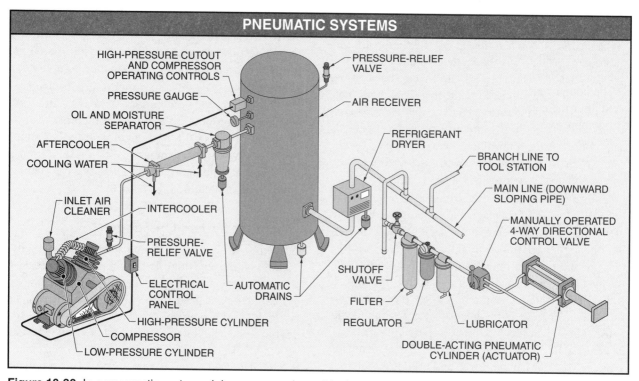

Figure 10-30. In a pneumatic system, air is compressed, used in the system, and exhausted to the atmosphere.

The operation of air compressors is similar to the operation of hydraulic pumps and refrigeration compressors. Air entering the compressor must be filtered. The air inlet should be located in a cool, dry place. The cooler the air drawn into the compressor, the more efficiently the compressor operates. Molecules of cool air are closer together than the molecules of warm air and are more easily compressed.

Air compressors include reciprocating, screw, and vane compressors. A *reciprocating compressor* is a compressor that uses pistons moving back and forth to increase fluid pressure. Reciprocating compressors are often driven by electric motors connected to the compressor by belts. Reciprocating compressors can be multi-staged to achieve high air pressures. A *multi-stage compressor* is a compressor that uses two or three cylinders, each with a progressively smaller diameter, to produce progressively higher pressures. Air enters the low-pressure cylinder that discharges through finned piping (intercooler) into a small high-pressure cylinder. An *intercooler* is a finned pipe heat exchanger that helps cool the air between compression stages. In the high-pressure cylinder, air is compressed again and discharged at a higher pressure.

Small reciprocating compressors with plastic piston rings and valves are also used to supply medical-quality, oil-free air. No oil is used in the compressor as plastic valves and rings do not require lubrication.

A *screw compressor* is a compressor that contains a pair of screw-like rotors that interlock as they rotate. Screw compressors are becoming more popular because they require less maintenance and introduce less oil into the system than do other compressors.

A *vane compressor* is a compressor that has multiple vanes located in an offset rotor. As the rotor rotates, its offset position allows the vanes to slide out and draw air from the inlet port. As the rotor continues to rotate, the volume between the vanes decreases, pushing the vanes into their slots in the rotor. The decreasing volume compresses the air and forces it out of the outlet port.

A pressure-relief valve and a high-pressure cutout are located on the outlet line of all air compressors. A pressure-relief valve on an air compressor serves the same purpose as a safety valve on a boiler. Other safety devices used on large compressors include a low oil pressure switch and/or a high air temperature switch. These devices stop the compressor before serious damage occurs. The compressor can be destroyed if there is not enough oil to lubricate its moving parts. Excessively high air temperature leaving the compressor indicates a problem with the compressor such as a coolant failure or a mechanical problem. These switches may be electrically or electronically operated and send signals to the compressor control system.

The components that clean and cool the air are located after the compressor in a pneumatic system. As it is compressed, the air becomes heated from the air molecules rubbing together frequently. Hot air is less efficient to store than cool air because it occupies a greater volume. After leaving the compressor, the air is cooled in an aftercooler. An *aftercooler* is a pipe-in-pipe heat exchanger in which water is used to cool compressed air. Heat is transferred from the hot air to the cool water. As the hot air cools, some of the moisture in the air condenses.

An *oil and moisture separator* is a device that removes oil and water droplets from a system by forcing compressed air to change direction quickly. The heavier oil and water droplets are separated, allowing clean, dry air to advance in the system. An automatic drain is located at the bottom of the oil and moisture separator. Compressed air from reciprocating compressors can carry oil from the crankcase that is used to lubricate the compressor valves. Oil may be unacceptable in medical or precision manufacturing and industrial applications that require clean, dry compressed air. Oil-free compressed air can be produced by a screw compressor or by specially designed reciprocating compressors.

Some pneumatic systems also use a refrigerant dryer to cool the compressed air. A *refrigerant dryer* is a device that uses a refrigeration process to lower the temperature of compressed air. The compressed air piping is routed through a small cooled space that cools the compressed air, causing the moisture to condense into an automatic water drain. Other systems use chemical filters that remove moisture by absorbing it in a desiccant. A *desiccant* is a drying agent that removes water vapor by absorption. Desiccant dryers are replaced or dried by heating when the desiccant is saturated with moisture.

Pneumatic systems commonly use a receiver. A *receiver* is an air tank that stores compressed air and allows it to cool before use. As the air cools, moisture condenses to the bottom of the receiver, where it is drained manually or with an automatic drain. The receiver is located where the compressor high-pressure cutout and ON/OFF pressure control are connected. Both devices have a similar function and operation as they do on boiler controls. The compressor high-pressure cutout senses pressure and opens an electrical circuit to shut down the compressor if pressure rises beyond a safe level. An ON/OFF pressure control operates an electric circuit that turns the compressor ON when the pressure is low (more air is needed), and turns the compressor OFF when the pressure is high. The receiver is also fitted with a mechanical pressure-relief valve to prevent an overpressure condition.

Clean, dry compressed air is directed through piping or tubing for use in the plant. Pneumatic system pipe must be plumbed with a downward slope so water in the lines flows into drains. The pipe must be large enough to allow airflow without causing a large pressure drop. The pipe must also be strong enough to withstand expected system pressures. If plastic pipe is used, it must be rated to withstand the expected pressures and designed for use in compressed air systems. Plastic pipe used for plumbing water systems is not designed to withstand the pressures in pneumatic systems. Piping systems are fitted with manual or automatic drains and shutoff valves for servicing system components.

Before the compressed air reaches the point of use, it is filtered again. If the compressed air is used to operate pneumatic tools, pressure may be reduced and lubricating oil added. These three functions can be accomplished by a filter, regulator, and lubricator (FRL) station. Air passes through a quick-closing shutoff valve, a filter, a pressure regulator that controls outlet air pressure, and a lubricator. A *lubricator* is a component that injects a mist of oil into the compressed air line for lubrication of pneumatic tools and internal motor parts. A quick disconnect attached to the outlet pipe allows quick attachment and removal of flexible air lines without losing air from the system. **See Figure 10-31.**

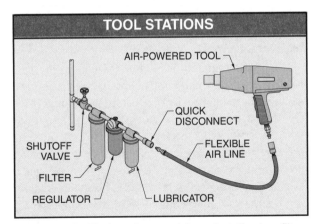

Figure 10-31. Tool stations contain a shutoff valve, filter, regulator, lubricator, and quick disconnect to enable clean, lubricated air to be easily connected to different air-powered tools.

Compressed air can be used to power pneumatic tools such as impact wrenches, nail guns, chippers, and grinders. Pneumatic tools are rugged and are not damaged by overloading. Pneumatic tools stall safely when overloaded and present no danger of electrical shock. In addition, pneumatic tools are lightweight and can be used in explosive hazard atmospheres.

Compressed air can be used for cleaning or drying hard-to-reach parts of machines or individual components. In this application, a flexible air line is fitted with a hand-operated nozzle. Cleaning operations require air pressure below 15 psi. Care must be taken to avoid possible injury and/or damage to the part being cleaned. Compressed air can drive small particles of metal, oil, moisture, and/or dirt through the skin and into the bloodstream. Some air nozzles are fitted with pressure-reducing valves to lower the air pressure to safe levels.

Warning: Proper eye, ear, and hand protection must be worn when using an air nozzle.

Pneumatic systems can use pressure and flow control valves to operate circuits similar to hydraulic circuits. Pneumatic component symbols are similar to hydraulic component symbols. **See Appendix.**

Pneumatic System Capacity Control

The demand for compressed air in a plant fluctuates as load demands change. The volume of compressed air delivered to a receiver must be controlled. Load demand is usually determined by the monitoring of the air pressure in the receiver. As load demand increases, more air flows from the receiver and pressure in the receiver decreases. As load demand decreases, less air flows from the receiver and pressure in the receiver increases.

The most common method for regulating pressure in a receiver for small compressors is by starting and stopping the compressor. The operation is similar to the ON/OFF control used on a boiler. The compressor stops when pressure in the receiver reaches the maximum set pressure and starts when pressure in the receiver reaches the minimum set pressure. Receiver pressure regulation is similar to pressure controls in boiler systems. In systems with dual compressors, one compressor is used to supply pressure for operating loads. The other compressor is set to operate when pressure in the receiver reaches a set minimum level. The compressors are operated alternately to distribute wear evenly.

Large compressors are designed to operate continuously because they are difficult and costly to start and stop repeatedly. The compressor continues to operate with pressure regulated by an unloader. An *unloader* is a device that allows the compressor to operate without adding pressure to the receiver. On a reciprocating compressor, the unloader holds the compressor suction valves open to allow air to flow freely in and out of the cylinder. This prevents air from being forced through the discharge valve to increase the air pressure in the system. The unloader unloads the suction side of the compressor when the system does not require any air to maintain its correct pressure range. An unloader on a screw compressor uses a slide valve at the compressor inlet. The slide valve opens and closes to allow more or less air into the compressor. Unloaders are operated by air pressure, electric signals, or electronic signals sent from the receiver to the compressor.

Compressors should always be started unloaded (no pressure against the discharge valves). This is often accomplished by bleeding off the air between the compressor discharge valves and by the use of a check valve installed after the compressor discharge. **See Figure 10-32.** A *bleed-off valve* is a valve that allows the pressure against the discharge valves to be released at startup. The valve is connected to the compressor crankshaft so it is open when the crankshaft is not turning and closed when the crankshaft is operating at near full speed. The bleed-off valve opens and air between the discharge valve and check valve escapes when the compressor stops. This ensures that there is no pressure working against the discharge valves at startup, allowing the compressor to start easily. Air leaving the cylinder exhausts through the bleed-off valve at low pressure until the compressor is at near full speed and the bleed-off valve closes. Air is then sent through the check valve to supply compressed air to the system.

PNEUMATIC SYSTEM MAINTENANCE

A pneumatic system is a pressurized system similar to a boiler or refrigeration system. A rupture in the system can cause serious injury and/or damage. All components in a pneumatic system require regular maintenance to prevent potential problems leading to ruptures.

System parts must be kept clean to provide efficient heat dissipation. Physical damage or corrosion damage must be prevented. This is especially important with the system receiver due to its internal pressure and large volume of air. Receiver and compressor surfaces are coated with protective paint by the manufacturer to prevent corrosion damage and efficiently dissipate heat. Improper and/or excessive coats of paint to any part of a pneumatic system can increase insulation, which prevents proper heat transfer. The manufacturer should be consulted for recommended coatings and procedures for pneumatic components.

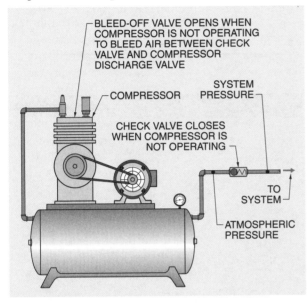

Figure 10-32. Compressors are started unloaded (no pressure against the discharge valves) by bleeding off the air between the compressor discharge valves and by the use of a check valve installed after the compressor discharge.

Ensuring the supply of clean, dry air is the most important task in pneumatic system maintenance. Filters must be changed according to manufacturer's recommendations. Drains and dryers must be routinely inspected for proper operation. Equipment is maintained according to manufacturer's recommendations. On large compressors, the lubrication and cooling systems should be monitored carefully. The correct type and amount of oil is critical to compressor operation. Equipment maintenance requirements are compiled and scheduled into the overall system maintenance procedures.

The amount and type of lubricant supplied to actuators is critical. Too much lubricant can clog an actuator. Too little lubricant can result in premature wear. The lubricant used must meet manufacturer's requirements. The mounting and alignment of all actuators should be checked regularly, especially heavily loaded pistons and motors. Actuators that are intermittently used or held for a long time in one position should be operated occasionally to ensure free movement.

TROUBLESHOOTING PNEUMATIC SYSTEMS

The most common problem in pneumatic systems is insufficient air pressure or airflow caused by leaks or disconnected or kinked lines. Leaks in a pneumatic system are costly and must be repaired promptly. Leaks can be located by listening for discharging air noise when the plant is quiet or by applying soapy water to the suspected location. The application of soapy water produces bubbles when applied to a leak. In addition to the noise of discharging air, clean areas are created and grooves are worn by escaping air. Leaks commonly occur at connection points in the system.

In some cases, the inner linings of flexible air lines collapse and are not visible on the surface. When checking for pressure or flow problems, work forward from the compressor or backward from the actuators. Couplings, valve operation, and pressure and flow readings are checked until the leak, disconnection, or blockage is located.

A common cause of pneumatic system problems is compressor controls set at incorrect pressure settings. Check the settings of pressure controls or pressure gauges on the compressor or receiver. If the compressor is working properly, check each component in sequence away from the compressor in the system piping. For example, a blocked filter at an FRL station restricts air flow to that portion of the system. Leaks can be located by their sound and by the application of soapy water. The actuators are tested if there are no obvious problems in the piping system.

To test an actuator, disconnect the load and energize the actuator. If the actuator functions normally without the load attached, the load may be too great, the actuator may have internal leaks, and/or there may be insufficient pressure to operate the load. The problem is probably the actuator or air supply components if the actuator does not operate with the load disconnected. Test for proper operation of components by working from the actuator back toward the compressor. Ensure that the actuator packing nuts are not overtightened. *Packing nuts* are components on an actuator that apply tension to the seals or packing to prevent leakage around the shaft. Plug-in pressure gauges may be used to trace pressure levels through a system if such couplings are available on the system piping.

> **TECH TIP**
> *Pneumatic systems are commonly used for food processing plant equipment because a leak or exhaust air does not contaminate the product.*

Air-Over-Oil Systems

Air pressure may be used to pressurize small hydraulic systems. A common example is an automobile lift in a service station. Rather than have a separate hydraulic system with a pump and valves, compressed air is used to pressurize hydraulic oil in an air/oil tank. **See Figure 10-33.** The pressurized hydraulic oil acts on the surface area of the cylinder, pressurizing the hydraulic oil. The hydraulic oil pressure acts on a large diameter hydraulic ram that raises the automobile.

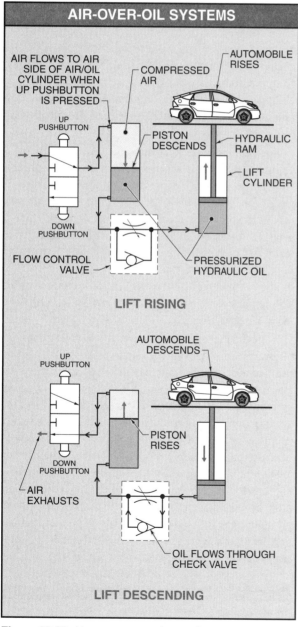

Figure 10-33. Air pressure may be used to pressurize small hydraulic systems in air-over-oil systems.

An *intensifier* is a device that converts low-pressure fluid power into high-pressure fluid power. An intensifier multiplies the force applied by sending pressurized air to a large diameter cylinder attached to a small piston that pressurizes a relatively small amount of hydraulic oil. **See Figure 10-34.** A large boost in system pressure is gained by having a large air piston attached to a small hydraulic ram. For example, air at 100 psi applied to a piston with a surface area of 5 sq in. exerts 500 lb of force in a straight line through the ram to its surface, pressurizing the oil in the small cylinder.

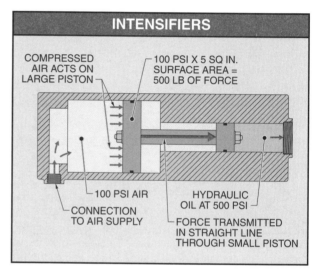

Figure 10-34. An intensifier multiplies the pressure applied by sending pressurized air to a large diameter cylinder attached to a small piston that pressurizes a relatively small amount of hydraulic oil.

PNEUMATIC SYSTEM SAFETY

Pneumatic system accidents are reduced by following basic safety rules. Pneumatic system safety rules include:

- Wear proper eye protection when working with pneumatic system components.
- Bleed off system air before opening system piping.
- Know the location of all shutoff valves.
- Never attempt to grasp a broken air line that is being whipped around by exhausting compressed air. If possible, shut OFF line pressure to stop the whipping. If the whipping must be stopped quickly, secure the air line from a safe distance with a long piece of wood.
- Use tools specifically designed for pneumatic service.

Troubleshooting

The two main activities of maintenance personnel are the proper maintenance of equipment and systems to prevent problems, and the effective troubleshooting of equipment to repair problems. Studying the results of preventive maintenance system operation leads to cost-effective operation of maintenance activities. Evaluation of the results of troubleshooting activities improves the effectiveness of troubleshooting.

MAINTENANCE AND TROUBLESHOOTING

Preventive maintenance (PM) is scheduled work required to keep equipment in peak operating condition. *Troubleshooting* is the systematic investigation of the cause of system problems in order to determine the best solution. The ability to troubleshoot effectively is a skill that combines technical expertise and logical and creative thought processes. This skill improves through experience with troubleshooting and evaluation of the causes and solutions of problems. If the causes of problems can be discovered, modifying the PM work on the equipment may prevent future problems.

Preventive maintenance and troubleshooting are often considered separate activities. There is a formal PM system with records and planned work while troubleshooting is performed as an as-needed activity. However, troubleshooting activities can generate extremely useful information that can be incorporated into the PM system. This information can be used to modify the procedures and amount of work required by the PM system in order to increase troubleshooting efficiency. When developing a PM system, the first step is to determine the PM tasks required. These tasks must be established and applied to the equipment and changes in equipment performance must be noted before the tasks can be modified to meet the needs of the plant.

Safety

Safety is the most important concern when performing maintenance and troubleshooting work. Wherever possible, de-energize and lock out the equipment according to all applicable national, regional, local, and plant safety procedures. When equipment must be operating as part of the troubleshooting process, ensure that personnel are completely safe before operating the equipment. Ensure that the equipment cannot be damaged by starting. When testing input devices such as pushbuttons and sensors, ensure that personnel cannot be injured by the sudden startup of the equipment.

Establishing PM Requirements

In the past, industrial maintenance was commonly based on repairing equipment after breakdown. This is no longer effective because of the complexity and high costs associated with modern industrial equipment. Today, equipment is maintained in order to prevent equipment breakdowns. Keeping modern equipment from breaking down requires the use of all information available. This information can be found in operator's manuals, service bulletins, online computer bulletin boards and databases, and telephone customer support services operated by equipment manufacturers and suppliers. Other sources of PM information include out-of-plant services, national codes and standards, trade magazines, and textbooks.

Once a piece of equipment has been surveyed and entered into the computerized maintenance management system (CMMS), maintenance personnel establish the PM tasks by deciding what must be done to keep the equipment working in peak operating condition. The first place to look for PM task recommendations is in the operation and maintenance manuals for each piece of equipment. The manuals should be obtained from the manufacturer as soon as possible if they are not available in the plant. The manuals offer specific PM recommendations. For example, a motor manufacturer's PM requirements include a three-month lubrication interval using a specific lubricant. **See Figure 11-1.**

Once the recommended PM tasks for a piece of equipment are known, they can be modified to meet plant conditions. For example, if the motor is running in an extremely hot location or is stopped and started frequently, the manufacturer may recommend a different lubricant or a more frequent lubrication interval. A manufacturer that provides a warranty with new equipment may specify the maintenance requirements that must be completed to maintain the warranty.

PM Care Level

PM care level is determined after the PM system has been operating for at least a year and the information can be analyzed. The level of PM care is the type and frequency of PM work a piece of equipment should receive. For example, an electric motor may be lubricated quarterly or semiannually. The cost of lubricating the motor more frequently may or may not justify the amount of money saved in prevented failures. For one motor, the cost savings may be negligible. For 500 motors, however, the cost savings may be substantial.

PM work must contribute to the efficient, profitable operation of a facility. Performing excessive PM work drives up the cost of maintenance operations. Performing too little PM work results in losses due to downtime and poor productivity. It is difficult to determine the exact cost/benefit of a PM system, though continuing analysis of PM work and results further refines the optimal level of PM care. The level of PM care can be thought of as the cost of PM work divided by the cost of failures.

Failures may include equipment downtime, quality disruptions that lead to the production of substandard product or reduced productivity, legal costs resulting from lawsuits or fines imposed by regulating agencies, and wasteful energy consumption. Only some of these costs can be measured with complete accuracy. For example, many companies know the exact cost of production downtime. These costs can be many thousands of dollars an hour. Costs of defective parts can be computed, but reduced productivity of office staff is difficult to determine. Legal costs saved because no problems arose are extremely difficult to determine. Costs and fines incurred by similar plants or facilities can give some idea as to the consequences of poor maintenance in these areas. For example, fines levied for accidental release of pollutants caused by the failure of pollution control equipment can be substantial. The resulting poor publicity can damage company reputation and sales of the product. Lawsuits launched by injured or sick employees can be expensive when legal costs are included.

A reliable measure of a PM system is the rate of breakdowns that require troubleshooting. The type and rate of breakdowns can serve as a guide to the amount of PM work a piece of equipment requires. The more frequently the equipment breaks down, the more PM care the equipment requires. For example, if the belt on a particular conveyor suffers a high level of rips and tears, the problem may be traced to a defective automatic belt adjuster designed to keep the conveyor belt tracking correctly. Inspection of the belt adjuster reveals that it failed due to lack of lubrication. This leads to increased inspection and cleaning of the automatic lubricator on the belt adjuster to reduce the incidence of belt damage. Because of this, all automatic lubricators are inspected more frequently. In this case, the increased cost of the maintenance inspections was more than offset by reduced production disruptions and the high cost of belt replacement. Collecting and analyzing troubleshooting information is a critical activity of the PM system and can also contribute to improved troubleshooting efficiency.

OPERATION AND MAINTENANCE MANUAL

6.17 DRIVE MOTOR LUBRICATION

Induction squirrel cage motors have antifriction ball or roller bearings front and rear. At extended intervals they require lubrication.

The periods between greasings of the motor bearings can vary, primarily with the severity of the service conditions under which the motor operates. As a general rule, the following applies:

Frequency of Lubrication-Normal Environments

Motor Size	Lubrication Interval
25-40 HP	3 Months (or 1000 hr)

NOTE: For severe duty - Dusty locations
                  - High Ambient
                    Temperatures

Reduce time intervals in preceding table to 1/2 the listed value.

Lubrication Procedure

⚠ **CAUTION**

Grease should be added when the motor is stopped and power disconnected.

When greasing, stop motor and remove inlet and outlet plugs. Inlet grease gun fittings and spring-loaded outlets are arranged at each end on the motor housing. Use a hand lever grease gun. Determine the quantity of grease delivered with each stroke of the lever. Add grease in the following quantity:

Motor Frame Size	Lubrication Amount	
	in³	oz
256-286	1.0	0.8
324-326	1.5	1.2

Do not expect grease to appear at the outlet. If it does, discontinue greasing immediately.

⚠ **CAUTION**

Overgreasing is a major cause of bearing and motor failure. Ensure dirt and contaminants are not introduced when adding grease.

Run motor for about ten minutes before replacing outlet plug. Certain TEFC motors have a spring relief outlet fitting on the fan end. If the outlet plug is not accessible at surface of hood, it is the spring relief type and need not be removed when greasing.

A major cause of motor bearing failure is overgreasing. The quantity of grease added should be carefully controlled. Small motors must be greased with a lesser amount of grease than large motors.

Recommended Motor Greases
(or equivalents)

Chevron SRI . . . . Standard Oil
                                        of California
Premium RB . . . . Texaco
Unirex N2 . . . . Exxon
Dolium R . . . . Shell
Rykon Premium . . . . American Oil

⚠ **CAUTION**

Never mix greases. Mixing greases can cause motor failure.

Figure 11-1. PM task recommendations are located in the operation and maintenance manuals for each piece of equipment.

Operational Maintenance and Troubleshooting

Maintenance personnel often operate equipment as part of their duties. This involves starting and stopping the equipment and making adjustments to keep it operating correctly. **See Figure 11-2.** Many of these functions are automated and computer-controlled, but some equipment cannot be completely automated. For example, an operator in an aluminum rolling plant might work from an automated control panel fitted with manual overrides should a problem develop. The operator may manually start the operation of a new roll of aluminum, then set the process to automatic operation. Operators may do their own troubleshooting or they may work with a partner, who makes the actual tests and repairs. After the tests and repairs are made, the operator tests the machine operation. This requires the technician to be knowledgeable in the correct operation of the machines requiring maintenance and/or troubleshooting.

Exxon Company

Figure 11-2. Maintenance personnel often operate equipment as part of their job duties.

Another common task is adjusting the timing of machine operations, such as a packaging machine where the product being wrapped must be at the correct place at the correct time for wrapping. Timing failures can damage the product and cause severe damage to equipment. Maintenance personnel also monitor the startup of equipment to ensure that the equipment is correctly calibrated and operating properly. Periodically, the technician makes adjustments and corrections to the equipment while it is operating. If the same equipment is used to make or wrap different products, maintenance personnel make adjustments to change the equipment operation to suit the different product. For example, product changeover often occurs in commercial bakeries. Maintenance personnel are always present when the machines are changed over and started because most problems occur at this time.

Other operations that may be part of the maintenance personnel's responsibility at a facility include the monitoring and adjusting of large systems such as cooling systems. The operation of such equipment is recorded in a logbook by the technician who takes regular equipment readings. **See Figure 11-3.** Logbook entries may be made hourly, several times each day, or weekly, depending on the size of the equipment and the amount of automatic monitoring equipment used.

While recording the readings in the logbook, maintenance personnel also look for signs of trouble that can be corrected before serious problems develop. Logbooks can be extremely useful in troubleshooting. For example, a large air-conditioning chiller requires frequent monitoring and logbook entries. If, during the course of taking readings, a technician notes a sudden rise in bearing temperature, the lubrication system is inspected immediately. If oil flow to the bearings

was disrupted, the bearings may fail quickly due to overheating. Such a failure is extremely dangerous and expensive to repair.

A less immediate problem may be a gradual lowering of the temperature difference between the water entering and leaving the chiller condenser. For example, the temperature difference between the water entering and leaving the condenser may change from 10°F to 8°F over a period of several months. This slow change may indicate dirty condenser tubes, which prevent the transfer of heat from the condenser to the cooling water. Such a problem causes a rise in the cost of operation and reduces the cooling capability of the system. This condition does not threaten the chiller with immediate, serious damage. Therefore, inspection and cleaning of the tubes may be scheduled for a convenient time when the chiller can be taken out of service. Dirty chiller condenser tubes are cleaned and measurements are taken to improve water treatment and testing to prevent the tubes from becoming dirty. All of these activities are recorded in the chiller equipment log so they can be studied if problems develop.

Large systems are often monitored and operated from a centralized control room using computerized equipment. **See Figure 11-4.** System conditions are monitored constantly and alarms are activated if problems arise. The alarms may be audible, indicator lights, or error codes. Technicians involved in these operations must be trained in routine and emergency plant operations. Experienced operators use alarms as a guide when solving operational problems. Complete dependence on automated alarm systems can lead to serious problems if unusual or complicated problems occur. Well-trained operators should know the location of all critical equipment and how to operate and repair the equipment in an emergency. Equipment operators often work with roving maintenance personnel who operate or make repairs to the equipment while the operator monitors system operation from the control room. Due to the increased power of computers, even small buildings or processes can be monitored electronically. The records taken can be analyzed to adjust plant efficiency and detect developing problems.

TECH TIP

Large facilities benefit from computerized maintenance management systems by centralizing maintenance activities, spare parts inventory, maintenance personnel, and purchasing activities.

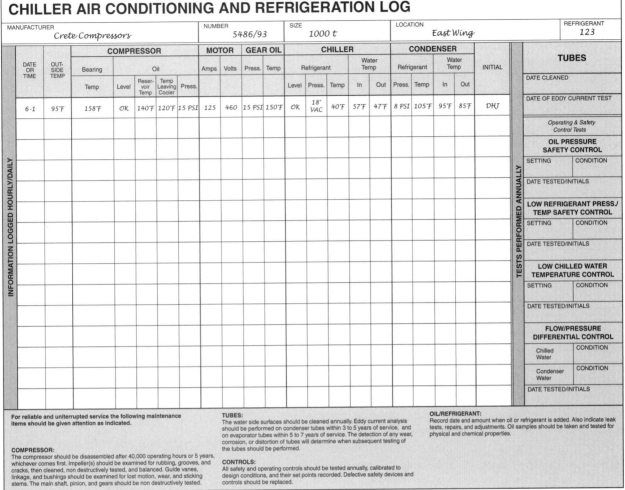

		COMPRESSOR					MOTOR		GEAR OIL		CHILLER					CONDENSER				INITIAL
DATE OR TIME	OUT-SIDE TEMP	Bearing	Oil				Amps	Volts	Press.	Temp	Refrigerant			Water Temp		Refrigerant		Water Temp		
		Temp	Temp	Level	Reservoir Temp	Temp Leaving Cooler	Press.				Level	Press.	Temp	In	Out	Press.	Temp	In	Out	
6-1	95°F	158°F	OK	140°F	120°F	15 PSI	125	460	15 PSI	150°F	OK	18" VAC	40°F	57°F	47°F	8 PSI	105°F	95°F	85°F	DHJ

CHILLER AIR CONDITIONING AND REFRIGERATION LOG

MANUFACTURER *Crete Compressors*   NUMBER *5486/93*   SIZE *1000 t*   LOCATION *East Wing*   REFRIGERANT *123*

INFORMATION LOGGED HOURLY/DAILY

TESTS PERFORMED ANNUALLY

TUBES

DATE CLEANED

DATE OF EDDY CURRENT TEST

Operating & Safety Control Tests

OIL PRESSURE SAFETY CONTROL

SETTING | CONDITION

DATE TESTED/INITIALS

LOW REFRIGERANT PRESS./ TEMP SAFETY CONTROL

SETTING | CONDITION

DATE TESTED/INITIALS

LOW CHILLED WATER TEMPERATURE CONTROL

SETTING | CONDITION

DATE TESTED/INITIALS

FLOW/PRESSURE DIFFERENTIAL CONTROL

Chilled Water | CONDITION

Condenser Water | CONDITION

DATE TESTED/INITIALS

For reliable and uninterrupted service the following maintenance items should be given attention as indicated.

COMPRESSOR:
The compressor should be disassembled after 40,000 operating hours or 5 years, whichever comes first. Impeller(s) should be examined for rubbing, grooves, and cracks, then cleaned, non destructively tested, and balanced. Guide vanes, linkage, and bushings should be examined for lost motion, wear, and sticking stems. The main shaft, pinion, and gears should be non destructively tested.

TUBES:
The water side surfaces should be cleaned annually. Eddy current analysis should be performed on condenser tubes within 3 to 5 years of service, and on evaporator tubes within 5 to 7 years of service. The detection of any wear, corrosion, or distortion of tubes will determine when subsequent testing of the tubes should be performed.

CONTROLS:
All safety and operating controls should be tested annually, calibrated to design conditions, and their set points recorded. Defective safety devices and controls should be replaced.

OIL/REFRIGERANT:
Record date and amount when oil or refrigerant is added. Also indicate leak tests, repairs, and adjustments. Oil samples should be taken and tested for physical and chemical properties.

Figure 11-3. Large system operation is recorded in a logbook.

The Foxboro Company

Figure 11-4. Large systems may be monitored and operated from a centralized control room.

Systems Troubleshooting

Troubleshooters deal with systems constantly. Troubleshooting by testing individual components solves most troubleshooting problems, but some problems, specifically complex problems, are difficult to solve in this way. Individual component testing narrows a troubleshooter's focus, but complex problems often require a broader view. This is because complex problems may lie in the relationship between parts of a large system or in other systems and not in individual components. Troubleshooters must consider the entire system when trying to identify and repair a problem.

Systems thinking is the consideration of an entire system and its interrelationships when troubleshooting. Before acting, experienced troubleshooters consider the whole system. For example, an intermittent computer printer problem could be related to occasionally high humidity in the air. A faulty flow control valve on the steam nozzles that add moisture to the air causes the occasional increase in humidity. This malfunction is caused by a buildup of scale in the pipe delivering the steam to the air. Occasionally, the flow control valve sticks too far open, delivering too much steam to the air, causing the intermittent printer problem. Focusing on only the printer prevents the troubleshooter from realizing the cause of the problem.

New system problems can also arise when correcting an old problem. For example, when trying to correct a drafty work area, the troubleshooter may reduce airflow to the area. Reducing airflow to one part of a building sends more air to the rest of the building. This could have a negative effect in other parts of the building, possibly creating hot or cold spots in the other areas. To prevent this, the troubleshooter may have to recalibrate the airflow to other parts of the building and possibly reduce the airflow to the entire building.

Systems may be open-loop or closed-loop control systems. **See Figure 11-5.** A *closed-loop control system* is a control system in which the result of an output is fed back into a controller as an input. An *open-loop control system* is a control system in which decisions are made based only on the current state of the system and a model of how it should work. Open-loop systems are less common than closed-loop systems. A common open-loop control system is used to warn of a sudden fall of outside air temperatures and then turn the heat ON in a building before it begins to cool because of the cold weather outside.

Most mechanical, electrical, and electronic systems are closed-loop systems. In a closed-loop system, each action by a component causes a reaction in the component following it. Eventually, the reaction works through the system to affect the component that initially caused the action.

Closed-loop systems become more complicated as their sophistication increases due to the increased number of input measurements and the sensitivity of electronic controls. For example, a basic closed-loop system may measure temperature. A more complex electronic control that measures and controls humidity may be added to the basic thermostat control because humidity contributes to human comfort. The system now has two measurement components to consider if the room is excessively hot or cold. There is a greater possibility that a fault will develop in two components than in a single component.

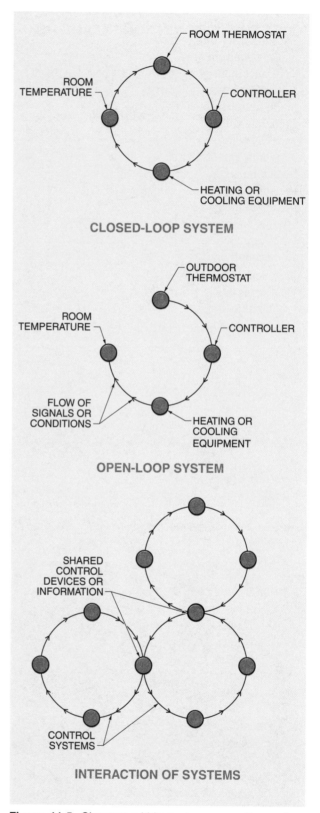

Figure 11-5. Changes within systems can alter system operation. Separate systems can interact, producing unexpected results.

Complex troubleshooting problems are also caused by the sensitivity of the inputs to control systems. For example, a controller may turn a device ON only when the controller input receives a signal of exactly 5.7 VDC. Such a device may be inadvertently turned ON by electric noise caused by other electrical or electronic systems that generate voltage signals in the input wires. Turning a motor ON whose power lines run near the signal wires may generate enough voltage to turn a controlled device ON when no signal is being sent by the input device. Conversely, if damage or poor connections change the resistance of the input wires, the voltage the controller receives is an incorrect signal. In this case, the device may not turn ON even when the input device is trying to send enough voltage to turn the device ON.

Troubleshooters also need an understanding of the many systems that interact in modern process and building controls. Systems may interact with other systems in ways that are not easily understood. Troubleshooters should work to understand the role a piece of equipment plays in a system and how other systems may interact with that system. It is common to find systems interacting with each other even if this is not shown on schematics. For example, if a building temperature is raised, heating bills rise, temperature-sensitive equipment may malfunction, and additional static electricity may be generated. While the increase in the heating bill is obvious and the malfunction of temperature-sensitive equipment could be easily traced to higher temperatures, the increase in static electricity may not be tied to the increase in building temperature. The increase in building temperature may lower relative humidity, enhancing the generation of static electricity that may cause random electronic failures if the static electricity is discharged into electronic components. These failures would appear to be random and have no obvious cause. Study by an experienced electronic technician would be required to identify the cause of the problem. Also, if a hundred new computer stations are added to a factory, the heat load on the building rises, requiring additional cooling capacity in climate-controlled areas. The electrical supply system may be strained by the additional demand. The harmonics generated by the computers may also cause severe power quality problems in the building electrical system.

Troubleshooting can be more complicated if time or distance separates the causes and effects of changes made in a system. For example, an adjustment made to a heating system for one area affects the temperature in another part of the building. The complaints do not arrive until four days later. The adjustment caused an immediate problem in the other part of the building and more problems when the complaints started arriving. Because of the time delay, the troubleshooter may not tie the adjustment in the heating system to the later complaints.

MAINTENANCE AND TROUBLESHOOTING RESOURCES

Troubleshooting involves using information gathered from operator's manuals, service bulletins, electronic monitoring systems, websites, manufacturers, machine operators, and advice from other technicians. Most manufacturers supply troubleshooting recommendations and symptom diagnostic guides with equipment. These materials should be organized for easy access. It is impossible to operate modern industrial equipment without using manufacturer's manuals and other resources.

Manufacturer's information is traditionally found in the operation and maintenance (operator's) manual. Updates to the manual may be distributed as service bulletins. The same information that is in the manuals and service bulletins is sometimes available on-line for use on-screen or to be printed in the plant as needed. Most companies operate websites that allow access to maintenance and other information using a computer connected to the internet.

Ludeca, Inc., representative of PRUEFTECHNIK AG.
Machine shafts that are properly aligned reduce vibration and add many years of service to equipment seals and bearings.

In addition, equipment sales personnel can be excellent sources of maintenance assistance. Equipment sales personnel may also have contact with manufacturer representatives or company engineers who have useful suggestions.

Trade journals and magazines are often excellent sources of maintenance information. Often, specific maintenance procedures are discussed in articles or case studies. The magazine's advertisements often contain manufacturer's contact numbers and may include manufacturer contact cards that allow maintenance personnel to obtain information from a variety of manufacturers.

General PM suggestions can be obtained from manuals for similar equipment. In addition, some manufacturers supply general maintenance suggestions. **See Figure 11-6.** For example, maintenance personnel may study general material produced by electrical equipment manufacturers. Many computerized maintenance management systems include general PM tasks. These suggestions are available as part of the software and can be used as general suggestions for PM tasks.

RECOMMENDED MAINTENANCE PROCEDURES

Hermetic Motors

Annually:

- Take insulation resistance test of stator windings. Values below 50 MΩ at an ambient temperature of 85°F or less may indicate moisture in the winding insulation.

- Inspect the contacts in the magnetic motor starter for signs of deterioration.

- Check all line and load side terminals for loose connections.

- Test control relays for proper timing sequence.

- Measure line voltage and current load for proper balance.

- Test motors which have been tripped by any protective devices. Do not restart them until the windings have been tested and the motor starter circuits have been examined to determine the reason for tripping.

Reciprocating Compressors

Annually:

- Sample oil for analysis. The results will indicate any need for a special service or maintenance activity.

- Check the crankcase heater circuit for operation.

- Test the low oil pressure cutoff switch, which should be within the time delay rating and at the pressure differential specified by the compressor manufacturer. Replace it if it fails to function properly.

Figure 11-6. Some manufacturers supply general maintenance suggestions, which can be used for equipment troubleshooting.

TECH TIP

Most maintenance software has the ability to record data and streamline inventory control, equipment history, job planning, and purchase requisition and purchase order generation.

Codes and standards may be used by state and local authorities when determining PM requirements. The National Electrical Manufacturers Association (NEMA) produces authoritative recommendations for establishing a PM system for industrial equipment in NEMA ICS 1.3, *Preventive Maintenance of Industrial Control and Systems Equipment*. Electrical maintenance guides are produced by the National Fire Protection Association (NFPA), publication NFPA 70B, *Electrical Equipment Maintenance*. PM requirements are also given in the American Society of Mechanical Engineers (ASME) boiler operating codes Section VI, *Recommended Rules for the Care and Operation of Heating Boilers*, and Section VII, *Recommended Rules for the Care of Power Boilers*. Electrical infrastructure regulations are found in the National Electrical Code® (NEC®).

The American Society of Heating, Refrigerating, and Air-Conditioning Engineers (ASHRAE®) codes cover topics such as refrigeration, indoor air quality, ventilation standards, building operation and maintenance, and energy efficiency. Some regional and local building and mechanical codes also specify maintenance activities. Local and national codes are usually available in libraries and state or local offices, or can be purchased along with guides to using and applying the codes. Some insurance companies specify the maintenance that must be completed on equipment that they insure.

Operator's Manuals

Troubleshooting and maintenance advice in operator's manuals is probably the most commonly used resource. For example, if a technician uses the incorrect flushing fluid, hydraulic system components may be damaged. Following the manufacturer's instructions is essential to avoid later problems.

Operator's manuals may contain troubleshooting information printed in chart form. The chart lists symptoms, possible causes, and suggestions on repairing the problem. **See Figure 11-7.** For example, if excessive noise is a problem in an air compressor, the troubleshooting chart indicates that V-belts may be slipping and need to be retensioned or replaced. The compressor bearings may be failing or the motor's rotor may be contacting the stator. If the belts are slipping, they may be badly worn and require replacement.

COMPRESSOR TROUBLESHOOTING CHART		
Symptom	Possible Causes	Suggestions
Compressor fails to start	115 V control voltage not available.	Check fuses. Check transformers and wiring connections.
		Inspect contactors.
		Rotate emergency stop button to disengage; press set button twice.
		Manually reset main motor overload relay; press set button twice.
		Check for defective sensor, bad sensor connection, or broken sensor wires.
Compressor shuts down	High air temperature.	Ensure that installation area has adequate ventilation.
		Ensure that cooling fan is operating. If not, reset circuit breaker inside starter box.
		Check coolant level. Add if required.
		Cooler cores dirty. Clean coolers.
	Aftercooler core dirty.	Inspect and clean.
	Enclosure panels not in place.	Install enclosure panels.
	No aftercooler on unit.	Install aftercooler.
	Drain line/drip leg incorrectly installed.	Slope drain line away from trap. Install drip leg.
	No refrigerated or desiccant dryer in air system.	Contact local distributor.
Excessive noise level	V-Belts slipping.	Adjust belt tension or replace belts.
	Compressor defective. (Bearing failure or rotor contact.)	Contact authorized distributor immediately, do not operate unit.
	Enclosure panels not in place.	Install enclosure panels.
	Loose component mounting.	Inspect and tighten.

Figure 11-7. Troubleshooting information may be printed in chart form.

Many operator's manuals provide excellent instructions for the regular maintenance activities required by the equipment. Often, such instructions are ignored because the job seems simple. For example, a manual may detail how to replace belts. Belt replacement is a common procedure. However, if a compressor belt drive has left-hand threads on the idler center shaft, the nut must be removed in the opposite direction from common right-hand threads. If this is not known, the technician may think the idler is stuck and apply damaging force in an attempt to remove it as if it is a right-hand thread. A mistake like this may be expensive to repair and can cause downtime. The required tools and recommended replacement parts may also be suggested in the operator's manual. Such information saves time when planning a job and ordering parts.

Troubleshooting information may also be presented as a flow chart. **See Figure 11-8.** The flow chart is read by beginning at the start ellipse, following the arrows, and answering yes or no to the questions in the diamonds. For example, if the answer to the question in the first diamond is no, the action is to turn power OFF then check to ensure the power is OFF. If the answer to the question is yes, the action is to check to ensure the power is OFF. The arrows are followed and the questions are answered by replying yes or no and following the respective paths that lead to the problem. The chart takes lengthy word descriptions and condenses them into a flow chart for quick problem solving.

Manuals for complicated equipment such as programmable logic controllers also present troubleshooting information as a flow chart. Sophisticated equipment such as a PLC could not be repaired without extensive use of the operator's manual. Operator's manuals should be used even for basic equipment. Effective troubleshooting of industrial equipment requires the use of all manuals and other forms of technical assistance.

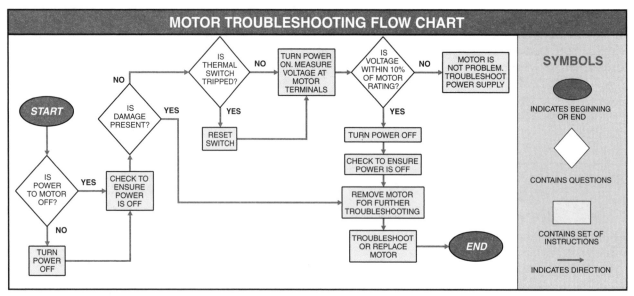

Figure 11-8. Troubleshooting information is often presented as a flow chart.

Troubleshooting Reports

Troubleshooting reports are patterned after the technical service bulletin system used in the automotive industry. After a new car model is introduced, information is gathered on its reliability and operational problems. All breakdowns or problems encountered in the dealer service facilities are recorded on a report form. Each report details one problem, its symptoms, causes, and repair procedures. As patterns of breakdowns or problems develop, technical service bulletins are distributed to all service facilities, where they are used by mechanics. If a mechanic has already diagnosed and solved a problem, another mechanic can save time by using this information to repair the same problem in another vehicle.

The technical service bulletin system works well in the highly organized automotive industry based on similar, widely distributed products such as automobiles. Because these conditions do not exist in most industrial settings, there are few service bulletin systems being operated by manufacturers of industrial equipment. However, the technical service bulletin system can be adapted for use within a plant by incorporating it into the plant PM system as a troubleshooting report. A *troubleshooting report* is a record of a specific problem that occurs in a particular piece of equipment, along with its symptoms, causes, and repair procedures. **See Figure 11-9.** Each troubleshooting report should include standard information such as the individuals who worked on the problem,

the department, the equipment identification number, the problem, symptoms, causes, repair procedures, and preventive maintenance action.

Individual maintenance personnel can evaluate troubleshooting reports to improve their own troubleshooting abilities. Troubleshooting reports that are incorporated into a plant PM system become part of the equipment history for each machine. If the cause of the problem is discovered, adjustments to the machine or modifications to the machine's PM work are made. Over time, these modifications reflect the needs of each plant or equipment situation.

A troubleshooting report is filled out for each breakdown or equipment problem immediately after the problem is solved. This information is filed manually or entered into a computer for future reference. The next time a particular machine requires troubleshooting, the technician reviews the machine's troubleshooting reports to learn if the symptoms of the current problem have occurred previously. If so, the technician uses the information on the troubleshooting reports to begin troubleshooting. Such information can result in a tremendous saving of troubleshooting time.

Over time, most equipment develops tendencies or problems that repeat. The use of troubleshooting reports enables maintenance personnel to develop plant procedures to deal with these repetitive problems. **See Figure 11-10.** For example, the troubleshooting and replacement of a defective PLC module can be written as a procedure. The technician follows the procedure when replacing the module.

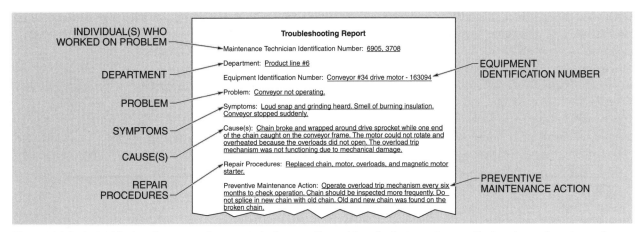

Figure 11-9. A troubleshooting report is a record of a specific problem that occurs in a particular piece of equipment.

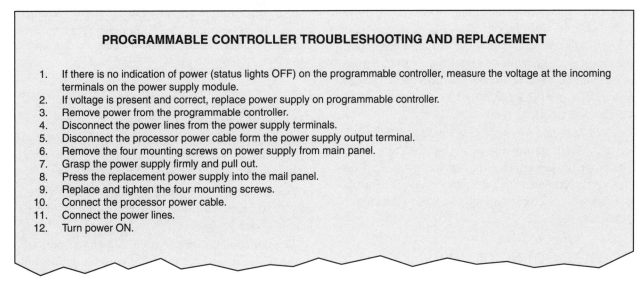

Figure 11-10. The use of troubleshooting reports enables maintenance personnel to develop plant procedures to deal with repetitive problems.

Electronic Monitoring Systems

Many industrial systems are equipped with electronic monitoring systems that display error codes. When a problem occurs in the system, numbers or words are displayed on a digital readout attached to the equipment or in a control room. The numbers or words are error codes that indicate the cause of the problem. The explanation of the code is found in the equipment manual and the technician follows instructions in the manual or tests for problems in the area indicated.

Some electronic monitoring systems are computerized. When a problem occurs, the monitoring system is accessed using a computer. The computer may be hand-held or a desktop model connected to the monitoring system. The error code is displayed on the computer screen followed by instructions for testing components and making repairs.

Out-of-Plant Services

Outside advice is often necessary for decision-making or work that requires expertise not found in the maintenance crew. In addition, installing new equipment or making major renovations may require companies with specialized tools and expertise. Outside companies can install new equipment or make major renovations more efficiently than a plant's maintenance crew, who must continue to maintain other plant operations. In addition, outside companies provide warranties for their work should problems occur during the equipment break-in period.

Outside experts, manufacturers, and distributors can be consulted by telephone, fax, or computer. Many manufacturers have toll-free numbers that provide access to service personnel. Sometimes, technical support services are part of a service contract and can be costly. Therefore, these resources must be used efficiently. When making calls to any outside support services for troubleshooting help, all symptoms should be written down, manuals and other service material should be nearby, and any computers should be ready to use. Portable phones allow for such calls to be made from the equipment location. All tools and test equipment should be within reach and ready to use because the service technician may need the results of tests and inspections. In some cases, service technicians can access the equipment's fault monitoring system using their own computer and run diagnostic tests remotely. Fax machines are commonly used to send parts lists, questions, and other written material or illustrations of equipment. Such faxed material should be easy to view and written in large, easy-to-read type to avoid communication problems.

Some manufacturers provide on-line information services where information is posted for those using the company's equipment. The service is accessed via the Internet, through which software or service information can be transferred or questions asked of the service technicians.

TROUBLESHOOTING PROBLEMS

The troubleshooting process investigates, isolates, remedies, and documents the problem. Troubleshooting reports provide valuable data when troubleshooting. This allows more efficient troubleshooting by maintenance personnel.

Troubleshooting Boiler Systems

Boilers supply steam or hot water for use in heating or production processes. Troubleshooting boilers requires quick action. Boiler problems can cause costly downtime and possibly result in extensive equipment damage and personal injury.

The most common problem is that the boiler does not start or restart. The burner control display may indicate the problem, such as failure of the pilot flame. If the pilot flame is out, a series of tests can indicate the cause of the flame failure. First, check the fuel supply. If the boiler is fueled by gas, the pressure may be too low or too high.

If the boiler uses fuel oil, the fuel may be too cold or at too low a pressure. The fuel filter may be clogged or the fuel lines may contain water or air.

Check the flame scanner to see if it is dirty or has loose connections. Check the ignition system for secure, tight connections and the proper source voltage. If a problem is suspected in the flame scanner or igniter, swap the component with a spare to see if that fixes the problem. Always purge the furnace of any accumulated fuel before attempting restart.

> **TECH TIP**
>
> *Some computerized maintenance management systems, when interfaced with the appropriate communication network, can monitor machine conditions from remote locations. This feature may be part of an integrated building automation and maintenance management system.*

Case Study: Boiler Flame Failure

A plant has three boilers in battery used to provide heat for an office and manufacturing facility. During warm weather conditions, one boiler operates and the other two are on standby. The operating boiler suddenly shuts down and sends an alarm message to the engineer's mobile phone.

The boiler is not firing, even though the pressure in the boiler is low enough to require the production of more steam. The engineer checks the control panel on the front of the boiler, and the digital error code indicates that the boiler has stopped because of a pilot flame failure during the light-off cycle. The maintenance engineer follows a step-by-step troubleshooting procedure to investigate, isolate, remedy, and document the problem. **See Figure 11-11.**

The maintenance engineer checks the high and low gas pressure interlocks on the gas line leading to the burner. The interlocks are not tripped, so the gas supply is not the cause of the failure. The engineer resets the boiler controller to start the light-off sequence, which operates normally until the time for the pilot light to ignite. No spark occurs to light the pilot light. The lack of the crackling sound that accompanies a normal spark ignition alerts the engineer to this problem. The engineer checks the connections between the ignition transformer and the electrode in the boiler's burner to ensure that the connections are snug.

BOILER TROUBLESHOOTING — FLAME FAILURE

Problem

A plant has three boilers in battery used to provide heat for an office and manufacturing facility. During warm weather conditions, one boiler operates, with the other two on standby. The operating boiler suddenly shuts down and sends an alarm to the maintenance technician's pager.

Troubleshooting Step 1 - Investigate

Operator not present - on automatic control.
No flame in boiler.
Steam pressure low.
Water level normal.
Burner control display readout - PILOT FLAME FAILURE.

Troubleshooting Step 2 - Isolate

Likely Causes	Action/Result
High or low gas pressure interlock.	Check reset on high and low pressure gas interlock - not tripped.
Dirty flame scanner.	Remove/inspect flame scanner - not dirty.
Ignition system failure.	Reset burner control and start light off - no igniter crackle heard/no spark observed through sight glass.
Loose igniter wire connections.	Tighten igniter wire connections, start light off - no change.
Faulty igniter wire.	Swap igniter wire from standby boiler, reset burner control, start light off - flame ignited. Original igniter wire damaged from overheating at connector.

Troubleshooting Step 3 - Remedy

Replace igniter wire. Obtain new igniter wire from inventory, reset burner control, start light off - flame ignited. Order replacement for inventory.

Troubleshooting Step 4 - Documentation

File reports. Complete troubleshooting report and notation in boiler and shift log.

PROCEDURE

Troubleshooting Report

Maintenance Technician Identification Number: 6704

Department: Heating Plant

Equipment Identification Number: Boiler #2 - 166734

Problem: Boiler would not fire.

Symptoms: Burner Control readout, "Pilot Flame Failure." No sound or sight of spark during pilot light off.

Cause(s): Open igniter wire. Connection at boiler front deteriorated from overheating.

Repair Procedures: Replaced igniter wire.

Preventive Maintenance Action: Check igniter wire connections on quarterly PM inspections.

DOCUMENTATION

Figure 11-11. Troubleshooting a boiler flame failure begins by investigating the condition of the equipment and isolating the likely causes.

The engineer tries to start the boiler again but it does not light. The digital readout indicates the same problem of pilot flame failure. The engineer decides to swap the igniter wire with one from a standby boiler because there is no visible sign of overheating and no smell of a burnout from the ignition transformer on the malfunctioning boiler. The firing cycle is restarted and the boiler fires correctly.

There is no replacement igniter in inventory, so the engineer immediately orders two, one to replace the one taken from the standby boiler and one to put in inventory. The parts are delivered by late afternoon, in time to bring the standby boiler online for the extra steam required for the evening's production and plant heating.

The maintenance engineer tests the damaged igniter wire using a DMM set to measure resistance and finds it open. The open was probably caused by the heating of the igniter wire as part of normal operation, although the problem may have been caused by a defect in the wire, causing high resistance that led to overheating. The engineer records the order for the spark igniter in the parts order log, fills out a troubleshooting report, and notes the problem in the boiler room log.

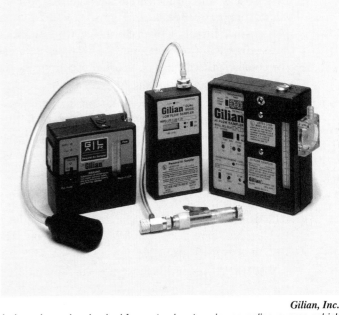

Gilian, Inc.

Indoor air can be checked for contaminants using sampling pumps, which collect vapors and particulates.

Troubleshooting Refrigeration Systems

The most common problem with a refrigeration or air conditioning system is that the temperature is incorrect, usually too hot. Begin troubleshooting this type of problem by taking the temperature at the thermostat, if there is one, and at several locations in the cooled space. If it is too high, check to see that cool air is reaching all portions of the cooled space. Remove any barriers to airflow. Check that there have been no unusual conditions such as cooler doors being left open or extremely high outside temperatures. These conditions can create too great a heat load and the system may not keep up even if it is working correctly.

Check the operation of the evaporator fans and the condition of the coils. Check the condenser for cleanliness and proper fan operation. Look for oil stains that might be signs of a refrigerant leak. If there are no obvious problems found during these visual inspections, then attach the gauge manifold to check system pressure and determine subcooling and superheating. Compare these readings to manufacturer's troubleshooting specifications.

Many refrigeration problems will present themselves as electrical problems. For example, a leaking system will become low on charge, causing the compressor to overheat. The compressor overloads will then open.

Case Study: Refrigeration System Leak

Several large walk-in coolers at a facility are excessively warm. Temperature readings confirm this and a visual investigation reveals nothing unusual with either cooled space evaporator. Questioning the workers who use the cooled space reveals no unusual occurrences. The compressor is not working and is very hot.

The maintenance engineer waits for the compressor to start. The compressor runs for a few minutes and then stops. The engineer moves into the cooled space and examines the sight glass near the evaporator when the compressor cycles ON. The sight glass is full of bubbles and appears to be empty at times. This indicates a low charge in the system. A review of the cooler maintenance records indicates that the evaporator has recently been cleaned but there have been no recent problems. The system is over 15 years old.

Because the product in the cooler will soon spoil and cannot be moved to another cooler, the engineer decides to charge the system to see if it operates correctly. The engineer uses 12 lb of R-22 refrigerant, which is the same refrigerant used in the system but is becoming expensive because it is being phased out. Once charged, the system works correctly. As required by environmental regulations and plant policy, the engineer notes the amount of refrigerant added in the PM system records and resumes looking for the leak.

The engineer first tests for leaks using an electronic refrigerant detector, but doesn't find a leak. **See Figure 11-12.** Using the fluorescent dye method, a leak is found in the evaporator, which was probably damaged during cleaning. This method uses a chemical dye added to the compressor oil. After circulating through the system, the dye illuminates leaks in system components when exposed to a special light.

By law, all large refrigeration system leaks must be repaired or the system must be replaced. Given the age of the system and the expense and coming scarcity of R-22 refrigerant, the chief maintenance engineer decides to investigate replacing the system. Several refrigeration system manufacturers are consulted, as are the manufacturers of environmentally safe refrigerants. After several weeks of discussion and planning, it is decided to replace the system during a planned production shutdown so production is not affected.

Figure 11-12. A refrigeration system that is not cooling properly may be low on refrigerant, which is checked by viewing the sight glass when the system is operating.

All actions regarding the system are recorded in case the records are needed to prove compliance with environmental regulations. All information regarding the purchase of the new system is retained for future reference because many of the plant's systems use old refrigerant and may require replacement in the near future. Due to the leak, the system inspection level is increased to prevent the system from being damaged beyond use before it is replaced. A warning to be careful when cleaning evaporators is added to all refrigeration system PM work orders because the recent cleaning may have caused the leak in the evaporator.

Troubleshooting PLCs

Programmable logic controllers are extremely reliable. When problems occur, the cause is usually in the input or output circuits. Troubleshooting these circuits individually can narrow down the source of the problem. All circuits should be checked for secure connections. The supply voltage to each control device and to the input modules should be measured. Monitor the status indicators (usually lights) to check that inputs are energized at the appropriate times. Outputs can be checked by placing the program in test mode. This mode prevents output devices from energizing, but still uses the output status indicators. As the program runs, monitor the actual status of the outputs against the expected status. This method can usually pinpoint problems down to individual inputs, outputs, or control devices.

It is less likely that the computer program is the source of the problem. When this occurs, the problem can be difficult to correct. Usually, the easiest way to remedy a faulty PLC program is to reload the original working copy of the program from the backup. This is why it is so important to maintain backup copies of all PLC programs. The cause of the faulty program may be data corruption from the computer or access to the program by unauthorized individuals. The computer hardware and access controls should be checked after reloading the program.

Case Study: Altered PLC Program

A PLC controls a pick-and-place robot that picks up a part to be fitted with a cover. The robot picks up the part, but does not move to place it into position to be covered. The arm does not extend and an alarm sounds. The maintenance engineer responds quickly and checks the mechanical operation of the arm. The arm is free to move. A check of the compressed air system supplying power for the arm reveals no obvious problem. The engineer decides to check the PLC inputs and outputs or the PLC program. **See Figure 11-13.**

The maintenance engineer finds the PLC wiring diagram and locates the output module controlling the arm. Other operations controlled by that output module are working correctly. The robot's movement starts when the correct input is pressed, but always stops at the same place and the output indicator light for the arm goes out. The engineer determines that the inputs and outputs are working correctly and accesses the PLC line diagram with a laptop computer. The arm operates correctly only when the particular output is forced.

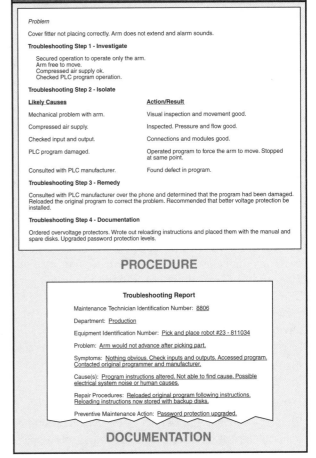

Figure 11-13. Troubleshooting a PLC begins by checking the inputs and outputs.

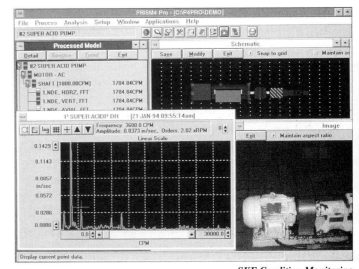

SKF Condition Monitoring
Condition monitoring software can assist in making accurate maintenance decisions.

The maintenance engineer gathers all the relevant prints and manuals and contacts the PLC manufacturer. The PLC manufacturer makes several suggestions for testing inputs and outputs but the problem is not corrected. The engineer then contacts the company that designed the robot and created its program. The programmer directs the maintenance engineer to access specific parts of the program. Finally, an alteration in the program is found in the instructions that advance the arm.

The engineer locates the backup discs containing the original program and, following instructions from the programmer, reloads the original program. The robot then operates correctly. The maintenance engineer works with the programmer to try to find the cause of the damage to the program, but no cause is determined. As a precaution, the voltage protective devices for the PLC are replaced and the grounding system is checked following instructions received from the PLC manufacturer and the robot designer. The level of password protection for the PLC is reviewed because someone accessing the program may have altered the program. The manuals and discs are returned to the correct locations. The steps for reloading a PLC program are included with the discs.

Troubleshooting Electrical Systems

Electrical systems can be quite complex, incorporating many electrical devices over a large area. Systematic troubleshooting can quickly narrow potential causes of electrical problems down to manageable subsystems or devices. Some electrical testing requires only a DMM, but others may require particular test equipment or electricians with special training, as in testing high voltage. Always be sure the correct test instrument is used for the desired test.

Electrical problems can be found by troubleshooting using the source, pathway, and load method. This is a troubleshooting method in which testing is started at the source of the system power, then moved through the control components in the pathway leading to the load. This method usually works well with problems in electrical, PLC, and fluid power systems. Depending on the situation, this method can also be reversed by working from the load to the source.

Intermittent problems are difficult to diagnose. This type of problem often requires more time to observe the system and record any short-term anomalies. An electrical test instrument with long-term recording functions is useful in this situation. The anomalies are compared to the normal operating conditions to help narrow down the possible causes. Performing this test on various parts of the system will isolate the source of the anomaly.

Damaged or destroyed motors are common electrical problems regardless of whether electrical or electronic controls operate the motor. Often, the only symptom is that a motor does not operate. If a motor will not start, the voltage supply coming into the control transformer and out of the transformer is checked. The transformer is replaced if it is bad. The fuse is checked and replaced if it is blown. A master stop could be stuck open or one or more of the OLs could be open. If one of the overloads is open, the maintenance engineer should try to determine why the motor was drawing enough current to cause the overload to open.

If a circuit contains one start/stop pushbutton station that operates multiple motors, none of which were working, the maintenance engineer would start by testing the source voltage. It is unlikely that multiple motors would fail at the same time, so the problem is probably in the voltage source.

If multiple motors are individually controlled and none of the motors will start, the problem is in the overall power supply. However, if only one motor will not energize, the problem must be in the controls for that motor, the stop and start buttons, the wiring connections, or the motor itself.

Case Study: Intermittent Electrical Problem

A high-speed automatic filling machine that fills cans with oil occasionally fails to fill a row of cans. The problem continues for several hours. The maintenance engineer watches the machine operate, but cannot see anything different in the mechanical motion of the filler from one cycle to the next. The engineer thinks an intermittent electrical problem is being caused by the rapid motion of the filler as it positions over the cans. The engineer inspects all parts of the machine as it operates, but does not see any obvious problem. **See Figure 11-14.**

The engineer attaches a DMM set to measure voltage to the two wires leading to the solenoid that controls the filling action. Normally, this reading is 0 V when the solenoid is not energized and 120 V when the solenoid is energized and the cans are being filled. As the technician watches, the reading suddenly flickers to about 80 V (instead of 120 V) and then returns to 0 V. Holding the DMM and watching the filler, the engineer notices that the flicker only occurs at a specific time in the cycle and when the cans are not filled.

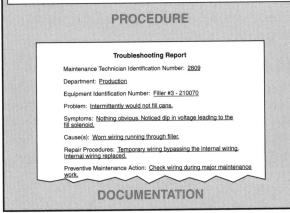

ELECTRICAL SYSTEM TROUBLESHOOTING — INTERMITTENT ELECTRICAL PROBLEM

Problem

Occasionally the oil filler does not fill a row of cans. Intermittent problem.

Troubleshooting Step 1 - Investigate

Observation shows no obvious mechanical problem.
Tested and observed mechanical and electrical operation of the filler solenoid.

Troubleshooting Step 2 - Isolate

Likely Causes	**Action/Result**
Mechanical problem.	Observation. Nothing obvious.
Electrical problem.	Observation and measurement of current and voltage in solenoid wire.
	Noticed occasional current and voltage dip in filler wires.
Damage to solenoid wiring.	Inspection. Found frayed wires.

Troubleshooting Step 3 - Remedy

Ran temporary wires outside the solenoid wiring conduit. Replaced the worn wires after the production run.

Troubleshooting Step 4 - Documentation

Filed troubleshooting report. Suggested increased inspection of all wiring on fillers.

PROCEDURE

Troubleshooting Report

Maintenance Technician Identification Number: 2809

Department: Production

Equipment Identification Number: Filler #3 - 210070

Problem: Intermittently would not fill cans.

Symptoms: Nothing obvious. Noticed dip in voltage leading to the fill solenoid.

Cause(s): Worn wiring running through filler.

Repair Procedures: Temporary wiring bypassing the internal wiring. Internal wiring replaced.

Preventive Maintenance Action: Check wiring during major maintenance work.

DOCUMENTATION

Figure 11-14. Intermittent problems are the most difficult problems to troubleshoot.

The engineer suspects that the filling solenoid is not energizing because it is not getting the correct voltage. By wiggling the wires leading to the solenoid, the engineer reproduces the voltage dip and prevents the cans from filling. The engineer suspects that the wires inside the filler are worn but not broken, or are occasionally touching the machine's frame to produce a high-resistance short to ground that is not enough to blow a fuse but causes a sudden voltage dip in the solenoid circuit.

The engineer knows that the wires cannot be repaired while the machine is working because the wires run inside the filler. The engineer runs two wires outside the filler, stops production for a few minutes, and connects the solenoid to the new wires. The problem is corrected when the filler restarts. The wires are secured temporarily until production ends and the internal wires can be replaced. When the faulty wires are removed, they appear worn and partially broken due to the motion inside the arm.

The problem is noted in a troubleshooting report that was entered into the computer because the plant has several of these machines that may develop similar problems.

Case Study: Electric Motor Failure

A conveyor in an industrial plant stops working. Questioning the machine operator reveals no unusual occurrences. The engineer finds no obvious problems and decides to test the electric motor operating the conveyor. **See Figure 11-15.**

The engineer ensures that all operators are clear of the conveyor and that its sudden startup would not damage product or endanger human safety. The engineer places a safety barrier around the work area and has another operator signal when everyone is clear.

ELECTRICAL SYSTEM TROUBLESHOOTING — ELECTRIC MOTOR FAILURE

Problem

Conveyor #17 stopped working.

Troubleshooting Step 1 - Investigate

No obvious problems and the nearby worker did not see or hear anything. No burnt insulation smell.
No visible mechanical problems with the conveyor or chain drive.

Troubleshooting Step 2 - Isolate

Likely Causes	**Action/Result**
Power supply or control circuitry.	Operated magnetic starter. Motor started. Problem is in the control circuitry.
Overloads tripped.	Tested overloads. Not tripped.
Magnetic motor starter coil.	No voltage change when energized.
Start and stop pushbuttons.	No voltage change when energized.
Control transformer.	No voltage.
Fuse.	Voltage test of fuse indicates the fuse is bad.

Troubleshooting Step 3 - Remedy

Replaced fuse and looked for cause of the electrical overload that blew the fuse. None found.

Troubleshooting Step 4 - Documentation

Note in log and monitor the conveyor for overload problems. Monitor motor for electrical problems.

PROCEDURE

Troubleshooting Report

Maintenance Technician Identification Number: 3893

Department: Assembly

Equipment Identification Number: Conveyor #17 - 311910

Problem: Motor not working.

Symptoms: Nothing obvious.

Cause(s): Blown fuse. No obvious cause.

Repair Procedures: None.

Preventive Maintenance Action: Monitor to see if fuse blows again.

DOCUMENTATION

Figure 11-15. Electrical circuit problems may be tested using the source, pathway, and load method of troubleshooting.

The engineer tests the motor operation first. The motor is tested manually by operating the magnetic motor starter. The motor starts, indicating that power is available to the motor. The engineer tests the overloads by resetting them. Overloads usually click into place if tripped. These overloads do not click into place, indicating that the overloads are not tripped. Based on these findings, the problem is in the control circuit.

The engineer tests the control circuit load (magnetic motor starter coil) using a DMM set to measure voltage. **See Figure 11-16.** On a good coil, the DMM should read 0 V and change to approximately 120 V when the start pushbutton is closed. The DMM reads 0 V whether the start pushbutton is open or closed. The engineer tests the control circuit pathway. The start pushbutton is tested using a DMM set to measure voltage. A good switch measures source voltage when the pushbutton is open and 0 V when the pushbutton is closed. The DMM reads 0 V when open and closed. The stop pushbutton is tested using a DMM set to measure voltage. A good stop pushbutton measures 0 V with the pushbutton contacts closed and source voltage with the contacts open and the start pushbutton held closed. The DMM reads 0 V when the contacts are open and closed.

The engineer tests the control circuit source. The secondary of the transformer is tested using a DMM set to measure voltage. The source voltage (120 V) is indicated if the transformer is good. The DMM reads 0 V, indicating that the transformer may be bad. The transformer fuse is tested using a DMM set to measure voltage. The fuse reads 0 V if good and source voltage if bad. The DMM reads 120 V, indicating that the fuse is bad (blown). This is confirmed with a resistance check of the fuse. The engineer replaces the fuse and operates the circuit. The circuit operates correctly.

After the fuse is replaced, the engineer inspects the circuit for signs of overheating or a short circuit that could have caused the fuse to blow. No obvious cause is found and the motor continues to operate effectively. The engineer notes the blown fuse in the troubleshooting report so that, if the fuse blows again, the circuit will be inspected more closely.

Troubleshooting Hydraulic Systems

Hydraulic systems can suffer problems produced by many causes. Hydraulic system problems could result from internal conditions such as dirty oil, from physical damage to components, or from electrical or electronic control circuits.

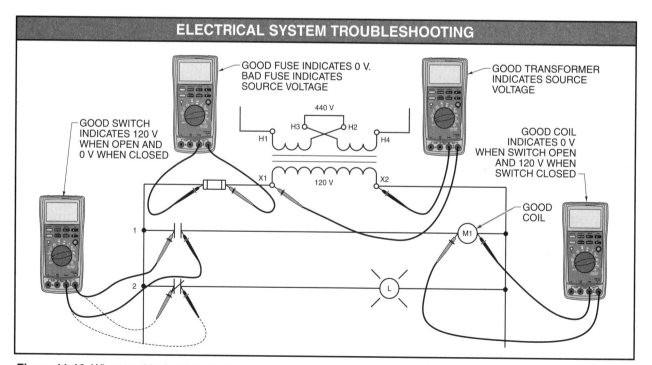

ELECTRICAL SYSTEM TROUBLESHOOTING

GOOD FUSE INDICATES 0 V.
BAD FUSE INDICATES
SOURCE VOLTAGE

GOOD TRANSFORMER
INDICATES SOURCE
VOLTAGE

GOOD SWITCH
INDICATES 120 V
WHEN OPEN AND
0 V WHEN CLOSED

440 V

GOOD COIL
INDICATES 0 V
WHEN SWITCH OPEN
AND 120 V WHEN
SWITCH CLOSED

GOOD
COIL

120 V

Figure 11-16. When troubleshooting electrical systems, each component is tested to locate the malfunctioning component.

A hydraulic piston rod reacts to changes in the direction of oil flow. The direction of oil flow is controlled by the position of the directional control valve, which can be operated by an electric solenoid or manual switches. When the valve is in the neutral position and flow stops, the piston will stay in position.

If a piston rod will not extend, the problem could be that the directional control valve is not moving because it is jammed, or because the valve operator (electric or manual) is not functioning properly. It could also be that the load the piston rod is attempting to move is too heavy for the system. In this case, the system might not be developing sufficient pressure to generate the force to move the load. The pump and the pressure relief valve should be checked. Dirt in a hydraulic system could have blocked oil flow or clogged pilot pressure lines, preventing component operation. The pressure relief valve setting could be incorrect. If adjustments are made, attach pressure gauges and flowmeters to the system to check that system pressure and flow specifications are not exceeded.

If the rod is moving, but at the incorrect speed, check the flow control valve and the pump. If the pump is not generating enough flow, the rod speed will be too slow. Restrictions in the hoses and tubing can also reduce flow and therefore speed. Maintenance personnel should also always look for kinked hoses and oil stains that indicate leaks.

Case Study: Hydraulic Press Failure

A hydraulic press is designed to bend and cut a piece of metal. The cutting action is controlled by a sequencing valve that operates after the metal is bent. A problem occurred when the press would bend but not cut the metal. The press operator demonstrates the problem but could not offer any suggestions. An examination of where the metal should be cut shows no marks, indicating that the cutter was not contacting the metal. The engineer obtains all the prints and some scrap metal to test in the machine. **See Figure 11-17.**

The engineer reviews the hydraulic prints. The source (pump) is not likely the problem because part of the circuit is working. The problem is likely in the branch of the circuit controlling the cutting action. The engineer makes a list of components and conditions to check including the cut piston condition, kinked lines feeding the piston, the operation of the sequence valve, and overall system pressure. The piston appears normal and the lines appear undamaged. While having the operator run a piece of scrap through the press, the engineer feels the sequence valve but detects no motion of the valve opening to allow oil flow to the cutter. After giving the valve a firm tap with the wooden end of a hammer, the press resumes normal operation.

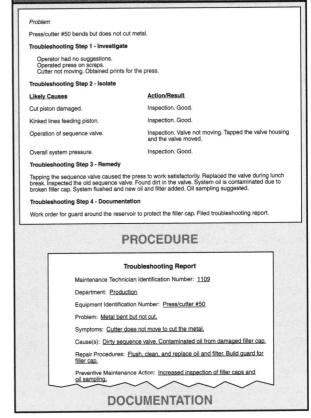

Figure 11-17. Hydraulic system fluid must be kept clean to ensure proper circuit operation.

The engineer watches the operation of the machine for several minutes and it malfunctions again. A tap on the sequence valve again restores normal machine operation. The operator continues work with the temporary fix while the engineer returns to the shop to obtain a new sequence valve. Later, the engineer locks out the press and replaces the sequence valve. The press operates correctly.

The engineer inspects the old sequence valve. Dirt is found in the valve, and the pilot line that operates the valve is partially plugged. The whole system is contaminated and requires cleaning.

The engineer returns to the press and finds that the filler cap on the reservoir is cracked and open to the atmosphere. Dirt has probably entered the reservoir during the daily cleaning of the press. The engineer studies the operator's manual and learns how to clean the system after such contamination. The engineer notes the exact flushing fluid to be used. A few days later during production downtime, the engineer drains and flushes the system and replaces the oil and filter.

The engineer suggests that a guard be built around the reservoir to prevent the filler cap from being hit and broken. All similar presses in the plant are inspected and eventually fitted with guards.

Troubleshooting Mechanical Systems

Mechanical troubleshooting requires patience and studying equipment from various angles while trying to visualize the forces acting on the equipment.

Bearings support shafts and allow them to rotate easily by reducing the friction of the rotation. Bearings are either lubricated and sealed when manufactured or can be lubricated manually on a regularly scheduled basis. Bearings can fail due to under- or overlubrication, dirt contamination, lubricant breakdown due to moisture contamination, or by using the incorrect lubricant. Overstressing the shaft can overload bearings. Slipping belts can cause the shaft to overheat, which in turn causes the bearings to overheat, leading to breakdown of the lubricant. Also, overloading of a mechanical system will cause the drive motor to overheat.

A motor can draw too much current in an attempt to maintain speed and torque, and trip the overloads that monitor current flow. If the motor current draw reduces when the bearings are lubricated, the problem is in the bearings, which must be replaced. The bearings can also be inspected using a mechanic's stethoscope. A low rumbling sound indicates that the bearings are worn. A high-pitched sound indicates poor lubrication and metal-to-metal contact.

To determine if the current overload results from the motor, its power supply, or the load, the motor is disconnected from the load and operated. If the current remains high, the problem is in the motor or its power supply. If current returns to normal, then the overload is in the load. For example, the load may be a conveyor that is carrying too much weight. The conveyor belt tension, tracking, and cleanliness should also be checked because these can also overload a motor.

Case Study: Gear Failure

A large mixer for mixing soup in a food processing plant develops a loud noise in the gearbox connected to the electric motor and mixing paddles. The operator calls the maintenance engineer. **See Figure 11-18.**

After hearing the loud noise, the engineer stops the mixer and quickly checks the gearbox oil. Metal shavings are present. The oil is drained and fresh gear oil is added. After production is over, the gearbox is replaced. The old gearbox is opened and inspected and the gears show signs of corrosion and broken teeth. Because the mixer is an old model, the engineer checks with the manufacturer to ensure that the replacement gearbox will work satisfactorily in this application. The manufacturer approves the use of the gearbox. The engineer prepares work orders to inspect the gearboxes on similar mixers.

MECHANICAL SYSTEM TROUBLESHOOTING – GEAR FAILURE

Problem

Loud noise from a motor driven gearbox driving a mixing paddle on a large mixer in a soup production plant.

Troubleshooting Step 1 - Investigate

Mixer gearbox made a loud noise when the mixer was started prior to mixing a new batch of soup.
Operator stopped the motor and called maintenance immediately.
Operated the mixer and the loud noise came from the gearbox.
The gearbox is an old model.

Troubleshooting Step 2 - Isolate

Likely Causes	Action/Result
Damaged gears.	Opened oil drain plug. Oil contained metal pieces.

Troubleshooting Step 3 - Remedy

Drained oil and replaced with fresh oil rather than stop production. The operator ran the mixer for one batch. A replacement gearbox was located in inventory and moved to the mixer. The gearbox was replaced after the batch was mixed. The damaged gearbox was opened and inspected and the oil analyzed. The gear teeth were badly damaged with signs of corrosion and broken teeth. Manufacturer was contacted to check the specifications for the mixer gearbox.

Troubleshooting Step 4 - Documentation

Troubleshooting report filed. Work orders written to inspect all similar motors and gearboxes. Gearbox maintenance and lubrication procedures reviewed to ensure the lubricants were not being mixed. Oil samples taken for analysis.

PROCEDURE

Troubleshooting Report

Maintenance Technician Identification Number: 68746

Department: Soup mixing

Equipment Identification Number: Mixer #45

Problem: Noisy gearbox. Damaged gears.

Symptoms: Loud noise.

Cause(s): Old equipment, overloaded gearbox, lubricating oil possibly mixed.

Repair Procedures: Replaced gearbox. Checked manufacturer's specifications. New gearbox is ok.

Preventive Maintenance Action: Inspect all similar motors and gearboxes. Review gearbox maintenance and lubrication requirements. Ensure the lubricants are not being mixed. Perform oil analysis.

DOCUMENTATION

Figure 11-18. Mixing lubricants can lead to mechanical equipment failure.

Troubleshooting HVAC Systems

Poor indoor air quality (IAQ) can cause serious health problems; the cause must be located and corrected quickly and safely. Prevention is the safest way to deal with IAQ problems. The best prevention for IAQ problems is effective preventive and predictive maintenance.

Indoor air quality problems are diagnosed by testing for progressively more complex causes. First, check the temperature and humidity in the space. Check for proper airflow in the space, not only from the supply ducts, but also throughout the space. Furniture and wall partitions can affect comfort by directing airflow away from occupants, creating areas where the air does not circulate and becomes stale. More detailed measurements are taken for carbon dioxide levels. High carbon dioxide levels indicate a lack of fresh air. These tests are best done over time to show how the air changes throughout the day. For instance, perhaps an error in the automatic program running the air handler closes the outside air dampers at a certain time each day. This kind of intermittent problem could cause discomfort for the occupants, but may only be found during an extended test. The areas near air inlets, both inside and outside, should be checked for any contaminants or chemical fumes that may be drawn into the system.

If these tests do not locate the cause of the problem, a more detailed investigation of the equipment and ductwork is required. Mold or other hard-to-find contaminants could be causing the occupants' health problems. At this point, it may be more efficient to bring in outside air quality experts to diagnose the problem. Specialized equipment may be necessary to test for and resolve the problem.

Case Study: Mold Problem

Several occupants in a section of a building become seriously ill with chest discomfort and allergic reactions. The building has a preventive maintenance program. When the problem becomes apparent, the maintenance engineer interviews all individuals who were sick and work in that area of the building and others who work in adjoining offices. **See Figure 11-19.**

There are no changes in activity in the rooms of the zone and no unusual events have occurred. Measurements indicate that airflow into the offices is normal, as are carbon dioxide levels. Due to the serious potential consequences of illnesses produced by IAQ problems,

Figure 11-19. Poor indoor air quality can cause serious health problems and must be located and corrected quickly and safely.

professional IAQ investigators are called to take specialized measurements of airborne chemicals. The presence of mold spores is suspected and tests verify that there is mold in the air in the work area.

The engineer assists the investigator in reviewing maintenance records and inspecting filters and ductwork. Mold is found growing in internal ductwork insulation. As a temporary solution, the ductwork is treated with anti-mold chemicals and the insulation liners are resealed. The replacement of the internally lined ducts with externally lined ducts is planned. The complete operation will take over a year and can be accomplished by moving tenants to empty parts of the building as their areas are renovated. During the procedure, all employees are kept informed of plans and activities. Air quality monitoring is increased throughout the building.

Appendix

WELD JOINTS AND POSITIONS				
BUTT	**LAP**	**T**	**EDGE**	**CORNER**
FLAT				
HORIZONTAL				
VERTICAL				
OVERHEAD				

ELECTRICAL SYMBOLS . . .

Lighting Outlets

OUTLET BOX AND INCANDESCENT LUMINAIRE — CEILING / WALL

INCANDESCENT TRACK LUMINAIRE

BLANKED OUTLET

DROP CORD

EXIT LIGHT AND OUTLET BOX. SHADED AREAS DENOTE FACES.

OUTDOOR POLE-MOUNTED FIXTURES

JUNCTION BOX

LAMPHOLDER WITH PULL SWITCH

MULTIPLE FLOODLIGHT ASSEMBLY

EMERGENCY BATTERY PACK WITH CHARGER

INDIVIDUAL FLUORESCENT LUMINAIRE

OUTLET BOX AND FLUORESCENT LUMINAIRE TRACK FIXTURE

CONTINUOUS FLUORESCENT LUMINAIRE

SURFACE-MOUNTED FLUORESCENT LUMINAIRE

Panelboards

FLUSH-MOUNTED PANELBOARD AND CABINET

SURFACE-MOUNTED PANELBOARD AND CABINET

Convenience Outlets

SINGLE RECEPTACLE OUTLET

DUPLEX RECEPTACLE OUTLET– 120 V

TRIPLEX RECEPTACLE OUTLET– 240 V

SPLIT-WIRED DUPLEX RECEPTACLE OUTLET

SPLIT-WIRED TRIPLEX RECEPTACLE OUTLET

SINGLE SPECIAL-PURPOSE RECEPTACLE OUTLET

DUPLEX SPECIAL-PURPOSE RECEPTACLE OUTLET

RANGE OUTLET — R

SPECIAL-PURPOSE CONNECTION — DW

CLOSED-CIRCUIT TELEVISION CAMERA

CLOCK HANGER RECEPTACLE — C

FAN HANGER RECEPTACLE — F

FLOOR SINGLE RECEPTACLE OUTLET

FLOOR DUPLEX RECEPTACLE OUTLET

FLOOR SPECIAL-PURPOSE OUTLET

UNDERFLOOR DUCT AND JUNCTION BOX FOR TRIPLE, DOUBLE, OR SINGLE DUCT SYSTEM AS INDICATED BY NUMBER OF PARALLEL LINES

Busways

SERVICE, FEEDER, OR PLUG-IN BUSWAY — B B B

CABLE THROUGH LADDER OR CHANNEL — C C C

WIREWAY — W W W

Switch Outlets

SINGLE-POLE SWITCH — S

DOUBLE-POLE SWITCH — S_2

THREE-WAY SWITCH — S_3

FOUR-WAY SWITCH — S_4

AUTOMATIC DOOR SWITCH — S_D

KEY-OPERATED SWITCH — S_K

CIRCUIT BREAKER — S_{CB}

WEATHERPROOF CIRCUIT BREAKER — S_{WCB}

DIMMER — S_{DM}

REMOTE CONTROL SWITCH — S_{RC}

WEATHERPROOF SWITCH — S_{WP}

FUSED SWITCH — S_F

WEATHERPROOF FUSED SWITCH — S_{WF}

TIME SWITCH — S_T

CEILING PULL SWITCH — S

SWITCH AND SINGLE RECEPTACLE — S

SWITCH AND DOUBLE RECEPTACLE — S

A STANDARD SYMBOL WITH AN ADDED LOWERCASE SUBSCRIPT LETTER IS USED TO DESIGNATE A VARIATION IN STANDARD EQUIPMENT — $a.b$ / $a.b$ / $S_{a.b}$

...ELECTRICAL SYMBOLS

Commercial and Industrial Systems

PAGING SYSTEM DEVICE

FIRE ALARM SYSTEM DEVICE

COMPUTER DATA SYSTEM DEVICE

PRIVATE TELEPHONE SYSTEM DEVICE

SOUND SYSTEM

FIRE ALARM CONTROL PANEL — FACP

Signaling System Outlets for Residential Systems

PUSHBUTTON

BUZZER

BELL

BELL AND BUZZER COMBINATION

COMPUTER DATA OUTLET

BELL RINGING TRANSFORMER — BT

ELECTRIC DOOR OPENER — D

CHIME — CH

TELEVISION OUTLET — TV

THERMOSTAT — T

Underground Electrical Distribution or Electrical Lighting Systems

MANHOLE — M

HANDHOLE — H

TRANSFORMER— MANHOLE OR VAULT — TM

TRANSFORMER PAD — TP

UNDERGROUND DIRECT BURIAL CABLE

UNDERGROUND DUCT LINE

STREET LIGHT STANDARD FED FROM UNDERGROUND CIRCUIT

Above-Ground Electrical Distribution or Lighting Systems

POLE

STREET LIGHT AND BRACKET

PRIMARY CIRCUIT

SECONDARY CIRCUIT

DOWN GUY

HEAD GUY

SIDEWALK GUY

SERVICE WEATHERHEAD

Panel Circuits and Miscellaneous

LIGHTING PANEL

POWER PANEL

WIRING—CONCEALED IN CEILING OR WALL

WIRING—CONCEALED IN FLOOR

WIRING EXPOSED

HOME RUN TO PANELBOARD
Indicate number of circuits by number of arrows. Any circuit without such designation indicates a two-conductor circuit. For a greater number of conductors indicate as follows: ⫫ (3 conductors)
⫫⫫ (4 conductors)

FEEDERS
Use heavy lines and designate by number corresponding to listing in feeder schedule

WIRING TURNED UP

WIRING TURNED DOWN

GENERATOR — G

MOTOR — M

INSTRUMENT (SPECIFY) — I

TRANSFORMER — T

CONTROLLER

EXTERNALLY-OPERATED DISCONNECT SWITCH

PULL BOX

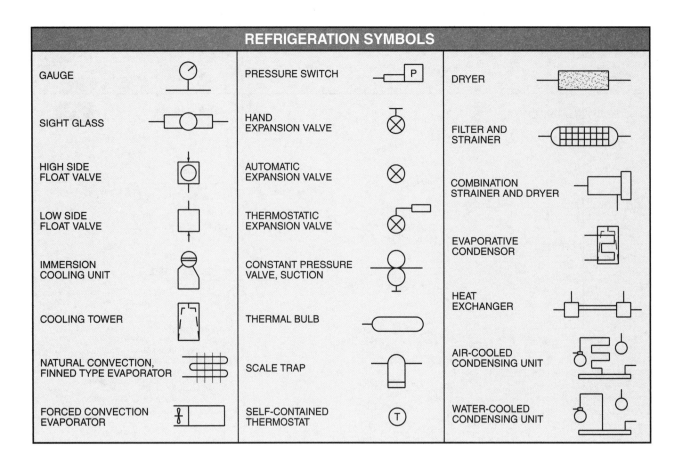

REFRIGERATION SYMBOLS		
GAUGE	PRESSURE SWITCH	DRYER
SIGHT GLASS	HAND EXPANSION VALVE	FILTER AND STRAINER
HIGH SIDE FLOAT VALVE	AUTOMATIC EXPANSION VALVE	COMBINATION STRAINER AND DRYER
LOW SIDE FLOAT VALVE	THERMOSTATIC EXPANSION VALVE	EVAPORATIVE CONDENSOR
IMMERSION COOLING UNIT	CONSTANT PRESSURE VALVE, SUCTION	HEAT EXCHANGER
COOLING TOWER	THERMAL BULB	AIR-COOLED CONDENSING UNIT
NATURAL CONVECTION, FINNED TYPE EVAPORATOR	SCALE TRAP	
FORCED CONVECTION EVAPORATOR	SELF-CONTAINED THERMOSTAT	WATER-COOLED CONDENSING UNIT

ROUTINE BOILER MAINTENANCE PROCEDURES*		
Daily	**Record in Boiler Room Log Daily**	**Weekly**
Check boiler water level Blow down water column Blow down gauge glass Blow down low water fuel cutoff Blow down feedwater regulator Clean fuel oil strainers Check combustion in burner	Makeup water used Flue gas temperature Water supply and return temperature Gas pressure Fuel oil pressure and temperature Steam pressure Extraordinary conditions	Check burner control linkage Check safety controls and interlocks Check alarms and indicating lights Check operating and limit controls Check for extraordinary vibration leaks, noise, or unusual conditions
Monthly	**Semi-Annually**	**Annually During Boiler Inspection**

Check for flue gas leaks
Inspect burner operation
Inspect for hot spots
Hand test safety valves
	Clean low water fuel cutoff
	Clean fuel oil heater
	Inspect refractory
	Clean air cleaner and/or fuel oil
	  separator
	Clean air pump coupling alignment
		Clean water side and fire side
		  surfaces
		Clean breeching
		Check safety valve operation

* water treatment program to be followed as required

BOILER ROOM LOG

TIME	NO. BOILER IN OPERATION	BOILER STEAM PRESSURE	FLUE TEMP (°F)	FURNACE DRAFT (IN IN. WG)	NO. BURNERS	BOILER FUEL METER READING	STEAM FLOW INTEGRATOR	WATER SOFTENER INTEGRATOR	WATER SOFTENER DIAL METER	FEEDWATER TEMP (°F)	OIL IN TANK (GAL.)	OIL TEMP FROM TANK AT PUMP (°F)	RETURN OIL TEMP (°F)	SMOKE ALARM SIGNAL	OUTSIDE TEMP (°F)	DATE: ENGINEER'S SIGNATURE
9 AM																
10																
11																
12																
1 PM																
2																
3																
4																
5																
6																
7																
8																
9																
10																
11																
12																
1 AM																
2																
3																
4																
5																
6																
7																
8																
9																

BOILER ROOM DAILY RECORD

Actual lb steam made		Fuel burned (gal.)	
Heat added/lb steam		Heat value of fuel/gal.	
Equivalent lb steam made		Total heat in fuel (MBtu)	
Boiler pressure		Efficiency	
Feedwater temperature		Average BHP rated	
Quality of superheat steam		Average BHP developed (@800,000 Btu/day)	
Total heat added to steam (MBtu)		Average % rating developed	

SHIFT	NO. OF BOILER	% BOILER EFFICIENCY	% MAKEUP WATER	LB OF STEAM/LB OF OIL
NO. 1				
NO. 2				
NO. 3				

PLUMBING SYMBOLS

Pipe Fitting and Valve Symbols

	FLANGED	SCREWED	BELL & SPIGOT		FLANGED	SCREWED	BELL & SPIGOT		FLANGED	SCREWED	BELL & SPIGOT
BUSHING				REDUCING FLANGE				AUTOMATIC BYPASS VALVE			
CAP				BULL PLUG				AUTOMATIC REDUCING VALVE			
REDUCING CROSS				PIPE PLUG				STRAIGHT CHECK VALVE			
STRAIGHT-SIZE CROSS				CONCENTRIC REDUCER				COCK			
CROSSOVER				ECCENTRIC REDUCER				DIAPHRAGM VALVE			
45° ELBOW				SLEEVE				FLOAT VALVE			
90° ELBOW				STRAIGHT-SIZE TEE				GATE VALVE			
ELBOW— TURNED DOWN				TEE—OUTLET UP				MOTOR-OPERATED GATE VALVE			
ELBOW— TURNED UP				TEE—OUTLET DOWN				GLOBE VALVE			
BASE ELBOW				DOUBLE-SWEEP TEE				MOTOR-OPERATED GLOBE VALVE			
DOUBLE-BRANCH ELBOW				REDUCING TEE				ANGLE HOSE VALVE			
LONG-RADIUS ELBOW				SINGLE-SWEEP TEE				GATE HOSE VALVE			
REDUCING ELBOW				SIDE OUTLET TEE— OUTLET DOWN				GLOBE HOSE VALVE			
SIDE OUTLET ELBOW— OUTLET DOWN				SIDE OUTLET TEE— OUTLET UP				LOCKSHIELD VALVE			
SIDE OUTLET ELBOW— OUTLET UP				UNION				QUICK-OPENING VALVE			
STREET ELBOW				ANGLE CHECK VALVE				SAFETY VALVE			
CONNECTING PIPE JOINT				ANGLE GATE VALVE— ELEVATION				GOVERNOR-OPERATED AUTOMATIC VALVE			
EXPANSION JOINT				ANGLE GATE VALVE—PLAN							
LATERAL				ANGLE GLOBE VALVE— ELEVATION							
ORIFICE FLANGE				ANGLE GLOBE VALVE—PLAN							

HVAC SYMBOLS

Equipment Symbols	Ductwork	Heating Piping

Equipment Symbols

EXPOSED RADIATOR	
RECESSED RADIATOR	
FLUSH ENCLOSED RADIATOR	
PROJECTING ENCLOSED RADIATOR	
UNIT HEATER (PROPELLER)—PLAN	
UNIT HEATER (CENTRIFUGAL)—PLAN	
UNIT VENTILATOR—PLAN	
STEAM	
DUPLEX STRAINER	
PRESSURE-REDUCING VALVE	
AIR LINE VALVE	
STRAINER	
THERMOMETER	
PRESSURE GAUGE AND COCK	
RELIEF VALVE	
AUTOMATIC 3-WAY VALVE	
AUTOMATIC 2-WAY VALVE	
SOLENOID VALVE	

Ductwork

DUCT (1ST FIGURE, WIDTH; 2ND FIGURE, DEPTH)	12 X 20
DIRECTION OF FLOW	
FLEXIBLE CONNECTION	
DUCTWORK WITH ACOUSTICAL LINING	
FIRE DAMPER WITH ACCESS DOOR	FD AD
MANUAL VOLUME DAMPER	— VD
AUTOMATIC VOLUME DAMPER	
EXHAUST, RETURN, OR OUTSIDE AIR DUCT—SECTION	20 X 12
SUPPLY DUCT—SECTION	20 X 12
CEILING DIFFUSER SUPPLY OUTLET	20" DIA CD 1000 CFM
CEILING DIFFUSER SUPPLY OUTLET	20 X 12 CD 700 CFM
LINEAR DIFFUSER	96 X 6-LD 400 CFM
FLOOR REGISTER	20 X 12 FR 700 CFM
TURNING VANES	
FAN AND MOTOR WITH BELT GUARD	
LOUVER OPENING	20 X 12-L 700 CFM

Heating Piping

HIGH-PRESSURE STEAM	—— HPS ——
MEDIUM-PRESSURE STEAM	—— MPS ——
LOW-PRESSURE STEAM	—— LPS ——
HIGH-PRESSURE RETURN	—— HPR ——
MEDIUM-PRESSURE RETURN	—— MPR ——
LOW-PRESSURE RETURN	—— LPR ——
BOILER BLOWOFF	—— BD ——
CONDENSATE OR VACUUM PUMP DISCHARGE	—— VPD ——
FEEDWATER PUMP DISCHARGE	—— PPD ——
MAKEUP WATER	—— MU ——
AIR RELIEF LINE	—— V ——
FUEL OIL SUCTION	—— FOS ——
FUEL OIL RETURN	—— FOR ——
FUEL OIL VENT	—— FOV ——
COMPRESSED AIR	—— A ——
HOT WATER HEATING SUPPLY	—— HW ——
HOT WATER HEATING RETURN	—— HWR ——

Air Conditioning Piping

REFRIGERANT LIQUID	—— RL ——
REFRIGERANT DISCHARGE	—— RD ——
REFRIGERANT SUCTION	—— RS ——
CONDENSER WATER SUPPLY	—— CWS ——
CONDENSER WATER RETURN	—— CWR ——
CHILLED WATER SUPPLY	—— CHWS ——
CHILLED WATER RETURN	—— CHWR ——
MAKEUP WATER	—— MU ——
HUMIDIFICATION LINE	—— H ——
DRAIN	—— D ——

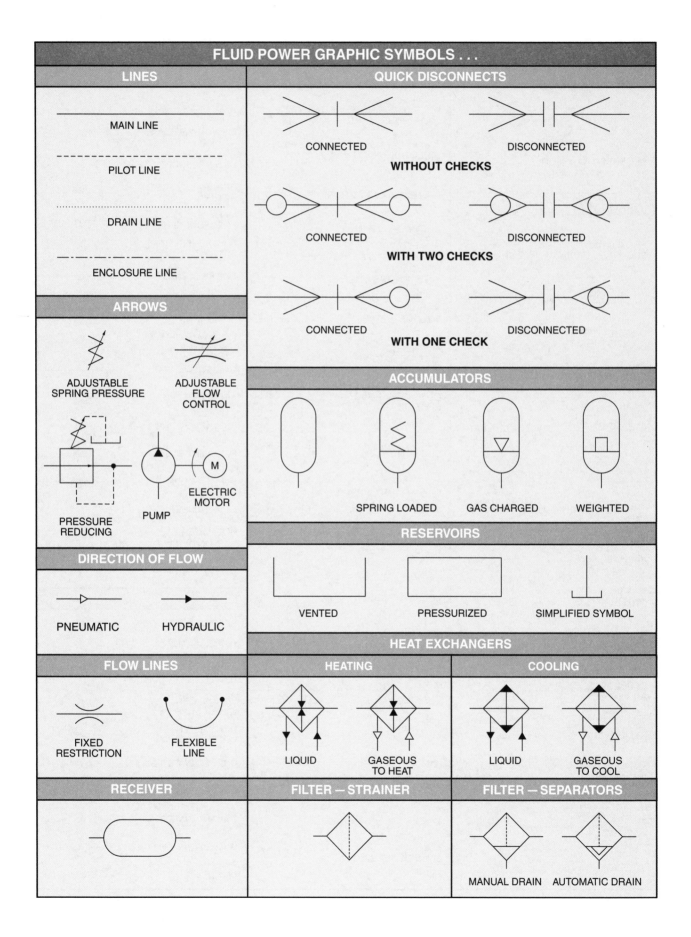

FLUID POWER GRAPHIC SYMBOLS . . .

LINES

MAIN LINE

PILOT LINE

DRAIN LINE

ENCLOSURE LINE

ARROWS

ADJUSTABLE SPRING PRESSURE

ADJUSTABLE FLOW CONTROL

PRESSURE REDUCING

PUMP

ELECTRIC MOTOR

DIRECTION OF FLOW

PNEUMATIC

HYDRAULIC

FLOW LINES

FIXED RESTRICTION

FLEXIBLE LINE

RECEIVER

QUICK DISCONNECTS

CONNECTED

DISCONNECTED

WITHOUT CHECKS

CONNECTED

DISCONNECTED

WITH TWO CHECKS

CONNECTED

DISCONNECTED

WITH ONE CHECK

ACCUMULATORS

SPRING LOADED

GAS CHARGED

WEIGHTED

RESERVOIRS

VENTED

PRESSURIZED

SIMPLIFIED SYMBOL

HEAT EXCHANGERS

HEATING		COOLING	
LIQUID	GASEOUS TO HEAT	LIQUID	GASEOUS TO COOL

FILTER — STRAINER

FILTER — SEPARATORS

MANUAL DRAIN

AUTOMATIC DRAIN

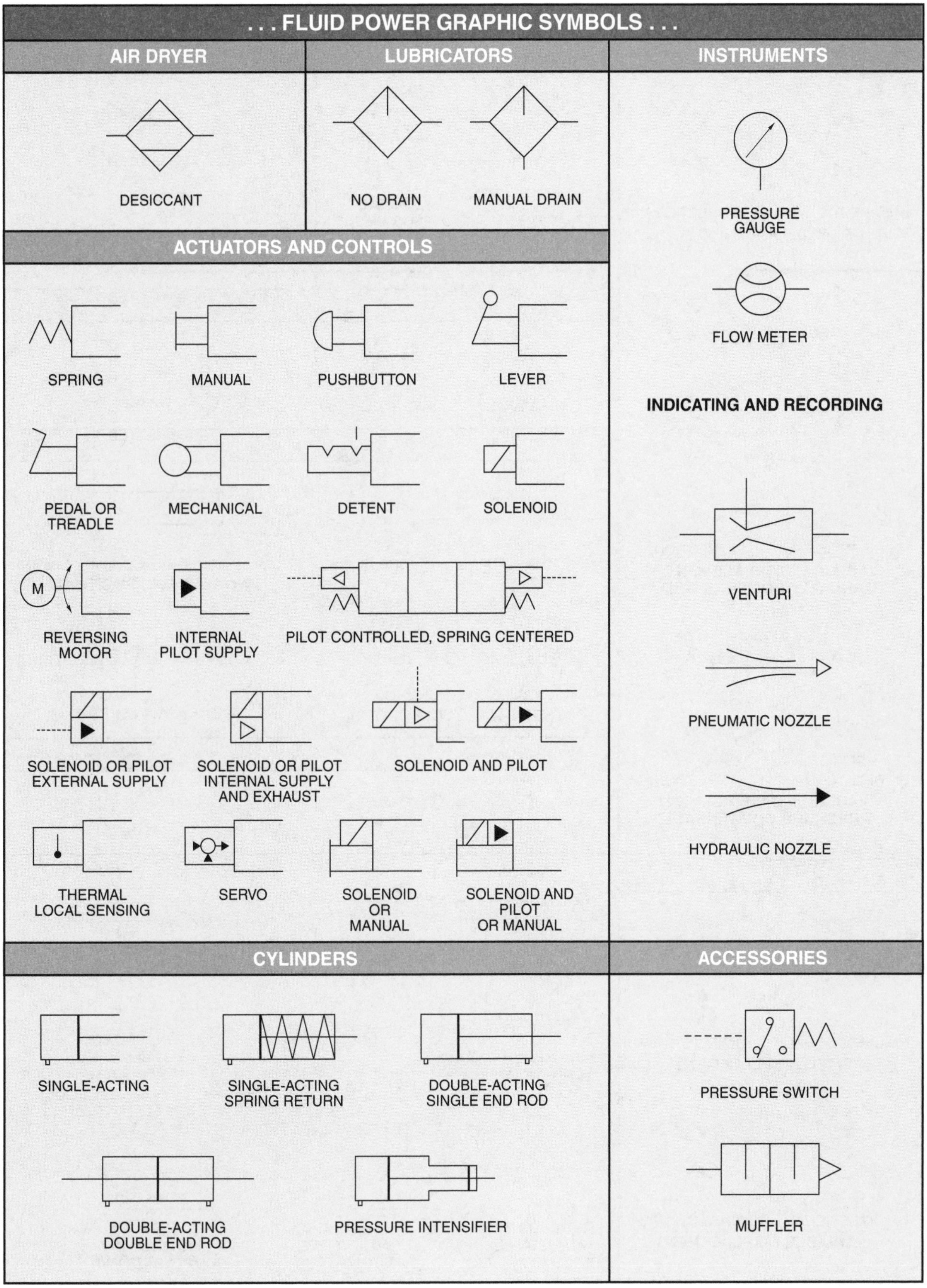

... FLUID POWER GRAPHIC SYMBOLS ...

AIR DRYER

DESICCANT

LUBRICATORS

NO DRAIN

MANUAL DRAIN

INSTRUMENTS

PRESSURE
GAUGE

FLOW METER

INDICATING AND RECORDING

VENTURI

PNEUMATIC NOZZLE

HYDRAULIC NOZZLE

ACTUATORS AND CONTROLS

SPRING

MANUAL

PUSHBUTTON

LEVER

PEDAL OR
TREADLE

MECHANICAL

DETENT

SOLENOID

REVERSING
MOTOR

INTERNAL
PILOT SUPPLY

PILOT CONTROLLED, SPRING CENTERED

SOLENOID OR PILOT
EXTERNAL SUPPLY

SOLENOID OR PILOT
INTERNAL SUPPLY
AND EXHAUST

SOLENOID AND PILOT

THERMAL
LOCAL SENSING

SERVO

SOLENOID
OR
MANUAL

SOLENOID AND
PILOT
OR MANUAL

CYLINDERS

SINGLE-ACTING

SINGLE-ACTING
SPRING RETURN

DOUBLE-ACTING
SINGLE END ROD

DOUBLE-ACTING
DOUBLE END ROD

PRESSURE INTENSIFIER

ACCESSORIES

PRESSURE SWITCH

MUFFLER

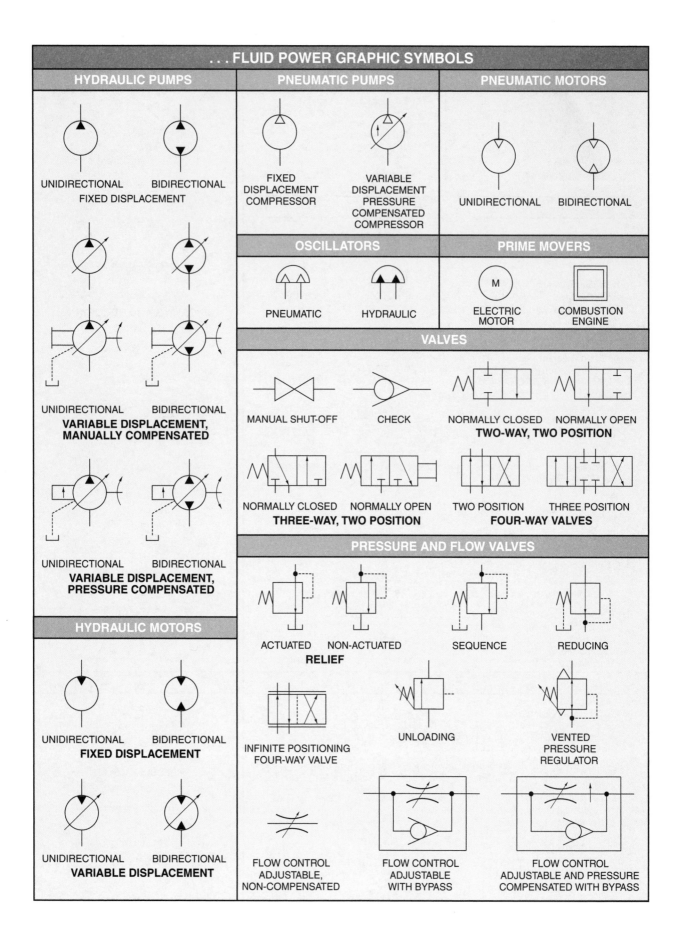

Glossary

A

abrasive wear: Wear that occurs when a hard surface rubs against a softer surface.

absorption system: A refrigeration system that uses the absorption of refrigerant by another chemical to facilitate heat transfer.

acceleration: The change in velocity.

accumulator: 1. In refrigeration systems, a metal container that catches refrigerant liquid that escapes from the evaporator coil before the refrigerant reaches the compressor. **2.** In fluid power systems, a container in which fluid is stored under pressure.

acrylic adhesive: A one-part UV or heat cure or a two-part adhesive that can be used on a variety of materials.

actuator: A component that produces motion from some other form of energy.

additive: A substance added to a lubricant to enhance a desired characteristic.

address: An identification number assigned to a specific input or output.

adhesive bonding: The joining of parts with an adhesive placed between mating surfaces.

adhesive wear: Wear that occurs when high spots on objects come in contact with each other, weld together, and tear away metal at the points of adhesion.

aerosol: A suspension of very fine solid particles or liquid droplets in air.

aftercooler: A pipe-in-pipe heat exchanger in which water is used to cool compressed air.

air compressor: A component that takes air from the atmosphere and compresses it to increase its pressure.

air contaminant: An undesirable element in the air.

air-cooled condenser: A condenser that uses air as the condensing medium.

air filter: A porous device that removes particles from air.

air handler: A device used to distribute conditioned air to spaces in a building.

alphabet of lines: A description of the various line types and their uses.

alternating current (AC): Current that reverses its direction of flow at regular intervals.

American National Standards Institute: A national organization that helps identify industrial and public needs for national standards.

ammeter: A test instrument that measures the amount of current in an electrical circuit.

ammonia system: A refrigeration system that uses ammonia as the refrigerant.

anaerobic adhesive: A one-part adhesive or sealant that cures by the absence of air displaced between mated parts.

antifriction bearing: A bearing that contains moving elements, which provide a low-friction support surface for rotating or sliding surfaces.

application drawing: A drawing that shows the use of a particular piece of equipment or product in an application.

arc blast: An explosion that may accompany an arc-fault event.

arc flash: An extremely high temperature discharge produced by an electrical fault in the air.

architectural drawing: A drawing that contains building information in the form of floor plans, elevation views, section drawings, and detail drawings.

arcing: The discharge of an electric current across an air gap.

arc rating: A PPE rating for the maximum amount of thermal energy the article can resist before exposing the wearer to second-degree burns.

arc tube: The light-producing element of an HID lamp.

area: The number of unit squares equal to the surface of an object.

armature: The moving part of a coil.

asperities: Microscopic peaks and valleys left over from the machining process.

atom: The smallest building block of matter that cannot be divided into a smaller unit without changing its basic character.

atomizing burner: A burner that uses steam, air, or fuel oil pressure to atomize fuel oil.

automatic expansion valve: A valve that is opened and closed by the pressure in the line ahead of the valve.

automatic switch: A switch that responds automatically to changes in a system.

axial flow fan: A fan that produces airflow parallel to the fan shaft.

axial load: A load applied parallel to the rotating shaft.

axial reciprocating pump: A pump that consists of pistons in a rotating cylinder block parallel to the drive shaft.

B

backing up: The process of saving computer data to more than one storage device so it is secure.

backlash: The amount of movement (play) between meshing gear teeth.

balanced draft: Mechanical draft from fans located both before and after the boiler.

balanced pump: A pump that has two inlets and outlets at opposite sides of the pump.

ballast: A transformer or solid-state circuit that limits current flow and supplies the high starting voltage for fluorescent and HID lamps.

barcode: A code consisting of a group of variously patterned bars and spaces that represent a certain number.

bearing: A machine component used to reduce friction and maintain clearance between stationary and moving parts.

belt: A flexible loop that connects two or more pulleys for transferring rotational power.

belt creep: The natural movement of the belt on the face of the pulley when it is subjected to changes in tension.

belt drive: A mechanical drive system that uses a belt and pulleys to transfer power between two shafts.

bending: Stress caused by forces acting perpendicular to the horizontal axis of an object.

bent-tube watertube boiler: A boiler that has shaped tubes surrounded by gases of combustion.

bevel gear: A gear with straight tapered teeth used in applications where shaft axes intersect.

biological attack: The deliberate release of illness-causing microorganisms.

biological hazard: A bacterium, virus, fungus, or other microorganism that can cause acute and chronic infections by entering the body directly or through breaks in the skin.

bleed-off valve: A valve that allows the pressure against the discharge valves to be released at startup.

block-and-jaw coupling: A mechanical-flexing coupling that contains two or more jaws and a floating block (spider).

bloodborne pathogen: A pathogenic (disease-causing) microorganism present in human blood that can infect and cause disease in individuals through exposure.

blowdown tank: A boiler accessory that collects water discharged through the bottom blowdown valves.

boiler: A closed metal container (vessel) in which water is heated to produce steam or heated water.

boiler horsepower: The power available from the evaporation of 34.5 lb of water per hour at a feedwater temperature of 212°F.

boiler lay-up: The preparation of a boiler for out-of-service status for an extended period of time.

boiler overpressure: A condition that occurs when a boiler is operating at or above its maximum allowable working pressure (MAWP).

boiler room log: A record of the temperatures, pressures, and fuel consumption of the boiler and accessories in the plant over a 24-hour period.

boiling point: The temperature at which a liquid changes to a gas.

bottom blowdown: The purging of boiler water from the bottom of the boiler to drain the boiler or remove settled sludge and sediment.

bottom blowdown valve: A valve used to release water from the bottom of a boiler in order to reduce the water level, remove sludge and sediment, reduce chemical concentrations, and/or drain the boiler.

Bourdon tube: A curved hollow tube closed on one end that straightens when steam pressure is applied in the tube.

brazing: A joining process that joins parts by heating the filler metal to temperatures greater than 840°F, but less than the melting point of the base metal.

breakdown maintenance: Unscheduled service on failed equipment that has not received any scheduled maintenance.

break-in period: The time just after installation when equipment achieves peak operating performance.

breeching: The duct connecting the boiler to the chimney.

bridge rectifier: A full-wave rectifier containing four diodes.

British thermal unit: The quantity of heat required to raise the temperature of 1 lb of water 1°F.

brittleness: The lack of ductility in a material.

broadband analysis: An vibration analysis that compares current condition readings with baseline condition readings to detect a wide variety of changes.

brownout: The drop in voltage by the power company to conserve power during times of peak usage or excessive loading of the power distribution system.

building automation system: A control system that uses digitally encoded messages shared on a common communication network to distribute information between control devices.

busway: A metal-enclosed distribution system of busbars available in prefabricated sections.

C

cage: An antifriction bearing component used to maintain the position and alignment of the rolling elements.

cam: A machine part that transmits motion using an irregular external or internal surface.

cam follower: A machine part that contacts the cam and moves in a designated path.

Canadian Standards Association: A Canadian government agency that tests products to verify conformance to national codes and standards.

capacitor: A device that stores electrical energy in an electrostatic field.

capacitor start-and-run motor: A 1φ motor that has capacitors in the starting and running windings.

capacitor-start motor: A 1φ motor that has a capacitor in the starting winding.

capillary tube: A long, thin tube that resists fluid flow, which causes a pressure decrease.

carrier frequency: The frequency of the short voltage pulses of varying length that simulate a lower fundamental frequency.

carryover: The inclusion of small water droplets in steam lines, which can cause water hammer.

cast iron boiler: A boiler that has modular cast iron sections that function similar to the watertubes in a watertube boiler.

caustic embrittlement: The accumulation of high alkaline elements that cause boiler metal corrosion.

cavitation: The process in which microscopic gas bubbles expand in a vacuum and suddenly implode when entering a pressurized area.

centrifugal compressor: A compressor that uses centrifugal force to increase fluid pressure.

centrifugal force: The outward force produced by a rotating object.

centrifugal pump: A pump that uses centrifugal force to increase the pressure of a fluid.

centrifugal switch: A switch that opens when a rotor reaches a certain preset speed and reconnects when the speed falls below a preset value.

centrifugal tension: The tension needed to offset the centrifugal force on the belt as it engages the pulley.

certified refrigerant technician: A person who has special knowledge and training, and has passed one or more EPA-approved tests in the charging, recovery, and recycling of refrigerants for air conditioning and refrigeration systems.

chain: A series of interconnected links that form a loop.

chain coupling: A mechanical-flexing coupling that includes two identical sprocketed hubs connected with a single-roller, double-roller, or silent chain.

chain drive: A synchronous mechanical drive system that uses a chain to transfer power from one sprocket to another.

chain grate stoker: A stoker that uses a rotating chain to advance coal into a furnace.

change of state: The process that occurs when enough heat is added to or removed from a substance to change it from one physical state to another.

charging: Adding refrigerant to a system.

check valve: A valve that allows flow in only one direction.

chemical attack: The deliberate release of a toxic gas, liquid, or solid that can poison individuals and the environment.

chemical hazard: A chemical in any form has toxic effects when inhaled, absorbed, or ingested.

chemical test: A metal identification test that uses chemicals that react when placed on certain metals.

chilled water system (chiller): A refrigeration system that cools water that is used to cool air.

chip test: A metal identification test that identifies metal by the shape of its chips.

circuit analyzer: A receptacle plug and test instrument that determines circuit wiring faults (reverse polarity or open ground), tests for proper operation of ground fault circuit interrupters (GFCIs) and arc fault breakers (AFCIs), and displays important circuit measurements (hot/neutral/ground voltage, impedance, and line frequency).

circuit breaker: An OCPD with a resettable mechanism that automatically opens the circuit when an overload condition or short circuit occurs.

clamp-on ammeter: A meter that measures current in a circuit by measuring the strength of the magnetic field around a single conductor.

cleaning tank: A tank used for cleaning parts in flammable solvents with a lid that automatically closes to contain flames during a fire.

clearance requirement: A specification of the operating space required around a piece of equipment in addition to its outline dimensions.

closed circuit: A circuit having a complete path for current flow.

closed feedwater heater: A feedwater heater in which steam and water do not come into direct contact.

closed-loop control system: A control system in which the result of an output is fed back into a controller as an input.

coal: A solid black fossil fuel.

code: A collection of regulations related to a particular trade or environment.

coil: A winding of insulated conductors arranged to produce a magnetic field.

color rendering: The appearance of a color when illuminated by a particular light source.

color test: A metal identification test that identifies metals by their color.

combustion controls: Boiler controls that optimize efficiency by regulating fuel supply, air supply, air-to-fuel ratio, and the removal of gases of combustion.

comfort: The condition that occurs when people cannot sense a difference between themselves and the surrounding air.

commercial watertube boiler: A watertube boiler that produces steam or hot water for commercial or medium-size applications.

complete combustion: Combustion of all fuel using an amount of air above the theoretical minimum.

compression: Stress caused by forces acting along the same axial line to squeeze an object.

compressor: A mechanical device that compresses gas.

computerized maintenance management system (CMMS): A software package that organizes preventive maintenance information and automatically generates reports, work orders, and other data for implementing and improving future maintenance activities.

condensate return tank: A boiler accessory that collects condensate returned from the steam system.

condenser: A heat exchanger that removes heat from high-pressure refrigerant vapor.

conduction: Heat transfer that occurs when heat is passed from molecule to molecule through a material.

conductivity meter: A test instrument used to determine total dissolved solids present in boiler water.

conductor: A material that has little resistance and permits electrons to move through it easily.

confined space: A space large enough that an employee can physically enter and perform assigned work, but has limited or restricted means for entry and exit, and is not designed for continuous employee occupancy.

connecting (master) link: A modified pin link that has a removable side bar and retainer.

constant volume system: An HVAC system that supplies a set amount of air at all times.

contact: The conducting part of a switch that operates with another conducting part to complete (close) or break (open) a circuit.

continuity: The presence of a complete path for current flow.

continuity tester: A test instrument that indicates if a circuit is open or closed.

continuous blowdown: The continuous purging of boiler water at a controlled rate to maintain the proper chemical concentration in the boiler water.

continuous monitoring: Equipment monitoring at all times.

controlled device: An HVAC device that is operated by signals from an HVAC control system.

convection: Heat transfer that occurs when currents circulate between warm and cool regions of a fluid.

conveyor belt: An extra wide flat belt made from rubber or synthetic material used for a conveyor system.

conveyor pulley: A long cylinder that supports one end of a conveyor belt.

conveyor system: A system used to transport material from one location to another.

cooling coil: A finned heat exchanger that removes heat from the air.

cooling tower: A chiller component that uses evaporation and airflow to cool water.

corrective work: The repair of a known problem before a breakdown occurs.

corrosion: The combining of metals with elements in the surrounding environment that leads to the deterioration of a material.

corrosive wear: Wear that occurs when acid or moisture deteriorates a surface.

coupling: A device used to connect a mechanical drive to a prime mover.

cross head: A metal frame that accommodates a threaded shaft and supports and positions the jaws of a puller.

crude oil: A mixture of semisolids, liquids, and gases formed from the remains of organic matter that has been changed by pressure and heat over millions of years.

current: The flow of electrons through a conductor.

current unbalance: The unbalance that occurs when current on each of the three power lines of a 3ϕ power supply or a 3ϕ load is not equal.

cushion cylinder: A cylinder that slows piston movement to provide a gentle stop.

cyanoacrylate adhesive: A one-part adhesive that cures instantly by reacting to trace surface moisture to bond mated parts.

cycle: One complete positive and negative alternation of a waveform over 360°.

cylinder: A component that converts fluid pressure into linear mechanical force.

D

damper: A movable plate that controls airflow.

data logging: The recording of measurements at specified intervals over a period of time.

dead short: A short circuit that opens the OCPD as soon as the circuit is energized.

deep vacuum: A vacuum between 400 μm and 700 μm.

deformation: A change in the shape of a material caused by stress.

defrosting: The process of removing frost or ice that builds up on evaporator coils.

desiccant: A drying agent that removes water vapor by adsorption.

dew point: The temperature below which moisture in the air begins to condense.

diagram: A graphic that shows the arrangement and relationship of objects, areas, or parts.

diaphragm coupling: A material-flexing coupling that includes two flanges separated by a floating member.

diaphragm gauge: A gauge that uses a flexible divider (diaphragm) to measure the pressure difference between the two sides of the diaphragm.

dielectric absorption test: An insulation resistance test that checks the absorption characteristics of wet or contaminated insulation.

diode: An electronic component that allows current to pass in only one direction.

direct current (DC): Current that flows in one direction only.

direct digital control system: A control system that uses low voltage and current to send variable signals from thermostats and controllers to controlled devices.

direct drive conveyor: A conveyor that has the material in direct contact with the belt.

direct drive fan: A fan that has the fan wheel attached directly to the motor shaft.

directional control valve: A valve that directs or prevents flow through selected passages.

disabling: The prevention of the activation of an output.

division point: A point in a refrigeration system where there is a significant pressure change.

doping: The altering of pure semiconductor material by adding small amounts of other elements.

draft: The difference in pressure between two points that causes air or gases to flow.

draft system: A system that regulates the flow of air to and from the boiler.

drawing: A graphic representation of an object or idea.

dryback firetube boiler: A boiler that has a refractory-lined chamber outside of the vessel that directs the gases of combustion from the furnace to the tube bank.

dry bulb temperature: A temperature reading that does not take into account the amount of water vapor in the air.

dry lay-up: Storage of a boiler with all water drained.

dry lubricant: A lubricant in non-liquid form.

ductility: The ability of a material to stretch, bend, or twist without breaking or cracking.

dust: A collection of minute solid particles suspended in the air or settled on surfaces.

duty cycle: The percentage of time a load or circuit is ON.

E

earmuff: A device worn over the ears to reduce the level of noise reaching the eardrum.

earplug: A compressible device inserted into the ear canal to reduce the level of noise reaching the eardrum.

efficiency: A measure of a device's useful output energy compared to its input energy.

elastic deformation: The ability of a stressed material to return to its original size and shape after being unloaded.

elastic limit: The maximum stress with which a material can be deformed and still return to its original shape.

elastomeric coupling: A material-flexing coupling that includes two flanged hubs connected and separated by elastomeric material.

electrical analysis: A type of analysis that uses electrical monitoring devices and/or test instruments to evaluate the performance of equipment and the quality of electrical power delivered to the equipment.

electrical circuit: The interconnection of conductors and electrical elements through which current flows.

electrical conductivity: The ability of a material to conduct the flow of electrons.

electrical noise: A disturbance that distorts the voltage and current sine wave.

electrical system: A system that produces, transmits, distributes, and delivers electrical power to satisfactorily operate electrical loads.

electric boiler: A boiler that produces heat with electrical resistance coils or electrodes.

electricity: The energy of the flow of electrons in a conductor.

electric operator: A device that uses electrical equipment to position or switch HVAC components.

electric-pneumatic switch: A switch powered by a solenoid that moves a plunger, opening or closing an air line.

electron: A subatomic particle that has a negative electrical charge of one unit.

electronic leak detector: A device that gives an audible tone or visual signal when the detector senses refrigerant.

electrostatic precipitator: A filtering system that electrostatically charges small airborne particles to remove them from the air stream.

emergency plan: A document that details procedures, exit routes, and assembly areas for facility personnel in the event of an emergency.

emergency work: Unscheduled service to correct an unexpected failure on equipment that receives regular maintenance.

enclosed belt variable-speed drive: A mechanical drive system that uses one or two split-half adjustable pulleys and an adjusting mechanism to change output speed.

energy: The capacity to do work.

energy audit: A comprehensive review of a facility's energy use and a report on ways to reduce the energy use through changes to buildings, equipment, and procedures.

enthalpy: The total heat contained in a substance.

epoxy adhesive: A two-part adhesive that cures when resin and hardener are combined.

ergonomic hazard: A physical task or body position that causes musculoskeletal stress or injury.

ergonomics: The study of the effects of job-related tasks and work-area arrangement on the health of the worker.

evacuation: The removal of all air from a refrigeration system before charging it with refrigerant.

evaporative condenser: A condenser that uses the evaporation of water from the outside surface of the coils to remove heat from refrigerant.

evaporator: A heat exchanger through which heat is transferred to low-pressure refrigerant liquid.

evaporator pressure regulator: A valve that allows two evaporators running from the same compressor to maintain different temperatures.

excess air: Air in the combustion process that is above the theoretical amount required to burn the fuel.

explosion: A blast that is caused by a bomb or the ignition of existing flammable material.

explosive range: The difference between the lower explosive limit and the upper explosive limit of combustible gases.

external boiler water treatment: The treatment of boiler water before it enters the boiler.

external gear pump: A pump that consists of a pair of meshing gears that seal with the pump housing.

extractor: A tool for removing broken bolts, studs, or screws.

F

fan: A mechanical device that creates airflow from the rotation of aerodynamic blades.

fan capacity: The volume of air a fan can move in a given period of time.

fatigue: Material failure caused by repeated bending stresses that exceed the material's limits.

feature information: Information about the relationships between equipment components and facility features.

feedwater: Water that is supplied to the boiler at the proper temperature and pressure.

feedwater heater: A boiler accessory that heats feedwater before it enters the boiler.

feedwater pump: A boiler accessory controlled by the feedwater regulator to supply feedwater to a boiler.

feedwater regulator: A boiler accessory that maintains the NOWL in a boiler by controlling the amount of feedwater pumped to the boiler.

feedwater system: A system that supplies water to the boiler at the proper temperature and pressure.

feedwater valve: A valve that controls the flow of feedwater from the feedwater pump to the boiler.

feeler gauge: A steel strip of a specific thickness.

fiber: A solid particle whose length is several times greater than its diameter.

field-erected boiler: A boiler that is constructed on-site.

file test: A metal identification test in which a file is used to indicate the hardness of a metal compared with that of the file.

filter-dryer: A combination filter and dryer located before the TXV that removes dirt and moisture from the refrigerant.

fin comb: A device used to straighten the edges of the thin metal that extend from a condenser or evaporator coil.

firebox boiler: A compact firetube boiler with a round top and flat sides.

firetube boiler: A boiler in which the hot gases of combustion pass through tubes surrounded by water.

first aid: Emergency care or treatment given to an injured or ill individual before professional medical assistance is available.

first class lever: A lever that has the fulcrum located between the resistance and the effort.

fixed displacement pump: A pump that moves a fixed amount of fluid during each revolution of the pump.

flame failure: A boiler condition when the flame in the boiler furnace has been unintentionally lost.

flame scanner: A safety device that senses if the pilot light and/or main flame are lit.

flammability hazard: The degree of susceptibility of materials to burning based on the form or condition of the material and its surrounding environment.

flange coupling: A coupling consisting of two hubbed flanges fastened together with mechanical fasteners (bolts).

flash protection boundary: The distance from live parts within which individuals could receive a second-degree burn due to an arc flash.

flat belt: A belt that has a rectangular cross-section.

flexible coupling: A coupling designed to join two shafts that have a small amount of misalignment.

flexible disk-ring coupling: A material-flexing coupling that includes two flanged hubs and flexible disks.

float thermostatic steam trap: A steam trap that has a float that opens and closes depending on the amount of condensate.

flooded evaporator: An evaporator that is full of liquid refrigerant.

floor plan: A view of a building level looking directly down from a horizontal cutting plane 5′ from the floor.

flow chart: A diagram that shows a logical sequence of troubleshooting steps for a given set of conditions.

flow control valve: A valve that regulates the volume of hydraulic oil flowing to components in a system.

fluid: A substance that tends to flow or conform to the outline of its container.

fluorescent lamp: An electric lamp that produces light as electricity passes through a gas.

foaming: 1. In boilers, the formation of steam bubbles trapped below the boiler water surface. **2.** In refrigeration systems, the formation of foam in a refrigerant/oil mixture due to the rapid evaporation of refrigerant.

force: Anything that changes or tends to change the state of rest or motion of a body.

forced draft: Mechanical draft from air pushed through the boiler with fans located in the front of the boiler furnace.

forcing: The activation of an input or output by a command entered on a PLC programming terminal keyboard rather than through an input device or the program functions.

fossil fuel: A fuel that is formed in the earth over millions of years from plant or animal remains.

fractional horsepower (FHP) belt: A V-belt designed for light-duty applications.

fracture test: A metal identification test that breaks the metal sample to check for ductility and grain size.

frequency: The number of waveform cycles per second.

friction bearing: A bearing consisting of a stationary bearing surface, such as machined metal or pressed-in bushings, that provides a low-friction support surface for rotating or sliding surfaces.

fuel oil: A liquid fossil fuel made from crude oil.

fuel system: A system that supplies fuel in the proper amount to the boiler.

full-body harness: A fall protection device that evenly distributes fall-arresting forces throughout the body to prevent further injury.

full-wave rectifier: A circuit containing multiple diodes that permit the positive and negative halves of an AC sine wave to pass.

fume: Vapor from volatilized solids that condenses in cool air.

fundamental frequency: The desired voltage frequency simulated by the varying ON/OFF pulses at a higher carrier frequency.

furnace: The combustion chamber of a boiler.

furnace explosion: The instantaneous combustion of flammable gases or vapors accumulated in the furnace.

fuse: An OCPD with a fusible link that melts and opens the circuit when an overload condition or short circuit occurs.

G

gasket: A rigid or semirigid pliable material placed between mating surfaces to prevent gas or liquid leakage.

gas laws: The predictable relationships between the volume, pressure, and temperature of a gas.

gas metal arc welding (GMAW): An arc welding process with a shielded gas arc between a continuous wire electrode and the weld metal.

gas tungsten arc welding (GTAW): An arc welding process in which a shielding gas protects the arc between a tungsten electrode and the weld area.

gate valve: A valve that controls flow by raising or lowering a wedge-like gate.

gauge glass: A boiler fitting that indicates the water level in the boiler.

gauge manifold: A set of valves and pressure gauges used to determine refrigerant system pressures and add or remove refrigerant.

gear: A toothed wheel that meshes with other toothed wheels to transfer rotational power.

gearbox: A sealed container that has an input shaft and an output shaft and houses at least one set of mating gears.

gear coupling: A mechanical-flexing coupling that includes two identical hubs with external gear teeth and a sleeve(s) with mating internal gear teeth.

gear drive: A synchronous mechanical drive system that uses the meshing of two or more gears to transfer motion from one shaft to another.

gear pump: A pump that consists of two meshing gears enclosed in a close-fitting housing.

gear ratio: The ratio of the number of teeth on the drive gear to the number of teeth on the driven gear.

general energy audit: A facility energy study that expands on the preliminary energy audit by collecting more information and performing a more detailed analysis.

generator: An electric device that converts mechanical energy into electrical energy by means of electromagnetic induction.

gerotor pump: A pump that consists of a rotating multiple-lobed inner rotor that meshes with lobes of a rotating outer rotor.

glazing: A slick polished surface caused by dirt and other debris being rubbed on the surface of a belt.

globe valve: A valve that controls flow by raising or lowering a circular disc.

grease dropping point: The temperature at which the thickening agent of a grease turns to a liquid.

grease fitting: A hollow tubular fitting used to direct grease to bearing components.

grease gun: A lubrication tool that is attached to a grease fitting to provide grease to a bearing.

grid coupling: A mechanical-flexing coupling that includes two identical hubs with axially cut slots along their perimeter and a wire grid.

ground circuit: A conducting connection between electrical equipment and the earth.

group relamping: The replacement of all lamps in a given area when they reach 60% to 80% of their rated average life.

guide link: The center link of a silent chain that has no teeth.

H

half-wave rectifier: A circuit containing a diode that permits only the positive half-cycles of an AC sine wave to pass.

hand-firing: The feeding of coal pieces into a furnace by hand.

hand tools: Tools that are used manually.

hardfacing: The application of surfacing material that has hardness properties needed to reduce wear damage.

hardness: The ability of a material to resist deformation by indentation.

harmonic distortion: The addition of voltage and/or current in a power line that is a multiple of the fundamental line frequency.

hazardous material: A substance that could cause injury to individuals or damage to the environment.

head end: The end where material exits a conveyor.

head pressure controller: A refrigeration system component that prevents the high-pressure side pressure from falling too low.

health hazard: The likelihood of a material to cause, either directly or indirectly, temporary or permanent injury or incapacitation due to an acute exposure by contact, inhalation, or ingestion.

heat: A form of energy identified by a temperature difference or change of state.

heater unit: A heating element used to open overload contacts for overcurrent conditions.

heating coil: A finned heat exchanger that adds heat to the air.

heating element: A conductor with an intentionally high resistance for producing heat when connected to an electrical power supply.

heating surface: The part of a boiler that has heat and gases of combustion on one side and water on the other.

heating value: The amount of heat energy per unit of fuel.

heating, ventilating, and air conditioning (HVAC) system: A system used to condition air by maintaining proper temperature, humidity, and air quality.

heat pump: A mechanical compression refrigeration system that can reverse the flow of refrigerant, switching between heating and cooling modes.

heat sink: A device that conducts and dissipates heat away from a component.

heavy wear: A severe wearing away of an object's cross-section.

helical gear: A gear with teeth that are cut following a line that spirals around the shaft axis.

helix: The curve traced on a cylinder or cone by a spiral.

hermetic compressor: A compressor that is completely sealed inside a welded case.

herringbone gear: A gear with two rows of helical teeth.

high-intensity discharge lamp: A lamp that produces light from an arc tube.

high-pressure boiler: A boiler that has an MAWP above 15 psi and over 6 BHP.

holding current: The minimum current required to keep an SCR conducting.

hole: An electrically positive space created by the missing electron in P-type semiconductor material.

horizontal return tubular boiler: A firetube boiler suspended over a furnace.

horsepower: A unit of power equal to 746 W or 550 lb-ft/sec.

hose: A flexible tube for carrying fluids under pressure.

hose coupling: A fitting that allows a line to be connected or disconnected by hand.

hot-gas defrost: Evaporator defrosting using hot gas from the compressor.

hot melt adhesive: A thermoplastic material applied in a molten state that cures to a solid state when cooled.

hydraulic filter: A component that traps dirt and contaminants present in hydraulic oil.

hydraulic motor: A device that converts hydraulic energy into rotation.

hydraulic system: A fluid power system that transmits energy in an enclosed space using a liquid (hydraulic oil) under pressure.

hydraulic tester: A device that measures the pressure, flow, and temperature of hydraulic oil in a system.

I

imbalance: Lack of balance.

implosion: An inward bursting.

improper phase sequence: The changing of the sequence of any two phases in a 3φ system or circuit.

incandescent lamp: An electric lamp that produces light by the flow of current through a tungsten filament inside a gas-filled, sealed glass bulb.

incident report: A document that details facts about an injury-causing accident in the facility.

inclined plane: A simple machine that allows force to be applied over a long, horizontal distance to move heavy loads vertically.

incomplete combustion: Combustion that occurs when not all fuel is burned, resulting in soot, smoke, and unburned gases.

indirect drive fan: A fan in which the fan motor is connected to the fan wheel through a belt and pulley arrangement.

indoor air quality (IAQ): A description of the type and quantity of contaminants in indoor air.

induced draft: Mechanical draft from air pulled through the boiler furnace with fans located in the breeching.

induction motor: A motor that rotates due to the interaction between the magnetic fields of the stator and rotor.

industrial hygiene: The science of anticipating, recognizing, evaluating, and controlling workplace conditions that may cause worker injury or illness.

industrial watertube boiler: A watertube boiler that produces steam or hot water for industrial process applications.

inert gas: A gas that does not readily combine with other elements.

inertia: The tendency of a physical body to persist in its state of rest or uniform motion until acted upon by an external force.

installation print: A print that outlines the general configuration and information needed to install a specific piece of equipment.

insulation spot test: A simple insulation resistance measurement.

insulation step voltage test: An insulation resistance test that puts electrical stress on internal insulation to reveal aging or damage not found during other motor insulation tests.

insulator: A material that has a high resistance.

integrated circuit: A circuit composed of multiple semiconductor devices, providing a complete circuit function, in one small semiconductor package.

intensifier: A device that converts low-pressure fluid power into high-pressure fluid power.

intercooler: A finned pipe heat exchanger that helps cool the air between compression stages.

internal boiler water treatment: The treatment of boiler water after it has entered the steam and water drum.

internal gear pump: A pump that consists of a small spur gear (drive gear) mounted inside a large internal spur gear (ring gear).

International Organization for Standardization: A nongovernmental, international organization that provides a worldwide forum for the standards-developing process.

interpersonal skills: Strategies and actions that allow a person to communicate effectively with other persons in a variety of situations.

inventory control: The organization and management of commonly used parts, vendors and suppliers, and purchasing records in the PM system.

inverted bucket steam trap: A steam trap in which steam enters the bottom and flows into an inverted bucket.

inverter: A device that changes DC voltage into AC voltage of any frequency.

investment-grade energy audit: A comprehensive facility-energy study that expands on the general energy audit by developing a dynamic model of the facility's energy-use characteristics.

isolated grounded circuit: A circuit that minimizes electrical noise by providing a separate grounding path.

isometric drawing: A pictorial drawing with the three axes 120° apart.

K

key: A removable part that provides a positive engagement between a shaft and a hub when mounted in a keyseat.

keyseat: A rectangular groove along the axis of a shaft or hub.

kinetic energy: Energy of motion.

L

ladder chain: A light-duty synchronous chain that is commonly used for actuating control functions of equipment, such as speed control.

latent heat: Heat energy that causes a change of state but no temperature change.

leakage current: The small amount of current flowing through a solid-state device when it is not conducting.

lever: A simple machine that consists of a rigid bar that pivots on a fulcrum with resistance and effort applied.

light emitting diode: A diode that produces light when current flows through it.

limited approach boundary: The distance from live parts within which a shock hazard exists.

linear bearing: A bearing designed to provide low-friction movement of a mechanical device that moves in a straight line.

linear motion: The movement of an object in a straight line.

line diagram: A diagram that uses lines and graphic symbols to show the logic of an electrical circuit.

link-type V-belt: A V-belt consisting of individual connected links.

liquid receiver: A storage tank for liquid refrigerant that is located after the condenser.

live steam: Steam that leaves the boiler directly without having its pressure reduced in process operations.

load shedding: The deliberate shutting down of equipment to reduce electrical use.

lockout: The use of locks, chains, or other physical restraints to prevent the operation of specific equipment.

logbook: A book or electronic file that documents all work performed during a shift and lists information needed to complete further work by maintenance personnel on other shifts.

lower explosive limit: The lowest concentration (air-fuel mixture) at which a gas can ignite.

low-pressure boiler: A boiler that has an MAWP of up to 15 psi.

low-pressure chiller: A chiller that uses an evaporator that operates in a vacuum.

low water fuel cutoff: A boiler fitting that shuts OFF the burner and trips an alarm if a low water condition occurs in the boiler.

lubricant: A substance that separates moving (bearing) surfaces to reduce the friction and/or wear between them.

lubricating grease: A semisolid lubricant consisting of a mixture of oil and thickening agents.

lubricating oil: A liquid lubricant having a mineral, synthetic, vegetable, or animal origin.

lubricator: A component that injects a mist of oil into the compressed air line for lubrication of pneumatic tools and internal motor parts.

M

magnetic motor starter: A specialized relay used to energize and de-energize a motor and includes motor overload protection.

magnetic test: A metal identification test that checks for the presence of iron in a metal by using a magnet to test for magnetism.

main steam line: A line that connects a boiler to the steam header.

main steam stop valve: A valve controls the flow of steam from the boiler.

makeup water: Water that is used to replace boiler water lost from leaks or from the lack of condensate returned to the boiler.

malleability: The ability of a material to be deformed by compression without developing defects.

manometer: A device that uses a liquid-filled tube to measure the difference in pressure between two locations.

manual switch: A switch that is operated by a person.

material-flexing coupling: A flexible coupling that uses a flexible material to accommodate shaft misalignment.

material safety data sheet (MSDS): Document containing hazard information about a certain chemical.

maximum allowable working pressure (MAWP): The recommended maximum pressure at which a boiler can safely be operated.

mechanical compression refrigeration: A refrigeration process that produces a refrigeration effect using a compressor and a pressure control device.

mechanical draft: Draft produced using fans.

mechanical drive system: A combination of mechanical components that transfer power from one location to another.

mechanical-flexing coupling: A flexible coupling with components that move or slide in relation to each other to accommodate shaft misalignment.

mechanical load: An external force applied to a body that causes stress in the material.

mechanical property: The characteristic of a material that describes its behavior under applied load.

mechanical switch: A switch that is operated by the movement of an object.

megohmmeter: A type of ohmmeter that is used to measure the very high resistances of insulating components.

melting point: The temperature point at which a solid changes to liquid.

membrane watertube boiler: A boiler that directs water through formed metal membrane tubes connected to upper and lower drums.

metering device: A component that controls the flow rate of refrigerant into an evaporator.

micron: An alternate term for a micrometer, or one millionth of a meter.

micron gauge: An electronic instrument that indicates the depth of vacuum in microns.

miter gear: A bevel gear used at right angles to transmit power at a 1:1 ratio.

momentary power interruption: A decrease in voltage to 0 V that lasts from 0.5 cycles up to 3 sec.

motor torque: Rotational force that is generated in a split-phase motor by the strength of and phase angle difference between the magnetic fields in the stator.

mounting dimension: A dimension used to locate holes or threads for mounting equipment with screws, studs, brackets, or clips.

mud drum: The lowest part of the water side of a watertube boiler and collects sludge and mud.

multimeter: A test instrument capable of measuring two or more electrical quantities.

multi-stage compressor: A compressor that uses two or three cylinders, each with a progressively smaller diameter, to produce progressively higher pressures.

N

narrowband analysis: A vibration analysis that focuses on specific vibration frequencies that correspond to equipment components or failure features.

narrow belt: A V-belt having a smaller cross-section and a higher profile than a standard belt.

National Electrical Manufacturers Association (NEMA®): A national trade association that provides information and develops standards concerning proper selection, rating, construction, testing, and performance of electrical equipment.

National Fire Protection Association® (NFPA®): A national standards organization that provides guidance in assessing fire-related hazards.

National Institute for Occupational Safety and Health (NIOSH): A national organization that acts in conjunction with OSHA to develop and periodically revise recommended exposure limits for hazardous substances or conditions in the workplace.

natural draft: Draft produced from the difference in air temperature inside and outside a chimney.

natural gas: Combustible fossil fuel gas found in pockets trapped underground.

neutral conductor: A wire that carries current from one side of the load to the grounded neutral bar in the circuit breaker panel.

neutral position: The position of a directional control valve when no controlling devices in the system are energized and there is no force causing the spool to move from the neutral position.

neutron: A subatomic particle that has no electrical charge.

noise reduction rating (NRR): The amount of the reduction of sound level (in decibels) provided by a hearing protection device.

noise suppressor: A device that reduces the random electrical signals (noise) on power lines.

nonferrous metal: A metal that does not contain iron.

nonreturn steam trap: A steam trap that removes condensate from steam lines and heat exchangers and delivers it to the condensate return tank.

non-synchronous drive system: A drive system that does not provide positive engagement between the drive and driven sides of the system.

nonthreaded fastener: A device that permanently joins or fastens parts together without threads.

non-time delay fuse: A fuse that opens the circuit almost instantly.

normally closed device: A device that is closed when the signal causing it to operate is absent or at its lowest level.

normally open device: A device that is open when the signal causing it to move is absent or at its lowest level.

N-type material: Semiconductor material with free electrons in its crystalline structure.

nuclear blast: An explosion with a damaging pressure wave, intense light and heat, and widespread radioactive material (fallout) that can contaminate the air, water, and ground surfaces.

nucleus: The dense center of an atom, consisting of protons and neutrons.

O

Occupational Safety and Health Administration (OSHA): A federal government regulatory agency that requires all employers to provide a safe environment for their employees.

OFF-delay timer: A timer that switches its contacts after being de-energized and then a preset time period elapses.

offset link: A chain link that is used to shorten or lengthen a chain and to connect two ends of a chain.

ohmmeter: A test instrument that measures resistance.

Ohm's Law: The relationship between voltage, current, and resistance in a circuit.

oil analysis: A predictive maintenance technique that detects and analyzes the presence of acids, dirt, fuel, and wear particles in lubricating oil.

oil and moisture separator: A device that removes oil and water droplets from a system by forcing compressed air to change direction quickly.

oil cup: An oil reservoir located on a bearing housing to provide lubrication to a bearing.

oil level sight glass: A window located on the compressor crankcase that indicates the level of oil in the compressor.

oil seal: A device used to contain oil inside a housing.

oil seal lip: An oil seal component that contacts the moving part of the equipment to prevent material from passing by the oil seal.

oil wick: An absorbent material which serves as a conduit for oil from the oil cup to the bearing surface.

ON-delay timer: A timer that switches its contacts after being energized and then a preset time period elapses.

one-shot timer: A timer that switches its contacts immediately and remain changed for the set period of time after the timer has received power.

open belt variable-speed drive: A mechanical drive system that uses two opposing cone pulleys linked together with a flat or V-belt to obtain variable speeds.

open circuit: A circuit having an incomplete path, which prevents current flow.

open feedwater heater: A feedwater heater in which steam and water come into direct contact (mix) to raise the temperature of the water.

open-loop control system: A control system in which decisions are made based only on the current state of the system and a model of how it should work.

operational short: A short circuit that causes the circuit to malfunction and may not open the OCPD.

operator: The device that is pressed, pulled, or rotated by the individual operating the circuit.

operator's manual: A document that contains instructions for the safe and efficient installation, operation, troubleshooting, and repair of equipment.

optocoupler: A device that converts electrical input signals to light signals and the light signals back to electrical signals on the output side.

outline dimension: A dimension that indicates the minimum space required to install the piece of equipment.

overcurrent condition: A condition that occurs when the amount of current flowing in a circuit exceeds the design limit of the circuit.

overcurrent protection device (OCPD): A device that automatically opens a circuit to prevent damage from a high-current condition.

overload: Breakage caused by one large sudden shock.

overvoltage: An increase in voltage of more than 10% above the normal rated line voltage that lasts longer than 1 min.

oxidation: The combination of metal and oxygen into metal oxides.

oxyacetylene welding (OAW): An oxyfuel welding process that uses oxygen mixed with acetylene.

oxyfuel welding (OFW): A welding process that produces heat from the combustion of a mixture of oxygen and a fuel gas.

P

package boiler: A boiler that is preassembled at the factory.

packing nuts: Components on an actuator that apply tension to the seals or packing to prevent leakage around the shaft.

panelboard: A wall-mounted distribution cabinet containing a group of overcurrent and short-circuit protection devices for branch circuits.

parallax error: A measurement error caused by an improper viewing angle of the pointer slightly above the scale.

parallel circuit: A circuit that has two or more components connected such that there are multiple paths (branches) for current flow.

partial fault: The malfunctioning of only a section or several sections of a machine.

particulate: A very small particle of solid or liquid matter.

perfect combustion: Combustion of all fuel using only the theoretical minimum amount of air.

periodic maintenance: Work completed at specific intervals.

permissible exposure limit (PEL): The OSHA estimate of the average airborne concentration of a substance to which workers may be exposed day after day without adverse effect.

permit-required confined space: A confined space that has specific health and safety hazards capable of causing death or serious physical harm.

personal protective equipment (PPE): Clothing and/or equipment worn by a worker to reduce the possibility of an injury.

phase sequence tester: A test instrument used to determine which of the 3ϕ power lines are powered and which power line is phase A, which is phase B, and which is phase C.

phase unbalance: The unbalance that occurs when 3ϕ power lines are more or less than 120° out of phase.

photodiode: A diode that produces current when absorbing light.

photoelectric switch: A solid-state sensor that can detect the presence of an object by means of a light beam.

phototransistor: A device that combines the effect of a photodiode and the switching capability of a transistor.

pH scale: A scale from 0 to 14 used to indicate the acidity or alkalinity of a solution.

physical hazard: A hazard caused by excessive levels of noise, vibration, illumination, temperature, and radiation.

pictorial drawing: A drawing that shows the length, height, and depth of an object in one view.

pilot control passage: A small pathway used to transmit a valve pressure signal.

pilot-operated valve: A valve controlled by a pressure signal in the pilot control passage connecting parts of the pressure-reducing valve.

pin: A cylindrical nonthreaded fastener that is placed into a hole to secure the position of two or more parts.

pinion gear: The smaller of two meshing gears.

pin link: A chain link that consists of two steel pins pressed into two side bars with matching holes.

pipe: A hollow cylinder of metal or other material of substantial wall thickness.

pitch: The distance between corresponding points on an adjacent pair of evenly spaced projections.

pitting: Localized corrosion that has the appearance of cavities (pits).

plant survey: A complete inventory and condition assessment of a facility's equipment and structure.

plastic deformation: The failure of a stressed material to return to its original size and shape after being unloaded.

plastic flow: The movement of material below the surface of an object under mechanical stress.

plot plan: A scaled drawing that shows the shape and size of a lot and the buildings on it.

pneumatic control system: A control system that uses compressed air to send variable signals from thermostats and controllers to controlled devices.

pneumatic-electric switch: A switch that uses air pressure to open or close a set of electrical contacts.

pneumatic operator: A device that uses air pressure to position HVAC components.

pneumatic system: A fluid power system that transmits energy using a gas (typically compressed air).

pneumatic thermostat: A device that converts temperature variations into a variable pneumatic signal.

polarity: The positive (+) or negative (–) state of an object.

polysulfide adhesive: A one- or two-part adhesive or sealant that cures by evaporation or catalyst.

polyurethane adhesive: A one- or two-part adhesive with excellent flexibility that cures by evaporation, catalyst, or heat.

poppet valve: A valve in which the valve seating element (poppet) pops open to allow flow in one direction.

positive displacement pump: A pump that has a seal between its inlet and outlet and moves a specific volume of fluid with each revolution.

potential energy: Stored energy a body has due to its position, chemical state, or condition.

power: The rate of doing work or using energy.

power distribution: The delivery of electrical power to where it is needed.

power formula: The relationship between power, voltage, and current in a circuit.

power supply: A device that provides the voltage required for the internal operation of a device.

power tools: Tools that are electrically, pneumatically, or hydraulically powered.

predictive maintenance (PDM): The monitoring of wear conditions and equipment operation characteristics for comparison against a predetermined tolerance to predict potential malfunctions or failures.

preliminary energy audit: An overview of a facility's major energy-consuming processes to identify only significant inefficiencies.

pressure: Force per unit area.

pressure control valve: A valve used to regulate pressure in a hydraulic system.

pressure cup: A pressurized grease reservoir that provides constant lubrication to a bearing.

pressure-reducing valve: A valve that limits the maximum pressure at its outlet, regardless of the inlet pressure.

pressure-relief valve: A valve that sets a maximum operating pressure level for a fluid-power circuit to protect the circuit from overpressure.

pressure-sequence valve: A pressure-operated valve that diverts flow to a secondary actuator while holding pressure on the primary actuator at a predetermined minimum value after the primary actuator completes its travel.

pressure switch: A switch operated by the amount of pressure acting on a diaphragm, bellows, or electronic element.

preventive maintenance (PM): Scheduled work required to keep equipment in peak operating condition.

preventive maintenance system: A system used to record and organize maintenance information, which is then used to make the decisions required to maintain the facility and equipment.

primary air: Air in the combustion process that regulates the rate of combustion.

primary coil: The transformer coil to which the incoming voltage is supplied.

printed circuit board: A thin plate of insulating material, such as fiberglass or phenolic, with conducting paths laminated to one or both sides.

process: A sequence of operations that accomplishes desired results.

programmable logic controller (PLC): A solid-state control device that is programmed and reprogrammed to automatically control an industrial process or machine.

prohibited approach boundary: The distance from live parts within which work is considered the same as making contact with the live part.

project work: Work on long-term projects that require advanced planning and more time than typical maintenance tasks.

protocol: A standardized set of rules and procedures for the exchange of digital information between two devices.

proton: A subatomic particle that has a positive electrical charge of one unit.

proximity sensor: A device that reacts to the nearness of a target without physical contact.

psychrometer: A thermometer that can measure both dry and wet bulb temperatures.

P-type material: Semiconductor material with empty spaces (holes) in its crystalline structure.

puller: A device used to remove gears, pulleys, sprockets, bearings, and couplings from a shaft or housing.

pulley: A simple machine consisting of a cylinder rotating freely on its axis, which uses a belt, rope, or chain to change the direction of an applied force or transmit rotational motion.

pulverizer: A mechanical device used to feed coal that has been ground to a fine powder.

pump: A mechanical device that causes fluid to flow.

pump-down control: A control system that uses a thermostat and pressure switch to control the temperature in the cooled space.

purge valve: A valve that removes unwanted gases from a system.

Q

qualified person: One with skills and knowledge of the relevant system and is trained to recognize and avoid related hazards.

R

race: The bearing surface of an antifriction bearing that supports the rolling elements.

rack gear: A spur gear with teeth spaced along a straight line.

radial flow fan: A fan that produces airflow perpendicular to the fan shaft.

radial load: A load applied perpendicular to the rotating shaft.

radial reciprocating pump: A pump that consists of pistons located perpendicular to the pump shaft.

radiation: Heat transfer from electromagnetic waves radiating outward from the source.

radiation threat: The use of conventional explosives to spread radioactive material over a targeted area.

ram-feed stoker: A stoker that uses a pushing motion and feeder blocks to advance coal into a furnace.

ramping: Evenly raising or lowering the voltage supply to a device.

random monitoring: Unscheduled equipment monitoring as required.

reactivity hazard: The degree of susceptibility of materials to explode or release energy by themselves or by exposure to certain conditions or substances.

receiver: An air tank that stores compressed air and allows it to cool before use.

receptacle tester: A test instrument that is plugged into a standard receptacle to determine if the receptacle is properly wired and energized.

reciprocating compressor: A compressor that uses pistons moving back and forth to increase fluid pressure.

reciprocating pump: A pump in which fluid flow is produced from pistons moving back and forth.

rectifier: An electrical component that converts AC to DC by allowing voltage and current to flow in only one direction.

recycle timer: A timer that switches its contacts open or closed repeatedly at a set interval once the timer has received power.

reduction: The loss or removal of oxygen from a material.

refractory: Material that retains its strength at very high temperatures.

refrigerant: A chemical that vaporizes (boils) at low temperatures.

refrigerant dryer: A device that uses a refrigeration process to lower the temperature of compressed air.

refrigerant oil: A specialized liquid lubricant for refrigeration systems.

refrigerant reclaiming: The process of cleaning heavily contaminated refrigerant.

refrigerant recovery: The process of removing refrigerant from a system and capturing it in a recovery cylinder, with no cleaning of the refrigerant.

refrigerant recycling: The process of removing dirt, oil, and moisture from lightly contaminated refrigerant that has been removed from a refrigeration system.

refrigeration: The process of moving heat from one area to another area by use of a refrigerant in a closed system.

refrigeration system: A closed system that controls the pressure and temperature of a refrigerant to regulate the absorption and rejection of heat by the refrigerant.

regenerative cylinder: A cylinder that has a piston rod sized to have a cross-sectional area equal to half of the piston area.

regulation: A rule made mandatory by a federal, state, or local government.

relative humidity: The amount of moisture in the air compared to the amount of moisture the air would hold if it were saturated.

relay: An electrical switch that is actuated by a separate circuit.

reservoir: A container for storing fluid under little or no pressure.

resistance: The opposition to the flow of electrons in a circuit.

respirator: A device that protects the wearer from inhaling airborne contaminants.

restricted approach boundary: The distance from live parts within which there is an increased risk of shock than at the limited approach boundary and further precautions are required.

reversing valve: A four-way valve that reverses the flow of refrigerant in a heat pump.

ridging: Plastic flow that occurs due to excessive loads in localized areas.

rigid coupling: A device that couples two shafts that are precisely aligned within a common frame.

rippling: Plastic flow that occurs from heavy loads, vibration, or improper lubrication.

rivet: A cylindrical metal pin with one preformed head that is deformed on the opposite side in order to hold parts together.

roller: A cylindrical device used to guide and support the conveyor belt.

roller chain: A synchronous chain that contains roller, pin, and connecting (master) links.

roller drive conveyor: A conveyor that has the material sitting on rollers that are rotated by the belt.

roller link: A chain link that consists of two bushings placed inside two rollers that are pressed into two side bars.

rolling and peening: Plastic flow that occurs due to excessive loads or impact loading.

rolling friction: Friction that occurs when a rolling device (roller or ball) moves on a stationary surface.

rotary cup burner: A burner that atomizes fuel oil using centrifugal force from the outer surface of a spinning cup.

rotor: The rotating part of an AC motor.

running winding: A coil that continues to operate after a 1ϕ motor is started.

rupture disc: A nonreusable device that bursts at a specific pressure.

S

safety valve: A fitting that prevents the boiler from exceeding its MAWP.

saturated liquid: Liquid at a certain pressure and temperature that vaporizes if the temperature increases.

saturated vapor: Vapor at a certain pressure and temperature that condenses if the temperature decreases.

saturation: The maximum amount of moisture that the air can hold at a specific temperature.

saturation temperature: The temperature at which a refrigerant changes state by vaporizing or condensing.

scale: The accumulation of calcium carbonate and magnesium carbonate on boiler heating surfaces.

scheduled maintenance: Work that is planned and scheduled for completion.

scheduled monitoring: Equipment monitoring at specific time intervals.

schematic diagram: A diagram that shows the electrical connections and functions of a specific circuit arrangement with graphic symbols.

screw: A simple machine consisting of a continuous spiral on a cylinder.

screw compressor: A compressor that contains a pair of screw-like rotors that interlock as they rotate.

screw-feed stoker: A stoker that uses a long rotating screw to advance coal into a furnace.

screw thread: A ridge of uniform section in the form of a helix on an internal or external surface of a cylinder or cone.

scroll fan: A radial flow fan contained in a scroll-shaped housing.

sealant: A product used to seal, fill voids, and waterproof parts.

sealed bearing: An antifriction bearing that is completely sealed so the lubricant stays inside the bearing and dirt is kept out.

secondary air: Air in the combustion process that controls combustion efficiency by controlling how completely the fuel is burned.

secondary coil: The transformer coil that supplies output voltage to the circuit.

secondary refrigeration system: A refrigeration system that uses a secondary coolant (usually antifreeze) to cool the cooled space.

second class lever: A lever that has the resistance located between the fulcrum and the effort.

sectional view drawing: A drawing that shows the internal features of an object.

semiconductor: A material with an electrical conductivity between that of a conductor (high conductivity) and an insulator (low conductivity).

semi-hermetic compressor: A sealed compressor that can be serviced through removable access plates.

sensible heat: Heat energy that can be measured by a change in temperature.

series circuit: A circuit that has two or more components connected such that there is only one path for current flow.

series/parallel circuit: A circuit that contains a combination of components connected in series and parallel.

service port: A service valve used on small refrigeration systems that is only opened or closed by fitting a gauge manifold hose onto the port.

service valve: A three-way, manually operated valve used to charge or remove refrigerant or monitor system pressure.

setpoint temperature: The temperature the HVAC system is set to maintain.

shear: Stress caused by parallel forces acting upon an object from opposite directions.

shielded metal arc welding (SMAW): An arc welding process in which the arc is shielded by the gases from the decomposition of a consumable electrode covering.

shim: A thin precision piece of material, usually metal, that provides an accurate spacing between two surfaces.

short circuit: An undesirable, low-resistance path for current to leave the normal current-carrying path through a load.

short cycling: The increase in the frequency of system operation due to improper feedback.

side bar: A steel plate with two precision holes used to connect two pins or two bushings.

sight glass: A refrigerant line fitting located before the TXV that contains a small glass window for observing refrigerant flow.

signature analysis: The visual comparison between two vibration frequency patterns (signatures) to detect differences.

silent chain: A synchronous chain that consists of a series of links joined together with a bushing and pin.

silicon controlled rectifier (SCR): A solid-state rectifier with the ability to rapidly switch heavy currents.

silicone: An adhesive or sealant that has excellent flexibility, resilience, and strength over a wide temperature range.

simple machine: Any device that transmits the application of a force into useful work.

sine wave: A periodic, symmetrical waveform that varies over time according to the trigonometric sine function.

single-phase motor: A motor designed to run on 1ϕ power.

single phasing: The complete loss of one phase on a 3ϕ power supply.

skiving: The removal of the cover from a hydraulic hose down to the reinforcement in order to attach a fitting.

slack-side tension: The tension on a belt when it is approaching the driven pulley.

sleeve coupling: A coupling consisting of a tube with an internal keyway and center hole that fits mating shafts.

sliding friction: Friction that occurs when one surface moves across another or both surfaces move in opposite directions.

slip: The movement of a belt on the face of the pulley when belt tension is too loose.

slugging: A condition in which liquid refrigerant enters a compressor and causes hammering.

snub roller: A roller used to guide (track) the conveyor belt on the conveyor pulleys.

soft foot: A condition that occurs when one or more machine feet do not make complete contact with the base plate.

soldering: A joining process that joins parts by heating the filler metal to temperatures up to 840°F, but less than the melting point of the base metal.

solenoid: An electric output device that converts electrical energy into a linear mechanical force.

solid-state device: An electronic component that switches or controls the flow of current in a circuit with no moving parts.

solid-state relay: An electronic switching device that has no moving parts.

solvent-base adhesive: A one-part adhesive with a rubber or plastic base that cures by solvent evaporation.

source voltage: Potential electrical energy available to do work.

spalling: The flaking off of a metal surface.

spark test: A metal identification test that identifies metals by the shape, length, and color of spark emitted from contact with a grinding wheel.

specific hazard: A designation on the NFPA Hazard Signal System that specifies special properties and hazards associated with a particular material.

spectrometer: A device that vaporizes materials and records the resulting light.

spider: A cross-shaped member having four ends providing pivot points.

spiral bevel gear: A bevel gear that has curved teeth, which provide smooth operation at high speeds.

split-phase motor: A 1φ AC motor that includes a starting and running winding in the stator.

spool directional control valve: A directional control valve that controls the flow of hydraulic oil to and from actuators.

spring coupling: A material-flexing coupling that uses the flexing of a spring to accommodate shaft misalignment.

sprocket: A wheel with evenly spaced teeth located around the perimeter that engage a chain.

spur gear: A gear with straight teeth cut parallel to the shaft axis.

stack effect: The rising of hot air up the center of a building through elevator shafts, stairwells, and service columns.

standard: A collection of voluntary rules developed through consensus and related to a particular trade, industry, or environment.

standard belt: A V-belt constructed from multiple cords, providing added strength.

starting winding: A coil that is energized temporarily at startup to create the torque required to start a 1φ motor rotating.

static pressure: The pressure exerted by airflow in a direction perpendicular to the flow.

stator: The stationary part of an AC motor.

status indicator: A light, number, or word that shows the condition of the components in the programmable logic controller.

steam header: The part of a steam heating system that distributes the steam to branches.

steam strainer: A boiler accessory that removes scale or dirt from steam.

steam system: A system that collects, controls, and distributes the steam produced in the boiler.

steam trap: A boiler accessory that removes air and condensate from steam lines and heat exchangers.

step-down transformer: A transformer in which the secondary coil has fewer turns of wire than the primary coil.

stepping: The pulsing of the voltage supply to a device.

step-up transformer: A transformer in which the secondary coil has more turns of wire than the primary coil.

stoker: A mechanical device used to feed coal pieces into a furnace.

straight-tube watertube boiler: A boiler that has straight tubes surrounded by gases of combustion.

strain: A material's deformation per unit length under stress.

stress: The internal effect of an external force applied to a solid material.

subcooling: The cooling of a substance to a temperature that is lower than its saturated temperature at a particular pressure.

superheat: Sensible heat added to a substance after it has turned to vapor.

superheat setting: The temperature difference between the point immediately after the TXV and the outlet of the evaporator.

supply air: The mixture of outside air and return air that is conditioned for use in a building.

surface blowdown: The purging of boiler water at the surface to remove floating impurities and foreign matter.

surface deterioration: The loss of a material due to one surface contacting another surface.

surface fatigue: Failure that occurs when repeated pressure exceeds the limits of a material.

surfacing: The process of applying a layer of material by welding or brazing to obtain the desired dimensions or surface properties.

surge suppressor: A device that limits the intensity of voltage transients that occur on a power distribution system.

sustained power interruption: A decrease in voltage to 0 V that lasts for more than 1 min.

swash plate: An angled plate in contact with the piston heads that moves the pistons in the cylinders of a pump.

switch: Any component that is designed to start, stop, or redirect the flow of current in a circuit.

switchboard: A piece of equipment in which incoming electrical power is broken down into smaller units for distribution throughout a building.

synchronous drive system: A drive system that provides a positive engagement between the drive and driven sides of the system.

system: A combination of components, units, or modules that are connected to perform work or meet a specific need.

systems thinking: The consideration of an entire system and its interrelationships when troubleshooting.

T

tagout: The attachment of a danger tag to the source of power to indicate that the equipment may not be operated until the tag is removed.

tail end: The end where material enters a conveyor.

tap: A tool for cutting internal threads.

technical society: An organization of technical personnel united by a professional interest.

temperature: A measurement of the intensity of heat.

temporary power interruption: A decrease in voltage to 0 V that lasts from 3 sec up to 1 min.

tensile member: Cording material that runs the entire length of the belt, increasing the tensile strength of the belt.

tensiometer: A device that measures the amount of deflection of a belt.

tension: Stress caused by forces acting along the same axial line to pull an object apart.

terminal unit: A device that is located close to the zone and heats or cools air flowing through it.

testing chart: A chart that lists possible causes of a problem along with testing procedures.

test instrument: An electrical measurement tool used to test the condition or operation of an equipment component or system.

test light: A test instrument with a bulb that is connected to two test leads to give a visual indication when voltage is present in a circuit.

therm: A unit of heat energy equivalent to 100,000 Btu.

thermal conductivity: The rate at which heat moves through a material.

thermal efficiency: The ratio of heat absorbed to the heat available.

thermal expansion: The change in volume of a material in relation to temperature.

thermal imager: An infrared-measuring device that displays an image based on temperature.

thermistor: A transducer that changes resistance in response to a change in temperature.

thermocouple: A device that produces electricity when two different metals that are joined together are heated.

thermodynamic steam trap: A steam trap that has a single movable disc that raises to allow the discharge of air and cool condensate.

thermography: A predictive maintenance procedure that uses heat energy emitted from operating equipment to analyze the status of moving components.

thermometry: The use of temperature-indicating or -measuring devices to quantify temperature or temperature changes.

thermostat: A temperature sensor inside a temperature-controlled space that sends signals to a control system in order to maintain a set temperature.

thermostatic expansion valve (TXV): A valve that uses temperature readings at the evaporator outlet to control the rate of refrigerant flow into the evaporator.

thermostatic steam trap: A steam trap that has a bellows filled with a fluid that boils at steam temperature.

third class lever: A lever that has the effort located between the fulcrum and the resistance.

threaded fastener: A device such as a bolt, screw, or nut that joins or fastens parts together with threads.

thread insert: A small coil with outside threads to hold the coil in a tapped hole and inside threads to fit a standard fastener.

thread lock coating: A liquid coating applied to a threaded fastener that prevents the loosening of assembled parts from vibration, shock, and/or chemical leakage.

three-way valve: A valve that has three ports and controls the flow of fluid from one to the other two.

threshold limit value (TLV®): An estimate of the average airborne concentration of a substance to which workers may be exposed day after day without adverse effect.

tie-down testing method: A testing method in which one test instrument probe is connected to a point on a circuit (L1 or L2) and the other probe is moved along the circuit to test sections sequentially.

tight-side tension: The tension on a belt when it is approaching the drive pulley.

time delay fuse: A fuse that opens a short circuit almost instantly, but allows small overcurrents to exist for short periods of time.

timer: A control device that uses a preset time period as a control function.

timing belt: A flat belt containing gear teeth.

tooth breakage: The removal of a gear tooth or part of a tooth.

torch test: A metal identification test that uses the application of heat to identify a metal by its color change, melting point, and behavior in the molten state.

torque: Rotational force.

torsion: Stress caused by forces acting in opposite twisting motions.

torsional dampening: A dampening process that smoothes torque fluctuations in a drive system.

total shutdown: The malfunction of an entire machine.

toughness: A combination of resistance to stress and ductility.

trace: A conducting path used to connect components on a PC board.

trade association: An organization that represents the producers of specific products.

transducer: A device used to convert physical parameters such as temperature, pressure, and weight into electrical signals.

transformer: An electric device that steps up or steps down alternating current.

transient voltage: Temporary, undesirable voltage spike, ranging from a few volts to several thousand volts and lasting from a few microseconds up to a few milliseconds.

transistor: A three-terminal semiconductor device that controls current flow depending on the amount of voltage applied to the base.

triac: An AC semiconductor switch that is triggered into conduction in either direction.

troubleshooting: The systematic investigation of the cause of system problems in order to determine the best solution.

troubleshooting report: A record of a specific problem that occurs in a particular piece of equipment, along with its symptoms, causes, and repair procedures.

try cock: A valve mounted on the water column that can be used to determine boiler water level.

tube bank: An assembly of tubes in a boiler.

two-way valve: A valve that has two ports and controls the flow of fluid between them.

U

ultrasonic analysis: Analysis that detects high-frequency vibrations to create an image or reading.

undervoltage: A drop in voltage of more than 10% (but not to 0 V) below the normal rated line voltage that lasts longer than 1 min.

Underwriters Laboratories Inc.® (UL): An independent organization that tests products to verify conformance to national codes and standards.

universal joint: A mechanical-flexing coupling that includes two yokes connected by a spider.

unloader: A device that allows the compressor to operate without adding pressure to the receiver.

unloading: 1. In refrigeration systems, the process of varying the amount of refrigerant pumped by a compressor. **2.** In fluid power systems, allowing a pump to run against little or no pressure.

unloading valve: A pressure control valve that directs hydraulic oil from the pump to the reservoir after system pressure has been reached.

unscheduled maintenance: Impromptu service that is required due to a failure.

upper explosive limit: The highest concentration (air-fuel mixture) at which a gas can ignite.

useful life: The period of time after the break-in period when most equipment operates as designed.

V

vacuum: Pressure lower than atmospheric pressure.

valence electron: An electron in the outermost shell of an atom.

valve: A device that controls the pressure, direction, or rate of fluid flow.

vane compressor: A compressor that has multiple vanes located in an offset rotor.

vane pump: A pump that contains vanes in an offset rotor.

variable air volume box: A terminal unit that varies the amount of air flowing into a zone.

variable displacement pump: A pump that moves a variable amount of fluid during each revolution of the pump.

variable frequency drive: A motor controller that is used to change the speed of AC motors by changing the frequency of the supply voltage.

variable-speed drive: A mechanical drive system that provides variable output speed without changing the speed of the drive motor.

variable volume system: An HVAC system that supplies varying amounts of air.

V-belt: A drive belt made from rubber or synthetic material that has a cross-section in the shape of a V.

V-belt strand: One V-shaped cross-section.

velocity pressure: The pressure of airflow in the direction of flow.

ventilation: The process of introducing fresh outdoor air into an indoor space.

vibration: Oscillating motion in response to a force.

vibration analysis: The monitoring of individual component vibration characteristics to analyze the component condition.

vibration cycle: Motion from a neutral position to the upper limit, from the upper limit to the lower limit, and from the lower limit back to the neutral position.

vibration displacement: The maximum range of motion from the upper limit to lower limit of the vibration cycle.

vibration frequency: The number of completed vibration cycles within a specified period of time.

vibration phase: The relationship between the peak of a vibration and a moment in time.

vibration velocity: The speed of travel from an extreme limit of the vibration displacement to the neutral position.

viscosity: The measure of the resistance of a fluid to flow.

visual and auditory inspection: The analysis of the appearance and sounds of operating equipment.

voltage: The amount of electrical pressure (electromotive force) that causes electrons to move in a circuit.

voltage drop: The reduction of circuit voltage caused by the resistance of circuit conductors and loads.

voltage indicator: A test instrument that indicates the presence of voltage when the test tip touches, or is near, an energized conductor or metal part.

voltage regulator: A device that provides precise voltage control to protect equipment from voltage fluctuations.

voltage sag: A drop in voltage of more than 10% (but not to 0 V) below the normal rated line voltage that lasts from 0.5 cycles up to 1 min.

voltage swell: A increase in voltage of more than 10% above the normal rated line voltage that lasts from 0.5 cycles up to 1 min.

voltage tester: A test instrument that indicates approximate voltage level and type (AC or DC).

voltage unbalance: The unbalance that occurs when the voltages of a 3ϕ power supply or the terminals of a 3ϕ load are not equal.

volume: The three-dimensional size of an object.

W

washer: A device used with threaded fasteners to distribute load, affect friction, prevent leakage, and/or ensure tightness.

water-base adhesive: A one-part adhesive that cures by water evaporation.

water column: A boiler fitting that reduces water turbulence in the gauge glass to provide an accurate water level reading.

water-cooled condenser: A condenser that uses water as the condensing medium.

water hammer: A banging caused by rapid water movement in steam lines.

water leg: The area around the furnace of a boiler that is filled with water to transfer heat from the furnace area.

watertube boiler: A boiler in which water passes through tubes surrounded by the hot gases of combustion.

way: A port into or out of a valve.

wear-out period: The period after the useful life of equipment when normal failures occur.

wear particle analysis: The study of the size, frequency, shape, and composition of wear particles present in the lubricating oil.

wedge: A simple machine that converts force applied to its blunt end into force that is perpendicular to its sloped surface.

welding: A joining process that fuses materials by heating them to melting temperature.

wetback firetube boiler: A boiler that has a water-cooled turnaround chamber that directs gases of combustion from the furnace to the tube bank.

wet bulb temperature: A temperature reading that takes into account the amount of water vapor in the air.

wet lay-up: Storage of a boiler filled with warm, chemically treated water.

wheel and axle: A simple machine consisting of a wheel and an axle fixed together along the same axis.

wire group: Any set of wires that are connected directly without being broken by a device such as a pushbutton, contact, or coil.

wiring diagram: A diagram that shows the electrical connections of a circuit.

work: The movement of an object by a force to a specific distance.

work order: A document that details specific maintenance tasks to be completed.

work priority: The order in which work should be done based on its importance.

worm: A screw thread that rotates the worm gear.

worm gear: A spur gear with specially cut teeth that are driven by a worm.

Z

zeolite: The mineral group including silicates of aluminum, sodium, and calcium that is used to soften water.

zone: A building subsection such as an auditorium, a warehouse, or a group of several rooms.

Index

USING THE *INDUSTRIAL MAINTENANCE* INTERACTIVE CD-ROM

Before removing the Interactive CD-ROM from the protective sleeve, please note that the book cannot be returned for refund or credit if the CD-ROM sleeve seal is broken.

System Requirements

To use this Windows®-compatible CD-ROM, your computer must meet the following minimum system requirements:

- Microsoft® Windows® 7, Windows Vista®, or Windows® XP operating system
- Intel® 1.3 GHz processor (or equivalent)
- 128 MB of available RAM (256 MB recommended)
- 335 MB of available hard disk space
- 1024 × 768 monitor resolution
- CD-ROM drive (or equivalent optical drive)
- Sound output capability and speakers
- Microsoft® Internet Explorer® 6.0 or Firefox® 2.0 web browser
- Active Internet connection required for Internet links

Opening Files

Insert the Interactive CD-ROM into the computer CD-ROM drive. Within a few seconds, the home screen will be displayed allowing access to all features of the CD-ROM. Information about the usage of the CD-ROM can be accessed by clicking on Using This Interactive CD-ROM. The Quick Quizzes®, Illustrated Glossary, Maintenance Forms, Flash Cards, Media Clips, and ATPeResources.com can be accessed by clicking on the appropriate button on the home screen. Clicking on the American Tech web site button (www.go2atp.com) accesses information on related educational products. Unauthorized reproduction of the material on this CD-ROM is strictly prohibited.